Algorithmic Number Theory

Volume I: Efficient Algorithms

Foundations of Computing
Michael Garey and Albert Meyer, editors

Complexity Issues in VLSI: Optimal Layouts for the Shuffle-Exchange Graph and Other Networks, Frank Thomson Leighton, 1983

Equational Logic as a Programming Language, Michael J. O'Donnell, 1985

General Theory of Deductive Systems and Its Applications, S. Yu Maslov, 1987

Resource Allocation Problems: Algorithmic Approaches, Toshihide Ibaraki and Naoki Katoh, 1988

Algebraic Theory of Processes, Matthew Hennessy, 1988

PX: A Computational Logic, Susumu Hayashi and Hiroshi Nakano, 1989

The Stable Marriage Problem: Structure and Algorithms, Dan Gusfield and Robert Irving, 1989

Realistic Compiler Generation, Robert Lee, 1989

Single-Layer Wire Routing and Compaction, F. Miller Maley, 1990

Basic Category Theory for Computer Scientists, Benjamin C. Pierce, 1991

Categories, Types, and Structures: An Introduction to Category Theory for the Working Computer Scientist, Andrea Asperti and Giuseppe Longo, 1991

Semantics of Programming Languages: Structures and Techniques, Carl A. Gunter, 1992

The Formal Semantics of Programming Languages: An Introduction, Glynn Winskel, 1993

Hilbert's Tenth Problem, Yuri V. Matiyasevich, 1993

Exploring Interior-Point Linear Programming: Algorithms and Software, Ami Arbel, 1993

Theoretical Aspects of Object-Oriented Programming: Types, Semantics, and Language Design, edited by Carl A. Gunter and John C. Mitchell, 1994

From Logic to Logic Programming, Kees Doets, 1994

The Structure of Typed Programming Languages, David A. Schmidt, 1994

Logic and Information Flow, edited by Jan van Eijck and Albert Visser, 1994

Circuit Complexity and Neural Networks, Ian Parberry, 1994

Control Flow Semantics, Jaco de Bakker and Erik de Vink, 1996

Algebraic Semantics of Imperative Programs, Joseph A. Goguen and Grant Malcolm, 1996

Algorithmic Number Theory, Volume I: Efficient Algorithms, Eric Bach and Jeffrey Shallit, 1996

Algorithmic Number Theory

Volume I: Efficient Algorithms

Eric Bach and Jeffrey Shallit

The MIT Press
Cambridge, Massachusetts
London, England

This book was set in Times Roman by Integre Technical Publishing Co., Inc., and was printed and bound in the United States of America.

Library of Congress Cataloging-in-Publication Data

Bach, Eric.
Algorithmic number theory / Eric Bach and Jeffrey Shallit.
p. cm. — (Foundations of computing)
Includes bibliographical references and index.
Contents: v. 1. Efficient algorithms.
ISBN 0-262-02405-5 (hc: alk. paper)
1. Number theory—Data processing. 2. Algorithms. I. Shallit, Jeffrey Outlaw. II. Title. III. Series.
QA241.B1085 1996
512′.72′015118—dc20 95-25458
CIP

To our parents

Contents

Series Foreword

Theoretical computer science has now undergone several decades of development. The "classical" topics of automata theory, formal languages, and computational complexity have become firmly established, and their importance to other theoretical work and to practice is widely recognized. Stimulated by technological advances, theoreticians have been rapidly expanding the areas under study, and the time delay between theoretical progress and its practical impact has been decreasing dramatically. Much publicity has been given recently to breakthroughs in cryptography and linear programming, and steady progress is being made on programming language semantics, computational geometry, and efficient data structures. Newer, more speculative, areas of study include relational databases, VLSI theory, and parallel and distributed computation. As this list of topics continues expanding, it is becoming more and more difficult to stay abreast of the progress that is being made and increasingly important that the most significant work be distilled and communicated in a manner that will facilitate further research and application of this work. By publishing comprehensive books and specialized monographs on the theoretical aspects of computer science, the series on Foundations of Computing provides a forum in which important research topics can be presented in their entirety and placed in perspective for researchers, students, and practitioners alike.

Michael R. Garey
Albert R. Meyer

Preface

This is the first volume of a projected two-volume set on algorithmic number theory, the design and analysis of algorithms for problems from the theory of numbers. This volume focuses primarily on those problems from number theory that admit relatively efficient solutions. The second volume will largely focus on problems for which efficient algorithms are not known, and applications thereof.

Prerequisites

We hope that the material in this book will be useful for readers at many levels, from the beginning graduate student to experts in the area. The early chapters assume that the reader is familiar with the topics in an undergraduate algebra course: groups, rings, and fields. Later chapters assume some familiarity with Galois theory. A good text is Herstein [1975].

We assume the reader is familiar with the analysis of algorithms, as treated, for example, in Aho, Hopcroft, and Ullman [1974], or Cormen, Leiserson, and Rivest [1990] and with the language of complexity theory, as treated, for example, in Garey and Johnson [1979]. However, we have tried to make our discussion of complexity theory relatively self-contained.

Finally, we assume the reader has had the equivalent of an undergraduate course in probability theory. A good text in this area is Feller [1968].

For some sections, a knowledge of more advanced mathematics will be useful. For example, complex analysis is needed for parts of Chapter 8, and some algebraic number theory will be useful for Chapter 9.

A typical graduate course for computer scientists might cover the following material: Chapter 1, Chapter 2, Chapter 4 (Sections 1–3), Chapter 5, Chapter 6 (Sections 1–5), Chapter 7 (Sections 1–4), Chapter 8 (Sections 1–5), Chapter 9 (Sections 1–5), and Chapters 11 and 12. Mathematicians may want to substitute Chapter 3 for Chapter 2.

Goals of This Book

As stated above, this book discusses the current state of the art in algorithmic number theory. This book is *not* an elementary number theory textbook, and so we frequently do not give detailed proofs of results whose central focus is not computational. Choosing otherwise would have made this book twice as long.

However, we firmly believe that a mere list of proved theorems is not particularly enlightening or useful. Therefore, we endeavor to prove as much as is feasible. Some proofs are left to the reader as exercises, with outlines suggested in Appendix A. And every theorem which is not proved in the text or left as an exercise has a reference in the "Notes" section that appears at the end of each chapter.

This book is also not intended solely as a practical guide for those who wish to implement number-theoretic algorithms. The variety of architectures of modern machines, and the profusion of languages and operating systems, frequently make specific remarks on running times useless. On the other hand, computational theory without the influence of practice seems artificial. Thus, the "Notes" section of each chapter contains remarks on practical implementations of the algorithms discussed.

We might mention several other books that cover approximately the same material. The first was Knuth [1969]; a second edition appeared as Knuth [1981]. More recently, there have appeared texts such as Riesel [1985a]; Kranakis [1986]; Koblitz [1987a]; Bressoud [1989]; H. Cohen [1993]; and Zippel [1993]. We hope that our book will be useful in addition to these, because of its somewhat different emphasis.

We restrict our attention in this book to those concepts and algorithms that relate to what we see as algorithmic number theory. Thus, we do not discuss some very interesting topics that seem more algebraic than number-theoretic: algorithms on finite groups, Gröbner bases, etc. We also restrict our attention to problems from *elementary* number theory: e.g., primality testing, factorization, discrete logarithm, etc. Thus there is virtually no overlap between the present work and the books of Zimmer [1972] and Pohst and Zassenhaus [1989].

In writing this book, we have found many opportunities to extend or refine known results. This is particularly true with regard to algorithms, many of whose running times were not completely worked out in the literature. As an alternative to listing all such improvements (which is clearly impractical), we give references to earlier results, such as are known to us, in the appropriate places.

Algorithm Descriptions

We write our algorithms in a "pseudo-code" notation similar to Pascal or C, which should be familiar to most readers. We do not provide a formal definition of this notation.

We do make the following observations: no explicit "begin/end" is used to denote the start and finish of a loop. Instead, the scope of loops is made clear through indentation. A "return" statement causes execution of a routine to terminate immediately. An "output" statement is identical to "return", except that execution continues with the next line; this allows a program to return multiple outputs at different places in the program.

Projected Contents of Volume II

Volume II: Intractable Problems will cover the following topics:

Acknowledgments

We would like to thank our thesis advisor, Manuel Blum, for having introduced us to algorithmic number theory and its applications.

Richard Carnes typed the course notes on which an earlier version of this book was based. Terrence Dick provided bibliographic and programming assistance.

We express our thanks to the staff members of the Mathematics Faculty Computing Facility at the University of Waterloo (in particular Anil Goel) for explaining some of the arcane features of the software used to write this book.

Thanks to Igor Khurgin, Arkady Kanevsky, and Valery Soloviev for help with Russian translation, to Jean-Paul Allouche for help with French translation, and to Christiane Whittington and Renate Scheidler for help with German translation.

Thanks to Dan Shanks for providing the reference to Robinson's factor stencils.

Thanks to Andrei Brodnik for bringing our attention to articles on maintaining partial sums of an array.

We would like to express our appreciation to the following people who read a preliminary version of our book and made many comments: John Brillhart, Gudmund Frandsen, Jerry Kuch, Arjen Lenstra, Hendrik W. Lenstra, Jr., Paul Beame, Kevin McCurley, Peter Montgomery, Evi Nemeth, Michael Somos, Rob Lambert, Stan Wagon, Andrew Odlyzko, Alfred Menezes, Louxin Zhang, Drew Vandeth, Will Gilbert, Joan Boyar, Donald Knuth, and Peter Høyer.

We are extremely grateful to many colleagues who have unselfishly given their time locating difficult-to-find papers and books for us. These include Stuart Kurtz, Howard Karloff, Hugh Williams, John Brillhart, and Jean-Paul Allouche. We wish to thank the library staff at the University of Waterloo, the University of Wisconsin, and Dartmouth College for their help. We also thank John Neu, Bibliographer in the History of Science department at the University of Wisconsin.

Of course, any mistakes are our responsibility alone. Please report mistakes by sending electronic mail to `shallit@graceland.uwaterloo.ca` or `bach@cs.wisc.edu`. Errata for the book can be found at

`http://math.uwaterloo.ca/~shallit`

which is the URL for Shallit's home page.

This work was supported in part by grants from the National Science Foundation (USA); Wisconsin Alumni Research Foundation, and the Natural Sciences and Engineering Research Council (Canada).

Eric Bach
Madison, Wisconsin

Jeffrey Shallit
Waterloo, Ontario

1 Introduction

The goal of this book is to provide a thorough introduction to the design and analysis of algorithms for problems from the theory of numbers.

The goal of algorithmic number theory is to *produce efficiently* the objects whose mere *existence* is proved in the classical theorems of number theory. For example, the famous proof of Euclid tells us that there exists a prime larger than any given number n; algorithmic number theorists ask: how can we quickly find such a prime? The fundamental theorem of arithmetic tells us that every number n can be written as the product of prime powers; algorithmic number theorists ask: can we quickly find this prime factorization of n? Even to state these questions precisely, we must introduce concepts from the theory of computation. We might say that algorithmic number theory is the marriage of an ancient subject, number theory, with a modern one: the theory of computational complexity.

Almost every problem we will discuss can be interpreted as finding the solution to an equation in a specified group, ring, or field. For example, factoring n can be viewed as finding the integer solutions to the equation $n = xy$, with $x, y > 1$. Thus the discipline of algorithmic number theory can be summarized succinctly as *finding solutions to equations, or proving their non-existence, making efficient use of resources such as time and space.* Implicitly included in this definition is the problem of *efficiently representing* the objects in question on a computer.

Unlike some other branches of mathematics, number theory has long been associated with computation. Roughly speaking, this association may be categorized as follows:

1. *Construction of tables:* Here the investigator may be concerned with obtaining evidence for conjectures. Many tables, computed by hand, can be found in the collected works of Gauss. (He conjectured the prime number theorem on the basis of a table of primes.) Tables giving the prime factorization of integers were constructed to higher and higher limits. To list just a few such tables, Cataldi published a table up to the limit of 800 in 1603, T. Brancker reached 100,000 in 1668, L. Chernac reached 1,020,000 in 1811, and D. N. Lehmer published a table up to 10,017,000 in 1909. Many investigators have constructed tables of periods of continued fractions for square roots, class numbers of algebraic number fields, etc. The celebrated conjectures of Birch and Swinnerton-Dyer, which relate the rank of an elliptic curve with the orders of the zeroes of certain L-functions, were based on extensive computations, as were the more recent heuristics on class groups of H. Cohen and H. W. Lenstra, Jr. Neither of these conjectures have been proved in their full generality.
2. *Gathering evidence for conjectures:* Goldbach, in a 1742 letter to Euler, conjectured that every even integer ≥ 4 can be expressed as the sum of two primes. This conjecture has recently been verified up to $4 \cdot 10^{11}$ by Sinisalo.

3. *Searching for counter-examples:* We give three well-known examples.

   Suppose $\lambda(n) = (-1)^r$, where r is the total number of primes that divide n, counted with multiplicity. Define $L(x) = \sum_{n \le x} \lambda(n)$. Polya conjectured that $L(x) \le 0$ for all $x \ge 2$. However in 1960, with the aid of a computer, R. S. Lehman showed that $L(906180359) = 1$.

   Euler had conjectured that there was no nontrivial integer solution to $a^5 + b^5 + c^5 + d^5 = e^5$, but in 1967, using a computer, L. J. Lander and T. R. Parkin found the counterexample $27^5 + 84^5 + 110^5 + 133^5 = 144^5$.

   If $\mu(n)$ denotes the well-known Möbius function (see Chapter 2), then define $M(x) = \sum_{n \le x} \mu(n)$. Then Mertens conjectured that $|M(x)| < \sqrt{x}$. However, using extensive computation, Odlyzko and te Riele showed in 1985 that this conjecture is false.

4. *Proofs by enumeration:* Some theorems may require enormous computation for their proofs, and the computer has made this task more feasible. (A classic non-number-theoretic example is the proof of the four-color conjecture by Appel and Haken.) We mention such a theorem of D. H. Lehmer: Every set of 7 consecutive integers greater than 36 contains either a prime or a multiple of a prime greater than 41.

   Fermat's "Last Theorem" is the conjecture that there exist no nontrivial integer solutions to $x^n + y^n = z^n$ for $n > 2$. (It is easy to see that it would suffice to prove the conjecture for prime exponents $n = p$.) Using a computer, D. H. Lehmer and E. Lehmer proved the truth of the conjecture for all prime exponents less than 2000. This bound has been improved to 4,000,000 by Buhler, Crandall, Ernvall, and Metsänkylä, using extensive computations. The *first case* of Fermat's "Last Theorem" states that there exist no solutions with $p \nmid xyz$. Using a computer, D. Coppersmith proved that the first case holds for all prime exponents $p \le 7.568 \times 10^{17}$. (Recently, A. Wiles and R. Taylor proved that Fermat's "Last Theorem" is true.)

   Waring conjectured that every positive integer is the sum of 19 fourth powers. This was recently proven by Balasubramanian, Deshouillers, and Dress, using extensive computations.

5. *Study of number-theoretic algorithms:* In this category we are more interested in the design and analysis of algorithms themselves than we are in the outcome of any specific computation. The categories above provide impetus for this study, especially when existing computational methods prove infeasible for problems of interest.

We may refer to the activities in the first four categories as *computational number theory*. The last category is the subject of this book: algorithmic number theory.

1.1 Number Theory and Complexity

Algorithmic number theory is interesting because there are problems at every level of complexity. Let us mention some examples, which are discussed more fully in the text:

1. There is no algorithm to decide whether an arbitrary Diophantine equation is solvable. (A *Diophantine equation* is a multivariate polynomial equation with integer coefficients, such as $x^3 + y^3 - z^3 = 3$. It is solvable if there exist *integer* values of the variables $x, y, z, \ldots$ satisfying the equation.)
2. Given positive integers a, b, c, determining if there is a positive integer $x < c$ such that $x^2 \equiv a \pmod{b}$ is $\mathcal{NP}$-hard. (Roughly, this means it is at least as difficult as many combinatorial problems which are believed to be hard.)
3. Factoring integers is believed to be intractable; for this problem no really fast algorithm is currently known to exist.
4. If a given integer is composite, we can quickly produce a proof of that fact with an algorithm that uses a source of random numbers. However, we do not know how to do this with a fast *deterministic* algorithm.
5. While we can do the basic operations of arithmetic quickly in parallel, there is no known fast parallel algorithm for computing the greatest common divisor.
6. Even the basic algorithms of arithmetic have not been completely analyzed. For example, we can form the product of two n-bit integers in $O(n(\log n)(\log\log n))$ steps on a random access machine (RAM), but the existence of a linear-time algorithm is still an open question.

The interest in algorithmic number theory is not confined to theoretical topics. Clark, Bannon, and Keller applied the discrete logarithm (see Chapter 12) to the problem of designing a fast hardware counter. There have also been proposals for parallelization of arithmetic hardware, based on the Chinese remainder theorem.

Number-theory algorithms form the backbone of computer algebra systems, which have become essential tools of the trade for physicists and engineers. For instance, if one wishes to compute using exact arithmetic, a greatest common divisor algorithm is required to reduce fractions to lowest terms. Less trivially, polynomial factoring algorithms use "local" methods based on finite and p-adic fields to construct factorizations over the integers.

Some of the best methods for the generation of pseudo-random numbers use either modular arithmetic or arithmetic in finite fields. Recently, G. Marsaglia and A. Zaman proposed a method for designing pseudo-random number generators with extremely long period. Determination of the period crucially depends on integer factorization, a problem which we address in Chapter 11. Truly random numbers are not efficiently predictable, but this is not a property shared by all pseudo-random number generators. Currently, there are fast methods which are easily inferable, and slower methods whose prediction is computationally equivalent to a problem such as integer factorization. These topics are discussed in more detail in Chapter 15.

Finally, cryptology relies increasingly on algebraic or number-theoretic principles. As the algebraist A. A. Albert remarked in 1941, "It would not be an exaggeration to state that abstract cryptography is identical with abstract mathematics." This new "applied number theory" has obvious military uses, and has been vigorously pursued by the U. S. Department of Defense and the National Security Agency. Regarding applications, G. H. Hardy said in 1940,

> There is one comforting conclusion which is easy for a real mathematician. Real mathematics has no effects on war. No one has yet discovered any warlike purpose to be served by the theory of numbers... and it seems very unlikely that anyone will do so for many years.

But this seems no longer to be true.

1.2 Number Theory and Computation: A Brief History

From earliest times, number theory has been associated with computation. Let us mention some events, without attempting to be inclusive.

The Euclidean algorithm (c. 300 B.C.) to compute the greatest common divisor of two integers has been called the "oldest nontrivial algorithm that has survived to the present day".

Eratosthenes (c. 250 B.C.) devised a "sieve" to create a list of prime numbers between 1 and n. Sieve methods are still being used in algorithmic number theory; for example, see Chapter 9.

The "extended Euclidean algorithm", which, given relatively prime integers x, y, finds integers a, b such that $ax + by = 1$, was apparently first given by Aryabhata in the Sanskrit astronomical work Aryabhatiya in the 5th-6th century A.D.

In the 7th century A.D., Brahmagupta gave an algorithm for obtaining new solutions to the equation $x^2 - Ny^2 = 1$ for certain values of N, once a particular solution is known. A method that works for all N was given by Jayadeva in the 11th century A.D.

Two statements by Fermat and Mersenne in the 17th century stimulated much interest in methods for factorization and primality testing. In 1640, Fermat observed that $2^{2^k} + 1$ is a prime for $0 \leq k \leq 4$, and conjectured it was prime for all $k \geq 0$; however, Euler showed in 1732 that 641 is a divisor of $2^{32} + 1$.

In 1644, Mersenne claimed that $2^p - 1$ was prime for $p = 2, 3, 5, 7, 13, 17, 19, 31, 67, 127, 257$, and for no other primes p less than 257. (The numbers $2^p - 1$ were of great interest because of their connection with *perfect numbers*; see Chapter 9 for more details.) It would not become feasible to test $2^{257} - 1$ for primality until almost three centuries later, so it is not surprising that Mersenne's list contained several mistakes: the numbers corresponding to $p = 67$ and $p = 257$ are not prime, and the list omitted the primes corresponding to $p = 61$ and $p = 89$. Numbers of the form $2^p - 1$ became known as *Mersenne numbers*.

Euler proved in 1772 that the Mersenne number $2^{31} - 1 = 2147483647$ was a prime.

In 1776, C. F. Hindenburg realized the sieve of Eratosthenes by using a "strip of thick paper with holes at proper intervals to show the multiples of p, for the successive primes p".

At the start of the nineteenth century, Gauss discussed the importance of two basic problems from algorithmic number theory in a quotation that has frequently been cited:

> The problem of distinguishing prime numbers from composite numbers and of resolving the latter into their prime factors is known to be one of the most important and useful in arithmetic. It has engaged the industry and wisdom of ancient and modern geometers to such an extent that it would be superfluous to discuss the problem at length. Nevertheless we must confess that all methods that have been proposed thus far are either restricted to very special cases or are so laborious and prolix that even for numbers that do not exceed the limits of tables constructed by estimable men, i.e. for numbers that do not yield to artificial methods, they try the patience of even the practiced calculator. And these methods do not apply at all to larger numbers. Even though the tables, which are available to everyone and which we hope will continue to be extended, are indeed sufficient for most ordinary cases, it frequently happens that the trained calculator will be sufficiently rewarded by reducing large numbers to their factors so that it will compensate for the time spent. Further, the dignity of the science itself seems to require that every possible means be explored for the solution of a problem so elegant and so celebrated.

Other writers were even more specific. For example, in 1874, W. S. Jevons said:

> Given any two numbers, we may by a simple and infallible process obtain their product, but it is quite another matter when a large number is given to determine its factors. Can the reader say what two numbers multiplied together will produce the number 8,616,460,799? I think it unlikely that anyone but myself will ever know, for they are two large prime numbers, and can only be rediscovered by trying in succession a long series of prime divisors until the right one be fallen upon. The work would probably occupy a good computer for many weeks, but it did not occupy me many minutes to multiply the two factors together.

(Today we might find this statement ironic, for Jevons himself had built a "logical machine" in 1869.)

There were also efforts to define a "good" or "direct" algorithm. For example, H. J. S. Smith said in an 1859 survey of number theory:

> The arithmetical solution of a problem should consist in prescribing a finite number of purely arithmetical operations (exempt from all tentative processes), by which all the numbers satisfying the conditions of the problem, and those only, are obtained.

While mechanical calculators were devised in the seventeenth century by Schickard, Pascal, and Leibniz, the eighteenth century marks the first application of mechanical aids to number-theoretic problems. Around 1776, C. F. Hindenburg and A. Felkel independently invented devices to aid in the construction of factor tables. A description of Felkel's machine, made of rods, is contained in the correspondence of J. H. Lambert. Felkel's machines were even offered for sale to the general public, although it seems unlikely there was much demand.

In June 1822, C. Babbage discussed an application of his "Difference Engine":

> With this machine I have constructed... a table from the singular formula $x^2 + x + 41$, which comprises amongst its terms so many prime numbers.

His machine could only display results, and not print them; Babbage employed a friend to write down the results. In a letter to Davy the next month, Babbage discussed the speed of his machine:

> [The Difference Engine] proceeded to make a table from the formula $x^2 + x + 41$. In the earlier numbers my friend, in writing quickly, rather more than kept pace with the engine; but as soon as four figures were required, the machine was at least equal in speed to the writer.

The letter also mentions a possibility for a future machine:

> I have also certain principles by which, if it should be desirable, a table of prime numbers might be made, extending from 0 to ten millions.

In a series of papers beginning in 1876, the French mathematician E. Lucas developed the first fast (today we would say "polynomial time") method of testing Mersenne numbers for primality. He said

> I have thus verified, but only once, I admit, that the number $A = 2^{127} - 1$ is a prime number; this number contains thirty-nine digits....

He pointed out that his test was particularly suited for computation using binary arithmetic.

Lucas may have used the "arithmometer", a simple mechanical calculator that could add and subtract, invented by Thomas de Colmar in 1820; he advocated the use of this machine in an 1878 paper. But apparently he was not completely confident of his calculations, for in 1876 he wrote, "It is with the aid of these theorems that I think I have demonstrated that the number $A = 2^{127} - 1$ is prime." By 1887, he was even less sure, for he said of $2^{61} - 1$, "It is the largest prime number currently known...." Lucas wrote extensively on mechanical aids to computation and was interested in the construction of machines to apply his primality tests. He wrote in 1876:

> I have devised the design of a mechanism that would enable one to find prime numbers having a *thousand* digits in the decimal system, and even a lot more.

Together with the French civil engineer H. Genaille, he planned to construct a simple version of this machine. Genaille reported in 1891 (the year of Lucas's untimely death)

> The "arithmetical piano" allows one to give a practical application of the method formulated by Mr. E. Lucas... for the verification of large prime numbers. By simply moving a few pegs, the problem of verification of prime numbers of the form $2^n - 1$ is almost always reduced to just several hours' work. This machine, which may succeed in automatically performing calculations of the greatest importance, will one day realize the goal of having a calculating machine that can perform the arithmetic operations on its own.

In 1912, however, A. Gérardin reported that "unfortunately, Mr. Genaille did not construct anything, and because the archives of the author [Lucas] do not contain anything on this subject, we must mourn the loss [of the machine]."

In 1883, the Russian priest I. M. Pervushin proved that $2^{61} - 1$ was prime, contrary to Mersenne's assertion. This was also proved independently by P. Seelhoff in 1886, and J. Hudelot in 1887. The latter used Lucas' test, requiring 54 hours.

In 1903, F. N. Cole factored $2^{67} - 1$. When E. T. Bell, the historian of mathematics, asked Cole how long it had taken him to find the factorization, he replied, "Three years of Sundays." Bell goes on to say

> At the October, 1903 meeting in New York of the American Mathematical Society, Cole had a paper on the program with the modest title *On the factorization of large numbers*. When the chairman called on him for his paper, Cole–who was always a man of very few words–walked to the board, and, saying nothing, proceeded to chalk up the arithmetic for raising 2 to the sixty-seventh power. Then he carefully subtracted 1. Without a word he moved over to a clear space on the board and multiplied out, by longhand,
>
> $$193{,}707{,}721 \times 761{,}838{,}257{,}287.$$
>
> The two calculations agreed.... For the first and only time on record, an audience of the American Mathematical Society vigorously applauded the author of a paper delivered before it. Cole took his seat without having uttered a word. Nobody asked him a question.

The story is an excellent example of the difference in complexity between *finding* a solution to an equation and *verifying* that a given solution is correct. The issue of the relative difficulty of these two problems has given rise to the theory of $\mathcal{NP}$-completeness. Nevertheless, today a microcomputer can find Cole's factorization in a few minutes.

We might credit H. C. Pocklington with the first explicit analysis of a number-theoretic algorithm in terms of "naive bit complexity." In a 1910 paper, he gave an algorithm to find

$\text{ord}_p\, a$, the least positive e such that $a^e \equiv 1 \pmod{p}$, when the prime factorization of $p-1$ is known, and remarked

> We notice that the labour required here is proportional to a power of the logarithm of the modulus, not to the modulus itself or its square root as in the indirect processes, and hence see that in the case of a large modulus the direct process will be much quicker than the indirect.

Perhaps Pocklington also deserves credit as the inventor of the randomized algorithm. In a 1917 paper, he gave an algorithm for computing square roots modulo a prime p which used approximately $(\log p)^3$ steps and observed "the labour varies roughly as the cube of the number of digits in the modulus, and so remains moderate for large moduli." His algorithm depended on finding a quadratic nonresidue of a special form by trial and error. He must have viewed this use of randomness as slightly unsatisfactory, for he explained in a footnote:

> We have to do this [find the nonresidue] by trial, using the Law of Quadratic Reciprocity, which is a defect in the method. But as for each value of u half the values of t are suitable, there should be no difficulty in finding one.

Apparently Pocklington also made use of mechanical aids to computation, for he writes of a method for primality testing in a 1914 paper, "It is also well adapted for use with the arithmometer."

In 1896, F. W. Lawrence outlined the plan of a sieving machine to aid in integer factorization. In 1912, both M. Kraitchik and A. Gérardin suggested constructing a sieving machine. A simple version of such a machine was built that same year by P. Carissan and later perfected by E. Carissan. It was exhibited at the Exposition de Machines à Calculer in Paris in June 1920. Carissan claimed that it could test 10-digit numbers for primality in between 2 and 15 minutes.

In 1914, T. E. Mason sketched the design of a mechanical device to apply Lucas' primality test to Mersenne numbers.

In 1925, A. J. C. Cunningham and H. J. Woodall published a set of tables of factorizations of $b^n \pm 1$ for small nonsquare b and n up to high powers. About this time, mechanical calculators with an automatic multiply and divide feature became generally available in the United States. Many investigators, including D. H. Lehmer, M. Kraitchik, and K. R. Isemonger worked on factorizations using desk calculators during this period.

The Cunningham-Woodall tables have been extended and corrected over the last sixty years; this ongoing work is now called the "Cunningham Project". In 1983, a new version of these tables was published, with J. Brillhart, D. H. Lehmer, J. L. Selfridge, B. Tuckerman, and S. S. Wagstaff, Jr., as authors; the second edition appeared in 1988. Many improvements in the algorithms for primality testing and factorization were motivated by the desire to fill the gaps in these tables.

Today, factoring a previously unfactored number from the Cunningham table is the real test for any new factoring method. Wagstaff periodically issues updates to the tables and lists of the "most-wanted" numbers. Currently the most wanted factorizations are those of $2^{569} - 1$, $6^{256} + 1$, and $2^{4096} + 1$.

In the 1920's, D. N. Lehmer invented the *factor stencil*, an aid to the factorization of large numbers. For all squarefree integers r with $|r| \leq 250$ a punched card was prepared. Each card had a large number of cells, each of which was assigned a prime p; a hole was punched if r is a quadratic residue of p. To factor a number N, one obtains several small quadratic residues of N, and then superimposes the corresponding punched cards. If all the holes corresponding to a particular prime are punched, then that prime is a possible divisor of N. Later these stencils were revised and extended by J. D. Elder to seven sets of cards with 800 primes each, allowing one to factor numbers up to 3,033,696,241. A similar stencil method was invented by R. M. Robinson to solve quadratic congruences.

D. H. Lehmer is one of the major figures in algorithmic number theory in the twentieth century–a pioneer in the development of both algorithms and machines. We mention just a few of his accomplishments: In 1926, he and his father, D. N. Lehmer, built a sieving device out of bicycle parts. This was "the first fully automatic machine to be used for factoring and primality testing" and could sieve 60 numbers a second. The machine was improved with the addition of a photoelectric cell in 1932, and could then scan 5000 numbers a second. By 1966, Lehmer had built a delay-line sieve that could scan one million numbers per second.

In 1930, he improved and made rigorous the test of Lucas for the primality of Mersenne numbers.

In July 1946, Lehmer was the first to apply an electronic digital computer (the ENIAC) to a number-theoretic problem. He computed a list of all primes $p < 4538791$ for which the multiplicative order of 2, modulo p, was less than 300. (See Chapter 5 for the definition of order.)

Lehmer initiated modern algorithmic number theory in a 1969 paper, *Computer technology applied to the theory of numbers*. In this paper, he distinguished between those problems that could be solved in polynomial time, such as computing the greatest common divisor, and those for which no polynomial-time algorithm was known, such as factorization. Lehmer was also the first to use parallelism to solve number-theoretic problems.

Let us return to our history of the application of machines to number theory. The first program ever run on a stored-program electronic digital computer was a routine to computer, via brute-force search, the largest prime factor of an integer–this was on the Mark I computer at Manchester University on 21 June 1948. Later, in June 1949, M. H. A. Newman, G. C. Tootill, and T. Kilburn wrote a routine to use Lucas' primality test to test the Mersenne numbers up to $2^{353} - 1$ for primality.

In 1952, A. Ferrier proved that $(2^{148} + 1)/17$ was prime using only a desk calculator. This still holds the record as the largest prime found without the use of a computer.

The first successful attempt to use an electronic computer to find new Mersenne primes was by R. M. Robinson in 1952. He used Lucas' test on the SWAC to discover that $2^p - 1$ was prime for $p = 521, 607, 1279, 2203$, and 2281. (The previous year, A. M. Turing had made an attempt to find new Mersenne primes, but was unsuccessful.)

J. L. Selfridge found factors of the Fermat numbers $2^{2^{10}} + 1$ and $2^{2^{16}} + 1$ on the SWAC in 1953.

Since 1953, many investigators have used computers to factor numbers from the Cunningham tables, or to test such factors for primality. We now know 33 Mersenne primes, the largest being $2^{859433} - 1$, a number of 258716 decimal digits.

More recently, investigators have used supercomputers and networks of smaller computers to factor numbers. J. A. Davis and D. B. Holdridge factored $(10^{71} - 1)/9$ on a Cray XMP in 1983. K. J. McCurdy and M. C. Wunderlich factored a 64-digit number on the MPP in 1986.

The first to use a distributed implementation of factoring algorithms was R. D. Silverman in 1986. He used the quadratic sieve algorithm (discussed in Chapter 11) on a network of Sun microcomputers at the MITRE corporation. Subsequently, A. K. Lenstra and M. Manasse used a parallel implementation of the quadratic sieve and 400 computers on three continents to split a 102-digit factor of $2^{391} - 1$ in 1988.

In 1992, A. K. Lenstra and D. Bernstein factored $2^{523} - 1$, a 158-digit number using the lattice sieve. In 1994, a team of eleven researchers from Amsterdam and Oregon factored the 162-digit number $(12^{151} - 1)/11$ using the number field sieve.

In 1981, J. W. Smith and Wagstaff designed a special-purpose computer, called the Extended Precision Operand Computer, or EPOC, to perform the continued-fraction factoring algorithm. They could factor a 70-digit number in about a month with this machine. H. Dubner and R. Dubner have designed their own special-purpose 200-chip microcomputer for number-theoretic calculations and have found many large primes, some as big as 7000 digits. W. G. Rudd, D. A. Buell, and D. M. Chiarulli are building a special processor for factoring large numbers, as are C. Pomerance, Smith and R. Tuler.

While computers were being applied to specific numbers, there was also increased interest in algorithm development and more rigorous analysis of algorithms.

In 1968, G. E. Collins popularized a measure of the efficiency of number-theoretic algorithms which we call *naive bit complexity*; this notion is discussed more fully in Chapter 3.

J. Brillhart and M. Morrison introduced the first (heuristic) subexponential-time algorithm for factoring in a 1975 paper. (It had been previously considered by Lehmer and Powers in 1931, but was rejected as unsuitable for hand computation with a desk calculator.) With it, they were able to factor $2^{2^7} + 1$ in 1970. In 1975, R. Schroeppel was the first to provide a heuristic analysis of the method's running time. In 1979, J. D. Dixon gave the first subexponential factoring algorithm that could be rigorously analyzed.

In the past twenty years, the development of algorithms in number theory has seen a flurry of activity–so much so that we cannot summarize the highlights here. This period has seen the increased use of concepts from analytic number theory, algebraic number theory, and algebraic geometry. Some of the appropriate chronology is discussed in Chapters 9-12.

With improvements in machines and algorithms, it is now possible to factor 100-digit numbers and test 1000-digit numbers for primality in a reasonable length of time.

We mention briefly some software packages designed especially for number-theoretic computations. Most systems for computer algebra, such as SAC-2, Macsyma, Maple, Mathematica, and Scratchpad, have routines for unlimited precision integer arithmetic, and so are suitable for simple number-theoretic computations. At Berkeley, P. Weinberger and D. H. Lehmer designed a FORTRAN package for number-theoretic computation. D. Cantor and Morrison at UCLA designed a number-theoretic package in assembly language, intended for use with PL/I. Wunderlich and Selfridge have designed an assembly language package for the IBM 360. More recently, H. Cohen and his collaborators C. Batut, D. Bernardi, and M. Olivier have written a package called PARI which is especially designed to optimize number-theoretic calculations. C. Hollinger and P. Serf developed a computer algebra system focusing on algebraic number theory.

1.3 Condensed History of the Theory of Computation

Having examined some of the history of algorithmic number theory, let us look at some events in the development of the theory of computation that relate to the subject of this book.

Hilbert's dream was to provide a formal procedure that could, in principle, give an algorithmic solution to every well-formulated problem in mathematics. At the International Congress of Mathematicians in 1900, Hilbert gave a celebrated list of problems. The tenth one asked for a general procedure to determine if an arbitrary polynomial equation

$$f(x_1, x_2, \ldots, x_n) = 0 \tag{1.1}$$

has a solution in integers. Part of the problem was to formalize what was meant by the term "general procedure".

In 1931, Gödel showed that any formal system strong enough to express statements in elementary number theory must contain statements that are undecidable within that system. In lectures given in 1934, Gödel identified mechanically calculable functions with the functions definable using general recursions *(Gödel's thesis)*.

At about the same time, A. Church introduced the λ-calculus, a mechanism for defining functions. He claimed *(Church's thesis)* that the set of λ-definable functions coincides with the set of effectively calculable functions, i.e. those computable in principle by any machine or person, with no limitations on time or space.

In 1936, Turing defined a formal model of computation, now known as the *Turing machine*. *Turing's thesis* says that a function is effectively calculable if and only if it is calculated by a Turing machine. He showed that the *halting problem* for Turing machines (given a Turing machine M and an input w, does M halt on input w?) is undecidable.

In fact, it can be shown that all these models define the same set of functions. (This is also called "Church's thesis".)

Some of the work of Gödel, Church, and Turing was anticipated by E. Post in the 1920's. In 1944, Post introduced several concepts of future importance in complexity theory. His goal was to study the relative computational difficulty of undecidable problems. To do this, he introduced formal notions of reducibilities between problems, and the idea of a "most difficult" or "complete" problem in a given class. (See Chapter 3.) This work, and that of his successors in the 1950's, set the style for present-day complexity theorists.

In 1964, A. Cobham observed that many functions could be computed in *polynomial time*, that is, time bounded by a polynomial in the size of the input. In 1965, J. Edmonds suggested that the terms "good algorithm" and "polynomial-time algorithm" were synonymous. In 1971, S. Cook defined the class $\mathcal{NP}$, and the notions of $\mathcal{NP}$-completeness and polynomial-time reducibility. Many natural combinatorial problems were shown to be $\mathcal{NP}$-complete by Cook and R. M. Karp.

In 1971, M. Davis, Yu. Matiyasevich, H. Putnam, and J. Robinson, using ideas that stem from Gödel and Turing, solved Hilbert's Tenth Problem by showing that a general procedure for determining the solvability of Diophantine equations did not exist. In modern terms, we would say that the solvability of (1.1) is undecidable.

A major development was the formal study of algorithms that used an unbiased source of random numbers ("coin flips"). For reasons that are not completely clear, such algorithms have proved to be especially useful in number theory and algebra. These algorithms were popularized by the 1974 result of R. Solovay and V. Strassen (published 1977). Their algorithm tested the primality of a given integer in polynomial time, with the unusual feature that the algorithm answered correctly if n was prime, but might err with very small probability if n was composite. M. O. Rabin obtained a similar algorithm in 1976, based on the thesis of G. L. Miller. Since then, many randomized algorithms for number-theoretic problems have been devised.

In 1976, W. Diffie and M. Hellman introduced the concept of *public-key cryptography*, where a message is encoded using a publicly available algorithm. Decryption of the message, however, would remain difficult unless a secret key were known. They also introduced algorithmic number theory into cryptology by constructing a key distribution scheme based on the discrete logarithm. The next year, R. L. Rivest, A. Shamir, and L. M. Adleman showed how to realize a public-key cryptosystem using another problem from number theory. Although many other schemes have been proposed (and some broken) since then, the RSA scheme (named for its inventors) still remains a viable method for sending messages securely. These inventions brought new interest in computational number theory from mathematicians, cryptographers, philosophers, and computer security specialists.

1.4 Notes on Chapter 1

A good discussion of factor tables and tables of primes can be found in Glaisher [1878a], D. N. Lehmer [1909, 1914], and D. H. Lehmer [1941]. Lebon [1907, 1908, 1912, 1914] began the construction of a table of primes up to 10^8, but apparently this was never completed. Also see Tarry [1907]. Currently, the largest table of primes is that of Bays and Hudson, up to 1.2×10^{12}. See Ribenboim [1988b].

For the verification of Goldbach's conjecture, see Sinisalo [1993].

For some famous conjectures inspired by tables, see Birch and Swinnerton-Dyer [1963, 1965] and H. Cohen and H. W. Lenstra [1984a, 1984b].

Discussion of the counterexamples mentioned in the text can be found in Lehman [1960] (also see Tanaka [1980]); Lander and Parkin [1967a]; and Odlyzko and te Riele [1985]. Elkies [1988] found the solution

$$(A, B, C, D) = (2682440, 15365639, 18796760, 20615673)$$

to $A^4 + B^4 + C^4 = D^4$.

The theorem on the prime divisors of 7 consecutive integers can be found in D. H. Lehmer [1964]. For machine-aided proofs of Fermat's last theorem up to various bounds, see D. H. Lehmer, Lehmer, and Vandiver [1954]; Wagstaff [1978]; Tanner and Wagstaff [1987]; Granville and Monagan [1988]; Tanner and Wagstaff [1989]; Coppersmith [1990]; Buhler, Crandall, and Sompolski [1992]; Buhler, Crandall, Ernvall, and Metsänkylä [1993]. For the recent proof of Fermat's "Last Theorem", see Wiles [1995]; Taylor and Wiles [1995].

For the proof of Waring's conjecture on fourth powers, see Balasubramanian, Deshouillers, and Dress [1986a, 1986b]; Deshouillers and Dress [1992].

D. H. Lehmer [1963] drew a distinction between the various uses of computers applied to number theory: a *result* is a statement about a finite set, such as "$2^{2^5} + 1$ is composite". A *theorem* is a statement about an infinite class. Finally, a *theory* is a collection of results and theorems related in "an aesthetically satisfactory way". Machines can easily be used to obtain results, sometime to prove theorems, but rarely to construct theories. Rather, computers are *tools* to aid mathematicians in the construction of theories. For an example of this, see Richards [1974].

There have been several conferences devoted to the use of computers in number theory. Older examples include Atkin and Birch [1971]; Birkhoff and Hall [1971]. A more recent example is Adleman and Huang [1994].

1.1

For the unsolvability of Diophantine equations, see, e.g., Matiyasevich [1993]. For the $\mathcal{NP}$-completeness of solutions to $x^2 \equiv a \pmod{b}$, see Manders and Adleman [1978].

For the use of the Chinese remainder theorem in hardware, see Svoboda [1962]; Szabo and Tanaka [1967]; and Orton, Peppard, and Tavares [1992]. For the application of the discrete logarithm to hardware, see Clark, Bannon, and Keller [1988].

For the pseudo-random number generator with long periods, see Marsaglia and Zaman [1991].

The quotes in this section come from Albert [1941] and G. Hardy [1940a].

1.2

The standard reference for the history of the theory of numbers is the encyclopedic work of Dickson [1919]. A discussion of the application of machines to factorization and primality testing, covering the period 1925 to date, can be found in Brillhart et al. [1983, 1988]. History of early computing devices can be found in Goldstine [1972] and Randell [1982].

The following surveys discuss the use of computers in number theory: E. Lehmer [1956]; Taussky [1956]; Vandiver [1958]; D. H. Lehmer [1963, 1968, 1969, 1971, 1974]; Swinnerton-Dyer [1967]; Fröberg [1968]; Zimmer [1972]; H. Cohen [1979]; Debnath [1982]; H. Williams [1982a]; Nicolas [1985a]; Churchhouse [1988a, 1988b]; Silverman [1991a].

The remark about the Euclidean algorithm comes from Knuth [1981].

The sieve of Eratosthenes was discussed by Nicomachus (see D'ooge, Robbins and Karpinski [1926]) roughly around 125 A. D..

The work of Aryabhata and Brahmagupta is discussed in Weil [1984].

Fermat's observation about the numbers $2^{2^n} + 1$ can be found in Fermat [1894, p. 206]. Euler's factorization of $2^{2^5} + 1$ is in Euler [1849, p. 2]. The American mental calculator Zerah Colburn claimed to have factored $2^{2^5} + 1$ in 1811, at the age of seven, "by the mere operation of his mind" [1833, p. 38].

For Mersenne's statement, see Mersenne [1644] and Dickson [1919, pp. 12–13]. Barlow [1811, p. 43] wrote that $2^{30}(2^{31} - 1)$ was the "greatest perfect number known at present, and probably the greatest that ever will be discovered; for, as they are merely curious without being useful, it is not likely that any person will attempt to find one beyond it." Archibald [1935], Drake [1971], and Heyworth [1982] have discussed possible reasons for Mersenne's conjecture, and Bateman, Selfridge, and Wagstaff [1989] have put forth a new conjecture that may be more plausible.

The primality of $2^{31} - 1$ is discussed in Euler [1849, p. 584].

For discussion of Hindenburg's use of the sieve of Eratosthenes, see Hindenburg [1776] and Dickson [1919, pp. 349].

For quotes on the difficulty of factoring, see Gauss [1801, Art. 329] and Jevons [1874, pp. 141]. For the logical machine, see Jevons [1870]; Akl [1980]; Gardner [1982]. Smith's work can be found in H. Smith [1859].

For the devices of Hindenburg and Felkel, see Hindenburg [1776]; Dickson [1919, pp. 349]; Glaisher [1878a]; Lambert [1785]; and Stephens and Williams [1990].

The difference engine is discussed in Babbage [1889]. The prime-generating polynomial $x^2 - x + 41$ was mentioned by Euler [1849, p. 584]; Legendre [1798] discussed $x^2 + x + 41$. The relationship between these polynomials and the class number of imaginary quadratic fields was explored by Frobenius [1912]; Rabinovitch [1912]; D. H. Lehmer [1936b]; Szekeres [1974]; and Ayoub and Chowla [1981]. Also see H. Cohn [1980, pp. 155–156]; Fendel [1985]; Ribenboim [1988a]; and Fung and Williams [1990].

Lucas' primality test was given in Lucas [1876a, 1876b, 1876c, 1877a, 1878a, 1878b, 1878d]. He discusses the construction of a machine to carry out his test in Lucas [1876b, 1876d, 1877a]; also see Genaille [1891]. The arithmometer is the subject of Lucas [1878b]. A discussion of Lucas' work and a bibliography can be found in Harkin [1957]. For the comment by Gérardin on Genaille's machine, see Gérardin [1912].

For the primality of $2^{61} - 1$, see Bouniakowsky [1884]; Imchenetzki and Bouniakowsky [1887]; Seelhoff [1886]; and Lucas [1887]. For biographical information about Pervushin, and a discussion of his work on $2^{61} - 1$, see Raik [1953].

For information about the early history of mechanical calculators, see Boys [1901]; Turck [1921]; Couffignal [1933]; Lilley [1942]; Comrie [1946]; and Taton and Flad [1963], just to name a few references.

The story about Cole is taken from Bell [1951]; also see Cole [1903].

Pocklington's work can be found in Pocklington [1910, 1914, 1917]. Galois had suggested [1830] that one should find an irreducible polynomial over a finite field by trial and error. But he was not as specific as Pocklington about the running time of his suggestion.

Lawrence's suggestion appears in Lawrence [1896]. For the sieve machines of Kraitchik and Gérardin, see Gérardin [1912]. The work of Carissan is described in Carissan [1920]; a more accessible reference is d'Ocagne [1922]. For the story of the machine's discovery, see Shallit, Williams, and Morain [1995]. As of this writing, Carissan's machine is located in the Conservatoire National des Arts et Métiers, Paris. Stephens and Williams [1990] discuss the history of sieving.

A device to perform Lucas' test is discussed in Mason [1914a, 1914b].

The original Cunningham tables are Cunningham and Woodall [1925]. The modern Cunningham tables are Brillhart et al. [1983] and [1988].

For information about the factor stencils, see D. N. Lehmer [1925, 1929]. For stencils for solving quadratic congruences, see Robinson [1940].

Lehmer's proof and improvement of Lucas' test can be found in D. H. Lehmer [1930a]. His 1946 computation of the multiplicative order of 2 modulo p is mentioned in Lehmer [1974].

For more information about the bicycle-chain sieve and its descendants, see D. N. Lehmer [1932] and D. H. Lehmer [1928b, 1932b, 1933a, 1934, 1966, 1980]. D. H. Lehmer's collected works have appeared as D. H. Lehmer [1981]. Patterson and Williams [1983] have developed a sieve machine that examines 133 million numbers per second.

Lehmer distinguished between "fast" and "slow" number-theoretic algorithms in D. H. Lehmer [1969]; also see D. H. Lehmer [1971, 1973]. For a discussion of parallelism in number-theoretic computation, see D. H. Lehmer [1976a].

An excellent biography of Alan Turing is Hodges [1983]. He mentions that in 1939 Turing designed, but never completed, a machine that would hunt for zeroes of the Riemann zeta function. The method is discussed in Turing [1945, 1953].

For the first program ever run on a stored-program electronic digital computer, see F. Williams [1975, p. 330] and Campbell-Kelly [1980, p. 134] (both these accounts contain typographical errors). For the unsuccessful search for Mersenne primes, see Hodges [1983, p. 392].

Ferrier's result can be found in Ferrier [1952].

The following are papers on the discovery of Mersenne primes by computer: Reid [1953]; Robinson [1954]; Gillies [1964]; Tuckerman [1971]; Slowinski [1979]; Noll and Nickel [1980]; I. Peterson [1988]; Colquitt and Welsh [1991].

For a discussion of the application of supercomputers to factoring, see J. Davis and Holdridge [1983, 1988], McCurdy and Wunderlich [1987] and Lerner [1988].

For a discussion of distributed implementations of factoring algorithms, see Caron and Silverman [1988].

Special processors for number theory are discussed in J. Smith and Wagstaff [1983]; Rudd, Buell, and Chiarulli [1984]; Dubner and Dubner [1986]; Pomerance, Smith, and Tuler [1988]; and Chudnovksy, Chudnovsky, Denneau, and Younis [1988].

For the use of microcomputers in number theory research, see Wagstaff [1990].

For software packages for number theory, see Wunderlich and Selfridge [1974]; Buell and Ward [1989]; Hollinger and Serf [1991].

1.3

Hilbert's original list of problems can be found in Hilbert [1900]. An English translation is Hilbert [1902]. The proceedings of a 1974 symposium devoted to Hilbert problems is Browder [1976].

For the work of Gödel, Church, Turing, and Post, see Gödel [1931]; Church [1932, 1933]; Turing [1936]; and Post [1944]. The history surrounding these developments is discussed in Kleene [1981] and Rosser [1984]. See M. Davis [1965] for a collection of papers dealing with undecidable problems.

For polynomial time, see Cobham [1964] and Edmonds [1965]. For $\mathcal{NP}$-completeness, see Cook [1971] and Karp [1972].

A good exposition of Hilbert's tenth problem can be found in M. Davis [1973]. Also see J. Jones and Matiyasevich [1991].

The following are some early examples of the analysis of randomized algorithms: Rabin [1976, 1980b], Solovay and Strassen [1977].

The RSA scheme was introduced in Rivest, Shamir, and Adleman [1978].

2 Fundamentals of Number Theory

This chapter has three purposes. The first is to recall some of the basic definitions and theorems of elementary number theory. This is done in the first three sections. Since we expect that the reader is familiar with some of this material, we do not always provide proofs, although some proofs are sketched as exercises, and references are given.

The second purpose is to state some results from analysis that can be used to estimate common sums and number-theoretic functions. These estimates will be valuable for bounding the running time of algorithms discussed in future chapters.

Finally, we provide a brief review of some notions from modern algebra that provide a very useful context for thinking about number-theoretic algorithms.

We start by giving some definitions and stating some of the fundamental computational problems in algorithmic number theory. We will see later that some of these problems seem to be "easy," while others are believed to be "hard."

2.1 Notation, Definitions, and Some Computational Problems

In this section, we state some of the basic definitions and theorems from elementary number theory. We contrast the statements of these theorems with six classical problems from algorithmic number theory.

We denote the ring of integers by the symbol $\mathbb{Z}$, the rational numbers by the symbol $\mathbb{Q}$, the real numbers by the symbol $\mathbb{R}$, and the complex numbers by the symbol $\mathbb{C}$. If z is a complex number, then $\Re(z)$ and $\Im(z)$ denote its real and imaginary part, respectively.

The *greatest integer* function or *floor* function $\lfloor x \rfloor$ maps every real number x to the greatest integer $\leq x$. Thus $\lfloor \pi \rfloor = 3$. Similarly, the *least integer* function, or *ceiling* function $\lceil x \rceil$ maps every real number x to the least integer $\geq x$. Thus $\lceil \pi \rceil = 4$.

We say that an integer m is a *divisor* of an integer n if there exists a third integer a such that $n = am$. If m is a divisor of n, then we write $m \mid n$. For example, $2 \mid 12$.

We say that d is a *common divisor* of m and n if $d \mid m$ and $d \mid n$. For example, 2 is a common divisor of 12 and 30.

The *greatest common divisor* of two integers m and n is a non-negative integer d that is a common divisor of m and n, such that if e is any common divisor of m and n, then $e \mid d$. We write $d = \gcd(m, n)$. Therefore, for $(m, n) \neq (0, 0)$, we see that $\gcd(m, n)$ is actually the *numerically greatest* common divisor, although it has been defined without reference to magnitude. For example, $6 = \gcd(12, 30)$. We define $\gcd(m, 0) = \gcd(0, m) = |m|$ for all m. We say two integers m, n are *relatively prime* if $\gcd(m, n) = 1$; in this case, we sometimes write $m \perp n$.

We say f is a *common multiple* of m and n if $m \mid f$ and $n \mid f$. The *least common multiple* of m and n is a non-negative integer f such that if e is any common multiple of m and n, then $f \mid e$. We write $f = \operatorname{lcm}(m, n)$. We define $\operatorname{lcm}(m, 0) = \operatorname{lcm}(0, m) = 0$ for all m.

PROBLEM 1 Given two positive integers, compute their greatest common divisor and least common multiple.

Problem 1 is discussed in more detail in Chapter 4.

A *prime* is an integer $p > 1$ whose only positive divisors are 1 and p. A number $n > 1$ which is not a prime is said to be *composite*. For example, 2, 3, 5, 7 are primes, but $91 = 7 \cdot 13$ is composite. The number 1 is neither prime nor composite; it is a *unit*.

THEOREM 2.1.1 There are infinitely many primes.

Proof This famous proof is due to Euclid: Assume there are only finitely many primes, say $p_1, p_2, \ldots, p_n$. Then $1 + p_1 p_2 \cdots p_n$ is divisible by a prime different from $p_1, p_2, \ldots, p_n$, a contradiction. ■

PROBLEM 2 Given a positive integer, determine whether it is prime.

Problem 2 is discussed in more detail in Chapter 9.

The number of primes less than or equal to x is frequently abbreviated $\pi(x)$. We write

$$\pi(x) = \sum_{p \le x} 1.$$

(This formula uses a convention, which we will use throughout this book, that summations over the variable p refer to primes only.)

PROBLEM 3 Given an integer x, compute the exact value of $\pi(x)$.

(Note: it is known (see Chapter 8 for more details) that

$$\lim_{x \to \infty} \frac{\pi(x)}{x/\log x} = 1;$$

this is called the *Prime Number Theorem*. Thus $x/\log x$ gives us a good *approximation* for the value of $\pi(x)$. Problem 3, however, asks us to compute the *exact* value of $\pi(x)$.)

Problem 3 is discussed further in Chapter 9.

THEOREM 2.1.2 (THE FUNDAMENTAL THEOREM OF ARITHMETIC) Every positive integer can be expressed as a product of nontrivial powers of distinct primes

$$n = p_1^{e_1} p_2^{e_2} \cdots p_k^{e_k},$$

and up to a rearrangement of the factors, this *prime factorization* is unique.

For a proof, see Exercise 6. Note that the integer 1 is given by the empty product of primes.

PROBLEM 4 Given a positive integer n, determine its prime factorization.

Problem 4 is discussed in Chapter 11.

Let n be a positive integer, and let a and b be arbitrary integers. Then we write $a \equiv b \pmod{n}$ when $n \mid a - b$; this notation is read as "a is congruent to b, modulo n". If n is understood, we simply write $a \equiv b$. It is easy to verify that $\equiv$ is an equivalence relation on $\mathbb{Z}$ and hence partitions $\mathbb{Z}$ into a set of n disjoint equivalence classes, which we write as $\mathbb{Z}/(n)$.

Frequently it will be useful to identify a canonical *representative* for each such equivalence class. A simple choice for n positive is the least non-negative residue. We abbreviate this choice as $a \bmod n$; this unparenthesized "mod" is a *function* defined as follows:

$$a \bmod n = \begin{cases} a, & \text{if } n = 0; \\ a - n\left\lfloor \frac{a}{n} \right\rfloor, & \text{otherwise.} \end{cases} \tag{2.1}$$

Note that this definition is meaningful for all integers a and n.

THEOREM 2.1.3 (FERMAT) If p is a prime and $\gcd(a, p) = 1$, then $a^{p-1} \equiv 1 \pmod{p}$.

For a proof, see Exercise 13.

PROBLEM 5 Given a, b, n, compute $a^b \bmod n$.

Problem 5 is discussed in Chapter 5.

The *Euler phi function*, $\varphi(n)$, is defined to be the number of positive integers $\leq n$ which are relatively prime to n. For example, $\varphi(10) = 4$. We write

$$\varphi(n) = \sum_{\substack{1 \leq k \leq n \\ \gcd(k,n)=1}} 1.$$

One application of $\varphi(n)$ is the following theorem, a generalization of Theorem 2.1.3:

THEOREM 2.1.4 (EULER) If n is a positive integer, and $\gcd(a, n) = 1$, then $a^{\varphi(n)} \equiv 1 \pmod{n}$.

For a proof, see Exercise 14.

PROBLEM 6 Given a positive integer n, compute $\varphi(n)$.

We will see later that Problems 1 and 5 admit "fast" solutions, while Problems 3, 4, and 6 seem to be "hard". The status of Problem 2 is too complicated to summarize quickly, and it is covered more thoroughly in chapter 9.

2.2 More Definitions

In this section, we give some more definitions of useful number-theoretic functions.

The *sum-of-divisors function*, $\sigma(n)$, is defined as follows:

$$\sigma(n) = \sum_{d \mid n} d.$$

(The notation for this sum follows the usual convention that $\sum_{d \mid n} f(d)$ denotes a sum over all positive divisors of n.) More generally, we have

$$\sigma_k(n) = \sum_{d \mid n} d^k.$$

We define $d(n) = \sigma_0(n)$, the number of positive divisors of the integer n.

Frequently we will need to express the highest power of a prime p which divides an integer n. This is written $\nu_p(n)$; formally, we have

$$p^{\nu_p(n)} \mid n \quad \text{but} \quad p^{\nu_p(n)+1} \nmid n.$$

If p^k divides n, and p^{k+1} does not, then we sometimes write $p^k \,\|\, n$. This definition can be generalized to the case where p is not necessarily prime: we write $m^k \,\|\, n$ if $m^k \mid n$ and $\gcd(n/m^k, m) = 1$.

If the prime factorization of n is $p_1{}^{e_1} \cdots p_k{}^{e_k}$, then we define $\omega(n) = k$, the number of distinct prime factors of n, and $\omega'(n) = e_1 + e_2 + \cdots + e_k$, the total number of primes dividing n.

Let y be a positive real number; we say an integer n is *y-smooth* if all of x's prime divisors are $\leq y$. For example, 36 is 3-smooth, but not 2-smooth. We let $\Psi(x, y)$ denote the number of integers in the interval $[1, x]$ which are y-smooth. For example $\Psi(10, 3) = 7$.

There are commonly-used notations for sums over certain functions of primes. For example, we have the following definitions:

$$\vartheta(x) = \sum_{p \leq x} \log p.$$

$$\psi(x) = \sum_{p^k \leq x} \log p = \log(\operatorname{lcm}(1, 2, 3, \ldots, \lfloor x \rfloor)).$$

(In this last equation, the summation is over all primes p and integers $k \geq 0$.)

Sums over primes are often expressed in terms of the *von Mangoldt lambda function*, defined by

$$\Lambda(n) = \begin{cases} \log p, & \text{if } n = p^k \text{ for some prime } p; \\ 0, & \text{otherwise.} \end{cases}$$

Thus, for example, $\psi(x) = \sum_{n \leq x} \Lambda(n)$.

Finally, $\zeta(s)$ denotes the *Riemann zeta function*; if s is a complex number with $\mathrm{Re}(s) > 1$, we have

$$\zeta(s) = \sum_{k \geq 1} \frac{1}{k^s}.$$

It is useful to know that $\zeta(2) = \pi^2/6$.

2.3 Multiplicative Functions and Möbius Inversion

Suppose the prime factorization of n is $\prod_{1 \leq i \leq k} p_i^{e_i}$. Exercise 7 asks you to prove that

$$\varphi(n) = \prod_{1 \leq i \leq k} (p_i - 1) p_i^{e_i - 1}. \tag{2.2}$$

Many number-theoretic functions can be written in a form similar to that of (2.2). If $f(mn) = f(m)f(n)$ whenever m and n are relatively prime, we say f is *multiplicative*. It can be shown that both $\varphi(n)$ and $\sigma(n)$ are multiplicative. A multiplicative function is completely determined by its values on the prime powers.

Another extremely useful multiplicative function is the *Möbius function*, defined as follows:

$$\mu(n) = \begin{cases} 0 & \text{if } n \text{ is divisible by a square} \neq 1; \\ (-1)^t & \text{if } n = p_1 \ldots p_t, \text{ the product of } t \text{ distinct primes.} \end{cases}$$

Note: $\mu(1) = 1$, since 1 is divisible by 0 distinct primes.

We say that n is *squarefree* if $\mu(n) \neq 0$. The utility of the Möbius function arises from the following theorem, called the *Möbius inversion formula*:

THEOREM 2.3.1 Suppose $g(n) = \sum_{d \mid n} f(d)$. Then $f(n) = \sum_{d \mid n} \mu(d) g(n/d)$.

To prove this theorem, we need two simple lemmas:

LEMMA 2.3.2 Let $f(n)$ be a multiplicative function. Then

$$g(n) = \sum_{d \mid n} f(d)$$

is also multiplicative.

Proof We need to show that if $\gcd(m, n) = 1$, then $g(m)g(n) = g(mn)$. So assume $\gcd(m, n) = 1$. Then

$$
\begin{aligned}
g(m)g(n) &= \left(\sum_{d \mid m} f(d)\right)\left(\sum_{e \mid n} f(e)\right) \\
&= \sum_{d \mid m}\sum_{e \mid n} f(d)f(e) \\
&= \sum_{d \mid m}\sum_{e \mid n} f(de) \\
&= \sum_{c \mid mn} f(c) \\
&= g(mn).
\end{aligned}
$$

■

LEMMA 2.3.3 $\sum_{d \mid n} \mu(d) = \begin{cases} 1 & \text{if } n = 1; \\ 0 & \text{otherwise.} \end{cases}$

Proof It can be easily verified that $\mu(n)$ is multiplicative. Then by the Lemma, $\sum_{d \mid n} \mu(d)$ is multiplicative. But $\sum_{d \mid p^k} \mu(d) = \mu(1) + \mu(p) = 0$. Thus $\sum_{d \mid n} \mu(d) = 0$ unless n has no prime divisors, i.e., $n = 1$. ■

We can now prove Theorem 2.3.1.

$$
\begin{aligned}
\sum_{d \mid n} \mu(d) g\left(n/d\right) &= \sum_{d \mid n} \mu(d) \sum_{c \mid (n/d)} f(c) \\
&= \sum_{d \mid n}\sum_{c \mid (n/d)} \mu(d)f(c) \\
&= \sum_{cd \mid n} \mu(d)f(c) \\
&= \sum_{c \mid n}\sum_{d \mid (n/c)} \mu(d)f(c) \\
&= \sum_{c \mid n} f(c) \sum_{d \mid (n/c)} \mu(d) \\
&= f(n).
\end{aligned}
$$

■

Some further examples are given in the exercises.

2.4 Notation: Big-O, Little-o, Big-Omega, Big-Theta

In this section, we introduce some useful notation for expressing the asymptotic growth rates of functions.

DEFINITION 2.4.1 We say that $f(n) = O(g(n))$ if there exist constants $c > 0$ and N such that $f(n) \leq cg(n)$ for all $n \geq N$.

We say $f(n) = o(g(n))$ if $\lim_{n\to\infty} f(n)/g(n) = 0$.

We say $f(n) = \Omega(g(n))$ if there exist constants c, N such that $f(n) \geq cg(n)$ for all $n \geq N$.

We say $f(n) = \Omega_\infty(g(n))$ if there exists a constant c such that $f(n) \geq cg(n)$ for infinitely many distinct positive integers n.

We say $f(n) = \Theta(g(n))$ if $f(n) = O(g(n))$ and $g(n) = O(f(n))$.

Finally, we say $f(n) \sim g(n)$ if $\lim_{n\to\infty} f(n)/g(n) = 1$.

Sometimes, we use these symbols with functions of more than one variable, which may take negative arguments. For example, if we say that $f(m, n) = O(|mn|)$, then we mean that there exist c, N such that $f(m, n) \leq c|mn|$ whenever $|m| \geq N$ or $|n| \geq N$.

2.5 Abel's Identity and Euler's Summation Formula

We assume that the reader is familiar with Stieltjes integrals.

Abel's identity is a powerful tool for estimating sums in terms of integrals.

LEMMA 2.5.1 Define $A(x) = \sum_{1\leq k\leq x} a(k)$ (note that $A(x) = 0$ for $x < 1$). If f has a continuous derivative on $[1, x]$, then

$$\sum_{1\leq k\leq x} a(k)f(k) = A(x)f(x) - \int_1^x A(t)f'(t)\,dt.$$

Proof Let $0 < y < 1$. Using Stieltjes integration, we see that

$$\begin{aligned}\sum_{y<k\leq x} a(k)f(k) &= \int_y^x f(t)\,dA(t) \\ &= f(x)A(x) - f(y)A(y) - \int_y^x A(t)\,df(t) \quad \text{(using integration by parts)} \\ &= f(x)A(x) - f(y)A(y) - \int_y^x A(t)f'(t)\,dt.\end{aligned}$$

Now, letting $y \to 1^-$, we get the desired result. ■

Euler's summation formula is a useful corollary:

COROLLARY 2.5.2 Let f be continuously differentiable on $[1, x]$. Then

$$\sum_{1\le k\le x} f(k) = f(1) - (x - \lfloor x\rfloor)f(x) + \int_1^x f(t)\,dt + \int_1^x (t - \lfloor t\rfloor)f'(t)\,dt.$$

Proof Put $a(k) = 1$ for all positive integers k; then $A(x) = \lfloor x\rfloor$. Then, using Abel's identity, we get

$$\sum_{1\le k\le x} f(k) = \lfloor x\rfloor\, f(x) - \int_1^x \lfloor t\rfloor\, f'(t)\,dt.$$

But, using integration by parts, we get

$$\int_1^x tf'(t)\,dt = xf(x) - f(1) - \int_1^x f(t)\,dt.$$

Combining these two equations gives the result. ■

Now let us look at some applications. Define $H_n = \sum_{1\le k\le n} 1/k$; H_n is sometimes called the n-th *harmonic number*. We have

THEOREM 2.5.3

$$H_n = \log n + \gamma + O\left(\frac{1}{n}\right),$$

where

$$\gamma = 1 - \int_1^\infty \frac{t - \lfloor t\rfloor}{t^2}\,dt \doteq .5772$$

is *Euler's constant*.

(We write $a \doteq b$ if the decimal expansion of a agrees with b, when rounded to the number of decimal places given in b.)

Proof By Corollary 2.5.2 with $f(x) = 1/x$, we get

$$\begin{aligned} H_n &= \int_1^n \frac{dt}{t} + 1 - \int_1^n \frac{t - \lfloor t\rfloor}{t^2}\,dt \\ &= \log n + \left(1 - \int_1^\infty \frac{t - \lfloor t\rfloor}{t^2}\,dt\right) + \int_n^\infty \frac{t - \lfloor t\rfloor}{t^2}\,dt \\ &= \log n + \gamma + O\left(\frac{1}{n}\right). \end{aligned}$$ ■

Another example of the use of Theorem 2.5.2 is the following estimate on the "average order" of $d(n)$, the number of divisors of n.

THEOREM 2.5.4

(a) Weak version:

$$\sum_{1\le n\le x} d(n) = x\log x + O(x)$$

(b) Stronger version (Dirichlet, 1849):

$$\sum_{1\le n\le x} d(n) = x\log x + (2\gamma - 1)x + O\left(\sqrt{x}\right)$$

Proof To prove part (a), we observe that $\sum_{1\le n\le x} d(n)$ is just the number of lattice points in the uv-plane between the axes $u = 0$ and $v = 0$ and the hyperbola $uv = x$. Thus we have

$$\begin{aligned}\sum_{1\le n\le x} d(n) &= \sum_{1\le d\le x}\ \sum_{1\le q\le x/d} 1\\ &= \sum_{1\le d\le x}\left\lfloor \frac{x}{d}\right\rfloor\\ &= \left(x\sum_{1\le d\le x}\frac{1}{d}\right) + O(x)\\ &= x\left(\log x + \gamma + O\left(\frac{1}{x}\right)\right) + O(x)\\ &= x\log x + O(x)\end{aligned}$$

Part (b) is left to the reader as Exercise 35. ■

2.6 Asymptotic Integration

Approximating a sum by the methods of the previous section sometimes leads to an integral that is difficult or impossible to evaluate in closed form. If one is interested in the asymptotic behavior of the integral, however, it is often easy to get good approximations.

One useful idea is the following: suppose we have an integral of the form $\int_a^x f(t)dt$, where $f(t) = s(t)t^{\mu}$; we suppose that $s(t)$ changes slowly compared to the power of t. Then it makes sense to try to approximate the integral by $s(x)x^{\mu+1}/(\mu+1)$, the rough idea being that $s(t)$ acts like a constant and should be pulled out of the integral. The next theorem formalizes this notion.

THEOREM 2.6.1 Let f be a positive real-valued function defined in $[a, \infty)$. Assume that f is continuously differentiable, and that $f'(x)/f(x) \sim \mu/x$ for some real number $\mu > -1$.

Then if $\mu \neq 0$, we have

$$\int_a^x f(t)\,dt \sim \frac{xf(x)}{\mu + 1}.$$

If $f'(x)/f(x) = o(1/x)$, then

$$\int_a^x f(t)\,dt \sim xf(x).$$

Proof We prove the first assertion; the second is left to the reader. By integrating f'/f, we see that $\log f(x) \sim \mu \log x$, that is, $f(x) = x^{\mu+o(1)}$. This guarantees that the integral diverges, so we have

$$\int_a^x f(t)\,dt \sim \int_a^x \frac{t}{\mu}\,df(t) = \frac{x}{\mu} f(x) - \frac{1}{\mu}\int_a^x f(t)\,dt + O(1).$$

Solving this for $\int_a^x f(t)\,dt$, we obtain the result. ■

As an example, we can use this theorem to approximate

$$\int_2^x \frac{dt}{\log t},$$

which cannot be evaluated in closed form. With $f(x) = 1/\log x$, we have $f'/f = -1/(x \log x)$, so that

$$\int_2^x \frac{dt}{\log t} \sim \frac{x}{\log x}.$$

We may continue this process to obtain an estimate of the error. If $\epsilon(x) = \int_2^x dt/\log t - x/\log x$, we have $\epsilon'(x) = 1/(\log x)^2$, from which it follows that $\epsilon(x) = \int_2^x dt/(\log t)^2 + O(1)$. Applying Theorem 2.6.1 again to this new integral, we get

$$\int_2^x \frac{dt}{\log t} = \frac{x}{\log x} + O\left(\frac{x}{(\log x)^2}\right).$$

2.7 Estimating Sums over Primes

In many situations, we wish to estimate $\sum_{p \le x} f(p)$, where the sum is taken over primes. Because roughly 1 out of every $\log t$ numbers near t is prime, it is usually reasonable to make the following approximations:

$$\sum_{p \le x} f(p) \approx \sum_{n \le x} \frac{f(n)}{\log n} \approx \int_2^x \frac{f(t)\,dt}{\log t}.$$

(We use $\approx$ to mean "approximate equality," which we do not define rigorously.) We now give a theorem from which estimates of this type can be rigorously proved.

THEOREM 2.7.1 Let f be continuously differentiable on an open interval containing $[2, \infty)$, and let $\pi(x) = \mathrm{li}(x) + \epsilon(x)$, where $\mathrm{li}(x) = \int_2^x dt/\log t$. Then if $x \geq 2$,

$$\sum_{p \leq x} f(p) = \int_2^x \frac{f(t)\,dt}{\log t} + \epsilon(x)f(x) - \int_2^x \epsilon(t)f'(t)\,dt.$$

Proof Choose $a < 2$. Writing the sum as a Stieltjes integral, we have

$$\sum_{p \leq x} f(p) = \int_a^x f(t)\,d\pi(t) = \int_a^x f(t)\,d(\mathrm{li}(t) - \epsilon(t)).$$

Observing that $d\ \mathrm{li}(t) = 1/(\log t)$ and using integration by parts on $\int_a^x f(t)\,d\epsilon(t)$, we see this equals

$$\int_a^x \frac{f(t)\,dt}{\log t} + [\epsilon(t)f(t)]_a^x - \int_a^x \epsilon(t)f'(t)\,dt.$$

Letting $a \to 2^-$ and observing that $\epsilon(a) \to 0$ completes the proof. ■

To use this theorem, we must estimate ϵ. According to the prime number theorem, $\epsilon(t) = o(t/\log t)$, and often it is not necessary to be more precise than this. For example, let us estimate

$$\sum_{p \leq x} p.$$

This is

$$\int_2^x \frac{t\,dt}{\log t} + x\epsilon(x) - \int_2^x \epsilon(t)\,dt;$$

the main term is asymptotic to $x^2/(2\log x)$, and each of the other terms is easily seen to be $o(x^2/\log x)$. So we have

$$\sum_{p \leq x} p \sim \frac{x^2}{2\log x}.$$

2.8 Basic Concepts of Abstract Algebra

In this section, we review some of the basic concepts of abstract algebra that will be used in this book. The reader who is unacquainted with this material should not expect to learn it here; rather, an introductory text, such as those mentioned in the Notes section, should be consulted first. This section may, however, prove useful to the reader whose command of algebra is a bit rusty.

2.8.1 Semigroups, Monoids, and Groups

Let G be a nonempty set and let $\cdot$ be a binary operation. If G is *closed* under $\cdot$, that is, if $a \cdot b \in G$ for all $a, b \in G$, and if further $\cdot$ obeys the *associative law*, $a \cdot (b \cdot c) = (a \cdot b) \cdot c$ for all $a, b, c \in G$, then $(G, \cdot)$ is said to be a *semigroup*.

If in addition G contains an *identity element* e such that $a \cdot e = e \cdot a = a$ for all $a \in G$, then $(G, \cdot)$ is said to be a *monoid*. For example, the set of binary strings under concatenation forms a monoid, with the empty string ϵ playing the role of the identity element.

Finally, if G also contains *inverses*, that is, if for each element $a \in G$ there exists an $a^{-1} \in G$ such that $a \cdot a^{-1} = a^{-1} \cdot a = e$, then $(G, \cdot)$ is called a *group*. For example, the set $\{0, 1\}$ forms a group under addition modulo 2.

If the binary operation $\cdot$ is understood, then we may abuse the notation and write simply that G itself is a group (or semigroup, or monoid, as appropriate). Furthermore, we will usually omit explicit use of the $\cdot$ in expressing products of elements, writing abc for $a \cdot b \cdot c$, and we will use 1 to represent the identity e.

If $|G|$ is finite, we say G is a *finite group*. If the binary operation $\cdot$ obeys the *commutative law* ($a \cdot b = b \cdot a$ for all $a, b \in G$), then we say G is *abelian*. Most, if not all, of the groups we will encounter in this book will be finite abelian groups. When the group is abelian, the group operation is commonly denoted by $+$ rather than $\cdot$; in this case, the identity element is denoted by 0.

A nonempty set $H \subseteq G$ is said to be a *subgroup* of G if $(H, \cdot)$ is itself a group. It is easy to see that the following two conditions are sufficient for $H \subseteq G$ to be a group: (i) $a \cdot b \in H$ for all $a, b \in H$, and (ii) $a^{-1} \in H$ for all $a \in H$. If H is finite, then the first condition alone suffices.

If S is a subset of G, then by $\langle S \rangle$ we mean the set of all elements of G representable as a (possibly empty) product of elements of S. We call $\langle S \rangle$ the *subgroup of G generated by S*. If $G = \langle S \rangle$, and $|S|$ is finite, then G is *finitely generated*. If $G = \langle a \rangle$ for some element $a \in G$, we say that G is *cyclic*. As an example, consider the integers (mod n); they form a cyclic group of order n under addition.

Suppose H is a subgroup of G. We write $a \equiv b \pmod{H}$ if $ab^{-1} \in H$; it is easy to verify that $\equiv$ is an equivalence relation. We define $Ha = \{ha : h \in H\}$. The set Ha is called a *right coset* of H in G. It is easy to see that $Ha = \{x \in G : x \equiv a \pmod{G}\}$. There is a one-one correspondence between any two right cosets of H in G, and G can be written as a disjoint union of right cosets. From this, we obtain *Lagrange's theorem:* if G is finite, and H is a subgroup of G, then $|H|$ divides $|G|$.

The *index* of H in G, written $(G : H)$, is defined to be the number of distinct right cosets of H in G. If G is finite, this is simply $|G|/|H|$.

The *order* of an element $a \in G$ is defined to be the least positive integer i such that $a^i = 1$; if no such integer exists, we take $i = \infty$. We write ord $a = i$. Now assume G is finite. By Lagrange's theorem we know that $|\langle a \rangle|$ divides $|G|$, and hence we conclude that ord a divides $|G|$. Furthermore, $a^{|G|} = 1$ for all $a \in G$.

The *exponent* of a group G, $\exp(G)$, is defined to be the least positive integer i such that $a^i = 1$ for all $a \in G$; if no such integer exists, we define $\exp(G) = \infty$.

We say a subgroup N of G is a *normal subgroup* of G if $gng^{-1} \in N$ for all $g \in G$, $n \in N$. By G/N we will mean the set of right cosets of N in G; if N is normal, this structure forms a group using the multiplication rule $(Na)(Nb) = N(ab)$. All of the subgroups of an abelian group are normal.

Let G and G' be groups, and let $\varphi : G \to G'$ be a map satisfying the condition $\varphi(ab) = \varphi(a)\varphi(b)$. Then φ is called a *homomorphism*. The *kernel* of a homomorphism φ is the set $K = \{x \in G : \varphi(x) = e'\}$, where e' is the identity in G'. It is not difficult to see that K is actually a normal subgroup of G.

If the homomorphism φ is one-one, then it is an *isomorphism*. Two groups G and G' are said to be *isomorphic* if there is an isomorphism from G onto G'. In this case, we write $G \cong G'$.

If φ is a homomorphism of G onto G' with kernel K, then G/K is a group that is isomorphic to G'.

If $G_1, G_2, \ldots, G_n$ are n groups, we define

$$G_1 \times G_2 \times \cdots \times G_n,$$

the *direct product* of the groups, to be the group

$$G = \{(g_1, g_2, \ldots, g_n) : g_i \in G_i\}.$$

Multiplication and inverses are computed componentwise; the identity element is $(e_1, e_2, \ldots, e_n)$, where e_i is the identity in group G_i.

The so-called *fundamental theorem of finite abelian groups* says that every finite abelian group is isomorphic to a direct product of cyclic groups of prime power order.

Two elements x, y in a group are called *conjugate* if there is another element z such that $x = zyz^{-1}$. This is an equivalence relation; the set of all elements conjugate to a given element is called its *conjugacy class*.

2.8.2 Rings

We now turn to rings. A nonempty set R, together with two binary operations $\cdot$ and $+$ is called a *ring* if $(R, +)$ is an abelian group, $(R, \cdot)$ is a semigroup, and the *distributive laws* are obeyed: $a \cdot (b + c) = a \cdot b + a \cdot c$, and $(b + c) \cdot a = b \cdot a + c \cdot a$. If in addition, $(R, \cdot)$ is a monoid, then R is called a *ring with unit element*. If the operation $\cdot$ obeys the commutative law, then R is called a *commutative ring*. For example, the integers form a commutative ring under ordinary addition and multiplication. Most, if not all, of the rings we will encounter in this book are commutative rings with unit element. If $(R, \cdot)$ forms a group, then R is called a *division ring*.

Let $a \in R$. If there exists a nonzero $b \in R$ such that $ab = 0$, then a is called a *zero-divisor*. If R is a commutative ring and possesses no zero-divisors, then it is called an *integral domain*.

An element $a \in R$ is said to be a *unit* if it possesses a multiplicative inverse in R. (Do not confuse *unit* with *unit element*!) For example, in the ring $\mathbb{Z}$, the units are 1 and -1. If $a = ub$, where u is a unit, then a and b are said to be *associates*. Let a be an element of R which is not a unit. If $a = bc$ for $b, c \in R$ implies that either b or c is a unit, then a is said to be an *irreducible* element.

If R is a ring with unit element, and R is an integral domain, then we say R is a *unique factorization domain* if every nonzero element in R is a unit or a product of a finite number of irreducible elements of R, and this decomposition is unique, up to order and associates. For example, $\mathbb{Z}$ is a unique factorization domain.

Let R be a commutative ring with unit element. By $R[X]$ (the *polynomial ring* in the indeterminate X over R) we mean the set of all formal expressions of the form

$$p(X) = a_nX^n + a_{n-1}X^{n-1} + \cdots + a_1X + a_0,$$

where $a_0, a_1, \ldots, a_n \in R$. If $a_n \neq 0$, the *degree* of the polynomial $p(X)$ is defined to be n; we write $\deg\ p = n$. The degree of 0 is defined to be $-\infty$. Addition, multiplication, and equality are defined in the usual way for polynomials. It is not hard to see that if R is an integral domain, so is $R[X]$. Furthermore, if R is a unique factorization domain, so is $R[X]$.

Let R and R' be rings, and let $\varphi : R \rightarrow R'$ be a map satisfying the two conditions $\varphi(a + b) = \varphi(a) + \varphi(b)$, and $\varphi(ab) = \varphi(a)\varphi(b)$ for all $a, b \in R$. Then φ is a *(ring) homomorphism*. The terms *isomorphism* and *isomorphic*, and the notation $R \cong R'$, are then defined in analogy with the usage in groups, above.

An *ideal* I of a ring R is a nonempty subset of R such that I is a subgroup of R (under addition), and for all $t \in I$ and $r \in R$, both tr and rt are in I. We define the *quotient ring* R/I to be the set of distinct cosets of I in R (considering I to be an additive subgroup of R). The ring operations in R/I are performed according to the rules $(a+I)+(b+I) = (a+b)+I$, and $(a + I)(b + I) = ab + I$.

An ideal $I \neq R$ of R is said to be *maximal* if whenever I' is an ideal of R such that $I \subseteq I' \subseteq R$, then either $I' = I$ or $I' = R$. If R is a commutative ring with unit element, and

I is an ideal of R, then I is maximal if and only if R/I is a field (see the definition of field below).

If R is a commutative ring, then an ideal I of R is said to be *prime* if for all $a, b \in R$, $ab \in I$ implies $a \in I$ or $b \in I$. Then I is a prime ideal of R if and only if R/I is an integral domain.

An ideal which is generated by a single element is said to be *principal*. A *principal ideal domain* (sometimes abbreviated *p.i.d.*) is an integral domain in which every ideal is principal. For example, $\mathbb{Z}$ is a principal ideal domain.

By R^* we mean the *group of units* of the ring R, that is, the group of elements which possess a multiplicative inverse.

If $R_1, R_2, \ldots, R_n$ are n rings, we define

$$R_1 \bigoplus R_2 \bigoplus \cdots \bigoplus R_n,$$

the *direct sum* of the rings, to be the ring

$$R = \{(r_1, r_2, \ldots, r_n) \ : \ r_i \in R_i\}.$$

This is analogous to the direct product of groups defined above. Multiplication and inverses are computed componentwise; the additive and multiplicative identities of R are formed from the additive and multiplicative identities of each R_i.

2.8.3 Fields

A *field* is a commutative division ring. Examples include $\mathbb{Q}$, $\mathbb{R}$, and $\mathbb{C}$. (Finite fields, such as $\mathbb{Z}/(p)$ for p a prime number, are discussed in detail in Chapter 6.) It is easy to verify that any *finite* integral domain must be a field. *Wedderburn's theorem* says that any finite division ring must be a field.

If K and L are fields, and $K \subseteq L$, then L is said to be an *extension* of K. The *degree* of L over K is defined to be the dimension of L when considered as a vector space over K, and is written $[L : K]$. The extension is *finite* if its degree is finite. If L is a finite extension of K and K is a finite extension of F, then L is a finite extension of F and $[L : F] = [L : K][K : F]$.

A typical way to generate field extensions is to *adjoin* an element α. If L is an extension of K, and $\alpha \in L$, then by $K(\alpha)$ we mean the smallest subfield of L containing both K and α. The field L does not appear in the notation, as it is usually obvious from the context. For example, $\mathbb{Q}(i)$ is the field of all numbers of the form $a + bi$, where $a, b \in \mathbb{Q}$; here i is a square root of -1, and $i \in \mathbb{C}$. The element α is said to be *algebraic* over K if it is a zero of a polynomial

$$f(X) = a_nX^n + a_{n-1}X^{n-1} + \cdots + a_1X + a_0$$

with $a_0, a_1, \ldots, a_n \in K$. If $f(X)$ is taken to be *monic* (i.e., the leading coefficient is 1), and of minimal degree, then f is unique; in this case it is called the *minimal polynomial* for α

over K. The zeroes of the minimal polynomial for α are called the *conjugates* of α over K. It can be shown that α is algebraic over K iff the index $[K(\alpha) : K]$ is finite.

Let K be a field and let $f(X) \in K[X]$. We say a nonzero polynomial $f(X)$ is *irreducible* over K if whenever $f(X) = a(X)b(X)$, with $a(X), b(X) \in K[X]$, then deg $a = 0$ or deg $b = 0$.

Another way to get an extension field of K is as follows: we take the polynomial ring $K[X]$ and form the quotient ring modulo the ideal generated by $f(X)$, where f is an irreducible polynomial over K. Then $K[X]/(f(X))$ is a field extension of K, and the degree of the extension is deg f. If α is a zero of $f(X)$, then in fact $K(\alpha)$ is isomorphic to $K[X]/(f(X))$.

A field K is said to be of *characteristic 0* if the equation $na = 0$, where $n \in \mathbb{Z}$ and $a \in K$ implies that $n = 0$ or $a = 0$. Otherwise there is a positive integer n such that $na = 0$ for all $a \in K$; in this case the field is of *finite characteristic*. The smallest such n is called the *characteristic* of the field. The characteristic is always a prime number.

By an *automorphism* of a field K, we mean a mapping σ from K onto K such that for all $a, b \in K$ we have $\sigma(ab) = \sigma(a)\sigma(b)$, and $\sigma(a + b) = \sigma(a) + \sigma(b)$.

Suppose K is a field extension of F. By the *Galois group* of K over F, we mean the group of all automorphisms σ of K which leave F fixed (i.e., $\sigma(a) = a$ for all $a \in F$). For example, the Galois group of $\mathbb{C}$ over $\mathbb{R}$ is of order two, containing the identity automorphism, and the automorphism that sends each complex number $a + bi$ to its complex conjugate $a - bi$.

2.9 Exercises

1. Suppose the prime factorization of m is $\prod_{1 \le i \le k} {p_i}^{e_i}$ and the prime factorization of n is $\prod_{1 \le i \le k} {p_i}^{f_i}$. Show that

$$\gcd(m, n) = \prod_{1 \le i \le k} {p_i}^{\min(e_i, f_i)}$$

and

$$\operatorname{lcm}(m, n) = \prod_{1 \le i \le k} {p_i}^{\max(e_i, f_i)}.$$

2. Suppose a, b, c are non-negative integers. Prove that $\gcd(ca, cb) = c \gcd(a, b)$ and $\operatorname{lcm}(ca, cb) = c \operatorname{lcm}(a, b)$.

3. Give a good definition of the greatest common divisor and least common multiple for rational numbers.

4. Suppose a and b are non-negative integers. Show that $\operatorname{lcm}(a, b) \gcd(a, b) = ab$.

5. Find a formula for $\operatorname{lcm}(a_1, a_2, \ldots, a_n)$ that is a good generalization of the one in the previous exercise. Check your formula on the case $a_1 = 10$, $a_2 = 6$, $a_3 = 15$.

6. Prove Theorem 2.1.2. (Hint: let n be the least integer with two distinct factorizations.)

7. Prove that equation (2.2) holds, i.e. that if

$$n = \prod_{1 \le i \le k} p_i^{e_i}, \text{ then } \varphi(n) = \prod_{1 \le i \le k} (p_i - 1) p_i^{e_i - 1}.$$

8. (Continuation.) Prove that $d(n) = \prod_{1 \le i \le k} (e_i + 1)$.

9. (Continuation.) Prove that, for $r \ge 1$, we have

$$\sigma_r(n) = \prod_{1 \le i \le k} \frac{p_i^{r(e_i+1)} - 1}{p_i^r - 1}.$$

10. Suppose p is a prime, e and r are positive integers, and $p \nmid r$. Show that if $n \mid p^e r$ and $n \nmid p^{e-1} r$, then $p^e \parallel n$.

11. Let a and b be positive integers. Show that there exist positive integers i, j such that $a^i = b^j$ if and only if there exist positive integers r, s, t such that $a = r^s$ and $b = r^t$.

12. Let p be a prime. Show that $\binom{p}{k} \equiv 0 \pmod{p}$ for $1 \le k \le p - 1$.

13. Prove Fermat's theorem: If p is a prime and $\gcd(a, p) = 1$, then $a^{p-1} \equiv 1 \pmod{p}$. (Hint: use the previous exercise to prove $a^p \equiv a \pmod{p}$.)

14. Prove Euler's theorem: If $\gcd(a, n) = 1$, then $a^{\varphi(n)} \equiv 1 \pmod{n}$. (Hint: show that if $x_1, x_2, \ldots, x_k$ are the integers between 1 and n that are relatively prime to n, then $ax_1, ax_2, \ldots, ax_k \pmod{n}$ is a permutation of $x_1, x_2, \ldots, x_k$.)

15. Prove that, if p is a prime, then

$$\nu_p(n!) = \sum_{k \ge 1} \left\lfloor \frac{n}{p^k} \right\rfloor.$$

16. Show that

$$\operatorname{lcm}(1, 2, 3, \ldots, n+1) = (n+1) \operatorname{lcm}\left(\binom{n}{0}, \binom{n}{1}, \ldots, \binom{n}{n}\right)$$

for all integers $n \ge 1$.

17. Show that $\varphi(n) = n \sum_{d \mid n} \mu(d)/d$. Hint: first, show $n = \sum_{d \mid n} \varphi(d)$. Then apply Möbius inversion.

18. Show that $\sum_{1 \le n \le x} \mu(n) \lfloor x/n \rfloor = 1$ for $x \ge 1$.

19. Show that if

$$g(x) = \sum_{1 \le k \le x} f(x/k)$$

for all real $x > 0$, then

$$f(x) = \sum_{1 \le k \le x} \mu(k) g(x/k).$$

Also prove the converse.

20. Use the previous exercise to show that

$$\vartheta(x) = \sum_{k \ge 1} \mu(k) \psi(x^{1/k}).$$

21. Show that

$$\sum_{1 \le n \le x} \frac{\mu(n)}{n^2} = \frac{6}{\pi^2} + O\left(\frac{1}{x}\right).$$

22. Show that

$$\sum_{1 \le n \le x} \varphi(n) = \frac{3}{\pi^2} x^2 + O(x \log x).$$

23. Show that $\log(x!) = x \log x - x + O(\log x)$.

24. Use Euler's summation formula to show that

$$\sum_{1 \le n \le x} \sigma(n) = \frac{\zeta(2)}{2} x^2 + O(x \log x).$$

25. Use Euler's summation formula to show that

$$\sum_{1 \le n \le x} \frac{\log n}{n} = \frac{(\log x)^2}{2} + A + O\left(\frac{\log x}{x}\right),$$

where A is a constant. Estimate the value of A.

26. Use Euler's summation formula to show that

$$\sum_{2 \le n \le x} \frac{1}{n \log n} = \log \log x + B + O\left(\frac{1}{x \log x}\right),$$

where B is a constant. Estimate the value of B.

In the next two exercises, $d_{m,k}(n)$ denotes the number of divisors of n which are congruent to $k \pmod{m}$.

27. (Cesàro.) Show that

$$\sum_{1 \le k \le n} (d_{2,1}(k) - d_{2,0}(k)) = n \log 2 + O(\sqrt{n}).$$

28. (E. Landau.) Let F_k denote the k-th Fibonacci number, defined by $F_0 = 0$, $F_1 = 1$, and $F_n = F_{n-1} + F_{n-2}$ for $n \geq 2$. Let α and β be the zeroes of the polynomial $X^2 - X - 1$, with $\alpha > 0 > \beta$. Show that

$$\sum_{j \geq 0} \frac{1}{F_{2j+1}} = \sqrt{5} \sum_{n \text{ odd}} \frac{d_{4,1}(n) - d_{4,3}(n)}{\alpha^n}.$$

Hint: use the fact that $F_n = (\alpha^n - \beta^n)/\sqrt{5}$, and a relationship between α and β.

29. Use the formula proved in Section 2.5 for $\sum_{k \leq n} d(k)$ and Exercise 26 to find an asymptotic formula for

$$\sum_{x_1 x_2 x_3 \leq n} 1.$$

30. (Continuation.) Guess and prove an asymptotic formula for

$$\sum_{x_1 x_2 \cdots x_k \leq n} 1.$$

31. Let $f(n)$ be the product of all positive divisors of n. Show $f(n) = n^{d(n)/2}$.

32. (Continuation.) Does $f(m) = f(n)$ imply $m = n$?

33. Prove the following:

$$\sum_{\substack{1 \leq k \leq n \\ \gcd(k,n)=1}} f(k) = \sum_{d \mid n} \mu(d) \sum_{1 \leq k \leq n/d} f(kd).$$

34. Give an example of two positive integer-valued functions $f(n)$ and $g(n)$ such that $f(n) \neq O(g(n))$ and $g(n) \neq O(f(n))$. Give another example where the functions are actually nondecreasing.

35. Prove part (b) of Theorem 2.5.4. Hint: use the symmetry in the hyperbola $xy = n$.

36. Estimate $\sum_{p \leq x} p^{-1/2}$.

2.10 Notes on Chapter 2

The following are recommended introductory texts on number theory: G. Hardy and Wright [1985]; Ireland and Rosen [1990]; Hua [1982].

2.1

The notation $\lfloor x \rfloor$ for the floor function and $\lceil x \rceil$ for the ceiling function was suggested by Iverson [1962]. For properties of the floor and ceiling, see R. Graham, Knuth and Patashnik [1989]. This book also popularized the notation $m \perp n$ for relative primality, although it had apparently been in use before then.

For seven other proofs of Theorem 2.1.1, see Ribenboim [1988b].

Theorem 2.1.2 was proved by Gauss [1801, Art. 16]. For previous results along these lines, see C. Goldstein [1992].

We caution the reader that many computer systems implement the mod function incorrectly. See, for example, Boute [1992].

Fermat stated Theorem 2.1.3 in 1640; see Fermat [1894, Vol. 2, p. 209]. For historical details, see Fletcher [1991]. Leibniz also seems to have proved this independently; see Vacca [1894]. The function $\varphi(n)$ was introduced by Euler [1760], where the generalization of Fermat's theorem was also proven.

The reader can find proofs of Theorems 2.1.2, 2.1.3, and 2.1.4 in G. Hardy and Wright [1985].

2.2

The function we have called $\omega'(n)$ is more often written as $\Omega(n)$, but since this conflicts with our asymptotic notation introduced in Section 2.4, we have avoided it.

The function $\sigma(n)$ has been extended to the Gaussian integers by Spira [1961].

2.3

The function $\mu(n)$ was defined by A. F. Möbius [1832]. The Möbius inversion formula was proved by R. Dedekind [1857, pp. 21, 25].

At even positive integers, the value of the zeta function is known. Euler proved that

$$\zeta(2n) = \frac{(2\pi)^{2n}(-1)^{n+1}B_{2n}}{2 \cdot (2n)!},$$

where B_{2n} denotes the $2n$-th Bernoulli number; see Edwards [1974].

2.4

Big-O notation was introduced by P. Bachmann [1894]. E. Landau [1909] invented the little-o notation. The big-omega notation was first used by G. Hardy and Littlewood [1914]; they used $\Omega(f(x))$ for what we have called $\Omega_\infty(f(x))$. Our use of big-omega seems more common in the computer science community. Big-theta was suggested by R. E. Tarjan and M. Paterson. For a discussion of these notations, see Knuth [1976a]. Also see Vitányi and Meertens [1984].

2.5

Readers not familiar with the Stieltjes integral may want to consult a good analysis text, such as Rudin [1964] or Apostol [1967]. Our treatment of Abel's identity is taken from Apostol [1976].

For the summation formula, see Euler [1732]. Euler's formula can be extended to give a much more detailed estimate than the one we have given; for example, see R. Graham, Knuth, and Patashnik [1989]. The resulting asymptotic series is often called the *Euler-Maclaurin* expansion.

It is possible, but not convenient, to compute Euler's constant using the expansion of H_n. A better strategy is to use the formula

$$\gamma = \sum_{k\geq 1} \frac{(-1)^{k-1} n^k}{k!\,k} - \log n + O(e^{-n}/n);$$

for this and similar methods, see Brent and McMillan [1980]. These authors calculated γ to 30, 100 decimal places. (For 3566 places see Sweeney [1963].)

Dirichlet's estimate is from Dirichlet [1849].

The *Dirichlet divisor problem* asks for the least number that can replace $1/2$ in the exponent of the error term in Theorem 2.5.4, part (b). It is known that $1/4$ cannot. (For a review of this problem, see Ivić [1985].)

2.6

The techniques of this section date back to work of du Bois-Reymond, who developed a heuristic *Infinitärkalkul*, that is, mechanical procedures for computing asymptotic approximations to integrals and derivatives. (See du Bois-Reymond [1870, 1872, 1879].) The theory was first put on a rigorous basis by G. Hardy [1910]. Theorem 2.6.1 appears in Dieudonné [1971, p. 81]. For a modern approach to the theory of asymptotic approximation, including more integration rules, see Chapter III of Dieudonné [1971], and the appendix of Bourbaki [1951].

2.7

Theorem 2.7.1 comes from Rosser and Schoenfeld [1962]. For other theorems useful in the estimation of prime number sums, see Landau [1909, section 55], and Kalecki [1964]. For bounding the error terms in such estimates, the results in Chapter 8 are very useful.

2.8

Readers unfamiliar with abstract algebra may be interested in consulting a text such as Herstein [1975]; Fraleigh [1976]; or Jacobson [1974]. Other books of interest include Albert [1961]; Birkhoff and Mac Lane [1977]; van der Waerden [1970]; and Lang [1965].

3 A Survey of Complexity Theory

When we develop algorithms to solve problems in number theory, we find that some problems can be solved by algorithms that run quickly, while others cannot be, at least by any known method.

Complexity theory gives us a precise language in which to make such statements; more exactly, it allows us to quantify an algorithm's use of scarce resources such as time and space. Thus, for example, we could prove statements about the number of steps required by the Euclidean algorithm for the greatest common divisor or the method of solving linear equations by Gaussian elimination.

For a given problem, we can, in principle, consider all possible algorithms for that problem, and compare their use of resources. We may also separate problems into *complexity classes*, which contain those problems that can be solved by *some* algorithm within a given resource bound. By choosing suitable bounds, we obtain many interesting complexity classes.

This chapter introduces the language and models necessary to discuss the complexity of number-theoretic algorithms.

3.1 Notation

Our complexity bounds will be phrased in terms of the size of the input. We define

$$\lg n = \begin{cases} 1, & \text{if } n = 0; \\ 1 + \lfloor \log_2 |n| \rfloor, & \text{if } n \neq 0. \end{cases}$$

Then $\lg n$ counts the number of bits in the binary representation of the integer n, excepting the sign bit. We will also denote the length of a character string x by $\lg x$.

We also recall the asymptotic notations defined in Chapter 2: O, o, and Ω. Roughly speaking, $f = O(g)$ means that f is at most proportional to g, $f = o(g)$ means that f/g tends to 0 for large arguments, and $f = \Omega(g)$ means that f is at least proportional to g. (See Section 2.4.)

3.2 The Notion of "Step"

To discuss the running time of a given algorithm, we need the notion of a *step*—the fundamental unit of computation. As with many other types of measurement, there is no single unit that is natural for all situations. For example, in analyzing sorting algorithms we might want to count the number of comparison steps. Early analyses of the Euclidean algorithm were concerned with bounding the number of "division steps" used. On the other hand, when we discuss calculations such as finding the determinant of a matrix, it seems natural (and certainly easy) to count each addition, subtraction, multiplication, or division as a single step.

But in these cases, a running time estimate may not correspond at all with what is observed when the algorithm is actually programmed on a computer. One difficulty, which is easy to handle, is the following: different computers have different operation times. For example, one computer might add two 32-bit integers in a microsecond, while another takes a millisecond. For this reason, it is usual to suppress constant factors in stating the running time of an algorithm; for such purposes, "big-O" notation is natural and convenient.

For example, a statement such as "the ordinary pencil-and-paper algorithm adds two n-bit integers in $O(n)$ steps", although asymptotic, nevertheless accurately describes the behavior of the algorithm on many different computers.

Big-O notation is useful for two other reasons. First, it lets us ignore small terms that occur in analyzing the running time of algorithms. For example, an algorithm that uses $n^2 + 3n + 2$ steps on an input of size n can be succinctly described as an $O(n^2)$ algorithm. Second, it lets us focus on the behavior of algorithms for large inputs, which is frequently (but not necessarily) a good indication of how they will behave in practice.

Our use of such notation is analogous to the use of quantitative theories by physical scientists: by suppressing details, it is frequently possible to give a crisp account of some phenomenon. However, the price paid for such simplifications is that the results have a limited range of applicability. The same is true of our estimates: it is possible that the details suppressed by a big-O estimate may be more important than the asymptotic result.

For the present purpose, a more important objection to assigning arithmetic operations unit cost is the following one. This assumption becomes unrealistic when the size of the inputs greatly exceeds the fixed word length of the computer, or when the required accuracy of the result demands extended-precision routines.

Further, if we assume that we can perform arithmetic on integers of arbitrary size in a single step, then we get theoretical results that contradict experience. For example, under such an assumption we could factor an integer in time proportional to its length (see Chapter 11). We know of no actual computer code for factoring with this property.

Thus to analyze our algorithms, we would like to find a unit that is both natural (in the sense that it is easy to work with) and practical (in the sense that it leads to accurate results). These criteria suggest that *it is useful to equate "step" with "bit operation"*.

We explain informally what this means. In principle, we deal only with variables that take the values 0 or 1, and perform the following logical operations upon these variables: conjunction ($\bigwedge$), disjunction ($\bigvee$) and negation ($\sim$). More complex objects are represented using these logical variables; for example, an integer may be represented in base 2, as a list of bits. The cost of a computation is then the number of logical operations performed. In this model, the size of operands affects the cost of even simple arithmetic operations. For example, $O(n)$ bit operations are needed to add two n-bit integers.

Counting bit operations as primitive is satisfying for both practical and theoretical reasons. Because all real computers are built using boolean circuits, the number of bit operations involved in a computation is clearly tied to the difficulty of its physical realization. It is also true that the actual cost of a multiprecise arithmetic operation is controlled by the input lengths. More importantly for this book, counting bit operations allows us to focus our attention on the cost of the arithmetic operations we use, and lets us ignore, to some extent, the cost of shuffling data. For these reasons, *unless otherwise stated, all statements made about complexity in this book will be phrased in terms of bit operations.*

Of course, real computers are not programmed one bit at a time. Nor would it be a good idea for us to describe all of our algorithms at the level of boolean operations. Rather, it seems preferable to analyze the basic arithmetic operations once and for all and base our estimates on the results of this analyses.

What we will actually do, then, is the following. In each of our algorithms, there will be a group, ring, or field (perhaps more than one) in which computation takes place. We will assign each operation in such a structure a certain cost, based on the number of bit operations involved in that operation. Then the total cost of an algorithm is the sum of the costs of the basic individual, ring, or field operations. Although any assignment of costs in such a scheme must ultimately be based on experience, we will take the ring of integers (with four operations) as basic, and derive costs, based on assumptions about integer arithmetic, for more complicated structures.

Table 3.1 provides our assumptions about the bit complexity of the four fundamental integer arithmetic operations, based on the ordinary pencil-and-paper algorithms. We have also listed the best known results in this regard, which are for a multi-tape Turing machine; we use the abbreviation

$$\mu(m, n) = \begin{cases} m(\lg n)(\lg\lg n), & \text{if } m \geq n; \\ n(\lg m)(\lg\lg m), & \text{otherwise.} \end{cases}$$

In our analyses, we will use the naive bounds, because we believe that results obtained through their use lead to accurate predictions. The reader will not go wrong thinking of

Table 3.1
Cost of Basic Arithmetic Operations

Operation	Naive Bit Complexity	Best Known
$a + b$	$\lg a + \lg b$	$O(\lg a + \lg b)$
$a - b$	$\lg a + \lg b$	$O(\lg a + \lg b)$
ab	$(\lg a)(\lg b)$	$O(\mu(\lg a, \lg b))$
$a = qb + r$	$(\lg q)(\lg b)$	$O(\mu(\lg q, \lg b))$

these bounds as axioms from which running times of more complicated algorithms may be derived.

We also point out one extra assumption that is commonly used, but not often stated: numbers are represented using some base $b \geq 2$, and the individual digits of a number may be accessed at no extra cost. For example, taking $b = 2$, if $n \approx 2^k$, then we can compute $\lfloor n/2^{k/2} \rfloor$, an approximate square root of n, in $O(\lg n)$ steps.

Our assumptions about bit complexity are also naive in the sense that they are not tied to any particular machine model. Using these assumptions, one can reason about algorithms in an informal manner, in much the same way as Church's thesis is used in recursive function theory to simplify the presentation of proofs.

For the moment, we will not present any detailed definition of "algorithm," although we do discuss a formal model for this purpose in Section 3.6. Readers familiar with computability theory may choose their favorite definition of this concept, although they should be warned that our naive assumptions about bit complexity may lead to results that differ slightly from those obtained via a formal model of computation. For readers not familiar with computability theory, we state that by an algorithm, we mean something akin to an ordinary computer program, in which operations may be done on arbitrary integers. By the running time of an algorithm, we mean the total number of bit operations involved in these integer operations, as given above.

3.3 Language Classes

It has also proved quite useful to study complexity theory "in the large," in the sense that running times and similar resource requirements are estimated roughly, certainly less precisely than a practitioner would like. Such a study has the advantage that the resulting theory is relatively insensitive to precise running time assumptions of the sort presented in the last section.

We now indicate the simplifications that are useful for such a study. First, although this book deals with algorithms whose inputs are numbers, we may if we wish consider these inputs to be strings from a finite alphabet. We do this, for example, by assuming that numbers are represented in binary notation. We can also, for the most part, restrict attention to algorithms that solve *decision problems*, that is, algorithms that take a string as input and output a 1 or 0 according to whether or not the string belongs to a particular set. In essence, such an algorithm decides whether or not its input has a specific form.

In this regard, the following definitions are useful. By an *alphabet* we mean any finite set, for example, $\{0, 1\}$. A *language* over a finite alphabet Σ is a set of strings whose symbols are chosen from Σ. We let Σ^* denote the set of all strings over Σ; thus, a language is just a subset of Σ^*. For example, PRIMES is a language over the alphabet {0, 1} consisting of those strings that are the binary representation of a prime number:

$$\texttt{PRIMES} = \{10, 11, 101, 111, 1011, 1101, 10001, \ldots\}.$$

If an algorithm takes a string as input and returns either 0 or 1, we may think of it as solving a decision problem for a particular language L. More precisely, such an algorithm takes a string x as input, and returns 1 if $x \in L$, and 0 otherwise. In this case, we say that the algorithm *accepts* the language L.

A *complexity class* is a collection of languages. We may rephrase many questions about the existence of certain algorithms as questions about complexity classes. For example, the collection of languages with decision algorithms forms a complexity class, which in mathematical logic is called the class of *recursive* sets. Then the negative solution of Hilbert's tenth problem may be summarized by saying that a certain set of strings encoding solvable Diophantine equations is not recursive.

Most of the complexity classes of interest to us originate in the notion of polynomial time computation. For this, we adopt the following definition.

DEFINITION 3.3.1 For a string x, let $\lg x$ denote the length of x. We say that a function f is *computable in polynomial time* if there is an algorithm to compute f that takes at most $p(\lg x)$ bit operations on input x, for some polynomial p.

The above definition is open to criticism because neither "algorithm" nor "computation step" has been formally defined. We remark, however, that these concepts can be made formal (we do this in Section 3.6), and our definition will be equivalent to the other definitions of polynomial time computation given in the literature. If the reader wishes, the notion of a polynomial-time computable function may be taken as primitive.

DEFINITION 3.3.2 We say that a language L is contained in the complexity class $\mathcal{P}$ if the characteristic function of L is computable in polynomial time.

For example, it is easy to see that the set

$$\texttt{EVEN NUMBERS} = \{0, 10, 100, 110, 1000, \ldots\}$$

of even natural numbers written in binary is contained in $\mathcal{P}$. (See Exercise 11.)

When we say that a set of integers is contained in $\mathcal{P}$, we really mean that a certain language, containing the binary encodings of numbers in that set, is in $\mathcal{P}$. We emphasize this point, because not doing so may lead one to incorrectly assign sets to $\mathcal{P}$ in certain cases. For example, a number n can be tested for primality by trying divisors up to $\sqrt{n}$, but this is not a polynomial time algorithm; its running time is exponential in the *length* of the input.

A very useful concept is that of *nondeterministic algorithm*. This is a decision algorithm that includes instructions allowing it to choose from a finite set of possibilities; it accepts the input if *at least one* sequence of possible choices could cause it to output a 1. (The other choices may cause the algorithm to output a 0, or run forever.) Informally, the running time of a nondeterministic algorithm is the minimum number of steps in any computation that outputs a 1, should one exist.

One may also think of a nondeterministic algorithm as one that first "guesses" a string indicating which choices would cause it to accept the input, and then attempts to "verify" that the guess is correct. The algorithm accepts the input if a correct, verifiable guess is possible. For example, a nondeterministic algorithm to recognize the set

$$\texttt{COMPOSITES} = \{100, 110, 1000, 1001, \ldots\}$$

of composite numbers written in binary may be specified as follows. If the input is x, the algorithm simply "guesses" a divisor d, $1 < d < x$, and then verifies that $d \mid x$.

For this example, and many others, verification of a correct guess is relatively easy. It is useful to have a complexity class containing languages with this property. Informally, a language is in the class $\mathcal{NP}$ if it can be accepted by a nondeterministic algorithm in polynomial time. Formally, we can define $\mathcal{NP}$ as follows:

DEFINITION 3.3.3

$$\begin{aligned} \mathcal{NP} = \{L \mid \exists \text{ a set } S \in \mathcal{P} \text{ and a polynomial } p \text{ such that} \\ x \in L \Longleftrightarrow \exists y, \lg y \le p(\lg x), \text{ and } (x, y) \in S\}. \end{aligned} \tag{3.1}$$

(By $S \in \mathcal{P}$, we mean that strings encoding the pairs in S can be recognized in polynomial time; for one possible coding scheme, see Exercise 12.) The string y should be thought of as a polynomial-length "proof" that $x \in L$, which may be checked by an algorithm that recognizes S. Note that $\mathcal{P} \subset \mathcal{NP}$; unfortunately it is not known if these classes are the same.

If C is a complexity class, we denote the complexity class obtained by complementing each of its members by co-C. The class co-$\mathcal{NP}$ therefore contains the languages for which membership can be quickly *disproved* by a nondeterministic algorithm. For example, $\texttt{PRIMES} \in$ co-$\mathcal{NP}$, because the set of all binary strings is the disjoint union of PRIMES, COMPOSITES, and OTHERS (strings not denoting numbers greater than 1). Because strings in the third category are easy to recognize, our nondeterministic polynomial time algorithm for recognizing composite numbers may be easily extended to one recognizing strings that do not denote a prime.

3.4 Reductions and $\mathcal{NP}$-Completeness

We all have an intuitive notion of what it means for one problem to be "at least as hard" as another. For example, computing the prime factorization of a number is at least as hard as testing it for primality, since the output of any algorithm for prime factorization may be used to tell if its input is prime.

For the moment, let us restrict our attention to decision algorithms, which have outputs in the set $\{0, 1\}$. Associated with each decision algorithm is the language of all strings corresponding to the output 1.

To formalize the intuitive notion of "relative hardness," we introduce the concept of a *reduction* from one language to another. Although the details vary from author to author, the basic idea is that a decision algorithm for a language is used as a "subroutine" in an algorithm to decide membership in another language.

DEFINITION 3.4.1 Let L_1 and L_2 be languages (that is, subsets of Σ^*). A computable function $f : \Sigma^* \rightarrow \Sigma^*$ is said to be a *(many-one) reduction* from L_1 to L_2 when $x \in L_1$ iff $f(x) \in L_2$. If f is computable in polynomial time, it is called a *polynomial time (many-one) reduction.*

We write $L_1 \leq_m L_2$ to indicate the existence of a many-one reduction from L_1 to L_2, and $L_1 \leq_m^{\mathcal{P}} L_2$ if the reduction is computable in polynomial time. (Note that these relations are transitive.) Roughly speaking, if $L_1 \leq_m^{\mathcal{P}} L_2$, then the decision problem for L_1 is "easier" than that for L_2, since an algorithm to decide membership for L_2 can be easily modified so as to decide membership for L_1.

It is a remarkable fact that $\mathcal{NP}$ contains certain languages that are as hard to recognize as any member of $\mathcal{NP}$; this leads to the notion of *$\mathcal{NP}$-completeness.*

DEFINITION 3.4.2 A language L is $\mathcal{NP}$-complete if $L \in \mathcal{NP}$ and for any $L' \in \mathcal{NP}$, we have $L' \leq_m^{\mathcal{P}} L$.

It is not hard to prove that $\mathcal{P} = \mathcal{NP}$ if and only if some (hence every) $\mathcal{NP}$-complete language is in $\mathcal{P}$. This indicates why $\mathcal{NP}$-complete problems are so interesting: they are all of roughly the same difficulty, and either no $\mathcal{NP}$-complete problem can be solved in polynomial time, or they all can.

The reader may object that most computational problems are not in fact decision problems. In most cases, however, this objection is easily handled by devising a decision problem that is roughly as difficult as the original problem. For example, the problem of factorization is to compute the prime factors of a given integer n. We may make this into a decision problem by considering the set

$$\texttt{FACTOR} = \{(n, k) \ : \ n \text{ has a nontrivial positive divisor} \leq k\}$$

Any algorithm to decide if a pair of numbers belongs to FACTOR may be used to factor n, by repeatedly performing a binary search for the smallest nontrivial positive divisor, which necessarily must be a prime. Thus the factorization problem and the associated decision problem are equally hard, in the sense that a polynomial-time algorithm for one leads to a polynomial-time algorithm for the other.

Thus when we say that a *problem* is $\mathcal{NP}$-complete, we really mean that the language for an associated decision problem is $\mathcal{NP}$-complete. In most cases, the specification of the associated decision problem, as well as the precise details of how instances of the decision problem are encoded, is omitted. However, we will require that all integers be coded in binary. (Actually, any base greater than or equal to 2 would do. The real goal is to forbid unary notation; see Exercise 22.)

There are now many known examples of $\mathcal{NP}$-complete problems, and this class has attained its present notoriety because it includes apparently intractable problems from diverse areas. We will indicate a few $\mathcal{NP}$-complete problems below; because of their diverse nature some definitions are required along with the examples.

DEFINITION 3.4.3 A *boolean formula* is either a variable chosen from the list $\{u_0, u_1, \ldots\}$, or an expression of the form $(\sim \varphi)$, $(\varphi \bigvee \psi)$, $(\varphi \bigwedge \psi)$, where φ and ψ are boolean formulas. It is said to be *satisfiable* if there is some assignment to the variables that makes the expression evaluate to 1.

EXAMPLE 1 The set

$$\texttt{SAT} = \{\varphi : \varphi \text{ encodes a satisfiable boolean formula}\}$$

is $\mathcal{NP}$-complete.

DEFINITION 3.4.4 A *clause* is a boolean formula of the form $\lambda_1 \bigvee \cdots \bigvee \lambda_k$, where each λ_i is a variable or its negation. (We omit parentheses in such descriptions.) A boolean formula is in *conjunctive normal form* if it has the form $\kappa_1 \bigwedge \cdots \bigwedge \kappa_n$, where each κ_i is a clause. It is in *3-conjunctive normal form* if each clause contains exactly 3 variables.

EXAMPLE 2 The set

$$\texttt{3-SAT} = \{\varphi : \varphi \text{ encodes a satisfiable formula in 3-conjunctive normal form}\}$$

is $\mathcal{NP}$-complete.

DEFINITION 3.4.5 A *partition problem* is a set of positive integers $\{a_1, \ldots, a_n\}$ expressed in binary notation; it is said to be *solvable* if there is a index set $I \subset \{1, \ldots, n\}$ such that $\sum_{i \in I} a_i = \sum_{i \notin I} a_i$.

EXAMPLE 3 The set

$$\texttt{PARTITION} = \{\pi : \pi \text{ encodes a solvable partition problem}\}$$

is $\mathcal{NP}$-complete.

There are many others.

Before going on, we indicate how to actually prove that a particular problem is $\mathcal{NP}$-complete. First, one has to show that a decision problem is in $\mathcal{NP}$; this is usually straightforward. Next, the definition requires one to show that *all* languages in $\mathcal{NP}$ reduce to the given problem. Because reductions can be composed, it suffices to reduce some known $\mathcal{NP}$-complete problem to the given one. This step usually requires some cleverness, and may involve subtleties that are not immediately apparent. For example, in proving that PARTITION is $\mathcal{NP}$-complete, it is crucial that the numbers not be given in unary, for otherwise the reduction will not be computable in polynomial time. (In fact, the unary version of PARTITION is in $\mathcal{P}$; see Exercise 29.)

Finally, we wish to point out the existence of other types of reductions. Although many-one reductions are convenient to work with, one can argue that a second type of reduction, called the *Turing reduction*, is closer to our intuitive ideas. We may informally contrast the notions of many-one and Turing reducibility as follows. A many-one reduction from L_1 to L_2 allows an algorithm to recognize an input in L_1 by returning the answer to a *single question* about membership in L_2. A Turing reduction, on the other hand, may use the answer to *many* questions about membership in L_2.

To explain this, we need to consider algorithms that take *sets* as inputs. Informally, this can be done as follows. Suppose A is a set of strings, possibly infinite. An algorithm is allowed, on demand, to obtain the answer to a question of the following sort: "is $x \in A$"? Further computations, including the posing of additional questions, may depend on the answer. We emphasize that there is to be no a priori decision about which questions are asked, nor any assumption that the decision problem for A is algorithmically solvable. If we like, we may think of the algorithm as using an external agent, called an *oracle*, to obtain the answers; for the purpose of computing running times, we assume that the oracle can decide if x belongs to A in $\lg x$ steps.

Although such a concept of "oracle computation" can be formalized, we do not do this here (but see Exercise 16). Rather, in line with our approach that polynomial time computation is primitive, we make the following definition.

DEFINITION 3.4.6 Let A be a set of strings. A function $f(x, A)$ is *computable in polynomial time* if there is an algorithm computing f, with an oracle for A, that takes at most $p(\lg x)$ bit operations on input x.

Note that the "size" of the set A—whatever that means—is not taken into account.

DEFINITION 3.4.7 Let L_1 and L_2 be languages. We say that L_1 is *Turing-reducible* to L_2 if there is a computable function $f : \Sigma^* \times 2^{\Sigma^*} \rightarrow \{0, 1\}$ for which

$$x \in L_1 \iff f(x, L_2) = 1.$$

If f is computable in polynomial time, we call f a *polynomial-time Turing reduction* from L_1 to L_2.

By analogy with our previous notation for reductions, we write $L_1 \leq_T L_2$ and $L_1 \leq_T^{\mathcal{P}} L_2$. Note that if $L_1 \leq_m^{\mathcal{P}} L_2$, then $L_1 \leq_T^{\mathcal{P}} L_2$; for sets in $\mathcal{NP}$, it is not known if the converse is true.

3.5 Randomized Complexity Classes

In many situations it is of interest to consider algorithms that have access to random samples from some distribution. The principal reason for this is that an algorithm may use an element of a certain type that, although difficult to explicitly construct, is plentiful. An algorithm that uses random choice to find the element may be more efficient than any known deterministic method for the same problem.

In all situations of practical interest, it suffices to endow such algorithms with the ability to generate random *bits*, that is, access to the result of a fair coin flip. However, some simplicity is gained by allowing them to contain instructions of the following kind: given a positive number b, generate a random sample from the uniform distribution on $\{0, \ldots, b-1\}$. We call algorithms that use such instructions *randomized algorithms*, and make the assumption that a randomized algorithm may obtain a random element of $\{0, \ldots, b-1\}$ in $\lg b$ steps.

We now ask what languages are accepted in polynomial time under this new assumption. To answer this question, we must be more precise; we find it convenient to define three new complexity classes for this purpose, which can be described informally as follows.

$\mathcal{RP}$ is the collection of languages L with the following property: there is a randomized algorithm accepting inputs in L with high probability, and always rejecting inputs not in L. The randomized algorithm is further required to run in polynomial time, no matter what sequence of random choices it makes. In this case we will also say that L is accepted with one-sided error.

$\mathcal{BPP}$ is the collection of languages L with the following property: there is a randomized algorithm accepting inputs in L with high probability, and rejecting inputs not in L with high probability. Again the algorithm must run in polynomial time, for all possible random choices. Sometimes we say that L is accepted with two-sided error.

$\mathcal{ZPP}$ is the collection of languages L with the following property: there is a randomized algorithm that accepts all inputs in L, rejects all inputs not in L, and runs in *expected* polynomial time.

We may give formal definitions of these complexity classes without appealing to the notion of randomized algorithm; although this is mathematically precise, the reader should consider how the definitions capture the intuitive ideas above. (See Exercise 15.)

DEFINITION 3.5.1

$$\begin{aligned} \mathcal{RP} = \{L \ : \ &\exists \text{ a set } S \in \mathcal{P} \text{ and a polynomial } p \text{ such that} \\ x \in L \Rightarrow \ &\text{for at least } 1/2 \text{ of all } y, \lg y \leq p(\lg x), \ \text{ we have } (x, y) \in S; \\ x \notin L \Rightarrow \ &\text{for all } y, \lg y \leq p(\lg x), \ \text{ we have } (x, y) \notin S\}. \end{aligned} \tag{3.2}$$

DEFINITION 3.5.2

$$\begin{aligned} \mathcal{BPP} = \{L \ : \ &\exists \text{ a set } S \in \mathcal{P}, \text{ and a polynomial } p \text{ such that} \\ x \in L \Rightarrow \ &\text{for at least } 3/4 \text{ of all } y, \lg y \leq p(\lg x), \ \text{ we have } (x, y) \in S; \\ x \notin L \Rightarrow \ &\text{for at least } 3/4 \text{ of all } y, \lg y \leq p(\lg x), \ \text{ we have } (x, y) \notin S\}. \end{aligned} \tag{3.3}$$

DEFINITION 3.5.3

$$\mathcal{ZPP} = \mathcal{RP} \cap \text{co-}\mathcal{RP}$$

The following containments hold among these complexity classes:

$$\mathcal{P} \subset \mathcal{ZPP} \subset \mathcal{RP} \subset \mathcal{NP} \cap \mathcal{BPP}.$$

No relation between $\mathcal{BPP}$ and $\mathcal{NP}$ is known, nor is it known if any of the containments given above are proper.

We can also classify the algorithms for accepting languages in randomized complexity classes, rather than the languages themselves. This has given rise to some colorful terminology.

We say that a randomized algorithm is a *Las Vegas* algorithm if its output is always correct; its running time, of course, may depend on the particular random choices made. By an *Atlantic City* algorithm, we mean one that is correct at least 3/4 of the time. Finally, an algorithm to recognize a language L is called a *Monte Carlo* algorithm if it identifies strings in L with probability at least 1/2, and never incorrectly assigns a string to L. (It will be seen that the numbers in these definitions are somewhat arbitrary; for example, we can replace 1/2 by any positive constant less than 1.)

Thus, if $L \in \mathcal{ZPP}$, there is a Las Vegas algorithm, running in expected polynomial time, to recognize strings in L. Similarly, there is a polynomial-time Monte Carlo algorithm for any L in $\mathcal{RP}$, and a polynomial-time Atlantic City algorithm for any L in $\mathcal{BPP}$.

We will see examples of these kinds of algorithms in Chapters 7 and 9.

3.6 A Formal Computational Model

Although most of the running-time analysis done in this book is based on a naive count of bit operations, it is useful to have a model of an actual computer on which computations could be performed. We will discuss one such model, the *polynomial-cost random access machine*, in detail, indicating both its connection with other models in the literature and the extent to which it mirrors our naive assumptions about bit complexity.

We call our machine polynomial-cost because each instruction is assigned a running time specified by a polynomial in the length of its data. For example, multiplying a by b uses $(\lg a)(\lg b)$ time units.

The machine consists of a finite program and a potentially unbounded memory. The memory is divided into cells, labeled $R_0, R_1, R_2, \ldots$, each of which can hold an integer. There is no bound on the number of cells that can be used, nor on the integers that a cell can hold. In addition, the machine has facilities for reading and writing integers; the inputs to the machine are listed in a sequence, to be read in order.

By a *program*, we mean a sequence of instructions, chosen from the list in Table 3.2. (The table also gives execution times, which we discuss later.) Most of the instructions should be self-explanatory. For example, the instruction $R_1 \leftarrow R_2 + R_3$ adds the numbers stored in cells 2 and 3 and stores the result in cell 1. An instruction of the form READ R_i causes R_i to take on the value of the next input number, and WRITE R_i causes R_i to be written as an output.

One unusual feature of the machine deserves comment: it has instructions that provide quick access to the digits of a number, written in base b. We assume that $b = 2$, although any larger number would serve as well. An instruction of the form $R_i \leftarrow$ DECODE R_j R_k has the following effect, assuming that r and s are the contents of R_j and R_k. The contents of memory cells $r, r+1, \ldots, r+s-1$ are interpreted as the base b digits of a number, which is then written into R_i. An instruction of the form $R_i \leftarrow$ LENGTH R_j computes the number of base-b digits in R_j and stores the result in R_i. Finally, an instruction of the form ENCODE R_i AT R_j "unpacks" the contents of R_i into s base-b digits, and stores these digits into memory cells $r, r+1, \ldots, r+s-1$, where r denotes the contents of R_j. In all such

Table 3.2
Instructions and Their Costs

Instr. No.	Instruction	Computation Cost
1	$R_i \leftarrow C$	1
2	$R_i \leftarrow R_j + R_k$	$\lg R_j + \lg R_k$
3	$R_i \leftarrow R_j - R_k$	$\lg R_j + \lg R_k$
4	$R_i \leftarrow R_j R_k$	$(\lg R_j)(\lg R_k)$
5	$R_i \leftarrow \lfloor R_j / R_k \rfloor$	$(\lg R_i)(\lg R_k)$
6	$R_i \leftarrow R_{R_j}$	$\lg R_j + \lg R_{R_j}$
7	$R_{R_i} \leftarrow R_j$	$\lg R_i + \lg R_j$
8	$R_i \leftarrow$ DECODE R_j R_k	$\lg R_j + R_k$
9	$R_i \leftarrow$ LENGTH R_j	$\lg R_j$
10	ENCODE R_i AT R_j	$\lg R_i + \lg R_j$
11	IF $R_i > 0$ GOTO m	$\lg R_i$
12	GOTO m	1
13	READ R_i	$\lg R_i$
14	WRITE R_i	$\lg R_i$
15	END	1

instructions, it is assumed that addresses are non-negative, and base-b digits are between 0 and $b - 1$.

We intend a random access machine to be an instantiation of a single algorithm; the reader may think of such a machine as analogous to an assembly language program, with multiprecise operations "built in." However, we emphasize that the program is an integral part of the machine; it is not stored in the memory.

By an *instantaneous description*, or *configuration*, of the machine, we mean a complete, but finite, description of the machine's state. This comprises the instruction to be executed next, the contents of all memory cells used so far, specifications of the inputs to be read next, and a list of outputs so far produced. (Memory cells not included in a configuration are supposed to contain 0.) By a *computation*, we mean a finite sequence of instantaneous descriptions, each following from its predecessor according to the meaning of the instructions. The computation is supposed to start with the first instruction in the program, and end with the "END" instruction.

The essential feature of our model is the way in which each computation is given a running time. We do this by assigning to each configuration a cost given by Table 3.2. For example, if a configuration specifies that $R_2 = 32$, $R_3 = 17$, and that $R_1 \leftarrow R_2 + R_3$ is to be executed next, we assign this configuration a cost of $\lg 32 + \lg 17 = 11$. The cost, or running time, of a computation is the sum of all these individual costs.

Here is a sample program, which computes the maximum of two integers. The reader may verify that the cost of this program is $O(\lg a + \lg b)$ steps on input (a, b).

```
READ R_1
READ R_2
R_3 ← R_1 − R_2
IF R_3 > 0 GOTO L1
WRITE R_2
GOTO L2
L1: WRITE R_1
L2: END
```

If we agree that by an algorithm, we mean a random access machine, it becomes possible to give formal definitions of the undefined concepts in Section 3.3. We fix once and for all a finite alphabet $\Sigma = \{0, \ldots, b-1\}$; we will assume $b = 2$, although any larger number would also do. The input and output strings are encoded as lists of integers between 0 and $b-1$, followed by a trailing -1. Then, by definition, a function from Σ^* to Σ^* is computable in polynomial time if the cost of computing this function is bounded by some polynomial in the length of the input.

Because the machine provides instructions to encode and decode integers in base-b format, this definition is equivalent to the following "naive" one, which is easier to use for number-theoretic problems. An algorithm runs in polynomial time if, for some polynomial p, its running time is at most $p(\lg x_1 + \lg x_2 + \cdots + \lg x_n)$ steps on the input $(x_1, x_2, \ldots, x_n)$.

By suitably extending our machine model, it is also possible to formalize the notion of algorithms that take sets as inputs; for this see Exercise 16.

We now give a critique of our model. First, the costs of our arithmetic instructions correspond to the naive complexity bounds in Table 3.2, and their suitability is open to question. For example, one might wish to assign multiplication a cost close to linear in the argument lengths, because asymptotically fast methods with this behavior are known. In response to this point, we argue that our costs simply represent the behavior that one is likely to observe, barring enormous inputs or heroic programming efforts. Further, the *raison d'être* of our model is to allow precise definitions of concepts such as "polynomial time algorithm"; within reason, these definitions are insensitive to a choice of cost function. Indeed, it can be proved formally that our definition of polynomial time computation is equivalent to the standard one based on multi-tape Turing machines. (See Exercise 19.)

A second issue is the relationship between running times obtained using our random access machine, which we may call "formalized bit complexity," and the naive bit complexity of Section 3.2. We have defined our machine so that these two estimates will be very close; however, discrepancies are possible. The essential reason for this is that naive complexity estimates ignore the cost of computing addresses and moving data, whereas formalized complexity estimates do not. For this reason, the latter may be higher, especially for algorithms that work with many relatively small data items. For example, the ordinary algorithm to multiply two $n \times n$ matrices with entries in $\{0, 1\}$ has naive bit complexity $\Theta(n^3)$, but formalized bit complexity $\Theta(n^3 \lg n)$, because n^3 addresses, each about $\lg n$ bits long, must be computed.

Indeed, even the assumption that accessing the k-th memory cell takes $O(\lg k)$ time may be questioned, because any real memory device with k cells will have a distance of $\Omega(k^{1/3})$ between some pair of cells. We must therefore give a caveat to the reader: although we believe that results using our models will be accurate and useful, they are not guaranteed to be so in all situations.

3.7 Other Resources

Although we primarily concern ourselves with running times in this book, it is of interest to consider other features of algorithms. The most interesting of these seem to be *space, randomness consumption,* and *depth*. We discuss these briefly in this section.

By the *space* used by an algorithm, we mean the maximum work space needed to store intermediate values, not counting the inputs and outputs. Informally, this may be estimated by a scheme similar to the one we have used for running times. We start by assuming that the integer x takes up $\lg x$ bits of space, and similarly assign sizes to elements of groups, rings, and fields derived from the integers. For example, we assign $\lg n$ bits of space to an integer modulo n. Then, to find the space needed by an algorithm on a certain input, we add up the sizes of all the elements it computes, taking care not to doubly count elements occupying the same storage location. We call this space accounting scheme *naive space complexity*; we use it when analyzing algorithms in this book.

As a practical matter, this suffices to estimate space consumption for the algorithms we are studying. However, it is useful to have a formal definition. We recall that algorithms are to be implemented using the machine model of Section 3.6, and that a computation, which traces a particular run of an algorithm, is a sequence of machine configurations. For any such computation, let R_m be the memory cell of highest index specified in any configuration. Then the space consumption of this computation is $\sum_{1 \le i \le m} S_i$, where S_i maximizes $\lg x$ over all integers x contained in R_i.

It is worthwhile to point out some features of this definition. First, an algorithm that uses memory cells in a sparse manner is charged for the unused space. Whether or not this is realistic depends on implementation details of a particular program. Second, our formalized definition agrees with a standard definition of space complexity based on Turing machines, and is useful for defining space complexity classes. However, our definition has certain anomalous features; for example, it is possible to design algorithms that run in polynomial time but use exponential space (see Exercise 21).

Using our formalized space measure, we define complexity classes, as follows. First, we say that a function $f : \Sigma^* \to \Sigma^*$ is *computable in polynomial space* if, for some polynomial p, there is an algorithm to compute f using space $p(\lg x)$ on all inputs x. A language is recognizable in polynomial space if its characteristic function is so computable. The most important complexity class from this point of view is the following one.

DEFINITION 3.7.1 $\mathcal{PSPACE}$ is the collection of all languages that are recognizable in polynomial space.

The complexity class $\mathcal{PSPACE}$ may be thought of as an upper limit for feasible computation. All the complexity classes discussed so far are contained within it. We can also identify certain problems as $\mathcal{PSPACE}$-complete, under polynomial-time reductions.

To give an example a few definitions are required. Consider logical formulas that are built out of variables, the constants 0 and 1, and predicates involving addition and multiplication. If all the variables are quantified (either existentially or universally), the formula is called a *sentence*. Such sentences are either true or false, depending on how the operations are interpreted; for example the sentence

$$\forall x\ \exists y\ (xy = 1)$$

is false in $\mathbb{F}_2$, the finite field of two elements. Without proof, we note the following.

EXAMPLE 4 The set

$$\texttt{F2} = \{\sigma : \sigma \text{ encodes a sentence that is true in } \mathbb{F}_2\}$$

is $\mathcal{PSPACE}$-complete.

It is also of interest to consider algorithms that use a very small amount of space, say logarithmic in the input size. This requires careful treatment, because our formalism charges for memory cells that hold the inputs, thereby making sublinear space consumption impossible. One way to handle this is the following. As before, all inputs and outputs are given in the form of strings over the fixed alphabet $\{0, \ldots, b-1\}$. We dedicate a special set of memory cells (say those with even indices) to the input values, and assume that the program first reads the inputs into these cells, and never changes them thereafter. Then space consumption is computed as before, but ignoring those cells containing input values.

By the *randomness consumption* of a computation, we mean the total number of random choices involved in that computation. Informally, we measure this as follows: if an algorithm chooses a random sample from $\{0, \ldots, b-1\}$, we say that it has used $\log_2 b$ bits of randomness. Unlike our other complexity measures, this is a real number, not an integer. It corresponds to the definition of entropy given in information theory, and is justified by the following fact. We can simulate a random choice from $\{0, \ldots, b-1\}$ with a fair coin;

on the average, $O(\log_2 b)$ coin flips will be required per choice. (See Exercise 24.) Just as with time and space, the randomness consumption of an algorithm is a random variable, and we can speak of its expected value. These informal remarks suffice for the purposes of this book; using the formalized notion of randomized algorithm given in Exercise 15, we can easily make them precise.

Finally, we may consider algorithms that do not use any branching instructions, or indirect addressing. In our machine model, we assume that such programs adhere to the following rigid format. Only the first five and the last three instructions of Table 3.2 are used. The program begins by reading in n input values into $R_0, \ldots, R_{n-1}$. Then there are k computation steps, in which cells $R_n, \ldots, R_{n+k-1}$ are assigned values, using constants or previously assigned cells. Finally, a subset of these $n+k$ cells used are written to the output in some order. Such algorithms are called *straight-line programs*. Given such a program, we inductively assign a level to each memory cell. Cells taking inputs or constants have level 0, and a cell depending on lower-numbered cells has a level 1 greater than the largest level of its arguments. The *depth* of the program is then the highest level of any memory cell appearing in the program; it indicates a lower limit on the running time of a parallel execution of the program.

For example, here is a straight-line program that on inputs (a, b) with $b \neq 0$, computes $a \bmod b$; it has depth 3.

READ R_0
READ R_1
$R_2 \leftarrow R_0/R_1$
$R_3 \leftarrow R_1 R_2$
$R_4 \leftarrow R_0 - R_3$
WRITE R_4

3.8 Parallel Complexity Classes

In this section we give a brief discussion of complexity classes based on parallel computation. There are many ways to do this; we have chosen a presentation based on boolean circuits. This fits in well with our general philosophy that complexity should be measured in terms of bit operations.

We first require some auxiliary concepts.

DEFINITION 3.8.1 A *directed graph* consists of a set V of *vertices*, and a set $E \subset V \times V$ of *edges*. If (v, w) is an edge, it is said to *connect* v to w. A directed graph is *acyclic* if there is no sequence of vertices $v_1, \ldots, v_m = v_1$, where some edge connects v_i to v_{i+1} for $i = 1, \ldots, m-1$.

DEFINITION 3.8.2 Let G be a directed graph. If v is a vertex of G, its *in-degree* is the number of vertices u such that $(u, v) \in E$.

We now consider directed acyclic graphs of the following type. Each vertex has in-degree 0, 1, or 2. Vertices of in-degree 0 are called *inputs*. The other vertices are labeled with boolean operations chosen from the set $\{\sim, \bigvee, \bigwedge\}$; we must use the label $\sim$ for all vertices of in-degree 1, and may choose either $\bigvee$ or $\bigwedge$ to label vertices of in-degree 2. Finally, certain vertices are distinguished as *outputs*.

Such a graph is called a *boolean circuit*; if it has n inputs and m outputs, it represents a function from $\{0, 1\}^n$ to $\{0, 1\}^m$ in a straightforward fashion. The number of vertices is called its *size*, and the number of edges in a longest path from an input to an output is called its *depth*. In this context we refer to a non-input vertex as a *gate*. (We remark that a boolean circuit can be specified by a straight-line "boolean program"—which we do not define—and that then the definition of depth given here agrees with that given in the last section.)

To discuss functions whose arguments have varying length, we introduce the notion of a *circuit family*. This is a sequence of circuits $C_0, C_1, C_2, \ldots,$ where C_n takes n inputs; it defines a function from $f : \{0, 1\}^* \rightarrow \{0, 1\}^*$ in the obvious way.

Many useful functions can be computed by circuit families with the following property: for polynomials p and q, C_n has size at most $p(n)$ and depth at most $q(\lg n)$. For example, the four basic arithmetic operations have this property. (See Exercise 27.)

In using these requirements to single out a class of functions, however, there is a technical difficulty. The function defined by a circuit family need not be computable; for instance, the output of C_n might be 0 or 1 depending on whether n belonged to a particular non-recursive set of integers. For this reason, we need to restrict the circuit families under discussion.

Our approach will be to demand that specifying the n-th circuit in the family is not too complex a task. To do this some preliminary definitions are required. The *description* of a boolean circuit is a string that lists each gate in the circuit, together with its predecessors and its type. (We also require that inputs and outputs be marked in some fashion.) We call a circuit family $C_1, C_2, \ldots$ *(log-space) uniform* if there is an algorithm (in the formal sense of Section 3.6) that takes as input the number n in unary, and returns the description of C_n using $O(\log n)$ space.

We now define the following class of functions.

DEFINITION 3.8.3 A function $f : \{0, 1\}^* \rightarrow \{0, 1\}^*$ is said to be in the class $\mathcal{NC}$ if it is given by a log-space uniform circuit family $C_0, C_1, \ldots$ and there are polynomials p and q such that C_n has size at most $p(n)$ and depth at most $q(\lg n)$.

It is known that any function in $\mathcal{NC}$ can be computed in polynomial time; whether the converse holds is unknown. Efforts to resolve this question have led to a theory of $\mathcal{P}$-completeness. (See Exercise 28.)

It has been proposed that the class $\mathcal{NC}$ contains exactly those functions that have "good parallel algorithms." The reason for this is the following. We may consider a circuit as an implementation schedule for a particular algorithm; the size of the circuit measures the total work required, and the depth measures the earliest time at which the outputs could be delivered. To say that a function is in $\mathcal{NC}$ is to say that it has a parallel algorithm that is both very fast and does a feasible amount of work.

For this reason, there has recently been great interest in showing problems to lie in $\mathcal{NC}$, and we would be amiss not to point out results of this sort that are relevant to number theory. However, it is only fair to point out that the identification of $\mathcal{NC}$ with functions that we can expect to compute quickly in parallel is controversial. The principal reasons for this are that the boolean circuit model does not quantify the difficulty of communicating results from gate to gate, and that even a polynomially bounded number of gates may be too large in practice. For these reasons, the reader should consider a proof that a certain function is in $\mathcal{NC}$ to be evidence in favor of its parallelizability, not a definitive judgement that this is so.

3.9 Exercises

Exercises 1–10 deal with naive bit complexity.

1. Give algorithms that compute the following functions in the times indicated; all arguments are positive.

   (a) $m \bmod n$, in time $O((\lg q)(\lg n))$, where $q = \lfloor m/n \rfloor$.

   (b) $\lfloor \sqrt{n} \rfloor$, in time $O((\lg n)^2)$.

   (c) $\lfloor n^{1/k} \rfloor$, where $2 \le k \le \log_2 n$, in time $O((\lg n)^2(\lg\lg n))$.

2. Let b and n be positive integers, with $1 < b \le n$. Show that the base-b digits of n can be computed using $O((\lg n)^2)$ bit operations, independently of b.

3. Show that if $m = m_1 m_2 \cdots m_k$, a product of k integers ≥ 2, then m can be computed using $O((\lg m)^2)$ bit operations, independently of k.

4. Consider the problem of determining if a number n is a perfect power; that is, given $n \ge 2$, we wish to know if there exist $a \ge 2$ and $k \ge 2$ such that $n = a^k$. Show that this problem can be solved using $O((\lg n)^3)$ bit operations, and that the same bound holds for finding the largest such value of a.

5. Suppose a and b are positive integers. Using the previous exercise, show how to determine in polynomial time whether there exist positive integers i, j such that $a^i = b^j$. Also show how to quickly determine whether $a^i = b^j$, where a, i, b, j are all given.

6. Use Exercise 2.23 and Exercise 3 above to show that $n!$ can be computed using $O(n^2(\lg n)^2)$ bit operations. What complexity results if $n!$ is computed as a product of its prime power factors?

7. Let F_k be the k-th Fibonacci number (introduced in Exercise 2.28). Show F_k can be computed using $O(k^2)$ bit operations.

8. Let $f(x)$ be a polynomial with non-negative integer coefficients, with at least one positive coefficient, and let n be a positive integer. Show how to determine in polynomial time whether the Diophantine equation $f(x) = n$ has a solution.

9. Let n be a positive integer. Show how to express n as a sum of $O(\lg \lg n)$ squares in polynomial time.

10. Design an algorithm that multiplies two n-bit numbers by recursively multiplying three smaller numbers and combining the results. Use this to show that the product, as well as the quotient with remainder, of two n-bit numbers can be found using $O(n^{\log_2 3})$ bit operations. (This requires the assumption that individual bits can be accessed.)

11. Show that `EVEN NUMBERS` is contained in $\mathcal{P}$, according to both naive bit complexity and the formalized model of Section 3.6.

12. In Section 3.3, the method for encoding pairs and other sequences of strings into one string was left unspecified. Describe how to do this.

13. There are many other machine models which we did not discuss. One option might be to use a random access machine, but limit each memory cell so that it can only hold integers of some fixed size. Discuss possible drawbacks of this model.

    (Hint: what problems arise when counting to an arbitrary limit?)

Exercises 14–16 deal with additional features that one could add to the machine of Section 3.6.

14. Suppose we amend our definition of a random access machine so that a program may also contain instructions of the form

$$\text{CHOICE } m_0 \; m_1;$$

such an instruction chooses one of the two instructions m_0, m_1 to perform next, in one step. We will only consider machines that have the following behavior: for any particular sequence of choices, either the machine halts and outputs a 1, or it runs forever. The *running time* of such a machine on a given input is the minimal cost of a halting computation on that input, should one exist.

Prove that a set is in $\mathcal{NP}$ (according to our definition) iff it is accepted by such a "nondeterministic machine" in polynomial time.

15. Similarly consider the effect of adding the following instruction type:

$$R_i \leftarrow \text{RANDOM } R_j.$$

The effect of such an instruction is to cause R_i to take on a value chosen uniformly at random from the set $\{0, \ldots, R_j - 1\}$ if $R_i > 0$, and 0 otherwise; it is supposed to take $\lg R_i$ steps.

Formalize the notions of a randomized algorithm's running time (this will be a random variable), and the probability that it recognizes an input string. Use these to clarify the informal definitions of $\mathcal{RP}$, $\mathcal{BPP}$, and $\mathcal{ZPP}$ in Section 3.5, and prove that they agree with the formal definitions (3.5.1–3.5.3).

16. Explain how to add an "oracle" instruction to the RAM model to allow the use of arbitrary sets as inputs.

17. Show how to determine the parity of an integer in polynomial time, using only addition, subtraction, and comparisons.

Exercises 18–20 assume that the reader is familiar with multi-tape Turing machines. (See Hopcroft and Ullman [1979].)

18. Sketch the design of multi-tape Turing machines that realize the bounds in the first column of Table 3.1.

19. Prove that a function is computable in polynomial time (in the sense of Section 3.6) if and only if it can be computed in a polynomial number of steps on a multi-tape Turing machine.

20. Compare the space complexity measure of our formalized model with that of a multi-tape Turing machine. In particular, show that a function is computable in polynomial space in our sense iff it can be computed in polynomial space on a Turing machine.

21. Consider a program that, on input n, writes 1 into memory cells $1, 2, 4, \ldots, 2^n$. What is its running time and space consumption?

22. Let b be an integer, with $b \geq 2$. For positive integers x, let $\lg_b x$ denote the length of x when expressed in base-b, so that $\lg x = \lg_2 x$. Show that $\lg_b x = \Theta(\lg_2 x)$. What happens if $b = 1$?

23. Prove that SAT $\leq_m^{\mathcal{P}}$ 3-SAT, in two stages. By introducing new variables to stand for the results of logical operations, show how the satisfiability problem can be reduced to the problem for formulas in conjunctive normal form. Finally, show how a clause containing more than three variables can be split into two clauses, each containing fewer variables.

24. Let α be a real number with $0 < \alpha < 1$. The following process will generate a 0-1 random variable with expected value α. Choose random bits $\beta_0, \beta_1, \ldots,$ and simultaneously generate $\alpha_0, \alpha_1, \ldots,$ the bits of α; stop when some $\beta_i \neq \alpha_i$, and output a 1 if $\beta_i < \alpha_i$, 0 otherwise.

    (a) Show that the expected number of coin flips needed by this process is 2.

    (b) Design a process that outputs a random sample from $\{0, \ldots, b-1\}$ and uses, on the average, at most $(\log_2 b) + 2$ bits per sample.

25. Let L be a language, and let $\epsilon > 0$. Consider a randomized algorithm such that, if $x \in L$, then x is accepted with probability at least $1/2 + \epsilon$, and if $x \notin L$, then x is accepted with probability at most $1/2 - \epsilon$. Show that if the algorithm is run n times, and a majority vote is taken, the probability of a wrong decision is at most $(1 - 4\epsilon^2)^{n/2}$.

26. Find a constant C with the following property: any recursive set can be recognized by a random access machine using C or fewer memory cells. (Therefore, counting the memory cells used by a program is not an interesting measure of its space complexity.)

27. Describe boolean circuit families that perform the following tasks with the costs indicated:

    (a) Add two n-bit numbers, with depth $O(\lg n)$.

    (b) Subtract two n-bit numbers, with the same depth.

    (c) Multiply two n-bit numbers, with the same depth.

    (d) Divide a $2n$-bit number by an n-bit number, yielding an n-bit quotient, with depth $O((\lg n)^2)$.

    In all cases, the circuits should have $O(n^k)$ gates for some $k > 0$.

28. Let L_1 and L_2 be languages. We say that L_1 *reduces to* L_2 *in logarithmic space* (written $L_1 \leq^{\mathcal{L}} L_2$) if there is a many-one reduction from L_1 to L_2, using $O(\lg n)$ space on inputs of length n. A language L is called *$\mathcal{P}$-complete* if $L \in \mathcal{P}$, and every other language in $\mathcal{P}$ is reducible to it in logarithmic space.

    (a) Show that if $L \leq^{\mathcal{L}} M$, and the characteristic function of M is in $\mathcal{NC}$, then the same is true of L. (Hence if any $\mathcal{P}$-complete language can be recognized with an $\mathcal{NC}$ algorithm, so can any language in $\mathcal{P}$.)

    (b) An instance of the *circuit value problem* consists of a boolean circuit with n inputs and one output, a list of input values, and a purported output value. Show that the language encoding circuits, inputs, and correct outputs is $\mathcal{P}$-complete.

29. The SUBSET SUM problem is the following: given a set of positive integers $A = \{a_1, \ldots, a_n\}$ and a target t, determine if there exists some index set $I \subseteq \{1, 2, \ldots, n\}$ such that $\sum_{i \in I} a_i = t$. Show that SUBSET SUM is $\mathcal{NP}$-complete. Further, show that SUBSET SUM can be solved in $O(n\, t \lg t)$ bit operations, if $a_i \leq t$ for all i.

30. The KNAPSACK problem is just like SUBSET SUM, except now we use multisets instead of sets. Clearly KNAPSACK $\in \mathcal{NP}$, and SUBSET SUM $\leq_m^{\mathcal{P}}$ KNAPSACK, so that KNAPSACK is $\mathcal{NP}$-complete. Give a polynomial-time reduction from KNAPSACK to SUBSET SUM.

31. The SUBSET PRODUCT problem has a specification similar to SUBSET SUM (see Exercise 29), except that now we want to know if there exists an index set $I \subseteq \{1, 2, \ldots, n\}$ such that $\prod_{i \in I} a_i = t$. Show that SUBSET PRODUCT is $\mathcal{NP}$-complete.

3.10 Notes on Chapter 3

3.2

Our "best known" bounds for multiplication and division are slight sharpenings of results due to Schönhage and Strassen [1971] and Cook (see Knuth [1981]). They may be found in G. Collins, Mignotte, and Winkler [1982]; this article also gives a slightly different estimate for the complexity of integer division.

The conventional wisdom about fast multiplication algorithms in practice seems to be the following. For numbers that are large, but not overwhelmingly so (a few hundred decimal digits), the divide-and-conquer algorithm of Karatsuba and Ofman [1962] outperforms ordinary multiplication; also see Maeder [1993]. For truly enormous numbers, one should use methods based on polynomial interpolation. For example, Comba [1989] suggests that the Toom-Cook method (see Knuth [1981]) be used for numbers with 10^3 to 10^6 decimal digits, and the fast Fourier transform thereafter. Also see Schönhage [1986]; Zuras [1993]. However, the recent book of Schönhage, Grotefeld, and Vetter [1994] argues that the fast Fourier method is practical even for numbers of a thousand digits or so.

See Brassard, Monet, and Zuffellato [1986] for a survey of large number arithmetic. See Kuechlin, Lutz, and Nevin [1991] for a comparison of several methods on microprocessors, and Weber [1990] and Fagin [1990] for practical aspects of parallel methods.

Our naive bit complexity is a variant of what is usually called "arithmetic complexity" in the literature. Both measures take into account only the arithmetic operations involved in an algorithm; the difference between our measure and standard arithmetic complexity is that we charge for an operation according to the lengths of the operands, whereas arithmetic complexity only charges one time unit.

3.3

Complexity theory has its roots in formal language theory (see Greibach [1981]), hence the terms "string," "language." Although time-bounded automata were first discussed by Yamada [1962], the idea of using them to define complexity classes was first expressed by Hartmanis and Stearns [1965]. We also note that complexity classes are usually required to satisfy some extra conditions; because we are only interested in specific classes, our general definition does no harm.

Cobham [1964] was the first to consider polynomial-time computable functions as such; as he pointed out, the class of such functions is resistant to changes in machine models and has several useful closure properties. Edmonds [1965] went further, and proposed that the informal notion of "good algorithm" be identified with the formal concept of polynomial time algorithm.

The concept of nondeterministic computation is apparently due to Rabin and Scott [1959]. The existence of "natural" $\mathcal{NP}$-complete problems was demonstrated by Cook [1971] and Levin [1973], independently. For compendia of $\mathcal{NP}$-complete problems, see Karp [1972] and Garey and Johnson [1979]. Sipser [1992] has recently given a history of the $\mathcal{P} = \mathcal{NP}$ question. Stockmeyer and Meyer [1973]; Gurari and Ibarra [1977]; Plaisted [1977,1978,1984,1985]; Manders and Adleman [1978]; Eppstein [1987]; and L. Heath [1990] gave $\mathcal{NP}$-complete problems drawn from number theory. Grigoriev, Karpinski, and Odlyzko [1992] showed that, under the assumption of the Generalized Riemann Hypothesis, nondivisibility of sparse multivariate polynomials belongs to $\mathcal{NP}$.

3.4

Oracles were introduced by Turing [1939]. The terms "many-one" and "Turing" reducibility are due to Post [1944]. In computer science, many-one and Turing reductions were popularized by Karp [1972] and Cook [1971]. For this reason, they are frequently called "Karp" and "Cook" reductions, respectively.

3.5

The earliest formal study of randomized algorithms seems to have been done by de Leeuw, Moore, Shannon, and Shapiro [1956]. Also see Santos [1969]. The classes $\mathcal{RP}$, $\mathcal{BPP}$, and $\mathcal{ZPP}$ were first investigated by Gill [1977]. Adleman [1978] showed that any language in $\mathcal{RP}$ could be recognized by a "non-uniform" deterministic algorithm running in polynomial time.

Although we use "Monte Carlo" in a precise technical sense, the name has long been associated with algorithmic procedures using random numbers. It was made famous in numerical analysis, the idea being apparently due to S. Ulam; see von Neumann [1947]. However, the idea goes back much further; see, for example, Kelvin [1901]. The concept of a "Las Vegas" algorithm was introduced by L. Babai (see D. Johnson [1984]). The name "Atlantic City" was first used by J. Finn [1982a].

See Welsh [1983] and Karp [1991] for surveys of randomized algorithms. Also see Motwani and Raghavan [1995].

3.6

Our polynomial-cost random access machine is based on the work of Cook and Reckhow [1973]. For recent results on the relationship of such models to Turing machines, see Katajainen, van Leeuwen, and Penttonen [1988].

The ideal computer model is both realistic (in the sense that running times for moderate inputs are accurate), and asymptotically fair (in the sense that devices with the claimed running times could be built). To some extent, these are contradictory goals. For discussions of the physical limitations to such a quest, see Sutherland and Mead [1977] and Vitányi [1988].

3.7

Apparently, there is no standard definition of space consumption for random access machines; our formal definition follows Cook (see Borodin [1973]). For a critique of various definitions, see Slot and van Emde Boas [1988]. Space complexity is surveyed in Michel [1992].

What we call `F2` is essentially the same as a language called `QBF` in the literature. It was shown $\mathcal{PSPACE}$-complete by Stockmeyer and Meyer [1973].

In Shannon's information theory [1948], the entropy of a distribution $(p_1, \ldots, p_n)$ is defined to be $-E(\log p_i)$. Using this language, we are assuming that each random choice has a randomness consumption equal to the entropy of the distribution from which it was selected.

Mansour, Schieber, and Tiwari [1989] showed that no straight-line program can compute $\lfloor\sqrt{n}\rfloor$.

3.8

All of the basic arithmetic operations on n-bit numbers can be done by circuit families of logarithmic depth, for which the best known size bounds are the following. For addition and subtraction, Ofman [1962] showed that $O(n)$ size is possible, and for multiplication, Schönhage and Strassen [1971] showed that $O(n(\lg n)(\lg\lg n))$ size is possible. Beame, Cook, and Hoover [1986] obtained logarithmic depth circuits for integer division, with size $O(n^4(\lg n)^3)$; they are not log-space uniform. For practical multiplication and division circuits, see Wallace [1964] and Anderson, Earle, Goldschmidt, and Powers [1967].

$\mathcal{NC}$—"Nick's class"—was first defined by N. Pippenger [1979]. $\mathcal{P}$-completeness was identified by Cook [1973]; for a list of such problems, see N. Jones and Laaser [1977].

4 The Greatest Common Divisor

One of the most fundamental calculations in number theory is that of the greatest common divisor of two integers. Let u and v be positive integers; it is easy to see (see Exercise 2.1) that if the prime factorization of u is $\prod_{1 \le i \le k} p_i^{e_i}$ and the prime factorization of v is $\prod_{1 \le i \le k} p_i^{f_i}$, then

$$\gcd(u, v) = \prod_{1 \le i \le k} p_i^{\min(e_i, f_i)}. \tag{4.1}$$

Unfortunately, a computation based on equation (4.1) appears to be intractable, since it uses the factorizations of u and v. Therefore, a different approach is necessary.

In this chapter we will examine three algorithms for the greatest common divisor: the Euclidean algorithm, the least-remainder Euclidean algorithm, and the binary algorithm. We discuss a modification known as the extended Euclidean algorithm. Finally, we discuss algorithms to compute gcd-free bases.

4.1 The Euclidean Algorithm

The Euclidean algorithm to compute the greatest common divisor of two integers is based on the following observation:

$$d \mid u \text{ and } d \mid v \quad \text{iff} \quad d \mid v \text{ and } d \mid u \bmod v. \tag{4.2}$$

(Prove this, using the definition in (2.1).) Thus $\gcd(u, v) = \gcd(v, u \bmod v)$.

Now suppose $u_1 > 0$. Then $u_1 > u_0 \bmod u_1$, so if we set $u_2 = u_0 \bmod u_1$, we see from (4.2) that $\gcd(u_0, u_1) = \gcd(u_1, u_2)$; furthermore, $u_1 > u_2$. We now continue this process, setting $u_3 = u_1 \bmod u_2$, and noting that $\gcd(u_1, u_2) = \gcd(u_2, u_3)$. This process cannot continue indefinitely, for $u_1 > u_2 > u_3 > \cdots$ and so eventually we must find $u_{k+1} = 0$. Since $\gcd(u_k, 0) = u_k$, we conclude that $\gcd(u_1, u_2) = u_k$.

For example, we have $\gcd(180, 146) = \gcd(146, 34) = \gcd(34, 10) = \gcd(10, 4) = \gcd(4, 2) = \gcd(2, 0) = 2$.

This process can be made into an algorithm as follows:

```
EUCLID(u, v)
    if (v = 0) then
            return(u)
        else
            return(EUCLID(v, u mod v))
```

We have proved

THEOREM 4.1.1 The algorithm EUCLID terminates on integer inputs u, v with $v > 0$, and returns $\gcd(u, v)$.

Although in our analysis we have assumed $v > 0$, by modifying line (2) of the algorithm to read

$$\text{(2) return } (|u|)$$

we get an algorithm that computes the greatest common divisor for *all* integers u and v.

The naive argument given above shows that the Euclidean algorithm terminates in $O(u)$ steps on input (u, v), which is not a polynomial time bound (in $\lg u$). However, we will show in the next section that the Euclidean algorithm runs in time polynomial in $\lg u$ and $\lg v$.

4.2 The Euclidean Algorithm: Worst-Case Analysis

We now examine the Euclidean algorithm on input (u, v), at each step keeping track of the quotients as well as the remainders. Let $u_0 = u$ and $u_1 = v$. Then we have

$$\begin{aligned} u_0 &= a_0 u_1 + u_2 \\ u_1 &= a_1 u_2 + u_3 \\ &\vdots \\ u_{n-2} &= a_{n-2} u_{n-1} + u_n \\ u_{n-1} &= a_{n-1} u_n, \end{aligned} \tag{4.3}$$

where $a_i = \lfloor u_i / u_{i+1} \rfloor$ for all i.

In the previous section, we showed that $u_n = d = \gcd(u, v)$. Notice that the last quotient, a_{n-1}, is always at least 2 (unless $n = 1$); for if $a_{n-1} = 1$ we would have $u_{n-1} = u_n$, and hence $u_{n-2} \bmod u_{n-1} = u_{n-1}$, a contradiction.

Let us call the integer n the number of *division steps* in the Euclidean algorithm; we define $E(u, v) = n$.

We are now ready for the goal of this section: proving a good upper bound on $E(u, v)$.

Let F_n denote the n-th *Fibonacci number*, defined by $F_0 = 0$, $F_1 = 1$, and $F_n = F_{n-1} + F_{n-2}$.

We have

LEMMA 4.2.1 Let u, v be integers such that $u > v > 0$, and the Euclidean algorithm on input (u, v) performs n division steps. Then $u \geq F_{n+2}$ and $v \geq F_{n+1}$.

Proof Let $u = u_0$ and $v = u_1$, and refer to the system of equations (4.3). We will prove by induction on n that $u_0 \geq F_{n+2}$ and $u_1 \geq F_{n+1}$.

This is true for $n = 1$. For then the entire Euclidean algorithm consists of one division, $u_0 = a_0u_1$, and since $u_0 > u_1$, to find the least u_0, u_1, we must take $u_1 = 1$, $a_0 = 2$, and hence $u_0 = 2$.

Now assume the statement is true for all $i < n$; we wish to prove it for n. Then the first step of the Euclidean algorithm sets $u_0 = a_0u_1 + u_2$, and we know that $E(u_1, u_2) = n - 1$. Thus $u_1 \geq F_{n+1}$ and $u_2 \geq F_n$ by induction. Then $u_0 \geq u_1 + u_2$; hence $u_0 \geq F_{n+2}$. ■

We now wish to determine the worst-case behavior of the Euclidean algorithm. More precisely, for each n, we would like to find the "smallest" input that forces the algorithm to perform n division steps. However, since the algorithm actually takes a *pair* of inputs (u, v), it is not immediately clear how to define the "smallest" input.

We will say that the pair (u, v) is *lexicographically less* than the pair (u', v') if $u < u'$, or if $u = u'$ and $v < v'$.

COROLLARY 4.2.2 (LAMÉ) Let u, v be integers such that $u > v > 0$, the Euclidean algorithm on input (u, v) performs n division steps, and (u, v) is lexicographically least among all pairs satisfying these properties. Then $(u, v) = (F_{n+2}, F_{n+1})$.

Proof From Lemma 4.2.1, we know that $u \geq F_{n+2}$ and $v \geq F_{n+1}$. To prove the corollary, we must verify that if $u_0 = F_{n+2}$ and $u_1 = F_{n+1}$, then in fact the Euclidean algorithm performs n division steps; actually, it produces the series of quotients $a_0 = 1$, $a_1 = 1$, $\ldots, a_{n-2} = 1$, $a_{n-1} = 2$. This amounts to proving that $\lfloor F_{k+2}/F_{k+1} \rfloor = 1$ and $F_{k-1} = F_{k+1} \bmod F_k$ for $k \geq 2$, which is left to the reader as Exercise 4. ■

The reader is asked to prove in Exercise 3 that

$$F_n = \frac{\alpha^n - \beta^n}{\sqrt{5}} \tag{4.4}$$

where $\alpha = (1 + \sqrt{5})/2$, $\beta = (1 - \sqrt{5})/2$.

From this, we prove

COROLLARY 4.2.3 Let $u > v > 0$. Then $E(u, v) < c_1 \log u + c_2 - 2 + c_3u^{-1}$, where $c_1 = 1/\log \alpha \doteq 2.08$, $c_2 = (\log \sqrt{5})/(\log \alpha) \doteq 1.67$, and $c_3 = 1/(\sqrt{5} \log \alpha) \doteq .93$. Also, $E(u, v) < c_1 \log v + c_2 - 1 + c_3v^{-1}$.

Proof Assume $u \geq F_{n+2}$. Then

$$u \geq \frac{\alpha^{n+2} - \beta^{n+2}}{\sqrt{5}} > \frac{\alpha^{n+2} - 1}{\sqrt{5}}.$$

Hence $(n + 2) \log \alpha \leq \log(1 + u\sqrt{5})$. Now we use the fact that $\log(x + 1) = \log x +$

$\log(1 + 1/x)$ and the estimate $\log(1 + 1/x) < 1/x$ to obtain

$$(n + 2) \log \alpha < \log(u\sqrt{5}) + 1/(u\sqrt{5}),$$

and we see $n < c_1 \log u + c_2 - 2 + c_3 u^{-1}$. Taking the contrapositive, we see that if

$$n \geq c_1 \log u + c_2 - 2 + c_3 u^{-1},$$

then $u < F_{n+2}$. To make the conclusion $u < F_{n+2}$ true, we choose

$$n = \left\lceil c_1 \log u + c_2 - 2 + c_3 u^{-1} \right\rceil .$$

From Lemma 4.2.1, if $u < F_{n+2}$, then $E(u, v) < n$. But $E(u, v)$ is an integer, so in fact $E(u, v) < c_1 \log u + c_2 - 2 + c_3 u^{-1}$.

The inequality $E(u, v) < c_1 \log v + c_2 - 1 + c_3 v^{-1}$ follows in a similar fashion. ■

Using Corollary 4.2.3, we easily get a polynomial-time bound on the bit complexity of the Euclidean algorithm on inputs u, v. For all the operations are on integers $\leq u$, and the algorithm performs $O(\lg u)$ divisions on integers with $O(\lg u)$ bits. This gives a time bound of $O((\lg u)^3)$.

With just slightly more work, however, we can improve this bound to $O((\lg u)(\lg v))$. Suppose $u = u_0$ and $v = u_1$ are the inputs, and refer to the system of equations (4.3). Now dividing u_i by u_{i+1} to get a quotient of a_i and a remainder of u_{i+2} can be done in $O((\lg a_i)(\lg u_{i+1}))$ bit operations. Thus the total bit complexity (disregarding constant factors) is

$$\begin{aligned}\sum_{0 \leq i \leq n-1} (\lg a_i)(\lg u_{i+1}) &\leq (\lg v) \sum_{0 \leq i \leq n-1} \lg a_i \\ &\leq (\lg v) \left(n + \log_2 \left(\prod_{0 \leq i \leq n} a_i \right) \right).\end{aligned}$$

Now from Corollary 4.2.3 we know $n = O(\lg u)$, and by Exercise 5, we have $a_0 a_1 \cdots a_{n-1} \leq u$; hence we conclude that the bit complexity is $O((\lg u)(\lg v))$. We have shown

COROLLARY 4.2.4 Let u, v be integers. We can compute the greatest common divisor of u and v using $O((\lg u)(\lg v))$ bit operations.

4.3 The Extended Euclidean Algorithm

The extended Euclidean algorithm takes as input a pair (u, v) and computes $d = \gcd(u, v)$ *and* a pair of integers (a, b) such that $au + bv = d$.

It is based on the following theorem:

THEOREM 4.3.1 Let u, v, and c be integers. Then the equation

$$au + bv = c$$

has a solution in integers (a, b) iff $\gcd(u, v) \mid c$.

Proof Let d be any integer such that $d \mid u$ and $d \mid v$. Then clearly $d \mid c$; hence $\gcd(u, v) \mid c$.

To prove the other direction, it clearly suffices to show that the equation $au + bv = \gcd(u, v)$ has a solution. This is done in the next theorem.

THEOREM 4.3.2 Let u and v be integers, and let $d = \gcd(u, v)$. Then there exist integers a and b such that

$$au + bv = d.$$

Proof We apply the Euclidean algorithm to $u = u_0$ and $v = u_1$, obtaining the sequence of quotients $a_0, a_1, \ldots a_{n-1}$ as in (4.3). Now we rewrite the system of equations (4.3) as follows:

$$\begin{bmatrix} u_0 \\ u_1 \end{bmatrix} = \begin{bmatrix} a_0 & 1 \\ 1 & 0 \end{bmatrix} \begin{bmatrix} u_1 \\ u_2 \end{bmatrix};$$

$$\begin{bmatrix} u_1 \\ u_2 \end{bmatrix} = \begin{bmatrix} a_1 & 1 \\ 1 & 0 \end{bmatrix} \begin{bmatrix} u_2 \\ u_3 \end{bmatrix};$$

$$\vdots$$

$$\begin{bmatrix} u_{n-2} \\ u_{n-1} \end{bmatrix} = \begin{bmatrix} a_{n-2} & 1 \\ 1 & 0 \end{bmatrix} \begin{bmatrix} u_{n-1} \\ u_n \end{bmatrix};$$

$$\begin{bmatrix} u_{n-1} \\ u_n \end{bmatrix} = \begin{bmatrix} a_{n-1} & 1 \\ 1 & 0 \end{bmatrix} \begin{bmatrix} u_n \\ u_{n+1} \end{bmatrix}.$$

where $d = u_n = \gcd(u_0, u_1)$ and $u_{n+1} = 0$.

Next, we combine these equations to get

$$\begin{bmatrix} u \\ v \end{bmatrix} = \begin{bmatrix} a_0 & 1 \\ 1 & 0 \end{bmatrix} \cdots \begin{bmatrix} a_{n-1} & 1 \\ 1 & 0 \end{bmatrix} \begin{bmatrix} d \\ 0 \end{bmatrix}.$$

Now define

$$M_k = \begin{bmatrix} b_k & c_k \\ d_k & e_k \end{bmatrix} = \begin{bmatrix} a_0 & 1 \\ 1 & 0 \end{bmatrix} \cdots \begin{bmatrix} a_k & 1 \\ 1 & 0 \end{bmatrix}.$$

Then we have

$$M_{n-1}^{-1}\begin{bmatrix} u \\ v \end{bmatrix} = \begin{bmatrix} d \\ 0 \end{bmatrix}.$$

Since $\det M_{n-1} = (-1)^n$, we see that

$$M_{n-1}^{-1} = (-1)^n \begin{bmatrix} e_{n-1} & -c_{n-1} \\ -d_{n-1} & b_{n-1} \end{bmatrix}.$$

Hence $e_{n-1}u - c_{n-1}v = (-1)^n d$, and we may take $a = (-1)^n e_{n-1}$, $b = (-1)^{n+1} c_{n-1}$. ■

The proof we have just given for the existence of a and b actually provides an efficient algorithm for calculating them. We have the following algorithm:

EXTENDED EUCLID(u, v)
{ Computes $d = \gcd(u, v)$ and a, b such that $au + bv = d$ }
$M \leftarrow \begin{bmatrix} 1 & 0 \\ 0 & 1 \end{bmatrix}$
$n \leftarrow 0$
while ($v \neq 0$) do
    $q \leftarrow \lfloor u/v \rfloor$ { compute a_n }
    $M \leftarrow M \begin{bmatrix} q & 1 \\ 1 & 0 \end{bmatrix}$
    $(u, v) \leftarrow (v, u - qv)$
    $n \leftarrow n + 1$
return($d = u$, $a = (-1)^n M_{22}$, $b = (-1)^{n+1} M_{12}$)

COROLLARY 4.3.3 Given integers u and v, we can find integers a, b such that

$$au + bv = d$$

in $O((\lg u)(\lg v))$ bit operations.

Proof The algorithm computes the matrices $M_0, M_1, \ldots M_{n-1}$. It is easy to see that

$$b_0 \leq b_1 \leq b_2 \cdots \leq b_{n-1} \leq u,$$

and similarly for the other entries of the M_i. Let us compute the number of bit operations required to compute the b_i, the other entries of the matrices M_i being handled similarly. Since $b_{i+1} = b_i a_{i+1} + c_i$ for $0 \leq i \leq n-2$, the total number of bit operations to compute all the b_i (disregarding constant factors) is bounded by

$$\sum_{0 \le i \le n-2} (\lg b_i)(\lg a_{i+1}) + (\lg b_i) + (\lg a_{i+1}) + (\lg c_i).$$

Using exercise 5 and Corollary 4.2.3, it is easy to see that this sum is $O((\lg u)(\lg v))$.

An example: let $u = 180$, $v = 146$. Then the quotients in the Euclidean algorithm are $a_0 = 1$, $a_1 = 4$, $a_2 = 3$, $a_3 = 2$, and $a_4 = 2$. We find $n = 5$ and

$$M_4 = \begin{bmatrix} 90 & 37 \\ 73 & 30 \end{bmatrix}.$$

Hence $a = -30$, $b = 37$, and $au + bv = 2$.

4.4 The Euclidean Algorithm and Continuants

Notice that, in the system of equations (4.3), each u_i is actually a multiple of $d = \gcd(u_0, u_1) = u_n$. For example,

$$u_{n-1} = a_{n-1}u_n = a_{n-1}d,$$

$$u_{n-2} = a_{n-2}u_{n-1} + u_n = (a_{n-2}a_{n-1} + 1)d$$

and

$$u_{n-3} = a_{n-3}u_{n-2} + u_{n-1} = (a_{n-3}a_{n-2}a_{n-1} + a_{n-3} + a_{n-1})d;$$

this suggests writing each u_i as the product of d and a polynomial in the a_i.

These polynomials are classical objects called *continuants*. Continuants are defined as follows:

DEFINITION 4.4.1

$$\begin{aligned} Q_0() &= 1 \\ Q_1(X_0) &= X_0 \\ Q_{n+1}(X_0, X_1, \ldots, X_n) &= X_n Q_n(X_0, X_1, \ldots, X_{n-1}) + Q_{n-1}(X_0, X_1, \ldots, X_{n-2}) \end{aligned} \tag{4.5}$$

for $n \ge 1$.

We will see below in Theorem 4.4.5 that

$$u_i = dQ_{n-i}(a_i, a_{i+1}, \ldots, a_{n-1}).$$

First, we prove some theorems on continuants. We will use the following shorthand notation: for $Q_{n+1}(X_0, X_1, \ldots X_n)$ we will write $Q[0, n]$.

The definition of continuants suggests viewing their calculation as a linear transformation. We have

$$\begin{bmatrix} Q[0,n] & Q[0,n-1] \\ Q[1,n] & Q[1,n-1] \end{bmatrix} = \begin{bmatrix} Q[0,n-1] & Q[0,n-2] \\ Q[1,n-1] & Q[1,n-2] \end{bmatrix} \begin{bmatrix} X_n & 1 \\ 1 & 0 \end{bmatrix} \tag{4.6}$$

for $n \geq 2$.

By iterating, we find

$$\begin{bmatrix} Q[0,n] & Q[0,n-1] \\ Q[1,n] & Q[1,n-1] \end{bmatrix} = \begin{bmatrix} X_0 & 1 \\ 1 & 0 \end{bmatrix} \begin{bmatrix} X_1 & 1 \\ 1 & 0 \end{bmatrix} \cdots \begin{bmatrix} X_n & 1 \\ 1 & 0 \end{bmatrix}. \tag{4.7}$$

This follows from (4.6) for $n \geq 2$. However, it is also true for $n = 1$ and even for $n = 0$, provided we define $Q[1, n-1] = 0$ for $n = 0$.

From this we can obtain many useful identities. For example

THEOREM 4.4.2

$$(-1)^{n+1} = Q[0,n]Q[1,n-1] - Q[0,n-1]Q[1,n].$$

Proof Take the determinant of both sides of (4.7). ■

THEOREM 4.4.3

$$Q_{n+1}(X_0, X_1, \ldots, X_{n-1}, X_n) = Q_{n+1}(X_n, X_{n-1}, \cdots, X_1, X_0).$$

Proof Take the transpose of both sides of (4.7). ■

THEOREM 4.4.4

$$Q[0,n] = Q[0,i]Q[i+1,n] + Q[0,i-1]Q[i+2,n]$$

for $0 \leq i \leq n-1$.

Proof Split the right side of (4.7) into two parts, one containing the product of the first $i+1$ matrices, and the other containing the product of the last $n-i$ matrices. ■

As we noted above, continuants express the relationship between the successive remainders and quotients in the Euclidean algorithm.

Let $u_0, \ldots, u_n$ and $a_0, \ldots, a_{n-1}$ be as in (4.3).

THEOREM 4.4.5 Suppose the Euclidean algorithm on inputs (u_0, u_1) takes n steps. Then

$$u_i = dQ_{n-i}(a_i, a_{i+1}, \ldots, a_{n-1})$$

where $d = \gcd(u_0, u_1)$.

Proof We will prove this by induction on i. The induction will proceed from $i = n - 1$ down, rather than the usual $i = 0$ up. The theorem is easily verified for $i = n$ and $i = n - 1$. Now assume it is true for all i such that $j \leq i \leq n$; we wish to prove it for $i = j - 1$. Then

$$\begin{aligned} u_{j-1} &= a_{j-1}u_j + u_{j+1} \\ &= a_{j-1}dQ_{n-j}(a_j, \ldots, a_{n-1}) + dQ_{n-j-1}(a_{j+1}, \ldots, a_{n-1}) \quad \text{(by induction)} \\ &= d(a_{j-1}Q_{n-j}(a_{n-1}, \ldots, a_j) + Q_{n-j-1}(a_{n-1}, \ldots, a_{j+1})) \quad \text{(by Theorem 4.4.3)} \\ &= dQ_{n-j+1}(a_{n-1}, \ldots, a_{j-1}) \quad \text{(by definition)} \\ &= dQ_{n-j+1}(a_{j-1}, \ldots, a_n). \end{aligned}$$

■

To conclude this section, we show that continuants can be computed quickly.

THEOREM 4.4.6 Let $a_0, \ldots, a_{n-1}$ be integers with $a_0 \geq 0$ and $a_i \geq 1$ for $1 \leq i \leq n - 1$. Set $u = Q_n(a_0, a_1, \ldots, a_{n-1})$ and $v = Q_{n-1}(a_1, \ldots, a_{n-1})$. Then we can compute u and v in $O((\lg u)(\lg v))$ bit operations.

Proof Essentially the same as the proof of Corollary 4.2.4. ■

4.5 Continued Fractions

In this section, we establish the relationship between the Euclidean algorithm and continued fractions, and discuss some properties of continued fractions. Some of this material will be used later in Chapter 11.

An expression of the form

$$X_0 + \cfrac{1}{X_1 + \cfrac{1}{X_2 + \cdots + \cfrac{1}{X_n}}} \tag{4.8}$$

is called a *continued fraction*; more precisely, a finite simple continued fraction. We usually abbreviate (4.8) as

$$[X_0, X_1, \ldots, X_n].$$

The X_i are called the *partial quotients*.

Although in our discussion the X_i will usually represent positive integers, for the moment it may be useful to think of the X_i as indeterminates or as representing arbitrary real numbers.

When x is a rational number, we can find a particular sequence of integers a_i such that $x = [a_0, a_1, \ldots, a_n]$ using the following algorithm, called the *continued fraction algorithm*. Given a rational number x, it produces a finite sequence of integers $(a_0, a_1, \ldots, a_n)$.

CFA(x)
{ returns finite or infinite sequence $(a_0, a_1, \ldots)$ }
$i \leftarrow 0$
$x_0 \leftarrow x$
$a_0 \leftarrow \lfloor x_0 \rfloor$
output(a_0)
while $(x_i \neq a_i)$ do

$\quad x_{i+1} \leftarrow \dfrac{1}{x_i - a_i}$

$\quad i \leftarrow i + 1$

$\quad a_i \leftarrow \lfloor x_i \rfloor$

$\quad$ output(a_i)

Note that we distinguish between the output of this algorithm and the rational function defined by (4.8).

THEOREM 4.5.1 CFA(x) returns a finite list $(a_0, a_1, \ldots, a_n)$ iff x is rational. In this case we have $x = [a_0, a_1, \ldots, a_n]$.

Proof Suppose CFA(x) returns a finite list, say $(a_0, a_1, \ldots, a_n)$. Then we claim that

$$x = [a_0, a_1, \ldots, a_n].$$

For let x_i, $i > 0$, be defined as in the function CFA. Then it is easy to prove by induction that

$$x_0 = [a_0, a_1, \ldots, a_{i-1}, x_i].$$

If CFA terminates and returns $(a_0, a_1, \ldots, a_n)$, we must have $x_n = a_n$. Hence

$$x = x_0 = [a_0, a_1, \ldots, a_{n-1}, a_n],$$

a rational number.

On the other hand, suppose x_i is rational, say $x_i = u_i/u_{i+1}$. Then $a_i = \lfloor x_i \rfloor = \lfloor u_i/u_{i+1} \rfloor$, and

$$x_{i+1} = \frac{1}{x_i - a_i}$$

$$= \frac{1}{u_i/u_{i+1} - \lfloor u_i/u_{i+1} \rfloor}$$
$$= \frac{u_{i+1}}{u_i - u_{i+1} \lfloor u_i/u_{i+1} \rfloor}$$
$$= \frac{u_{i+1}}{u_i \bmod u_{i+1}}.$$

If we represent the fraction u/v by the pair (u, v), then we see that the transformation that takes x_i to x_{i+1} maps the pair

$$(u_i, u_{i+1}) \quad \text{to} \quad (u_{i+1}, u_i \bmod u_{i+1}).$$

This is precisely the transformation of the Euclidean algorithm, which terminates on integer inputs, as we have already seen. We eventually find that $u_i \bmod u_{i+1} = 0$, which implies that $u_i/u_{i+1} = \lfloor u_i/u_{i+1} \rfloor$; i.e. $x_i = a_i$, which is the termination condition in the algorithm CFA. ■

Note that the last output of CFA, a_n, is always ≥ 2, except possibly when $n = 0$.

We have also proved

THEOREM 4.5.2 $\text{CFA}(a/b) = (a_0, a_1, \ldots, a_{n-1})$ iff the Euclidean algorithm for $\gcd(a, b)$ takes n division steps.

The next theorem relates continuants to continued fractions.

THEOREM 4.5.3

$$\frac{Q_{n+1}(X_0, X_1, \ldots, X_n)}{Q_n(X_1, X_2, \ldots, X_n)} = [X_0, X_1, \cdots, X_n]$$

Proof See Exercise 8. ■

Now fix a continued fraction $x = [a_0, a_1, \cdots, a_n]$. We define some useful abbreviations. We write

$$a_i' = [a_i, a_{i+1}, \ldots, a_n],$$

the i-th *complete quotient*.

We also write

$$p_i = Q_{i+1}(a_0, a_1, \ldots, a_i)$$

and

$$q_i = Q_i(a_1, a_2, \ldots, a_i).$$

The fraction p_i/q_i is referred to as the i-th *convergent*. These abbreviations are useful and frequently occur in the literature; for example, Theorem 4.4.2 is written $(-1)^{n+1} = p_n q_{n-1} - p_{n-1} q_n$.

Since

$$x = [a_0, a_1, \ldots, a_{i-1}, a_i'],$$

we have

$$x = \frac{a_i' p_{i-1} + p_{i-2}}{a_i' q_{i-1} + q_{i-2}}.$$

COROLLARY 4.5.4

$$\gcd(p_i, q_i) = 1.$$

Proof By Theorem 4.4.2, we have $p_i q_{i-1} - p_{i-1} q_i = (-1)^{i+1}$. Hence by Theorem 4.3.1, $\gcd(p_i, q_i) = 1$. ■

We will now prove that continued fraction expansions are essentially unique. We say a continued fraction $[a_0, a_1, \ldots a_n]$ is *regular* if the a_i are integers with $a_i \geq 1$ for $i \geq 1$.

THEOREM 4.5.5 If the two numbers represented by regular continued fractions

$$[a_0, a_1, \ldots, a_n], \quad [b_0, b_1, \ldots, b_m]$$

are equal, and $a_n > 1$ if $n > 0$, and $b_m > 1$ if $m > 0$, then $n = m$ and the sequences are identical.

Proof The reader can easily check that $a_i = \lfloor a_i' \rfloor$ for $0 \leq i \leq n$.

We have $x = a_0' = b_0'$. Thus $a_0 = b_0 = \lfloor x \rfloor$. Now suppose that the first i partial quotients in the continued fractions are the same:

$$x = [a_0, a_1, \ldots, a_{i-1}, a_i'] = [a_0, a_1, \ldots, a_{i-1}, b_i'].$$

If $i = 1$, then $a_0 + 1/a_1' = a_0 + 1/b_1'$; hence $a_1' = b_1'$ and so $a_1 = b_1$, as above. If $i > 1$, then we have

$$\frac{a_i' p_{i-1} + p_{i-2}}{a_i' q_{i-1} + q_{i-2}} = \frac{b_i' p_{i-1} + p_{i-2}}{b_i' q_{i-1} + q_{i-2}}$$

and so $(a_i' - b_i')(p_{i-1} q_{i-2} - p_{i-2} q_{i-1}) = 0$. But by Theorem 4.4.2, $p_{i-1} q_{i-2} - p_{i-2} q_{i-1} = (-1)^i$, so $a_i' = b_i'$. Hence $a_i = b_i$.

Now suppose that $n < m$. The argument of the previous paragraph shows that $a_i = b_i$ for $i \leq n$. Then

$$\frac{p_n}{q_n} = [a_0, a_1, \ldots, a_n] = [a_0, a_1, \ldots, a_n, b_{n+1}, \ldots, b_m] = \frac{b'_{n+1}p_n + p_{n-1}}{b'_{n+1}q_n + q_{n-1}},$$

so $p_n q_{n-1} - p_{n-1} q_n = 0$, a contradiction. Hence $n = m$. ■

We emphasize that when we write $x = [a_0, a_1, \ldots, a_n]$, we mean only that x equals the rational function (4.8) evaluated at the $n + 1$ inputs $a_0, a_1, \ldots, a_n$. We do *not* mean that the continued fraction algorithm produces the output $(a_0, a_1, \ldots, a_n)$ on input x; for this, we write $\text{CFA}(x) = (a_0, a_1, \ldots, a_n)$.

On the other hand, Theorem 4.5.5 allows us to speak of *the* regular continued fraction for a rational number. It ensures that if $u/v = [a_0, a_1, \ldots, a_n]$ for integers a_i satisfying the special conditions $a_i \geq 1$ for $i \geq 1$, and $a_n \geq 2$ for $n \neq 0$, then in fact $\text{CFA}(u/v) = (a_0, a_1, \ldots, a_n)$.

4.6 The Least-Remainder Euclidean Algorithm

The ordinary Euclidean algorithm proceeds by replacing the pair (u, v) with $(v, u \bmod v)$. However, it seems reasonable to consider choices different from $u \bmod v$; for example, we could replace the pair (49, 10) with $(10, -1)$ instead of (10, 9).

The *least-remainder Euclidean algorithm* replaces (u, v) with $(v, u \bmod v)$ or $(v, (u \bmod v) - v)$, depending on whether

$$|u \bmod v| \leq |v|/2 \quad \text{or} \quad |u \bmod v| > |v|/2.$$

This amounts to defining a new function

$$\text{round}(x) = \lceil x - 1/2 \rceil$$

and a new "mod" function

$$a \,\text{rmod}\, n \stackrel{\text{def}}{=} \begin{cases} a, & \text{if } n = 0; \\ a - n\,\text{round}(\frac{a}{n}), & \text{otherwise,} \end{cases} \tag{4.9}$$

and running the Euclidean algorithm with rmod in place of mod.

Similarly, by replacing occurrences of the floor function with the function round in the algorithm CFA, we get a different continued fraction algorithm, NICFA, in which both negative and positive quotients may appear. This new continued fraction algorithm is called the *nearest-integer continued fraction* algorithm. For example, $\text{NICFA}(7/19) = (0, 3, -4, 2)$.

THEOREM 4.6.1 Every rational number can be written uniquely as a nearest-integer continued fraction, $[a_0, a_1, \cdots, a_n]$, where the a_i are integers and

(1) $|a_i| \geq 2$ for $1 \leq i \leq n$;

(2) $a_n \neq -2$;

(3) if $a_i = -2$ for $i < n$, then $a_{i+1} \leq -2$; and

(4) if $a_i = 2$ for $i < n$, then $a_{i+1} \geq 2$.

Proof Left to the reader as Exercise 9.

We now prove that the least-remainder Euclidean algorithm uses fewer division steps in the worst case than the ordinary Euclidean algorithm.

Define $A_0 = 0$; $A_1 = 1$; and $A_n = 2A_{n-1} + A_{n-2}$ for $n \geq 2$.

THEOREM 4.6.2 (DUPRÉ) Let u, v be integers such that $u \geq v > 0$, the least-remainder Euclidean algorithm on input (u, v) performs n division steps, and (u, v) is lexicographically least among all pairs having these properties. Then $u = A_{n-1} + A_n$, $v = A_n$.

Proof Left to the reader as Exercise 11.

Exercise 10 asks you to prove that

$$A_n = \frac{\eta^n - \theta^n}{2\sqrt{2}},$$

where $\eta = 1 + \sqrt{2}$ and $\theta = 1 - \sqrt{2}$. Let $\lambda(u, v)$ be the number of division steps performed by the least-remainder Euclidean algorithm on input (u, v). Using the proof method of Corollary 4.2.3, we easily show the following:

COROLLARY 4.6.3 Let $u \geq v > 0$. Then $\lambda(u, v) < d_1 \log u + d_2 + d_3 u^{-1}$, where $d_1 = 1/\log \eta \doteq 1.13$, $d_2 = 1 + (\log 2\sqrt{2} - \log(\eta + 1))/(\log \eta) \doteq .79$, and $d_3 = 1/(2\sqrt{2} \log \eta) \doteq .40$.

The least-remainder Euclidean algorithm has another beautiful property: it requires no more division steps than any other Euclidean algorithm which chooses between a remainder of $u \bmod v$ or $(u \bmod v) - v$ at each step.

THEOREM 4.6.4 (KRONECKER-VAHLEN) Let $\lambda(u, v)$ denote the number of division steps used by the least-remainder Euclidean algorithm on input (u, v), and let $\kappa(u, v)$ denote the same quantity for any other Euclidean algorithm. Then $\lambda(u, v) \leq \kappa(u, v)$.

Proof We assume, without loss of generality, that $u > v > 0$. (It is easily verified that $\lambda(u, v) = \lambda(u, -v) = \lambda(-u, v) = \lambda(-u, -v)$.) Let us write $u = u_0$ and $v = u_1$.

To simplify the proof somewhat, we will rewrite the division steps of the least-remainder algorithm as follows:

$$
\begin{aligned}
u_0 &= a_0 u_1 + \epsilon_0 u_2 \\
u_1 &= a_1 u_2 + \epsilon_1 u_3 \\
&\vdots \\
u_{n-2} &= a_{n-2} u_{n-1} + \epsilon_{n-2} u_n \\
u_{n-1} &= a_{n-1} u_n
\end{aligned}
\tag{4.10}
$$

where $\epsilon_i = \pm 1$. This allows us to assume that the u_i's and a_i's are positive throughout.

LEMMA 4.6.5 Let $u_0 \geq 2u_1 > 0$. Then $\lambda(u_0, u_1) \leq \lambda(u_0, u_0 - u_1)$.

Proof This is easily verified for $2 \leq u_0 \leq 4$. Now assume it is true for all $u_0 < N$. We prove it for $u_0 = N$.

Let us write $u'_0 = u_0$ and $u'_1 = u_0 - u_1$, and write the least-remainder algorithm on (u'_0, u'_1) as follows:

$$
\begin{aligned}
u'_0 &= a'_0 u'_1 + \epsilon'_0 u'_2 \\
u'_1 &= a'_1 u'_2 + \epsilon'_1 u'_3 \\
&\vdots \\
u'_{n-2} &= a'_{n-2} u'_{n-1} + \epsilon'_{n-2} u'_n \\
u'_{n-1} &= a'_{n-1} u'_n
\end{aligned}
\tag{4.11}
$$

Case I: $u_0 \geq 3u_1$. In this case we find $a'_0 = 1$, $\epsilon'_0 = 1$, $u'_2 = u_1$, $a'_1 = a_0 - 1$, $\epsilon'_1 = \epsilon_0$, and $u'_3 = u_2$. Thus $\lambda(u'_2, u'_3) = \lambda(u_1, u_2)$, and so $\lambda(u_0, u_0 - u_1) = 1 + \lambda(u_0, u_1)$.

Case II: $2u_1 \leq u_0 < \frac{5}{2}u_1$. In this case we find $a_0 = 2$, $\epsilon_0 = 1$, $a'_0 = 2$, $\epsilon'_0 = -1$. If $u_2 = 0$, then $u'_2 = 0$ and hence $\lambda(u_0, u_1) = \lambda(u_0, u_0 - u_1) = 1$. If not, we have $u'_2 = u_0 - 2u_1 = u_2$, $a'_1 = a_1 + 1$, $\epsilon'_1 = \epsilon_1$, and $u'_3 = u_3$. Thus $\lambda(u'_2, u'_3) = \lambda(u_2, u_3)$, and hence $\lambda(u_0, u_1) = \lambda(u_0, u_0 - u_1)$.

Case III: $\frac{5}{2}u_1 \leq u_0 < 3u_1$. In this case we find $a_0 = 3$, $\epsilon_0 = -1$, $a'_0 = 2$, $\epsilon'_0 = -1$. Then $u_2 = 3u_1 - u_0$, and $u'_1 = u_0 - u_1 = 2u_1 - u_2$. Then $u'_2 = u_0 - 2u_1 = u_1 - u_2$. Now $\lambda(u'_0, u'_1) = 1 + \lambda(u'_1, u'_2)$, and $\lambda(u'_1, u'_2) = \lambda(2u_1 - u_2, u_1 - u_2) = \lambda(u_1, u_1 - u_2)$. Hence $\lambda(u_0, u_0 - u_1) = 1 + \lambda(u_1, u_1 - u_2)$. But $u_1 < N$, so the induction hypothesis applies to $\lambda(u_1, u_1 - u_2)$, and so we find $\lambda(u_1, u_2) \leq \lambda(u_1, u_1 - u_2)$. Since $\lambda(u_0, u_1) = 1 + \lambda(u_1, u_2)$, we conclude $\lambda(u_0, u_1) \leq \lambda(u_0, u_0 - u_1)$.

This concludes the proof of the lemma. ■

Now we can finish the proof of the Kronecker-Vahlen theorem. We prove it by induction on u_0. It is easily verified in the case $u_1 \mid u_0$, so we may assume $u_1 \nmid u_0$. It is easily verified for $u_0 = 3$. Now assume true for all $u_0 < N$, and we will prove it for $u_0 = N$.

Let $u_0 = a_0 u_1 + \epsilon_0 u_2$ be the start of the least-remainder Euclidean algorithm, and let $u_0 = a_0' u_1 + \epsilon_0' u_2'$ be the start of any other algorithm. If $u_2 = u_2'$, then by induction we have $\lambda(u_1, u_2) \le \kappa(u_1, u_2)$ and we are done.

Otherwise $a_0' = a_0 + \epsilon_0$, $\epsilon_0' = \epsilon_0$, and so $u_2' = u_1 - u_2$. In this case, we have $\kappa(u_1', u_2') \ge \lambda(u_1', u_2')$, by the induction hypothesis. But by the lemma $\lambda(u_1', u_2') \ge \lambda(u_1, u_2)$. Hence we conclude $\kappa(u_0, u_1) \ge \lambda(u_0, u_1)$, as desired. ■

We conclude this section by defining generalized continued fraction expansions, which include the ordinary and nearest-integer continued fractions as special cases.

Consider any function $f : \mathbb{R} \rightarrow \mathbb{Z}$ with the following properties:

1. $f(x + j) = f(x) + j$, for any $x \in \mathbb{R}$, $j \in \mathbb{Z}$;
2. $|f(x) - x| < 1$.

Such functions f are called *integer functions*.

Define

$$a \bmod_f n = \begin{cases} a, & \text{if } n = 0; \\ a - nf(\frac{a}{n}), & \text{otherwise,} \end{cases} \tag{4.12}$$

Note that $|u \bmod_f v| < v$ for $v \neq 0$. Hence a modified Euclidean algorithm, where $\bmod_f$ replaces mod, always terminates on input (u, v) and produces $\gcd(u, v)$.

Furthermore, if $f(x)$ replaces $\lfloor x \rfloor$ in the CFA algorithm, we get a generalized continued fraction expansion, in which positive and negative terms may appear.

These ideas are continued further in the exercises.

4.7 The Binary gcd Algorithm

A "non-Euclidean" method for calculating the greatest common divisor was discovered by J. Stein; it is frequently called the *binary gcd algorithm*.

It is based on the following facts:

(a) If u and v are both even, then

$$\gcd(u, v) = 2 \gcd\left(\frac{u}{2}, \frac{v}{2}\right). \tag{4.13}$$

(b) If u is even and v is odd, then

$$\gcd(u, v) = \gcd\left(\frac{u}{2}, v\right). \tag{4.14}$$

(c) If u and v are both odd, then

$$\gcd(u, v) = \gcd(|u - v|/2, v). \tag{4.15}$$

These ideas lead to the following algorithm:

```
BINARY GCD(u, v) {binary gcd algorithm}
{inputs are u, v > 0}
(1)  g ← 1 {g holds powers of 2 in gcd(u, v)}
(2)  while (u mod 2 = 0) and (v mod 2 = 0) do
(3)       u ← u/2
(4)       v ← v/2
(5)       g ← 2g   {remove powers of 2}
{now at least one of u, v is odd}
(6)  while (u ≠ 0) do
(7)       if (u mod 2 = 0) then u ← u/2
(8)       else if (v mod 2 = 0) then v ← v/2
(9)       else {both odd}
(10)           t ← |u − v|/2
(11)           if u ≥ v then u ← t else v ← t
               { replace larger of u, v with |u−v|/2 }
(12) return (g · v)
```

THEOREM 4.7.1 Let $u, v > 0$ be integers. The function BINARY GCD(u, v) computes the greatest common divisor of u and v in $O((\lg uv)^2)$ bit operations.

Proof First, let us prove that the algorithm terminates. It is clear that the first loop must eventually terminate. We claim in each pass through the loops on lines (2)–(5) and (6)–(11), the non-negative quantity uv is decreased by a factor of at least 2. This is clearly true for each iteration of the first loop, which divides both u and v by 2. The second loop divides u by 2 or v by 2, whichever is even; if both are odd, then it replaces the larger of u and v with $|u - v|/2$. This proves the claim, and in fact shows that in at most $1 + \log_2 uv$ passes through the loops, we find $uv = 0$. Hence either $u = 0$ or $v = 0$. But v can never be set to 0; if it were, it would have to occur in lines (10)–(11); this implies $u = v$, and so the algorithm would have set u to 0, not v. Hence u must eventually be set to 0, and the second loop terminates.

The correctness of algorithm BINARY GCD can be verified using the identities (4.13)–(4.15).

Now, as we have seen, the algorithm terminates in at most $1 + \log_2 uv$ passes through the loop. Each of the loop steps (a division by 2; a subtraction) can be done in $O(\lg uv)$ bit operations, and the complexity bound follows. ■

The worst case of the binary gcd algorithm is discussed in Exercise 26. It is possible to "extend" the binary algorithm to solve $au + bv = \gcd(u, v)$, as we did for the Euclidean algorithm in section 4.3; see Exercise 30.

4.8 Constructing a gcd-Free Basis

In this section, we introduce the concept of a *gcd-free basis*, a useful but little-known tool for the construction of number-theoretic algorithms. To motivate this concept, consider the following problem: given six positive integers, a, b, c, i, j, k, efficiently determine whether

$$a^i b^j = c^k.$$

We cannot simply multiply a^i and b^j, and see if the result is c^k, for a^i, b^j, and c^k cannot be computed in polynomial time (in $\lg abcijk$). If we knew the prime factorizations of $a, b,$ and c, we could solve the problem efficiently by comparing the exponents for each prime divisor of abc. However, there is no efficient algorithm known for factorization.

One way to solve this problem is to express a, b, and c not as the product of primes, but as the product of relatively prime pieces. The advantage to this method is that such a "factorization" can be completed in polynomial time.

Let $A = (a_1, a_2, \ldots, a_m)$ be a nonempty list of m positive integers, not necessarily distinct. Let $B = \{b_1, b_2, \ldots, b_n\}$ be a set of integers, each ≥ 2. We say B is a *gcd-free basis* for A if

(a) $b_i \perp b_j$ for all $i \neq j$; and

(b) there exist mn non-negative integers e_{ij} such that $a_i = \prod_{1 \leq j \leq n} b_j^{e_{ij}}$ for all i, $1 \leq i \leq m$.

For example, $\{2, 7, 15\}$ is a gcd-free basis for the list (4, 30, 14, 49). Note that $\emptyset$, the empty set, is a gcd-free basis for the list (1).

One reason for the importance of the concept of the gcd-free basis is the following:

THEOREM 4.8.1 Let the set B be a gcd-free basis for $A = (a_1, a_2, \ldots, a_m)$. Then each element of A can be expressed uniquely as the product of non-negative powers of elements of B (up to the order of the factors).

Proof Suppose there were two distinct factorizations of a_i, as follows:

$$a_i = b_1^{e_{i1}} \cdots b_n^{e_{in}} = b_1^{f_1} \cdots b_n^{f_n}.$$

Then

$$b_1^{e_{i1}-f_1} \cdots b_n^{e_{in}-f_n} = 1.$$

If the two factorizations were truly different, this product must contain at least one term with a negative exponent, and at least one term with a positive exponent. Then, by rearranging, we would have

$$\prod_{i \in S} b_i^{g_i} = \prod_{j \in S'} b_j^{g_j}.$$

Choose any term $b_i^{g_i}$ on the left-hand side with $g_i \neq 0$, and let p be a prime dividing b_i. Then p must divide some $b_j^{g_j}$ on the right-hand side; hence $\gcd(b_i, b_j) \geq p$, a contradiction. ■

COROLLARY 4.8.2 Let $B = \{b_1, b_2, \ldots, b_n\}$ be a gcd-free basis for $A = (a_1, a_2, \ldots, a_m)$. Then there exist non-negative integers $u_1, u_2, \ldots, u_n$ and $v_1, v_2, \ldots, v_n$ such that

$$\gcd(a_1, a_2, \ldots, a_m) = \prod_{1 \leq i \leq n} b_i^{u_i}$$

and

$$\operatorname{lcm}(a_1, a_2, \ldots, a_m) = \prod_{1 \leq i \leq n} b_i^{v_i}.$$

Not only is there "unique factorization" as a product of elements of B—we can even find this factorization efficiently:

THEOREM 4.8.3 Suppose $A = (a_1, \ldots, a_m)$ is a list of integers, each ≥ 2, and $B = \{b_1, \ldots, b_n\}$ is a gcd-free basis for A. Then we can compute the mn non-negative integers e_{ij} such that

$$a_i = \prod_{1 \leq j \leq n} b_j^{e_{ij}}$$

using $O((\lg c)^2 + (\lg c)(\lg d))$ bit operations, where $c = a_1 a_2 \cdots a_m$, and $d = b_1 b_2 \cdots b_n$.

Proof We compute the e_{ij} by trial division: for each a_i, we attempt to remove a factor of b_j as many times as possible. Details can be found in the BASIS-FACTOR algorithm below.

```
BASIS-FACTOR(A, B)
(1) for i ← 1 to m do
(2)     t ← a_i
(3)     for j ← 1 to n do
(4)         e_ij ← 0
(5)         while b_j | t do
(6)             t ← t/b_j
(7)             e_ij ← e_ij + 1
(8) return(e)
```

Correctness of the algorithm follows from Theorem 4.8.1.

To determine the running-time of BASIS-FACTOR, we compute the cost for all the divisions in lines (5) and (6). We leave it to the reader to verify that the total cost for lines (5) and (6) majorizes the cost for the other lines.

The cost of lines (5) and (6) for each pair (a_i, b_j) is, from our definition of naive bit complexity,

$$O((e_{ij} + 1)(\lg a_i)(\lg b_j)).$$

Hence the total cost associated with lines (5) and (6) is, ignoring the constant implicit in the $O()$,

$$\begin{aligned}
\sum_{i,j}(e_{ij} + 1)(\lg a_i)(\lg b_j) &\le \sum_{i,j}(e_{ij} + 1)(1 + \log_2 a_i)(1 + \log_2 b_j) \\
&\le 4\sum_{i,j}(e_{ij} + 1)(\log_2 a_i)(\log_2 b_j) \\
&= 4\sum_{i,j} e_{ij}(\log_2 a_i)(\log_2 b_j) + 4\sum_{i,j}(\log_2 a_i)(\log_2 b_j) \\
&= 4\sum_{i,j}(\log_2 a_i)(\log_2 b_j^{e_{ij}}) + 4\sum_{i,j}(\log_2 a_i)(\log_2 b_j) \\
&= 4\sum_{i}(\log_2 a_i)\sum_{j}\log_2 b_j^{e_{ij}} + 4\sum_{j}(\log_2 b_j)\sum_{i}\log_2 a_i \\
&= 4\sum_{i}(\log_2 a_i)^2 + 4(\log_2 c)\sum_{j}\log_2 b_j \\
&\le 4(\lg c)^2 + 4(\lg c)(\lg d).
\end{aligned}$$

This completes the analysis of BASIS-FACTOR. ■

Our next task is to show how to efficiently compute a gcd-free basis for a list of integers $A = (a_1, a_2, \ldots, a_m)$. First, we discuss the case where $m = 2$.

Consider the following algorithm, which "refines" a partial factorization of $n = ab$ into relatively prime pieces:

```
REFINE(a, b)
    g ← gcd(a, b)
    if g = 1 then return(a, ∅, b)
        else (ℓ1, S1, r1) ← REFINE(a/g, g)
            (ℓ2, S2, r2) ← REFINE(r1, b/g)
            return(ℓ1, S1 ∪ S2 ∪ {ℓ2}, r2)
```

We first prove that this algorithm always terminates:

THEOREM 4.8.4 The algorithm REFINE always terminates, and on input $(a, b) \neq (1, 1)$, it makes at most $2 \log_2 ab - 2$ additional recursive calls to itself.

Proof By induction on ab. Clearly the algorithm halts on input $(1, 1)$. For the rest of the proof, we assume $ab \geq 2$. If $ab = 2$, then the input is $(1, 2)$ or $(2, 1)$, and no recursive calls are made.

Now assume the result is true for all a, b with $1 < ab < c$; we will prove the result for $ab = c$.

If the input is $(c, 1)$ or $(1, c)$, the result clearly holds. Otherwise, $a, b > 1$. Let $R(c)$ denote the maximum number of recursive calls made on input (a, b), as a, b range over all positive integers such that $c = ab$; a priori $R(c)$ could be infinite. Then we have

$$R(c) \leq 2 + \max_{\substack{ab=c \\ a,b\neq 1}} (R(a) + R(b)).$$

By induction, all quantities appearing in the sum are finite, so $R(c)$ is finite. Also, we have

$$\begin{aligned} R(c) &\leq 2 + \max_{\substack{ab=c \\ a,b\neq 1}} (R(a) + R(b)) \\ &\leq 2 + \max_{\substack{ab=c \\ a,b\neq 1}} (2\log_2 a + 2\log_2 b - 4) \\ &\leq 2\log_2 c - 2. \end{aligned}$$

This completes the proof. ∎

THEOREM 4.8.5 If, on input (a, b), the algorithm REFINE returns an output (ℓ, S, r), then

(i) $\ell \mid a$ and $r \mid b$;

(ii) each element of S divides $\gcd(a, b)$;

(iii) the set $(S \cup \{\ell, r\}) - \{1\}$ is a gcd-free basis for (a, b); and

(iv) $\gcd(\ell, b) = \gcd(r, a) = 1$.

Proof

(i) We show that $\ell \mid a$. If the algorithm terminates on line 2 without any recursive calls, this is clear. Otherwise by induction, $\ell \mid a/g$ in line 3, and hence $\ell \mid a$. In a similar fashion, $r \mid b$.

(ii) We show that each element of S divides $\gcd(a, b)$. If the algorithm terminates on line 2, this is trivially true. Otherwise, by induction, each element of S_1 divides $g = \gcd(a, b)$. From the result in the previous paragraph, $r_1 \mid g$, and ℓ_2 and each element of S_2 divides r_1. Therefore each element of $S = S_1 \cup S_2 \cup \{\ell_2\}$ divides g.

(iii) In order to show that

$$S' = (S \cup \{\ell, r\}) - \{1\}$$

is a gcd-free basis for (a, b), we first show that $\ell \perp r$. This is clear if the algorithm terminates on line 2. Otherwise, by a result above, $\ell_1 \mid a/g$, $r_2 \mid b/g$, and therefore $\gcd(\ell, r) = \gcd(\ell_1, r_2) \mid \gcd(a/g, b/g) = 1$.

We now introduce a useful notational convention: if q is an integer and S is a set, by $q \perp S$ we mean q is relatively prime to each member of S. Using this convention, we now show $\ell \perp S$. This is clearly true if the algorithm terminates on line 2, for then $S = \emptyset$. Otherwise, by induction we have $\ell_1 \perp S_1$. Also by induction, all elements of S_2 must divide b/g, but $\ell_1 \mid a/g$, so $\ell_1 \perp S_2$. By the result above, $\ell_1 \perp r_1$, and $\ell_2 \mid r_1$, so $\ell_1 \perp \ell_2$. In a similar fashion, we can show $r \perp S$.

Next, we show that the elements of S are pairwise relatively prime. By induction, the elements of S_1 are pairwise relatively prime, and so are the elements of S_2. Let $s_1 \in S_1$, and $s_2 \in S_2$; then $s_1 \mid a/g$ and $s_2 \mid b/g$. Therefore $s_1 \perp s_2$, since $\gcd(a/g, b/g) = 1$. Next, $\ell_2 \perp S_1$, since $\ell_2 \mid r_1$, and $r_1 \perp S_1$ by induction. Finally, $\ell_2 \perp S_2$, by induction.

Now we show that a and b can be expressed as the product of (powers of) elements of $S \cup \{\ell, r\}$. If the algorithm terminates at line 2, this is clear. Otherwise, $a = (a/g)g$, and by induction, each of these can be expressed as the product of elements of $S_1 \cup \{\ell_1, r_1\}$; also, r_1 can be expressed as the product of elements of $S_2 \cup \{\ell_2, r_2\}$. A similar proof applies to b. It follows that $(S \cup \{\ell, r\}) - \{1\}$ is a gcd-free basis for $A = (a, b)$.

Our next task is to show that $\ell \perp b$. If the algorithm terminates on line 2, this is clearly true. Otherwise, by a result above, $\ell_1 \mid a/g$. Now $\gcd(\ell_1, b) \mid \gcd(a/g, b)$, and $\gcd(a/g, b) \mid g$. Therefore $\gcd(\ell_1, b) \mid g$. But in line 3, $\ell_1 \perp g$ by induction. Hence $\ell_1 \perp b$.

Similarly, we can show $r \perp a$. Again, this is trivially true if the algorithm terminates on line 2. Otherwise, $r_2 \perp r_1$ by induction, and $r_2 \perp \ell_1$ and $r_2 \perp S_1$ by the results above.

Since $(\ell_1, S_1, r_1) = \text{Refine}(a/g, g)$, we know that a/g and g can be expressed as the product of elements of $S_1 \cup \{\ell_1, r_1\}$. Therefore $r_2 \perp a/g$ and $r_2 \perp g$. It follows that $r_2 \perp a$.

This completes the proof of the properties of the output of the algorithm. ■

We now show that the algorithm Refine is surprisingly efficient:

THEOREM 4.8.6 On input (a, b), the algorithm Refine uses $O((\lg ab)^2)$ bit operations.

Proof From Corollary 4.2.4, we know that the cost for computing $\gcd(a, b)$, a/g, and b/g, measured in bit operations, is $O((\lg a)(\lg b))$. Provided $a \neq 1$ and $b \neq 1$, this cost is bounded by a constant times $2(\log a)(\log b)$, and using this figure greatly simplifies the analysis. It does not, however, correctly handle the case where $a = 1$ or $b = 1$. Thus, we need to add the cost of computing gcd's when one or both of the arguments to Refine is 1. However, by Theorem 4.8.4, we know that no more than $2\log_2 ab - 2$ recursive calls are made to Refine on input $(a, b) \neq (1, 1)$, and it is easy to see that no recursive calls are made when the input is $(1, 1)$. Hence the total cost of handling inputs where one factor is 1 is $O((\log ab)^2)$.

To obtain the desired time bound, then, we are free to charge $2(\log a)(\log b)$ for the cost of computing $\gcd(a, b)$, a/g, and b/g. Let $T(c)$ be the cost for computing a gcd-free basis (not including the cost of dealing with inputs containing 1) for (a, b), where $c = ab$. We claim that $T(c) \leq (\log c)^2$, and we prove this inequality by induction on c. It is clearly true when $c = 1$. For $c > 1$, we have

$$\begin{aligned} T(c) &\leq T(a/g \cdot g) + T(r_1 \cdot b/g) + 2(\log a)(\log b) \\ &\leq T(a) + T(b) + 2(\log a)(\log b) \\ &\leq (\log a)^2 + (\log b)^2 + 2(\log a)(\log b) \\ &= (\log a + \log b)^2 \\ &= (\log c)^2, \end{aligned}$$

and the result is true by induction. ■

We now consider the problem of efficiently constructing a gcd-free basis for three or more numbers. Consider the following algorithm:

```
GCD-Free Basis(a_1, a_2, ..., a_m)
  if m = 1 then
      if a_1 = 1 then
          return(∅)
      else
```

```
            return({a_1})
    else if m = 2 then
            (ℓ, S, r) ← REFINE(a_1, a_2)
            return((S ∪ {ℓ, r}) − {1})
    else (t_1, t_2, ..., t_s) ← GCD-FREE BASIS(a_1, a_2, ..., a_{m−1})
            T ← ∅
            r_0 ← a_m
            for i ← 1 to s do
                (ℓ_i, S_i, r_i) ← REFINE(t_i, r_{i−1})
                T ← T ∪ S_i ∪ {ℓ_i}
            return(T ∪ {r_s} − {1})
```

THEOREM 4.8.7 Algorithm GCD-FREE BASIS correctly computes a gcd-free basis for input $(a_1, a_2, \ldots, a_m)$. It uses $O((\lg a_1 a_2 \cdots a_m)^2)$ bit operations.

Proof Left to the reader as exercise 16. ■

4.9 Exercises

1. Give a non-recursive version of the Euclidean algorithm.

2. Show how to compute $\operatorname{lcm}(u, v)$ in $O((\lg u)(\lg v))$ bit operations.

3. (de Moivre.) Show that $F_n = (\alpha^n - \beta^n)/\sqrt{5}$, where F_n is the n-th Fibonacci number, and $\alpha = (1 + \sqrt{5})/2$, $\beta = (1 - \sqrt{5})/2$.

4. Complete the proof of Theorem 4.2.2 by showing that $\lfloor F_{k+2}/F_{k+1} \rfloor = 1$ and $F_{k-1} = F_{k+1} \bmod F_k$ for all $k \geq 2$.

5. Suppose the Euclidean algorithm computes the quotients $a_0, a_1, \ldots, a_{n-1}$ on input (u, v) with $u, v > 0$. Prove that $a_0 a_1 \cdots a_{n-1} \leq u$ and $a_1 a_2 \cdots a_{n-1} \leq v$.

6. Suppose

$$\begin{bmatrix} p_n & p_{n-1} \\ q_n & q_{n-1} \end{bmatrix} = \begin{bmatrix} a_0 & 1 \\ 1 & 0 \end{bmatrix} \begin{bmatrix} a_1 & 1 \\ 1 & 0 \end{bmatrix} \cdots \begin{bmatrix} a_n & 1 \\ 1 & 0 \end{bmatrix}.$$

   Show that $p_n/q_n = [a_0, a_1, \ldots, a_n]$.

7. We can generalize simple continued fractions to the case where the numerators are not necessarily 1. Suppose

$$\begin{bmatrix} p_n & p_{n-1} \\ q_n & q_{n-1} \end{bmatrix} = \begin{bmatrix} X_0 & 1 \\ 1 & 0 \end{bmatrix} \begin{bmatrix} X_1 & 1 \\ Y_1 & 0 \end{bmatrix} \cdots \begin{bmatrix} X_n & 1 \\ Y_n & 0 \end{bmatrix}.$$

Then show that

$$\frac{p_n}{q_n} = X_0 + \cfrac{Y_1}{X_1 + \cfrac{Y_2}{X_2 + \cdots + \cfrac{Y_n}{X_n}}}$$

8. Prove Theorem 4.5.3.

9. Prove Theorem 4.6.1.

10. Show that if $A_0 = 0$, $A_1 = 1$, and $A_n = 2A_{n-1} + A_{n-2}$, then

$$A_n = \frac{(1+\sqrt{2})^n - (1-\sqrt{2})^n}{2\sqrt{2}}.$$

11. Prove Theorem 4.6.2.

12. Show that $\gcd(m_1, m_2, \ldots, m_k)$ and $\text{lcm}(m_1, m_2, \ldots, m_k)$ can be computed in polynomial time.

13. Let a and b be positive integers with $a > b$. We wish to determine if there exist positive integers i, j such that $a^i = b^j$. Consider the following algorithm for this problem: first test if $b \mid a$. If not, return "no". Otherwise, replace (a, b) with $(a/b, b)$ if $a \geq b^2$, and $(b, a/b)$ if $a < b^2$. If eventually a pair $(a', 1)$ is reached, return "yes". Show that this algorithm uses $O((\lg a)^2)$ bit operations.

14. Show that two "randomly chosen" positive integers are relatively prime with probability $6/\pi^2$. More formally, show that

$$\frac{1}{n^2} \sum_{1 \leq i \leq n} \sum_{1 \leq j \leq n} i \perp j = \frac{6}{\pi^2} + O\left(\frac{\log n}{n}\right),$$

where $i \perp j$ is defined to be 1 if $\gcd(i, j) = 1$, and 0 otherwise. Hint: the result of exercise 2.22 may prove useful.

15. Show that there does not exist a polynomial in two variables $f(x, y)$ such that $f(m, n) = \gcd(m, n)$ for all positive integers m, n.

16. Prove Theorem 4.8.7. (Hint: prove that if the algorithm computes a gcd-free basis $\{b_1, \ldots, b_n\}$ on input $(a_1, \ldots, a_m)$, then the number of bit operations used (exclusive of

those used to deal with 1's encountered in the course of the algorithm) is bounded by a constant times

$$(\log a_1 a_2 \cdots a_m)^2 - \sum_{1 \le j \le n} (\log b_j)^2.)$$

17. Suppose we have a frictionless rectangular billiard table, m integer units long by n integer units wide. Show that if a ball is started rolling at a 45° angle from a corner A, then it will eventually strike one of the table's corners. Further, the point on the table's longest side that is nearest A (but not A) on the ball's path is exactly $2\gcd(m, n)$ units from the origin.

18. Following the analysis of section 4.2, show that if $d = \gcd(u, v)$, $u \ge v > 0$, then $\gcd(u, v)$ can be computed in $O((\lg u/d)(\lg v))$ bit operations.

19. Prove that if $a_0, a_1, \ldots$ is an infinite sequence of integers with $a_i \ge 1$ for $i \ge 1$, then

$$\lim_{n \to \infty} [a_0, a_1, \ldots, a_n] \tag{4.16}$$

exists. We abbreviate (4.16) by $[a_0, a_1, \ldots]$.

20. Let x be an irrational number, and suppose $\text{CFA}(x) = (a_0, a_1, \ldots)$. Show that $x = [a_0, a_1, \ldots]$.

The next three exercises deal with the properties of integer functions. See the end of Section 4.6 for the definition.

21. Show that $\lfloor x + a \rfloor$ is an integer function if $0 \le a < 1$. Similarly, show that $\lceil x - a \rceil$ is an integer function if $0 \le a < 1$.

22. (Continuation.) Show that there are as many integer functions as there are subsets of the half-open interval $[0, 1)$.

23. (Continuation.) Show that if $f(x)$ is an integer function, then so is its *dual* $\hat{f}(x) = -f(-x)$. Also show that $\hat{f} \ne f$.

24. Show that the binary gcd algorithm does *not* use $O((\lg u)(\lg v))$ bit operations.

25. Consider the binary gcd algorithm. Call the step that replaces the larger of u and v with $|u - v|/2$ a *reduction* step. Let the number of reduction steps performed on input (u, v) be denoted by $R(u, v)$. Show that $R(u, v) \le \lfloor \log_2(u + v) \rfloor$.

26. Use the previous exercise to prove the following: let $u \ge v > 0$, and let (u, v) be the lexicographically least pair such that the binary gcd algorithm performs n reduction steps on input (u, v). Then $u = 2^{n-1} + 1$, $v = 2^{n-1} - 1$ for $n \ge 3$.

27. (G. B. Purdy.) Consider the following variant of the binary algorithm for computing $\gcd(u, v)$: We may assume u and v are not both even.

    (a) If u is even, then set $u \leftarrow u/2$. ("shift")

    (b) If v is even, then set $v \leftarrow v/2$. ("shift")

    (c) Otherwise, u and v are both odd, and we set $(u, v) \leftarrow ((u + v)/2, (u - v)/2)$. ("reduction")

    We repeatedly perform steps (a)–(c) in that order. If one of u, v is 0, we stop, returning the other number as the gcd.

    Prove that this algorithm always terminates and returns the correct result.

28. (Continuation) Let u, v be odd and positive. Denote the total number of steps (a) and (b) performed on input (u, v) by $P_s(u, v)$, and denote the total number of steps (c) performed by $P_r(u, v)$. First show that $P_r(u, v) = O(\lg uv)$. Second, show that $P_s(u, v) = O((\lg uv)^2)$. Third, show that there exist infinitely many inputs (u, v) with $P_s(u, v) = \Omega((\lg uv)^2)$. Finally, show that $P_s(u, v) = P_s(v, u)$, and $P_r(u, v) = P_r(v, u)$.

29. Use Dirichlet's theorem (Theorem 8.4.1) and the extended Euclidean algorithm to prove the following:

    Let r be a positive rational number, not an integer. Then there exists an infinite number of primes in the sequence $(\lfloor rn \rfloor)_{n \geq 1}$.

30. Develop an "extended" binary gcd algorithm; i.e. one that computes integers a, b such that $au + bv = \gcd(u, v)$, using only additions, subtractions and "shifts" (multiplications and divisions by 2). Hint: maintain a 2×2 integer matrix M_k such that

$$M_k \begin{bmatrix} u \\ v \end{bmatrix} = \begin{bmatrix} u_k \\ v_k \end{bmatrix},$$

    where u_k and v_k denote the values of the variables u and v after k steps through the algorithm.

(More applications of the extended Euclidean algorithm.)

31. Let a and b be relatively prime and positive. Show how to express every $n > ab$ in the form $n = xa + yb$ where x and y are *positive* integers. Your method should run in polynomial time.

32. A positive integer n is called *abundant* if $\sigma(n) > 2n$. Show that every multiple of an abundant number is abundant.

33. (Continuation.) Use Exercises 31 and 32 to show that every integer greater than 20161 can be written as the sum of two abundant numbers.

34. (Continuation.) Use Exercise 31 to show that every number greater than 11 can be written as the sum of two composite numbers.

35. Consider the algorithm which expresses a real number x, $0 < x \leq 1$ in the form

$$x = \frac{1}{a_1} - \frac{1}{a_1 a_2} + \frac{1}{a_1 a_2 a_3} - \cdots$$

where the a_i form a *strictly* increasing sequence of positive integers, and the expansion may or may not terminate. If the expansion does terminate with

$$\frac{(-1)^{n+1}}{a_1 a_2 \cdots a_n}$$

as the last term, then we impose the additional restriction $a_{n-1} < a_n - 1$.

Design an algorithm that given x, produces the a_i. Be sure your algorithm has the proper condition for termination.

36. (Continuation.) Prove that the algorithm of the previous exercise terminates if and only if x is rational.

37. (Continuation.) Let $b_1, b_2, \ldots, b_n$ be specified positive integers that are strictly increasing. Prove that

$$\Pr[a_1 = b_1, a_2 = b_2, \ldots, a_n = b_n] = \frac{1}{b_1 b_2 \cdots b_n (b_n + 1)}.$$

(Here we consider the input x to be uniformly distributed, and Pr denotes the probability of an event occurring.)

38. (Continuation.) Using the previous result, show the following result on conditional probability:

$$\Pr[a_{n+1} = k \mid a_n = j] = \frac{j+1}{k(k+1)}.$$

39. (Continuation.) Using the result in the previous exercise, show that

$$\Pr[a_n = k] = \frac{\left[k \atop n \right]}{(k+1)!}$$

where $\left[{k \atop n}\right]$ is a Stirling number of the first kind, defined by

$$\prod_{0 \le i \le k-1} (X + i) = \sum_{0 \le n \le k} \left[{k \atop n}\right] X^n.$$

40. Let r, s be integers with $r > s > 0$. Let $s_1 = s$ and define $s_{k+1} = r \bmod s_k$ for $k \ge 1$. Prove that $s_n = 0$ for some n, and that $\gcd(r, s) \mid s_{n-1}$. What is the relationship between this algorithm and the algorithm discussed in Exercise 35?

41. (Continuation.) Let $s_1, s_2, \ldots, s_n$ be as in the previous exercise, and define $\alpha(r, s) = n$. Find good upper and lower bounds for $\alpha(r, s)$ and $(1/r) \sum_{1 \le s \le r} \alpha(r, s)$.

42. Define

$$A_n = \sum_{0 \le k \le n} (F_{2^k})^{-1},$$

where F_j is the j-th Fibonacci number. Show that for $n \ge 3$ the continued fraction expansion of A_n is given by

$$[2, 2, \overbrace{1, \ldots, 1}^{2^n - 5}, 2].$$

Conclude that $\lim_{n \to \infty} A_n = \frac{7 - \sqrt{5}}{2}$.

43. Show that $\sum_{1 \le i, j \le n} \gcd(i, j) = \frac{1}{\zeta(2)} n^2 \log n + O(n^2)$.

44. Show that $\sum_{1 \le i, j \le n} \operatorname{lcm}(i, j) = \frac{\zeta(3)}{4\zeta(2)} n^4 + O(n^3 \log n)$.

45. Let $a \ge 3$, an integer. Find the continued fraction expansion for the number $\sum_{n \ge 0} a^{-2^n}$.

46. (Floyd and Knuth.) An *addition machine* is a computing device with a finite number of registers. It has only six instructions: input(x); $x \leftarrow y$; $x \leftarrow x + y$; $x \leftarrow x - y$; if ($x \ge y$) then goto(label); and output(x). Show how to compute $x \bmod y$ using $O(\lg(x/y))$ operations in this model.

47. (Continuation.) Show how to compute $\gcd(x, y)$ using $O(\lg n)$ operations on an addition machine, where $n = \max(x, y)/\gcd(x, y)$.

48. (Golomb.) Let $n = \prod_{1 \le i \le k} p_i^{e_i}$. Define $\gamma(n)$ to be the number of distinct ways to write $n = a^b$ with a, b positive integers. Show that $\gamma(n) = d(\gcd(e_1, e_2, \ldots, e_k))$, where $d(n)$ denotes the number of divisors function.

4.10 Notes on Chapter 4

It is not known if the gcd of n-bit inputs can be computed in parallel with a polynomial number of processors and depth significantly less than $O(n)$; this question was first raised by Cook [1985]. Kannan, Miller, and Rudolph [1987] showed how to compute the gcd on n bit inputs in $O(n(\lg\lg n)/\lg n)$ parallel time using $O(n^2(\lg n)^2)$ processors. Chor and Goldreich [1990] improved this to $O(n/\lg n)$ time and $O(n^{1+\epsilon})$ processors. Also see Sorenson [1994]. Adleman and Kompella [1988] showed how to compute the extended gcd in polylogarithmic depth and subexponential size.

Mansour, Schieber, and Tiwari [1991a] gave a $\Omega(\log\log n)$ lower bound (in the algebraic decision tree model) for any algorithm to compute the gcd using the four field operations, mod, and comparisons. A related result is in Mansour, Schieber, and Tiwari [1991b]. Also see Deng [1989]; Lin-Kriz and Pan [1992].

A. Zavrotsky invented an analogue computer for finding the greatest common divisor based on the result in Exercise 17; see Gardner [1971]; Zavrotsky [1961].

4.1

In the literature, the greatest common divisor of u and v is frequently written (u, v), a notation we have avoided in this book.

The Euclidean algorithm can be found in Euclid's *Elements*, Book VII, Proposition 2; for example, see Heath [1956]. For an extensive discussion of the Euclidean algorithm, see Bachmann [1902, pp. 99–153]. For algorithms to compute the gcd of more than two inputs, see Rosser [1942]; Ficken [1943]; Weinstock [1960]; Blankinship [1963]; Bradley [1970]; Waterman [1977]; and Majewski and Havas [1994]; Havas, Majewski, and Matthews [1995]. Barlow [1811, p. 22] compared computing the greatest common divisor of two or more numbers via the prime factorization of the given numbers; he wrote

> ... this method [via factorization] is, however, rather theoretical than practical, being by no means so ready in application as the rule generally given for this purpose in books of arithmetic.

In practice, the Euclidean algorithm may be sped up considerably by using the technique of *loop unrolling*; for more on this, see Bentley [1982].

P. Cohn [1961] proposed a very general form of the Euclidean algorithm.

4.2

Theorem 4.2.2 is traditionally credited to Lamé [1844]. However, seven years earlier it appeared without proof in a paper of Léger [1837]. Also, de Lagny [1733, pp. 363–364] discussed the "simplest" rational numbers having a continued fraction of a given length.

Finck [1841, pp. 44–45] observed that if $u > v > 0$, then $u \bmod v < u/2$. From this it follows, he remarked, that the number of steps $E(u, v)$ performed by the Euclidean algorithm is bounded above by $2 \log_2 v + 1$. This analysis is reproduced in Finck [1842] and Finck [1843, pp. 57–58]. Other remarks about Lamé's theorem can be found in Finck [1845]; von Grosschmid [1911]; Grossman [1924]; Thoro [1964a, 1964b]; V. Harris [1970b]; Meijer [1992]. See Shallit [1994] for the history of early analyses of the Euclidean algorithm.

Lamé's theorem was originally stated as follows: the number of division steps in the Euclidean algorithm on input (u, v) is bounded above by 5 times the number of decimal digits in $\min(u, v)$.

Theorem 4.2.4 was first proved by G. Collins [1968]. G. Collins [1974a] also obtained similar results on the average bit complexity of the Euclidean algorithm.

D. H. Lehmer [1938] proposed an interesting method of speeding up the Euclidean algorithm by doing most of the divisions with single precision numbers. Also see Jebelean [1993c]. However, the binary gcd algorithm appears to be significantly faster in practice. Sorenson [1995] analyzed a variant of Lehmer's method, and proved it uses $O(n^2/\log n)$ bit operations on inputs of n bits.

The average-case complexity of the Euclidean algorithm (counting the number of division steps) was first discussed by Heilbronn [1969]. Further work was done by Dixon [1970, 1971]; Tonkov [1974]; Kilian [1983]; and G. Norton [1990]. Letting

$$t(n) = \frac{1}{\varphi(n)} \sum_{\substack{1 \leq k \leq n \\ \gcd(k,n)=1}} E(n, k),$$

Porter [1975] showed that

$$t(n) = \frac{12 \log 2}{\pi^2} \log n + C + O(n^{-1/6+\epsilon}).$$

Knuth [1976b] proved the following expression for C:

$$C = \frac{6 \log 2}{\pi^2}(3 \log 2 + 4\gamma - 24\pi^2 \zeta'(2) - 2) - \frac{1}{2}.$$

More recently, Hensley [1992] has shown that the number of division steps done by the Euclidean algorithm over all pairs (u, v) with $0 < u \leq v \leq x$ is asymptotically normally distributed, with mean close to $12(\log 2)\pi^{-2} \log x$.

Plankensteiner [1970] counted the number of pairs (u, v) for which $E(u, v) = k$.

A subtractive version of the Euclidean algorithm was analyzed by Yao and Knuth [1975]; Pitteway and Castle [1988]; Zheng [1994]. Also see Knopfmacher [1992]. Brun [1964] discussed the relationship between the subtractive algorithm and musical theory.

Using the Schönhage-Strassen fast multiplication method, Knuth [1971] gave an algorithm for the gcd that uses $O(n(\lg n)^5 \lg \lg n)$ steps on n bit integer inputs. This was improved by Schönhage [1971] to $O(n(\lg n)^2 \lg \lg n)$. A generalization was given by Moenck [1973]. See Schönhage, Grotefeld, and Vetter [1994] for a discussion of a practical implementation of this method.

An integral domain D in which there exists a map $\delta : D \to \mathbb{N}$ such that if a, b are nonzero elements of D, then there exist q, r in D such that $a = qb + r$ with $\delta(r) < \delta(b)$, is called *Euclidean*. A well-known theorem says that all Euclidean domains have unique factorization. For example, $\mathbb{Z}$ and

$\mathbb{Z}[i]$ (the Gaussian integers) are Euclidean domains. Samuel [1971] and P. Cohn [1991] discussed the properties of Euclidean domains. For the Euclidean algorithm in real quadratic fields, see Chatland [1949]; Chatland and Davenport [1950]. Also see Kaltofen and Rolletschek [1989] and Rolletschek [1990].

Hurwitz [1887] gave a Euclidean algorithm for $\mathbb{Z}[i]$; also see Caviness and Collins [1976]; Shallit [1979b]. The analogue of the Kronecker-Vahlen theorem for $\mathbb{Z}[i]$ was given by Rolletschek [1986].

Other papers on computational aspects of the gcd include Nievengloski [1849]; Daykin [1970]; Shea [1973]; V. Harris [1974]; Schaefer [1975]; van Trigt [1978]; Abramov [1979]; Marcus [1981]; Epasinghe [1985]; Bojanczyk and Brent [1987]; C. Purdy and Purdy [1990a, 1990b]; Yang [1990]. Mays [1987] discusses some variations on the Euclidean algorithm.

4.3

The first explicit solution to $au + bv = d$ appears in the Sanskrit astronomical work Aryabhatiya, of the 5th–6th century A.D.; see Datta and Singh [1938] or Weil [1984].

Blum and Kannan [1989, 1995] showed how to design a "self-checking" program for the extended gcd.

4.4

Continuants were introduced by Euler [1762]. Our presentation, based on 2×2 matrices seems to have originated with Hurwitz; see Hurwitz and Kritikos [1986]. It was rediscovered by Frame [1949] and Kolden [1949], independently. The term "continuant" was invented by T. Muir [1874]. Also see Muir [1878].

4.5

The material in this section is based on G. Hardy and Wright [1985]. For the history of continued fractions, see Brezinski [1991].

Many authors have been concerned with generalizing the continued fraction algorithm to dimensions $n > 2$. We list just a few of the relevant references here: L. Bernstein [1971]; Ferguson and Forcade [1979, 1982]; Brentjes [1981]; Ferguson [1986, 1987]; Hastad, Just, Lagarias and Schnorr [1989]; and Lagarias [1994].

4.6

The least-remainder algorithm was first analyzed by Binet [1841, p. 454]. Later remarks were given by Nievengloski [1845]. Theorem 4.6.2 is due to Dupré [1846]. Also see Brown and Duncan [1971].

Theorem 4.6.4 was stated by Kronecker [1901, p. 118]; it was first proved by Vahlen [1895]. Our presentation is based on Uspensky and Heaslet [1939]. Lazard [1977] generalized the situation of Theorem 4.6.4 to cover the case where *any* remainder is chosen at each step of the algorithm.

A relationship between the ordinary Euclidean algorithm and the least-remainder algorithm was discussed by Goodman and Zaring [1952]. Also see T. Moore [1992].

4.7

The binary gcd algorithm was invented by J. Stein [1967]. Using a heuristic model, it was analyzed by Knuth [1969, 1981] and Brent [1976a]. Similar algorithms were discussed by Brent, Kung, and Luk [1983]; Brent and Kung [1983, 1985]; G. Purdy [1983]; Norton [1985]; Yun and Zhang [1986]; Norton [1987]; Mandelbaum [1988]; Parikh and Matula [1991]; Shallit and Sorenson [1994]. Sorenson [1994] developed generalizations of the binary algorithm, called the k-ary gcd algorithm, and proved a running time of $O(n^2/\log n)$ on n-bit inputs. A similar algorithm was studied by Jebelean [1993b]; also see Weber [1995]. A cross between the Euclidean algorithm and the binary gcd algorithm was proposed by V. Harris [1970a]. In Alagić and Arbib [1978, pp. 223–226], a polynomial-time binary gcd algorithm is transformed through the magic of top-down program design into an exponential-time algorithm.

The simplicity of the operations used in the binary gcd algorithm make it the method of choice on real computers. For example, J. Sorenson (personal communication) found that the binary method was between 10 and 25 times as fast as the Euclidean algorithm on a DECstation 3100, on inputs of 10 to 1000 digits in size. Also see the article of Norton [1985], which found a similar, though less dramatic improvement on large inputs. However, different results were obtained by Jebelean [1993a].

4.8

Stieltjes [1890, §19] appears to have been the first to study an algorithm similar to our gcd-free basis algorithm. He was interested in expressing $\text{lcm}(a, b) = AB$, where A and B are integers satisfying the conditions $A \perp B$, $A \mid a$, and $B \mid b$. (A and B can also be computed efficiently with a gcd-free basis.) Later, Dedekind [1897] studied factorization by repeated use of the Euclidean algorithm. However, his construction took exponential time.

Several authors have considered constructing a gcd-free basis for polynomials; see Collins [1974b]; P. S. Wang [1980]. Also see von zur Gathen [1984b, 1986b] and Kaltofen [1985a, 1985b].

For such algorithms on integers, see Bach, Miller, and Shallit [1984, 1986]. Theorems 4.8.6 and 4.8.7 were first proved by Bach, Driscoll, and Shallit [1990, 1993]. The proof we have presented is due to H. W. Lenstra, Jr. Our presentation of some of the material in this section, particularly Theorem 4.8.5, is based on a manuscript of D. Bernstein [1992].

Other applications of the gcd-free basis construction can be found in H. W. Lenstra [1992]; Ge [1993]; Buchmann and Lenstra [1994].

5 Computing in $\mathbb{Z}/(n)$

In this chapter we introduce the finite ring $\mathbb{Z}/(n)$, the ring of integers modulo n, and the finite group of its invertible elements, $(\mathbb{Z}/(n))^*$. We discuss some of the basic algorithms for computing in these structures: the power algorithm, the Chinese remainder theorem, and algorithms for computing the Legendre and Jacobi symbols.

5.1 Basics

Recall from section 2.8.2 the following notation: if R is a ring, then R^* denotes the multiplicative group of invertible elements (units).

$\mathbb{Z}/(n)$ is a commutative ring, with the equivalence class of 1 playing the role of the identity. Its group of units is $(\mathbb{Z}/(n))^*$.

The least positive exponent e such that $a^e \equiv 1 \pmod{n}$ is called the *order* of a modulo n. We write $e = \text{ord}_n\, a$.

We now recall the theorem of Fermat-Euler, discussed in Chapter 2. In Exercise 14 we gave an elementary proof; for the sake of variety, we give another proof here.

THEOREM 5.1.1 Let $a \in (\mathbb{Z}/(n))^*$, and let $\varphi(n)$ be the Euler phi function. Then

$$a^{\varphi(n)} \equiv 1 \pmod{n}. \tag{5.1}$$

Proof By group theory we know that the order of any element divides the order of the group. But the order of $(\mathbb{Z}/(n))^*$ is $\varphi(n)$. ■

Unless otherwise stated, all algorithms in this section take as inputs the canonical representative $a \bmod n$ of an equivalence class, modulo n, and return canonical representatives as output. See equation (2.1).

5.2 Addition, Subtraction, Multiplication

THEOREM 5.2.1 We can add and subtract elements of $\mathbb{Z}/(n)$ in $O(\lg n)$ bit operations. We can multiply elements of $\mathbb{Z}/(n)$ using $O((\lg n)^2)$ bit operations.

Proof We only need to observe that if $0 \le a, b < n$, then

$$(a+b) \bmod n = \begin{cases} a+b, & \text{if } a+b < n; \\ a+b-n, & \text{if } a+b \ge n. \end{cases}$$

Thus no division is necessary for modular addition. In a similar fashion, no division is necessary for modular subtraction. The bound for multiplication modulo n is clear. ■

5.3 Multiplicative Inverse

THEOREM 5.3.1 Let $d = \gcd(a, n)$. Then the congruence $ax \equiv b \pmod{n}$ has a solution x iff $d \mid b$. The solution is unique, modulo n/d.

Proof Suppose x_0 is a solution. Then $ax_0 - b = kn$ for some k. Thus $ax_0 - kn = b$. Since $d \mid a$ and $d \mid n$, we see that $d \mid b$.

Suppose $d \mid b$. Then by Theorem 4.3.1 we can find integers x_1, y_1 such that $ax_1 + ny_1 = d$. But this implies that $ax_1c + ny_1c = b$, for $c = b/d$. Now put $x_0 = x_1c$. Then $ax_0 \equiv b \pmod{n}$.

To prove the last sentence of the theorem, suppose x_0 and x_1 are both solutions to $ax \equiv b \pmod{n}$. Then $(a/d)x_0 \equiv (a/d)x_1 \pmod{n/d}$. Since $\gcd(a/d, n/d) = 1$, we can multiply by $(a/d)^{-1} \pmod{n/d}$ to obtain $x_0 \equiv x_1 \pmod{n/d}$. ■

COROLLARY 5.3.2 Let $d = \gcd(a, n)$. If $d \mid b$, we can solve $ax \equiv b \pmod{n}$ using $O((\lg n)^2)$ bit operations. In particular, if $d = 1$, by setting $b = 1$ we can find the multiplicative inverse of $a \pmod{n}$ using $O((\lg a)(\lg n))$ bit operations.

Proof This follows immediately from Theorem 4.3.2. ■

We need a notation for the canonical representative of the multiplicative inverse of $a \pmod{n}$, where $\gcd(a, n) = 1$. By abuse of notation, we will use $a^{-1} \bmod n$.

5.4 The Power Algorithm

In later chapters we will need to evaluate $a^e \bmod n$. Repeated multiplication by a gives an algorithm that uses $\Omega(e(\lg n)^2)$ bit operations, which is not satisfactory for large e.

To improve the running time, we introduce one of the most important algorithms in computational number theory: the "power algorithm" or "binary method" for exponentiation. This algorithm computes $a^e \bmod n$, using $O((\lg e)(\lg n)^2)$ bit operations.

It is based on the following observation:

$$a^e = \begin{cases} a \cdot a^{e-1}, & \text{if } e \text{ is odd} \\ (a^{e/2})^2, & \text{if } e \text{ is even.} \end{cases}$$

From this we can easily derive a recursive algorithm for $a^e \bmod n$:

POWER(a, e, n)
$\{a, e, n$ are integers with $e \geq 0, n \geq 2\}$
if $e = 0$ then return (1)

```
else if e mod 2 = 0 then
    t ← POWER(a, e/2, n)
    return (t^2 mod n)
else
    t ← POWER(a, e − 1, n)
    return (at mod n)
```

THEOREM 5.4.1 Let a, e, n be integers with $n \geq 2$, $e \geq 0$, and $0 \leq a < n$. Let $s_2(e)$ denote the number of 1's in the binary expansion of e. Then the algorithm POWER computes $a^e \bmod n$, using $(\lg e) - 1$ squarings modulo n and $s_2(e)$ multiplications by a, modulo n. Hence POWER uses $O((\lg e)(\lg n)^2)$ bit operations to compute $a^e \bmod n$.

Proof The proof is by induction on e. Clearly the result is true for $e = 0$. Now assume it is true for all $f < e$; we wish to prove it for e. If e is odd, say $e = 2k + 1$, then the algorithm first computes a^{2k} (which takes $(\lg 2k) - 1$ squarings and $s_2(2k)$ multiplications by a, by induction) and then multiplies by a. Thus $(\lg 2k) - 1 = (\lg(2k + 1)) - 1$ squarings and $s_2(2k) + 1 = s_2(2k + 1)$ multiplications by a are performed. Similarly, if e is even, say $e = 2k$, then the algorithm first computes a^k and squares the result. Thus by induction $\lg k = (\lg 2k) - 1$ squarings and $s_2(k) = s_2(2k)$ multiplications are performed.

Finally, by observing that $O(\lg e)$ squarings or multiplications of numbers of size $O(\lg n)$ are performed, the time bound of $O((\lg e)(\lg n)^2)$ bit operations follows. ■

It is easy to give two iterative versions of POWER; one based on reading the bits of the exponent from right to left, and the other from left to right. See Exercise 3.

Note that the power algorithm works in any semigroup, not just the abelian multiplicative semigroup of integers modulo n. For example, by replacing multiplication modulo n with integer addition, we get a polynomial-time algorithm for multiplication.

The power algorithm can also be useful in many instances where its applicability is not immediately obvious. For example, let us consider the problem of computing the k-th Fibonacci number (mod n), where k is large. If we use the fact that $F_k = F_{k-1} + F_{k-2}$, then an iterative approach uses $O(k)$ additions of numbers smaller than n, which is not a very efficient method. We can rephrase the problem in terms of matrices if we observe that

$$\begin{bmatrix} F_{k+1} & F_k \\ F_k & F_{k-1} \end{bmatrix} = \begin{bmatrix} F_k & F_{k-1} \\ F_{k-1} & F_{k-2} \end{bmatrix} \begin{bmatrix} 1 & 1 \\ 1 & 0 \end{bmatrix}.$$

Hence by induction, we have

$$\begin{bmatrix} F_{k+1} & F_k \\ F_k & F_{k-1} \end{bmatrix} = \begin{bmatrix} 1 & 1 \\ 1 & 0 \end{bmatrix}^k.$$

Now if we replace multiplication of integers in the power algorithm with multiplication of 2×2 matrices, we see that we can compute $F_k \bmod n$ using $O((\lg k)(\lg n)^2)$ bit operations.

Other applications of the power algorithm can be found in Exercises 5–9.

We note that the power algorithm is most useful when the representation of a^e is not significantly larger than that of a. This is the case in the examples above, where all intermediate results are reduced modulo n before continuing.

5.5 The Chinese Remainder Theorem

In this section we discuss the computational complexity of the Chinese remainder theorem, the fundamental tool for converting between a single congruence and a system of congruences with smaller moduli.

DEFINITION 5.5.1 We say that $m_1, m_2, \ldots, m_k$ are *relatively prime in pairs* iff $\gcd(m_i, m_j) = 1$ for $i \neq j$.

THEOREM 5.5.2 (CHINESE REMAINDER THEOREM, VERSION 1) Let $m_1, m_2, \ldots, m_k$ be positive integers, relatively prime in pairs. Let $m = m_1 m_2 \cdots m_k$. Then, given integers $x_1, x_2, \ldots, x_k$, there is an integer x such that

$$x \equiv x_i \pmod{m_i},$$

and this integer x is uniquely determined, modulo m.

Proof Suppose we could find integers e_i, $1 \leq i \leq k$ such that

$$e_i \equiv \begin{cases} 1 \pmod{m_i}, & \\ 0 \pmod{m_j}, & \text{if } j \neq i. \end{cases} \tag{5.2}$$

Then if

$$x \equiv \sum_{1 \leq i \leq k} x_i e_i,$$

we would have $x \equiv x_i \pmod{m_i}$ for $1 \leq i \leq k$.

How do we find such integers e_i? Suppose we define

$$f_i = \frac{m}{m_i}.$$

Then $f_i \equiv 0 \pmod{m_j}$ for $j \neq i$. Now let $g_i \equiv f_i^{-1} \bmod m_i$; the inverse exists because the m_j are relatively prime in pairs. If we let $e_i = f_i g_i$, then e_i satisfies equation (5.2).

Thus we have found an integer x satisfying the given congruences.

To see that such an integer is uniquely determined modulo m, let x and y both satisfy the congruences. Then $x - y \equiv 0 \pmod{m_i}$ for $1 \le i \le k$. Since the m_i are pairwise relatively prime, $x \equiv y \pmod{m}$. ∎

COROLLARY 5.5.3 Given moduli $m_1, m_2, \ldots, m_k > 1$, relatively prime in pairs, and integers $x_1, x_2, \ldots, x_k$, we can find a solution to the congruences $x \equiv x_i \pmod{m_i}$, $1 \le i \le k$, using $O((\lg m)^2)$ bit operations, where $m = m_1 m_2 \cdots m_k$.

Proof We note that the restriction to moduli greater than 1 is to avoid the case where, for example, all the moduli are 1 and there are many of them. In this case, $m = m_1 m_2 \cdots m_k = 1$, and we could not solve the congruences (or even examine all of them) within the stated time bound.

To achieve the desired complexity bound, note that the solution given above consists of (i) computing m and $f_i = m/m_i$; (ii) computing $g_i = f_i^{-1} \bmod m_i$; (iii) computing $e_i = f_i g_i$ and (iv) computing $x = \sum_i x_i e_i$.

By Exercise 3.3, we can compute m using $O((\lg m)^2)$ bit operations. It costs $O((\lg m)(\lg m_i))$ to compute f_i; summing this from $i = 1$ to k gives us $O((\lg m)(k + \log m)) = O((\lg m)^2)$, since $k = O(\log m)$. Similar results hold for (ii)–(iv). ∎

The previous version of the Chinese remainder theorem can be regarded as a statement about solutions to certain equations. The next version is a *structure* theorem.

THEOREM 5.5.4 (CHINESE REMAINDER THEOREM, VERSION 2) Let $m_1, m_2, \ldots, m_k$ be positive integers, relatively prime in pairs. Then

$$\mathbb{Z}/(m_1 m_2 \cdots m_k) \cong \mathbb{Z}/(m_1) \oplus \mathbb{Z}/(m_2) \oplus \cdots \oplus \mathbb{Z}/(m_k).$$

Proof Let $f_i : \mathbb{Z} \to \mathbb{Z}/(m_i)$ be the map that takes n to its equivalence class $n \bmod m_i$. We construct a map from $\mathbb{Z}$ to $\mathbb{Z}/(m_1) \oplus \mathbb{Z}/(m_2) \oplus \cdots \oplus \mathbb{Z}/(m_k)$ as follows: $f(n) = (f_1(n), f_2(n), \ldots, f_k(n))$. It is easily checked that f is a ring homomorphism.

What is the image of f? We have $(a_1, a_2, \ldots, a_k) = f(n)$ iff $f_i(n) = a_i$ for $1 \le i \le k$. By the first version of the Chinese remainder theorem, we know that such an n always exists. Thus f is onto.

What is the kernel of f? We have $f(n) = 0$ iff $n \equiv 0 \pmod{m_i}$ for $1 \le i \le k$, iff $n \equiv 0 \pmod{m_1 m_2 \cdots m_k}$. Thus the kernel of f is the ideal $(m_1 m_2 \cdots m_k)$.

Hence the two rings are isomorphic. ∎

This version of the Chinese remainder theorem can be interpreted in terms of "data structures". We can represent an integer i in the range $0 \le i < m = m_1 m_2 \cdots m_k$ in two ways: (i) as an integer modulo m, or (ii) as a list of integers, modulo each m_j. It is easy to see that we can convert from representation (i) to representation (ii) using $O((\lg m)^2)$ bit operations, and by Corollary 5.5.3, we can convert from representation (ii) to representation (i) in the same time bound.

The Chinese remainder theorem can be generalized to cover the case where the moduli are not necessarily relatively prime, as follows:

THEOREM 5.5.5 The system of congruences

$$x \equiv x_i \pmod{m_i},$$

$1 \le i \le k$ has a solution iff $x_i \equiv x_j \pmod{\gcd(m_i, m_j)}$ for all $i \neq j$. If the solution exists, it is unique $\pmod{\operatorname{lcm}(m_1, m_2, \ldots, m_k)}$.

Proof Suppose $x \equiv x_i \pmod{m_i}$ and $x \equiv x_j \pmod{m_j}$. Then clearly $x \equiv x_i \pmod{\gcd(m_i, m_j)}$ and $x \equiv x_j \pmod{\gcd(m_i, m_j)}$, and we conclude $x_i \equiv x_j \pmod{\gcd(m_i, m_j)}$.

To prove uniqueness of the solution, note that if $x \equiv x_i \pmod{m_i}$ for $1 \le i \le k$ and $y \equiv x_i \pmod{m_i}$ for $1 \le i \le k$, then $x - y \equiv 0 \pmod{m_i}$; hence $x - y$ is a multiple of $\operatorname{lcm}(m_1, m_2, \ldots, m_k)$.

To complete the proof, we must show that a solution exists. Gauss gave a constructive method, which works by successively reducing a pair of congruences to a single congruence. Suppose we have

$$x \equiv x_1 \pmod{m_1}$$

$$x \equiv x_2 \pmod{m_2}.$$

Then $x = x_1 + tm_1$ for some t. Then $x_1 + tm_1 \equiv x_2 \pmod{m_2}$; hence $tm_1 \equiv x_2 - x_1 \pmod{m_2}$. Now let $d = \gcd(m_1, m_2)$; then $d \mid x_2 - x_1$. Thus

$$t\frac{m_1}{d} \equiv \frac{x_2 - x_1}{d} \pmod{m_2/d}.$$

By Theorem 5.3.1, this congruence has a unique solution $t \equiv t_1 \pmod{m_2/d}$. Substituting $t = t_1 + jm_2/d$ in $x = x_1 + tm_1$, we find $x = x_1 + t_1m_1 + jm_1m_2/d$. Hence $x \equiv x_1 + t_1m_1 \pmod{\operatorname{lcm}(m_1, m_2)}$. By repeating this construction $k - 1$ times, we find the solution to a system of k simultaneous congruences. ■

COROLLARY 5.5.6 Let $m_1, m_2, \ldots, m_k$ be integers, each ≥ 2, and define $m = m_1m_2 \cdots m_k$, and $m' = \operatorname{lcm}(m_1, m_2, \ldots, m_k)$. Given the system S of congruences $x \equiv x_i \pmod{m_i}$, $1 \le i \le k$, we can determine if S has a solution, using $O((\lg m)^2)$ bit operations, and if so, we can find the unique solution $\pmod{m'}$, using $O((\lg m)^2)$ bit operations.

Proof Left to the reader as Exercise 4. ■

Although implementing the Chinese remainder theorem can be done in polynomial time, it turns out that a similar problem, which we might call the "anti-Chinese remainder theorem", is $\mathcal{NP}$-complete:

THEOREM 5.5.7 The following decision problem is $\mathcal{NP}$-complete: given k and a list of k pairs

$$(x_1, m_1), (x_2, m_2), \ldots, (x_k, m_k),$$

determine if there exists x such that $x \not\equiv x_i \pmod{m_i}$ for $1 \le i \le k$.

Proof Note that we do *not* require that the m_i be relatively prime in pairs.

First, note that the problem is in $\mathcal{NP}$, as a nondeterministic algorithm can guess such an x and verify it satisfies the k incongruences in polynomial time. If such an x exists, it satisfies $0 \le x < \text{lcm}(m_1, m_2, \ldots, m_k)$, and hence $\lg x$ is bounded by a polynomial in the length of the input.

To show the problem is $\mathcal{NP}$-complete, we use reduction from `3-SAT`. Let the variables be $y_1, y_2, \ldots, y_t$, and let F be a formula which is the conjunction of n clauses of the form $C_i = (z_{a_i} \bigvee z_{b_i} \bigvee z_{c_i})$, where z_{a_i} is either y_{a_i} or $\overline{y_{a_i}}$, and the same for b_i and c_i. We encode a truth assignment to the variables as both a bit vector $(e_1, e_2, \ldots, e_t)$ and an integer x via the correspondence $x \equiv e_i \pmod{p_i}$, where p_i is the i-th prime number. (From Theorem 8.8.4, we know that $p_t < t(\log t + \log\log t)$, for $t \ge 6$, so we can determine the first t primes via brute force in time polynomial in t.)

Now define a_i' to be 0 if $z_{a_i} = y_{a_i}$, and 1 if $z_{a_i} = \overline{y_{a_i}}$, and similarly for the sequences b_i' and c_i'. We define a certain integer x_i for each clause C_i, $1 \le i \le n$, as follows:

$$x_i \equiv a_i' \pmod{p_{a_i}}$$

$$x_i \equiv b_i' \pmod{p_{b_i}}$$

$$x_i \equiv c_i' \pmod{p_{c_i}}$$

We can find such an integer x_i in the range $[0, p_{a_i} p_{b_i} p_{c_i})$ in polynomial time by the Chinese remainder theorem.

Finally, our system S of linear incongruences is given by

$$x \not\equiv 2 \pmod{3}$$

$$x \not\equiv 2, 3, 4 \pmod{5}$$

$$\vdots$$

$$x \not\equiv 2, 3, \ldots, p_t - 1 \pmod{p_t}$$

$$x \not\equiv x_1 \pmod{p_{a_1} p_{b_1} p_{c_1}}$$

$$x \not\equiv x_2 \pmod{p_{a_2} p_{b_2} p_{c_2}}$$

$$\vdots$$

$$x \not\equiv x_n \pmod{p_{a_n} p_{b_n} p_{c_n}}$$

Note that there are $n + \sum_{2 \le p \le p_t} p - 2 = O(n + t^3)$ incongruences.

Now we prove that the formula F is satisfiable if and only if this system of incongruences has a solution. The first few incongruences are needed to ensure that $x \equiv 0, 1 \pmod{p_i}$. Now let an assignment be $(y_1, y_2, \ldots, y_n) = (y_1', y_2', \ldots, y_n')$. Then clause C_i is satisfied if and only if $(y_{a_i}', y_{b_i}', y_{c_i}') \ne (a_i', b_i', c_i')$, which by our construction means $x \not\equiv x_i \pmod{p_{a_i} p_{b_i} p_{c_i}}$. ■

5.6 The Multiplicative Structure of $(\mathbb{Z}/(n))^*$

In this section we examine the multiplicative group $(\mathbb{Z}/(n))^*$. We start with a simple lemma:

LEMMA 5.6.1 Suppose $R_1, R_2, \ldots, R_k$ are rings and $S = R_1 \oplus R_2 \oplus \cdots \oplus R_k$. Then $S^* = R_1{}^* \times R_2{}^* \times \cdots \times R_k{}^*$.

Proof Suppose $u \in S$ is a unit. Then there is a $v \in S$ such that $uv = (1, 1, \ldots, 1)$, the identity element for S. Hence if $u = (u_1, u_2, \ldots, u_k)$ and $v = (v_1, v_2, \ldots, v_k)$, we have $u_i v_i = 1$ for $1 \le i \le k$. Thus u_i is a unit for each i.

Conversely, if u_i is a unit for each i, then there exists v_i such that $u_i v_i = 1$ for each i. Then $u = (u_1, u_2, \ldots, u_k)$ is a unit, since if $v = (v_1, v_2, \ldots, v_k)$, then $uv = (1, 1, \ldots, 1)$. ■

We now apply this lemma to the multiplicative structure of $(\mathbb{Z}/(n))^*$. Let the prime factorization of n be

$$n = p_1{}^{e_1} p_2{}^{e_2} \cdots p_k{}^{e_k}.$$

Then from Theorem 5.5.4 we have

$$\mathbb{Z}/(n) \cong \mathbb{Z}/(p_1{}^{e_1}) \oplus \mathbb{Z}/(p_2{}^{e_2}) \oplus \cdots \oplus \mathbb{Z}/(p_k{}^{e_k}).$$

Then from the preceding lemma we have

$$(\mathbb{Z}/(n))^* \cong (\mathbb{Z}/(p_1{}^{e_1}))^* \times (\mathbb{Z}/(p_2{}^{e_2}))^* \times \cdots \times (\mathbb{Z}/(p_k{}^{e_k}))^*.$$

Thus to determine the multiplicative structure of $(\mathbb{Z}/(n))^*$, it suffices to determine the structure of $(\mathbb{Z}/(p^e))^*$ for each prime power $p^e \,\|\, n$.

We say a group G is *cyclic* if it is generated by a single element. In the case where $G = (\mathbb{Z}/(n))^*$, such a generating element is frequently called a *primitive root*.

THEOREM 5.6.2 Let p, e be integers with p prime and $e \geq 1$. Then $(\mathbb{Z}/(p^e))^*$ is an abelian group with $p^{e-1}(p-1)$ elements. If $p \geq 3$, then $(\mathbb{Z}/(p^e))^*$ is a cyclic group. If $p = 2$, then $(\mathbb{Z}/(p^e))^*$ is cyclic for $e \leq 2$. For $e > 2$, $(\mathbb{Z}/(p^e))^*$ is the direct product of two cyclic groups, one of order 2 and the other of order 2^{e-2}; it is generated by the elements -1 and 5.

Proof See Exercises 14–21.

THEOREM 5.6.3 Let n be an integer > 1. Then $(\mathbb{Z}/(n))^*$ has a primitive root iff $n = 2, 4, p^e, 2p^e$, where $e \geq 1$ and p is an odd prime.

Proof We know from Theorem 5.6.2 that $(\mathbb{Z}/(2^e))^*$ is cyclic if $e = 1$ or $e = 2$. We also know from Theorem 5.6.2 that $(\mathbb{Z}/(p^e))^*$ is cyclic if p is an odd prime. Also, since $(\mathbb{Z}/(2p^e))^* = (\mathbb{Z}/(2))^* \times (\mathbb{Z}/(p^e))^*$, we see that $(\mathbb{Z}/(2p^e))^*$ is cyclic.

Now assume n is not of the required form. Then (i) $n = 2^e$, for $e > 2$; or (ii) we can write $n = rs$, where $\gcd(r, s) = 1$ and $r, s > 2$. In the first case, we know from Theorem 5.6.2 that $(\mathbb{Z}/(2^e))^*$ is not cyclic, as it has at least two elements of order 2. In the second case, we know that $(\mathbb{Z}/(n))^* \cong (\mathbb{Z}/(r))^* \times (\mathbb{Z}/(s))^*$; and $\varphi(r)$ and $\varphi(s)$ are both even. Hence again $(\mathbb{Z}/(n))^*$ has at least two elements of order 2, and thus is not cyclic. ■

5.7 Quadratic Residues

DEFINITION 5.7.1 Suppose m and n are positive integers. Suppose a is an integer and $\gcd(a, n) = 1$. Then we say a is an m-th *power residue* (mod n) iff $x^m \equiv a \pmod{n}$ is solvable.

THEOREM 5.7.2 Suppose $(\mathbb{Z}/(n))^*$ is cyclic and $\gcd(a, n) = 1$. Then a is an m-th power residue (mod n) iff $a^{\varphi(n)/d} \equiv 1 \pmod{n}$, where $d = \gcd(m, \varphi(n))$.

Proof Let g be a generator for $(\mathbb{Z}/(n))^*$, and let $a \equiv x^m \pmod{n}$. Then we can write $a \equiv g^b$ and $x \equiv g^y$. Then $g^{my} \equiv g^b \pmod{n}$, which implies that $my \equiv b \pmod{\varphi(n)}$. By Theorem 5.3.1, this implies that $d \mid b$. Hence $a^{\varphi(n)/d} \equiv g^{b\varphi(n)/d} \equiv 1 \pmod{n}$.

Conversely, suppose $a^{\varphi(n)/d} \equiv 1 \pmod{n}$. Then $g^{b\varphi(n)/d} \equiv 1 \pmod{n}$; hence $\varphi(n) \mid b\varphi(n)/d$ and so $d \mid b$. By Theorem 5.3.1, we can find integers e, f such that $em + f\varphi(n) = b$, and so $(g^e)^m \equiv g^b \equiv a \pmod{n}$. ■

A second power residue (mod n) is called a *quadratic residue*. If $\gcd(a, n) = 1$ and a is not a quadratic residue, then it is called a *quadratic nonresidue* (mod n). For example, the quadratic residues of 7 are 1, 2, and 4, and the quadratic nonresidues are 3, 5, and 6.

COROLLARY 5.7.3 (EULER'S CRITERION) Let p be an odd prime and let $\gcd(a, p) = 1$. Then a is a quadratic residue modulo p if

$$a^{(p-1)/2} \equiv 1 \pmod{p},$$

and a is a quadratic nonresidue modulo p if

$$a^{(p-1)/2} \equiv -1 \pmod{p}.$$

We use the term *quadratic character* to denote the classification of a number as a residue or nonresidue. Corollary 5.7.3 gives an $O((\lg p)^3)$ algorithm for determining the quadratic character of a number a modulo p. However, in Section 5.9 we will see that an $O((\lg p)^2)$ algorithm is possible.

COROLLARY 5.7.4 There are $(p-1)/2$ quadratic residues and $(p-1)/2$ quadratic nonresidues, modulo an odd prime p.

Proof By Exercise 16, we know that $X^{(p-1)/2} \equiv 1 \pmod{p}$ has exactly $(p-1)/2$ solutions. ■

We will see in Chapter 7 that quadratic nonresidues are very useful in number-theoretic algorithms. This corollary gives us a random polynomial time algorithm for finding a quadratic nonresidue: since exactly half of all elements of $(\mathbb{Z}/(p))^*$ are nonresidues, an element picked at random from $(\mathbb{Z}/(p))^*$ will be a nonresidue with probability $1/2$. Thus with an expected number of 2 choices, we will find a nonresidue. Assuming for the moment the result of Theorem 5.9.3, we have proved:

COROLLARY 5.7.5 We can find a quadratic nonresidue mod p with a Las Vegas algorithm that uses $O((\lg p)^2)$ expected bit operations.

5.8 The Legendre Symbol

In this section, we define a special symbol, called the *Legendre symbol*. This symbol is essentially a homomorphism from $(\mathbb{Z}/(p))^*$ to $\{-1, 1\}$.

Let a be an integer, and let p be an odd prime. The Legendre symbol $\left(\frac{a}{p}\right)$ is defined as follows:

$$\left(\frac{a}{p}\right) = \begin{cases} 1, & \text{if } a \text{ is a quadratic residue} \pmod{p}; \\ -1, & \text{if } a \text{ is a quadratic nonresidue} \pmod{p}\ ; \\ 0, & \text{if } p \mid a\ . \end{cases}$$

THEOREM 5.8.1 Let p, q be odd primes. Then

(1) $\left(\frac{a}{p}\right) \equiv a^{(p-1)/2} \pmod p$; in particular, $\left(\frac{-1}{p}\right) = \begin{cases} 1 & \text{if } p \equiv 1 \pmod 4; \\ -1 & \text{if } p \equiv 3 \pmod 4. \end{cases}$

(2) $\left(\frac{a}{p}\right)\left(\frac{b}{p}\right) = \left(\frac{ab}{p}\right)$;

(3) $\left(\frac{a}{p}\right) = \left(\frac{b}{p}\right)$ if $a \equiv b \pmod p$;

(4) $\left(\frac{a^2}{p}\right) = 1$ if $p \not| \, a$;

(5) $\left(\frac{2}{p}\right) = (-1)^{(p^2-1)/8}$; in other words, $\left(\frac{2}{p}\right) = \begin{cases} 1 & \text{if } p \equiv 1, 7 \pmod 8; \\ -1 & \text{if } p \equiv 3, 5 \pmod 8. \end{cases}$

(6) If $p \neq q$, $\left(\frac{p}{q}\right)\left(\frac{q}{p}\right) = (-1)^{\frac{p-1}{2}\frac{q-1}{2}}$. In other words, $\left(\frac{p}{q}\right) = \pm\left(\frac{q}{p}\right)$, with the $+$ sign employed unless $p \equiv q \equiv 3 \pmod 4$.

Proof

(1) This follows immediately from Corollary 5.7.3.

(2) $\left(\frac{a}{p}\right)\left(\frac{b}{p}\right) \equiv a^{(p-1)/2}b^{(p-1)/2} \equiv (ab)^{(p-1)/2} \equiv \left(\frac{ab}{p}\right) \pmod p$.

(3) This follows from definition of quadratic residue.

(4) $(a^2)^{(p-1)/2} \equiv a^{p-1} \equiv 1 \pmod p$.

(5) See Exercise 31.

(6) This is a very famous theorem called the *quadratic reciprocity law*. It was first proved by Gauss in 1796, although both Euler and Legendre had tried to prove it. We have outlined one proof in Exercise 45. ■

Using the power algorithm and part (1) of this theorem, we can compute the Legendre symbol $\left(\frac{a}{p}\right)$ using $O((\lg p)^3)$ bit operations. However, we can do better, as we will see in the next section.

5.9 The Jacobi Symbol

Jacobi generalized the Legendre symbol to lower entries that are odd, but not necessarily prime. This symbol, called the *Jacobi symbol*, is also written $\left(\frac{a}{n}\right)$.

Jacobi's generalization is perhaps not the most obvious generalization of the Legendre symbol, but does have the virtue of being multiplicative and easy to compute.

DEFINITION 5.9.1 Let n be odd with prime factorization $p_1^{e_1} p_2^{e_2} \cdots p_k^{e_k}$. Then

$$\left(\frac{a}{n}\right) = \left(\frac{a}{p_1}\right)^{e_1} \left(\frac{a}{p_2}\right)^{e_2} \cdots \left(\frac{a}{p_k}\right)^{e_k}. \tag{5.3}$$

In particular, $\left(\frac{a}{1}\right)$ is defined to be 1 for all integers a.

THEOREM 5.9.2 Let m, n be positive odd integers, and let a, b be integers. Then the Jacobi symbol $\left(\frac{a}{n}\right)$ has the following properties:

(1) $\left(\frac{ab}{n}\right) = \left(\frac{a}{n}\right)\left(\frac{b}{n}\right)$;

(2) $\left(\frac{a}{mn}\right) = \left(\frac{a}{m}\right)\left(\frac{a}{n}\right)$;

(3) $\left(\frac{a}{n}\right) = \left(\frac{b}{n}\right)$ if $a \equiv b \pmod{n}$;

(4) $\left(\frac{-1}{n}\right) = (-1)^{(n-1)/2}$;

(5) $\left(\frac{2}{n}\right) = (-1)^{(n^2-1)/8}$;

(6) If $\gcd(a, n) = 1$, and a is odd and positive, then

$$\left(\frac{a}{n}\right)\left(\frac{n}{a}\right) = (-1)^{\frac{a-1}{2}\frac{n-1}{2}}. \tag{5.4}$$

Proof Properties (1) and (3) follow from the corresponding properties of the Legendre symbol.

Property (2) follows from the definition of the Jacobi symbol.

To prove property (4), we note that if $n_1, n_2, \ldots, n_k$ are odd, then

$$\sum_{1 \le i \le k} (n_i - 1)/2 \equiv (n_1 n_2 \cdots n_k - 1)/2 \pmod{2}. \tag{5.5}$$

Property (5) is proved similarly: if $n_1, n_2, \ldots, n_k$ are odd, then

$$\sum_{1 \le i \le k} (n_i^2 - 1)/8 \equiv (n_1^2 n_2^2 \cdots n_k^2 - 1)/8 \pmod{2}.$$

Property (6) is proved similarly; see Exercise 32. ■

We point out that it is *no longer necessarily true* that if $\left(\frac{a}{n}\right) = 1$, then a is a quadratic residue, modulo n. For example, $\left(\frac{8}{15}\right) = 1$, but 8 is not a quadratic residue, modulo 15.

The properties of the Jacobi symbol given above enable us to compute $\left(\frac{a}{n}\right)$ using $O((\lg a)(\lg n))$ bit operations. Without loss of generality let us assume $0 < a < n$. We write

$$a = 2^e \cdot n'$$
$$n = qn' + a',$$

where $e \geq 0$, n' is odd, and $0 \leq a' < n'$. Using properties (3), (5), and (6) of the Jacobi symbol, we see

$$\left(\frac{a}{n}\right) = \left(\frac{a'}{n'}\right)(-1)^s$$

where $s = e \cdot \frac{n^2-1}{8} + \frac{n-1}{2} \cdot \frac{n'-1}{2}$. Note that this holds *even if* $\gcd(a, n) > 1$, for then both sides are 0.

We also point out that it is not necessary to compute s itself explicitly, since only the quantity $(-1)^s$ is used. It is easy to see that $(-1)^s$ depends only on e modulo 2, n' modulo 4, and n modulo 8.

Now since $a' < n' \leq a < n$, we have reduced the problem of computing $\left(\frac{a}{n}\right)$ to a problem with smaller inputs. If we continue in this fashion, eventually $a' = 0$, and the process terminates. Note that $\left(\frac{0}{n}\right) = 1$ if $n = 1$, and 0 otherwise.

This idea is the basis for the following algorithm:

```
JACOBI(a, n)
{0 < a < n; n odd}
t ← 1
while (a ≠ 0) do
        while (a mod 2 = 0) do
            a ← a/2
            if (n mod 8 = 3) or (n mod 8 = 5) then t ← −t
        interchange(a, n)
        if (a mod 4 = 3) and (n mod 4 = 3) then t ← −t
        a ← a mod n
if (n = 1) then return(t) else return(0)
```

THEOREM 5.9.3 Let n be an odd non-negative integer, and a be any integer. Then the Jacobi symbol $\left(\frac{a}{n}\right)$ can be computed using $O((\lg a)(\lg n))$ bit operations.

Proof Without loss of generality, we may assume that $a > 0$; for otherwise, $\left(\frac{a}{n}\right) = (-1)^{(n-1)/2}\left(\frac{-a}{n}\right)$. Furthermore, we may assume $a < n$; for otherwise, $\left(\frac{a}{n}\right) = \left(\frac{a \bmod n}{n}\right)$, and we can compute $a \bmod n$ in $O((\lg a)(\lg n))$ bit operations.

Thus $0 < a < n$, and we may use the algorithm JACOBI. The correctness of the algorithm follows from Theorem 5.9.2. We claim that JACOBI uses $O((\lg a)(\lg n))$ bit operations. To see this, let us model the computation as follows: write $a_0 = a$, $n_0 = n$, and

$$\begin{aligned} a_0 &= 2^{e_1} n_1 \\ n_0 &= q_1 n_1 + a_1 \\ a_1 &= 2^{e_2} n_2 \\ n_1 &= q_2 n_2 + a_2 \\ &\vdots \\ a_{k-1} &= 2^{e_k} n_k \\ n_{k-1} &= q_k n_k + a_k, \end{aligned} \tag{5.6}$$

where the n_i are odd and $a_k = 0$. Note that $n_0 > a_0 \geq n_1 > a_1 \geq n_2 > \cdots > a_{k-1} \geq n_k > a_k$. First, we claim that k, the number of division steps, satisfies $k = O(\lg n)$. For if q_i is odd, then a_i must be even; hence $e_{i+1} \geq 1$ and so $n_{i+1} \leq a_i/2$. On the other hand, if q_i is even, then $q_i \geq 2$; hence $n_i \leq n_{i-1}/2$. Thus, no matter what the parity of q_i is, we have $n_{i+1} < n_{i-1}/2$. It follows that $n_{2j} < n2^{-j}$, and so $k = O(\lg n)$.

The total cost of performing the algorithm is, by Eqs. (5.6), given by

$$\sum_{1 \leq i \leq k} (e_i + 1)(\lg a_{i-1}) + (\lg q_i)(\lg n_i).$$

But, as in the proof of Theorem 4.4.6, we have $q_1 q_2 \cdots q_k \leq n$, and so

$$\sum_{1 \leq i \leq k} (\lg q_i)(\lg n_i) \leq (\lg n_1)(k + \log q_1 q_2 \cdots q_k) = O((\lg a)(\lg n)).$$

Similarly, it is easy to see $2^{e_1 + e_2 + \cdots + e_k} \leq a$, and hence

$$\sum_{1 \leq i \leq k} (e_i + 1)(\lg a_{i-1}) \leq (\lg a)(k + \sum_{1 \leq i \leq k} e_i) = O((\lg a)(\lg n)).$$

This completes the proof. ■

5.10 Exercises

1. Let G be an group, and suppose $a^e = 1$ in G, where 1 denotes the group identity. Show that ord a divides e.

2. (Continuation.) Suppose $a^e = a^f = 1$. Show that $a^d = 1$, where $d = \gcd(e, f)$.

3. Give two non-recursive versions of the algorithm POWER.

4. Prove Corollary 5.5.6.

5. Show how to compute $1 + x + x^2 + \cdots + x^{n-1} \pmod{m}$, using $O((\lg n)(\lg m)^2)$ bit operations. Here m is an arbitrary integer, not necessarily a prime.

6. Pseudo-random numbers are frequently generated by an algorithm called a *linear congruential generator*. Here we choose a modulus M (whose factorization is unknown), a multiplier a, an increment c, and a seed X_0. Successive pseudo-random numbers are generated by the congruence

$$X_{k+1} \equiv aX_k + c \pmod{M}.$$

Show how to compute X_n in time polynomial in $\lg n$ and $\lg M$.

7. (G. Miller.) Show how to compute the n-th digit in the base b representation of $1/a$ using $O((\lg b)(\lg a)) + O((\lg n)(\lg a)^2)$ bit operations.

8. Suppose you are given a prime p and a complete factorization of $p - 1$. Show how to compute the order of an element a in $(\mathbb{Z}/(p))^*$ using $O((\lg p)^4/(\lg\lg p))$ bit operations.

9. We say a sequence $(A_n)_{n\geq 0}$ satisfies a *linear recurrence* of order k if

$$A_n = \sum_{1\leq i\leq k} c_i A_{n-i}$$

for $n \geq k$. Suppose the constant coefficients $c_1, c_2, \ldots, c_k$ and the starting values $A_0, A_1, \ldots, A_{k-1}$ are all integers. Show how to compute $A_n \pmod{m}$ using $O(k^3(\lg n)(\lg m)^2)$ bit operations. Can you improve this to $O(k^2(\lg n)(\lg m)^2)$ bit operations?

10. Suppose one wants to compute a^e for an integer a with e large. Assuming that it takes exactly $(\log u)(\log v)$ steps to multiply u and v, show that a^e can be computed using $\frac{e(e-1)}{2}(\log a)^2$ steps by repeated multiplication.

11. (Continuation.) Using the same model as in the previous exercise, let $T(e)(\log a)^2$ be the number of steps required to compute a^e using the binary method. Prove that for $e \geq 1$, we have $T(2e) = T(e) + e^2$ and $T(2e+1) = T(e) + e^2 + 2e$. Conclude that the power algorithm computes a^e in $(e^2/3 + O(e))(\log a)^2$ steps.

12. (Continuation.) Let $\operatorname{oddpart}(n) = n/2^{\nu_2(n)}$. Show that

$$T(n) = \left(\sum_{1\leq k\leq n} \operatorname{oddpart}(k)\right) - s_2(n)$$

where the function $s_2(n)$ counts the number of 1's in the binary representation of n.

13. Suppose a is a fixed positive integer, and that it costs exactly uv to multiply a^u by a^v. Show that the power algorithm gives the optimal way to compute a^n, among all algorithms that multiply two previously computed powers of a at each step.

The next 5 exercises are devoted to showing that $(\mathbb{Z}/(p))^*$ is a cyclic group:

14. Show that if $f(X)$ is a polynomial of degree d over a field k, then $f(X) = 0$ has at most d distinct roots in k.

15. Show that if p is a prime, then the congruence $X^{p-1} \equiv 1 \pmod{p}$ has exactly $p - 1$ solutions. (Hint: factor $X^{p-1} - 1$ over the finite field $\mathbb{F}_p$.)

16. Show that if $d \mid p - 1$, then the congruence $X^d \equiv 1 \pmod{p}$ has exactly d solutions. (Hint: show that $X^d - 1$ divides $X^{p-1} - 1$ and use the previous exercise.)

17. Suppose G is an abelian group and a and b are elements with $\operatorname{ord} a = m$, $\operatorname{ord} b = n$, and $\gcd(m, n) = 1$. Show that $\operatorname{ord} ab = mn$.

18. Show that $(\mathbb{Z}/(p))^*$ is cyclic. (Hint: factor $p - 1 = \prod_i {p_i}^{e_i}$. Use Exercise 16 to construct, for each i, an element x_i that is of order ${p_i}^{e_i}$. Then use Exercise 17 to construct an element of order $p - 1$.)

19. Suppose p is odd and $p \nmid a$. Show $\operatorname{ord}(1 + ap) = p^{e-1}$ in $(\mathbb{Z}/(p^e))^*$.

20. Show that if p is odd, then $(\mathbb{Z}/(p^e))^*$ is cyclic. (Hint: show that if g is a generator for $(\mathbb{Z}/(p))^*$, then either g or $g + p$ is a generator for $(\mathbb{Z}/(p^e))^*$.)

21. Show that $(\mathbb{Z}/(2^e))^*$ is cyclic for $e \leq 2$. Show that for $e > 2$, $(\mathbb{Z}/(2^e))^*$ is the direct product of two cyclic groups, one of order 2 and one of order 2^{e-2}.

22. Let p be an odd prime. Give an example where g, $1 < g < p$, is a generator for $(\mathbb{Z}/(p))^*$, but not a generator for $(\mathbb{Z}/(p^2))^*$.

23. Show that if $(\mathbb{Z}/(n))^*$ is cyclic, then there are $\varphi(\varphi(n))$ primitive roots modulo n.

24. Count the number of solutions to $x^2 = x$ in the ring $\mathbb{Z}/(n)$.

25. Count the number of solutions to $x^3 = x$ in the ring $\mathbb{Z}/(n)$.

26. Show that if x is of order a in $(\mathbb{Z}/(m))^*$ and of order b in $(\mathbb{Z}/(n))^*$, and $\gcd(m, n) = 1$, then x is of order $\operatorname{lcm}(a, b)$ in $(\mathbb{Z}/(mn))^*$.

27. Let G be a finite cyclic group with n elements, written multiplicatively, and let g be a generator for G. Prove that if $r = g^e$, then r is of order $n/\gcd(n, e)$.

28. Let G be a finite cyclic group with n elements. Consider a subgroup $H = \langle a_1, a_2, \ldots, a_k \rangle$. Show that

$$|\langle a_1, a_2, \ldots, a_k \rangle| = \operatorname{lcm}_{1 \le i \le k} \operatorname{ord} a_i.$$

Show, by providing a counterexample, that the result may not be true when G is not a cyclic group.

29. Let p be an odd prime, let e be a positive integer, and let $a \in (\mathbb{Z}/(p^e))^*$. Show that $(\operatorname{ord}_{p^e} a)/(\operatorname{ord}_p a)$ is equal to p^f for some f, $0 \le f < e$.

30. Prove *Gauss's Lemma*: Consider the following set S_p of representatives, modulo p:

$$S_p = \{-(p-1)/2, -(p-3)/2, \ldots, -1, 0, 1, 2, \ldots, (p-1)/2\};$$

the set S_p is called the set of *least residues*. Define μ to be the number of least residues of

$$a, 2a, 3a, \ldots, \frac{p-1}{2}a$$

that are negative. Show that $\left(\frac{a}{p}\right) = (-1)^{\mu}$.

31. Using the previous exercise, prove Theorem 5.8.1, part (5).

32. Prove part (6) of Theorem 5.9.2.

33. Prove that there are infinitely many primes congruent to 1, modulo 4. (Hint: Assume there are only finitely many such primes, $p_1, p_2, \ldots, p_k$ and consider the number $(2p_1p_2 \cdots p_k)^2 + 1$.)

34. Prove that there are infinitely many primes congruent to 7, modulo 8. (Hint: mimic the proof in the previous exercise and consider $(4p_1p_2 \cdots p_k)^2 - 2$.)

35. Let $n = 2^e p_1^{e_1} p_2^{e_2} \cdots p_k^{e_k}$ be the prime factorization of n, and suppose that $\gcd(a, n) = 1$. Show that $x^2 \equiv a \pmod{n}$ is solvable iff the following conditions are satisfied:

    (a) If $e = 2$, then $a \equiv 1 \pmod{4}$.

    (b) If $e \ge 3$, then $a \equiv 1 \pmod{8}$.

    (c) For each i we have $a^{(p_i-1)/2} \equiv 1 \pmod{p_i}$.

36. Show that if a is a square modulo all but finitely many primes, then it is a rational square (the square of an integer).

37. Suppose R is a commutative ring. Show that if p is prime, then $(a + b)^p \equiv a^p + b^p \pmod{pR}$. Prove the same result for any number of summands.

38. (R. Haddad.) Use the Chinese remainder theorem to prove that there are infinitely many integer solutions to $x^3 + y^4 = z^5$ with x^3, y^4, z^5 distinct integers. (Hint: let $w = 2^a 3^b$ for some non-negative integers a, b and observe that $w + 2w = 3w$. Now rig the exponents a and b so that w is a cube, $2w$ is a fourth power, and $3w$ is a fifth power.)

39. Besides the ordinary absolute value function, the rational numbers have *ultrametric* absolute value functions (one for each prime p), defined in the following way: if $x = mp^e/n$, with m and n relatively prime to p, then

$$|x|_p = p^{-e}.$$

(We also define $|0|_p = 0$.)

(a) Verify that the ultrametric absolute values satisfy the triangles inequality, by proving the stronger inequality $|x + y|_p \le \min\{|x|_p, |y|_p\}$.

(b) What is the complexity (in bit operations) of computing $|\cdot|_p$?

(c) The *Artin-Whaples approximation theorem* states that for any finite set of absolute values (ultrametric or ordinary) $|\cdot|_1, \ldots, |\cdot|_n$, any rational numbers $x_1, \ldots, x_n$, and any $\epsilon > 0$, there is an x such that

$$|x - x_i|_i < \epsilon, \qquad i = 1, \ldots, n.$$

Explain why this is a generalization of the Chinese remainder theorem.

(d) (Watrous.) Find a polynomial-time algorithm to compute the x satisfying the conditions of the approximation theorem.

40. Consider the following decision problem: given a set $A = \{(a_1, b_1), \ldots, (a_n, b_n)\}$ of pairs of non-negative integers and an integer k, $1 \le k \le n$, determine whether there exists a subset $A' \subseteq A$ of cardinality k and an integer x such that $x \equiv a_i \pmod{b_i}$ for each $(a_i, b_i) \in A'$. Show that this problem is $\mathcal{NP}$-complete.

41. (M. O. Rabin). Suppose Alice and Bob each have an n bit message, M_A and M_B. They believe their messages are identical, but wish to verify this, with high probability, using only a small amount of communication. In particular, n is so large that merely transmitting the entire message will not do.

They choose k primes $p_1, p_2, \ldots, p_k$ uniformly at random from the set of the first $2n$ primes, and they check if $M_A \equiv M_B \pmod{p_i}$ for $1 \le i \le k$. Show that if $M_A = M_B$, then $M_A \equiv M_B \pmod{p_i}$ for all i, while if $M_A \neq M_B$, then with probability at least $1 - 2^{-k}$, $M_A \not\equiv M_B \pmod{p_i}$ for some i.

42. (M. O. Rabin.) As in the previous exercise, suppose the k moduli are chosen not from the set of the first $2n$ prime numbers, but instead from the first n^2 positive integers. Show that if n is sufficiently large, then the probability of error is again 2^{-k}. (The estimates in Chapter 8 may be useful.)

43. (Factorization of Mersenne numbers.) Let p be an odd prime. Show that if q is a positive divisor of $2^p - 1$, then

    (a) $q \equiv 1 \pmod{p}$;

    (b) $q \equiv \pm 1 \pmod{8}$.

    Now use the above results to show that $2^{13} - 1$ and $2^{17} - 1$ are primes. Also find factors of the following numbers: $2^{23} - 1$, $2^{29} - 1$, $2^{37} - 1$, $2^{43} - 1$.

44. (Factorization of Fermat numbers.) Show that if q is a positive divisor of $F_k = 2^{2^k} + 1$, then $q \equiv 1 \pmod{2^{k+1}}$, and if $k \geq 2$, then $q \equiv 1 \pmod{2^{k+2}}$.

45. For this exercise you will need a knowledge of abstract algebra.

    Prove the law of quadratic reciprocity by showing each of (a)–(d) is equivalent to the statement immediately following it.

    Let p and q be distinct odd primes in $\mathbb{Z}$.

    (a) p is a square $\pmod{q}$.

    (b) left multiplication by p gives an even permutation of $\mathbb{Z}/(q)$.

    (c) the discriminant of $X^q - 1$ is a square in $\mathbb{Z}/(p)$. (Here the discriminant of a monic polynomial $f(x)$ is defined to be $\prod_{i<j}(\alpha_i - \alpha_j)^2$, where the α_i are the roots of f.)

    (d) $(-1)^{(q-1)/2} \cdot q$ is a square $\pmod{p}$.

46. Eisenstein suggested the following algorithm for computing the Jacobi symbol $\left(\frac{a}{n}\right)$, for a, n odd and positive. We write $n = qa + \epsilon r$ where $q = \lfloor n/a \rfloor$ or $q = \lceil n/a \rceil$, whichever is even, and $\epsilon = \pm 1$. Then r is odd, and we can use the reciprocity law and the formula for $\left(\frac{-1}{r}\right)$ to continue the algorithm with (r, a) in place of (a, n).

    Prove that this algorithm can take exponential time on certain inputs.

47. Lebesgue suggested the following algorithm for computing $\left(\frac{a}{n}\right)$, for a, n odd and positive. We write $n = qa + \epsilon 2^e r$ where $\epsilon 2^e r = n \operatorname{rmod} a$, and rmod is the least-remainder

function defined in Section 4.6. Here $\epsilon = \pm 1$ and r is odd and positive. We then use the reciprocity law, and the formulas for $\left(\frac{-1}{r}\right)$ and $\left(\frac{2}{r}\right)$ to continue the algorithm with (r, a) in place of (a, n). Prove the analogue of Theorem 4.2.2 for this algorithm.

48. Prove the analogue of Theorem 4.2.2 for the algorithm JACOBI.

49. (van Lint.) Consider the $n \times n$ matrix $M = M(n, a)$ defined as follows:

$$M_{ij} = \begin{cases} 2, & \text{if } i + j \le a + 1 \text{ or } i + j \ge 2n - a + 1; \\ 0, & \text{if } |j - i| \ge a + 1; \\ 1, & \text{otherwise.} \end{cases}$$

Show that $\det M(n, a) = (2a + 1)(-1)^{a(n-1)}\left(\frac{n}{2a+1}\right)$.

50. Define $f(a, 1) = a$ and $f(a, k + 1) = a^{f(a,k)}$. Show that, for a fixed positive integer n, the sequence $(f(a, i) \bmod n)_{i \ge 1}$ is eventually constant.

51. In analogy with the algorithm BINARY GCD given in section 4.7, design an efficient "binary" algorithm to compute the Jacobi symbol.

52. Show how to compute the Jacobi symbol $\left(\frac{a}{n}\right)$ efficiently from the continued fraction expansion of a/n. Conclude that $\left(\frac{a}{n}\right)$ can be computed in $O((\lg n)(\lg \lg n)^2(\lg \lg \lg n))$ bit operations.

53. (Nagasake, Shiue, and Ho.) Show that the system of congruences $x \equiv x_i \pmod{m_i}$, $1 \le i \le k$, where the are m_i positive integers, relatively prime in pairs is equivalent to the single congruence

$$\left(\sum_{1 \le i \le k} b_i f_i \right) x \equiv \sum_{1 \le i \le k} x_i b_i f_i \pmod{m},$$

where the b_i's are arbitrary integers with $\gcd(b_i, m_i) = 1$, $m = m_1 m_2 \cdots m_k$, and $f_i = m/m_i$ for $1 \le i \le k$. What is the bit complexity of a method to compute x based on this observation?

5.11 Notes on Chapter 5

5.2

For a discussion of hardware and software techniques intended to speed up multiplication (mod n), see Head [1980]; Norris and Simmons [1981]; Blakley [1983]; Brickell [1983]; P. Montgomery [1985]; Kukihara [1985]; P. Baker [1987]; Gibson [1988]; Kawamura and Hirano [1988]; Kukihara [1989]; Walter and Eldridge [1990]; Morita [1990a, 1990b]; Shand, Bertin, and Vuillemin [1990];

Alia and Martinelli [1991]; Even [1991]; Eldridge [1991]; Walter [1991, 1992, 1995]; N. Takagi [1991, 1992, 1993]; N. Takagi and Yajima [1992]; Skavantzos and Rao [1992]; Rao and Skavantzo [1992]; Iwamura, Matsumoto, and Imai [1993a]; Kornerup [1993].

5.3

From the results of Knuth [1971] and Schönhage [1971], it follows that we can compute $a^{-1} \bmod n$ for $1 \leq a < n$ in $O((\lg n)(\lg\lg n)^2(\lg\lg\lg n))$ steps. This method is likely to be unsuitable for practical use. However, a dissenting view can be found in Schönhage, Grotefeld, and Vetter [1994].

A major unsolved problem is to determine the parallel complexity of computing inverses in $(\mathbb{Z}/(n))^*$. It is not known to be in $\mathcal{NC}$, nor is it known to be $\mathcal{P}$-complete.

5.4

The power algorithm or binary method for exponentiation goes back to at least 200 B.C.; see Datta and Singh [1935, p. 76]. A similar method can be used to multiply two numbers and is sometimes known as "Russian peasant multiplication." The binary method is mentioned in the following 18th century works: Euler [1761]; Lagrange [1769]; Legendre [1798]. Pocklington [1910] may have been the first to remark explicitly that the binary method for computing $a^e \pmod n$ uses $O(\log e)$ multiplications $\pmod n$.

Depending on the cost of a multiplication, the binary method may not necessarily be the cheapest. Other algorithms may be faster for the symbolic computation of the power of a polynomial. See, for example, Gentleman [1972], Fateman [1974a, 1974b], and McCarthy [1977, 1986].

Willoner and Chen [1981] discussed computing $a^e \pmod n$ efficiently for many values of a, where n is fixed. Brickell, Gordon, McCurley and Wilson [1993] showed how to calculate g^n for $n < N$ efficiently.

The constant factor in the time bound for exponentiation mod n can be reduced by "k-ary" methods and "addition chains". See, for example, Scholz [1937]; Brauer [1939]; Utz [1953]; Erdős [1960]; Gioia, Subbarao, and Sugunama [1962]; Straus [1964]; Cottrell [1973]; Thurber [1973a, 1973b, 1976, 1993]; Schönhage [1975]; Vegh [1975]; Lipton and Dobkin [1976]; Pippenger [1976]; Yao [1976]; van Leeuwen [1977]; Gioia and Subbarao [1978]; Dobkin and Lipton [1980]; Pippenger [1980]; Downey, Leong, and Sethi [1981]; Olivos [1981]; Knuth [1981, Sect. 4.6.3]; Semba [1983]; Volger [1985]; Tsai and Chin [1987]; Bergeron, Berstel, Brlek and Duboc [1989]; Subbarao [1989]; Bos and Coster [1990]; Morain and Olivos [1990]; Eğecioğlu and Koç [1990]; Yacobi [1991a]; Brlek and Mallette [1992]; Brlek, Castéran, Habsieger, and Mallette [1995]; Sauerbrey and Dietel [1993]; H. W. Lenstra [1993]; Aiello and Subbarao [1993]; and Bergeron, Berstel and Brlek [1994]. In de Melo and Svaiter [1995], the authors examine the basic properties of addition-subtraction-multiplication chains.

The following question is open: given integers a, b, i, can one compute the i-th bit of a^b in polynomial time? In von zur Gathen [1984a, 1987b] the question is attributed to A. Borodin.

There is no known $\mathcal{NC}$ algorithm for computing $a^e \pmod{n}$ in parallel. Von zur Gathen [1984a, 1987b] has given such an algorithm in the case where the modulus has only small prime factors. (See also Zeugmann [1992].)

For methods to compute $a^e \pmod{n}$ in hardware, see Findlay and Johnson [1990]; Orup and Kornerup [1991]; Sauerbrey [1993].

Kompella and Adleman [1991] have shown how to design an efficient "checker" for a modular exponentiation program.

A lower bound for modular exponentiation was proved by Meidanis [1991].

For efficient methods to compute Fibonacci numbers, and terms of linear recurrences, see Escott [1900]; J. Miller and Spencer Brown [1966]; Shortt [1978]; Wilson and Shortt [1980]; Urbanek [1980]; Gries and Levin [1980]; Pettorossi [1980]; Er [1983]; Protasi and Talamo [1989]; and Capocelli, Cerbone, Cull, and Holloway [1990]; Maeder [1991]. The most efficient method has been given by Fiduccia [1985].

The following problem is still open: given an order-k linear recurrence $S(i)$ and an integer n, can one determine in polynomial time whether there exists j such that $S(j) = n$? Partial results were obtained by Kannan and Lipton [1986]. The problem also is open when considered modulo p, as it is a generalization of the discrete logarithm. See Chapter 12.

5.5

A special case of the Chinese remainder theorem was given between 280 and 437 A.D. by Sun Tsŭ. See Needham [1959, pp. 33–34, 119–120]. See Dickson [1919, V. II, pp. 57–64] for early work on this theorem. Kangsheng [1988] has discussed the history of the Chinese remainder theorem. Also see Ku and Sun [1992]. For six proofs of the Chinese remainder theorem, from the ancients to the present day, see P. Davis and Hersh [1980, pp. 187–195].

Our constructive proof of the Chinese remainder theorem in Theorem 5.5.2 is from Gauss [1801, Art. 36]. The generalized Chinese remainder theorem, Theorem 5.5.5, is from Gauss [1801, Art. 32].

Theorem 5.5.7, the "anti-Chinese remainder theorem", is from Stockmeyer and Meyer [1973].

For a systolic array to do Chinese remaindering, see Koç and Cappello [1989].

5.7

Euler's criterion gives an $O((\lg p)^3)$ algorithm for determining the quadratic character of a, modulo p. Gauss, in writing about this method, said [1801, Art. 106]: , "... as soon as the numbers we are examining are even moderately large this criterion is practically useless because of the amount of calculation involved." It is unclear why Gauss held this opinion. It seems unlikely that Gauss did not know the binary method of exponentiation, as the method appears in Euler [1761] (a paper that Gauss recommends) and was known by his contemporaries (see Lagrange [1769], Legendre [1798]). Note that the $O((\lg p)^2)$ algorithm of section 5.9 is superior to the algorithm based on Euler's criterion, even for small numbers.

Burgess [1957, 1963a] proved that the least quadratic nonresidue of a prime p is bounded by $p^{1/(4\sqrt{e})+o(1)}$. We note that $a := 1/(4\sqrt{e}) \doteq .151633$. Burgess's result proves the existence of a *deterministic* algorithm to find a quadratic nonresidue using $O(p^{a+\epsilon})$ bit operations for every $\epsilon > 0$. This appears to be the best known unconditional result. For more on this problem, including algorithms that run quickly assuming the Extended Riemann hypothesis, see Chapter 8.

For an application of quadratic residues to hashing, see Maurer [1968] and Radke [1970].

5.8

For a proof of the quadratic reciprocity law, see Exercises 5.45 and 7.26. A short proof can also be found in Frame [1978]. Many proofs can be found in Bachmann [1902] and Pieper [1978]. For a detailed historical treatment, see Cooper [1926]. Rousseau [1994] gave a simple proof of the reciprocity law.

For Legendre's symbol, see Legendre [1830, Vol. I, Art. 135].

5.9

The Jacobi symbol was first discussed in Jacobi [1837]. Our algorithm is essentially that of A. Cobham [1966] and H. Williams [1980]. Eisenstein [1844a] suggested an algorithm that does not run in polynomial time, and Lebesgue [1847] suggested an efficient method of calculation based on the least-remainder algorithm. For the worst-case analysis of these algorithms, see G. Collins and Loos [1982]; Shallit [1990].

Shallit and Sorenson [1993] gave a "binary" method for the Jacobi symbol that seems to be quite efficient in practice.

There are other ways to generalize Legendre's symbol, such as *Kronecker's symbol*; see Hua [1982, Sect. 12.3]. Also see Cartier [1970] and Estes and Pall [1973].

Using a result of Schönhage [1971] on the rapid calculation of continued fractions, it is possible to compute $\left(\frac{a}{n}\right)$, where $|a| < n$, in $O((\lg n)(\lg\lg n)^2(\lg\lg\lg n))$ bit operations, but this method is unlikely to be competitive in practice. See Exercise 52.

Meidanis [1991] gave a lower bound for the problem of computing the Jacobi symbol.

6 Finite Fields

In this chapter we discuss finite fields and the ring of polynomials in one variable over a finite field. There are surprising analogies between the latter ring and the ring of integers $\mathbb{Z}$. In fact, many of the algorithms from elementary number theory transfer over with a minimum of fuss, once the proper definitions are set up. We would like to emphasize that this analogy is *computational* as well; under a proper interpretation, the same algorithms have the same running times.

As fruitful as it is, this analogy is not perfect. In particular, there is no known fast algorithm for factorization in $\mathbb{Z}$, whereas polynomials over finite fields can be factored quickly, if randomization is used. In this chapter we cover problems that *can* be efficiently solved deterministically: arithmetic in finite fields, the Euclidean algorithm, continued fractions, and computation of characters. In addition, we discuss the structure of various groups and rings associated with finite fields. The problems of factoring polynomials over finite fields and constructing extensions of finite fields are discussed in the next chapter.

6.1 Basics

We will take some results from algebra as known (proofs will be sketched in the exercises). Recall that a *field* is a commutative ring in which every nonzero element is invertible. If k is a field, then the polynomial ring in one indeterminate with coefficients from k, denoted by $k[X]$, is a unique factorization domain. An element of this polynomial ring is a unit if and only if it is a nonzero constant polynomial, and every polynomial in $k[X]$ has a factorization into irreducible polynomials that is unique up to constant factors.

In this chapter we are concerned with fields k that are *finite*. The simplest examples of such fields are the rings $\mathbb{Z}/(p)$ when p is prime, called *prime fields*. We will let $\mathbb{F}_p$ denote such a finite field; for example $\mathbb{F}_2 = \{0, 1\}$. Evidently any two fields with p elements are isomorphic, since an isomorphism must map 1 to 1; thus the notation $\mathbb{F}_p$ is unambiguous.

By the process of root adjunction, we can take a finite field and produce a larger one. The following theorem shows how this is done.

THEOREM 6.1.1 Let k be a finite field, and let f be an irreducible polynomial of degree n in $k[X]$. Then the ring $k[X]/(f)$ is a finite field with $|k|^n$ elements.

Proof We first show that $k[X]/(f)$ is an integral domain. Suppose there were nonzero polynomials g and h of degree less than n, for which $gh \equiv 0 \bmod f$. By unique factorization, one of them would be a multiple of f, which is impossible. Therefore, $k[X]/(f)$ is a finite integral domain, and necessarily a field (see Exercise 2). By counting coefficients, we see that it has $|k|^n$ elements. ∎

Any finite field must have p^n elements for some prime p, as it must contain $\mathbb{F}_p$ as a subfield, and is a vector space of finite dimension over $\mathbb{F}_p$. We now show the converse.

THEOREM 6.1.2 Let p be a prime number and let n be a positive integer. Then there is a field with p^n elements.

Proof We may assume $n > 1$. There is a finite field k, containing $\mathbb{F}_p$, containing all the roots of $X^{p^n} - X = 0$. This can be constructed by successively adjoining zeroes of irreducible factors of $X^{p^n} - X$, as in Theorem 6.1.1. Within k, the zeroes of $X^{p^n} - X$ form a subfield, as can be readily checked by computation. It remains to show that $X^{p^n} - X$ has distinct zeroes. Thus let α be a zero of this polynomial. Then

$$((X - \alpha)^{p^n - 1} - 1)(X - \alpha) = (X - \alpha)^{p^n} - (X - \alpha) = X^{p^n} - X$$

so α is a simple zero. ■

Although we will not prove it here, any two finite fields with q elements are isomorphic (see Exercise 3), so we can unambiguously use $\mathbb{F}_q$ to denote such a field when considering its algebraic properties. However, one may wish to distinguish different implementations of $\mathbb{F}_q$, say, for reasons of efficiency. When such distinctions are being made we call the various realizations *models* of $\mathbb{F}_q$.

For our purposes, it is sufficient to adopt the following formal definition. A model of $\mathbb{F}_{p^n}$ consists of a prime p, a basis for $\mathbb{F}_{p^n}$ as a vector space over $\mathbb{F}_p$, and a multiplication table (consisting of n^3 elements of $\mathbb{F}_p$) relative to this basis. To fully specify a model of $\mathbb{F}_{p^n}$ in this way requires $\Theta(n^3 \lg p)$ bits. In practice, however, we can usually do this with much less information. For example, the pair (p, f), where p is a prime and f is a monic irreducible polynomial of degree n in $\mathbb{F}_p[X]$, determines a model whose basis is $\{1, X, \ldots, X^{n-1}\}$, and whose multiplication table gives us $X^{i+j} \bmod f$ for $1 \le i, j < n$. We will refer to models using a shorthand that should be clear from context, for example "the model $\mathbb{F}_p(\alpha)$" means the model whose basis is the powers of α. (See the Notes for more discussion of these points.)

When stating running time bounds for algorithms in general finite fields, we assume, unless otherwise noted, that our models of $\mathbb{F}_{p^n}$ are of the form $\mathbb{F}_p[X]/(f)$, where f is an irreducible polynomial over $\mathbb{F}_p$ of degree n:

$$f(X) = X^n + a_{n-1}X^{n-1} + \cdots + a_1 X + a_0.$$

In such a model, an element of $\mathbb{F}_{p^n}$ can be represented as a polynomial of degree $n - 1$ with coefficients in $\mathbb{Z}/(p)$, i.e.,

$$c_{n-1}X^{n-1} + \cdots + c_1 X + c_0.$$

In discussing algorithms for $\mathbb{F}_{p^n}$, we will assume that the characteristic p is explicitly given, say, in binary.

We add two field elements by adding the corresponding polynomials; subtraction is similar. To multiply two field elements, we multiply them as polynomials, then reduce powers of X greater than X^n using the rule

$$X^n = -a_{n-1}X^{n-1} - \cdots - a_1X - a_0.$$

To invert a nonzero field element a in polynomial time, one might use the formula $a^{-1} = a^{p^n-2}$, which follows from Lagrange's theorem: the order of a subgroup divides the order of the group. However, a more efficient procedure will be explained in Section 6.4.

We now give an example of a non-prime finite field. Let $f(X) = X^2 - 2$, which is irreducible over $\mathbb{F}_5$, since 2 is a quadratic nonresidue modulo 5. Modulo f, X has the same properties as $\sqrt{2}$, so the elements of $\mathbb{F}_{25}$ can be represented in the form $a + b\sqrt{2}$, with $a, b \in \mathbb{F}_5$. Multiplication is done in the usual way, for example, we have

$$(3 + \sqrt{2})^2 = 9 + 6\sqrt{2} + (\sqrt{2})^2 = 1 + \sqrt{2}.$$

In this field we can invert elements as in high-school algebra:

$$\frac{1}{1+\sqrt{2}} = \frac{1-\sqrt{2}}{(1-\sqrt{2})(1+\sqrt{2})} = -1 + \sqrt{2}.$$

Later in this chapter we will give bounds on the number of bit operations used for various algorithms in finite fields and their polynomial rings. To avoid undue complications for fields of prime power order, which are themselves polynomial rings, we first discuss bounds for the number of field operations, then use these bounds to get estimates for bit operations.

6.2 The Euclidean Algorithm

In this section we discuss the Euclidean algorithm for polynomials, and the closely related continued fraction algorithm. Our running time bounds in this section count *field operations*. For f in $k[X]$, we define

$$\operatorname{dg} f = \begin{cases} 1, & \text{if } f = 0; \\ 1 + \deg f, & \text{if } f \neq 0. \end{cases}$$

Since f has $\operatorname{dg} f$ coefficients, $\operatorname{dg} f$ is roughly equal to $\deg f$. However, we use this function instead of the degree to avoid treating constant polynomials specially. We also use the convention that $\deg 0 = -\infty$.

With the usual methods of algebra, we can perform arithmetic operations on polynomials in $k[X]$. The following theorems give complexity bounds for these operations.

THEOREM 6.2.1 Let u and v be polynomials in $k[X]$. Then $u + v$ and $u - v$ can be computed using $O(\text{dg}\, u + \text{dg}\, v)$ operations in k, and uv can be computed using $O((\text{dg}\, u)(\text{dg}\, v))$ operations.

Proof Left to the reader. ∎

THEOREM 6.2.2 Let u and v be polynomials in $k[X]$, with $v \neq 0$. Then there are unique polynomials q and r with

$$u = qv + r$$

such that $\deg r < \deg v$. These polynomials can be computed using $O((\text{dg}\, q)(\text{dg}\, v))$ operations in k.

Proof Left to the reader. ∎

Making an analogy with the integers, we let $u \bmod v$ denote the remainder after division of u by v. This allows us to write the Euclidean algorithm exactly as done in Chapter 4. In particular, if $u_0 = u$ and $u_1 = v$, with $v \neq 0$, then

$$\begin{aligned} u_0 &= a_0 u_1 + u_2 \\ u_1 &= a_1 u_2 + u_3 \\ &\vdots \\ u_{n-2} &= a_{n-2} u_{n-1} + u_n \\ u_{n-1} &= a_{n-1} u_n \end{aligned}$$

where $u_i = u_{i-2} \bmod u_{i-1}$ for $i = 2, \ldots, n$. The final divisor u_n is a greatest common divisor of u and v, in the sense that u_n divides both u and v, and any other polynomial with this property divides u_n.

THEOREM 6.2.3 Let u and v be two nonzero polynomials in $k[X]$. Then the Euclidean algorithm finds a greatest common divisor of u and v using $O((\text{dg}\, u)(\text{dg}\, v))$ operations in k.

Proof We imitate the proof of Corollary 4.2.4. All the quotients a_i are nonzero except possibly the first; if $a_0 = 0$, then the first step of the algorithm just interchanges the arguments. Therefore we may as well assume that none of the quotients are zero, and in this case the algorithm uses

$$\sum_{0 \leq i < n} (\text{dg}\, a_i)\,(\text{dg}\, u_{i+1}) \leq \text{dg}\, u_1 \sum_{0 \leq i < n} (1 + \deg a_i) \leq \text{dg}\, u_1 \left(n + \deg \prod_{1 \leq i \leq n-1} a_i \right)$$

operations, which is $O((\text{dg}\, u)(\text{dg}\, v))$. ∎

Each iteration of the Euclidean algorithm produces a new polynomial that is a linear combination of its two predecessors. Alternatively, each polynomial is a linear combination of its two successors, and this relationship can be expressed precisely as

$$\begin{bmatrix} u_i \\ u_{i+1} \end{bmatrix} = \begin{bmatrix} a_i & 1 \\ 1 & 0 \end{bmatrix} \begin{bmatrix} u_{i+1} \\ u_{i+2} \end{bmatrix}.$$

Using the continuant notation from Section 4.4 we find that

$$\begin{bmatrix} u \\ v \end{bmatrix} = \begin{bmatrix} Q_n(a_0, \ldots, a_{n-1}) & Q_{n-1}(a_0, \ldots, a_{n-2}) \\ Q_{n-1}(a_1, \ldots, a_{n-1}) & Q_{n-2}(a_1, \ldots, a_{n-2}) \end{bmatrix} \begin{bmatrix} d \\ 0 \end{bmatrix},$$

where d is a gcd of u and v. As explained in the proof of Theorem 4.3.2, this allows us to express d as a linear combination of u and v; if

$$a = (-1)^n Q_{n-2}(a_1, \ldots, a_{n-2})$$

and

$$b = (-1)^{n+1} Q_{n-1}(a_0, \ldots, a_{n-2})$$

then $d = au + bv$.

THEOREM 6.2.4 Let u and v be two nonzero polynomials in $k[X]$. Then the extended Euclidean algorithm finds d, a greatest common divisor of u and v, and polynomials a and b with $au + bv = d$, using $O((\mathrm{dg}\, u)(\mathrm{dg}\, v))$ operations in k.

Proof Recall that all the quotients a_i are nonzero except possibly the first; if $a_0 = 0$, then we can just consider the algorithm with u and v interchanged. Therefore assume that $a_0 \neq 0$; then the recurrence relation

$$Q_{i+1}(a_0, \ldots, a_i) = a_i Q_i(a_0, \ldots, a_{i-1}) + Q_{i-1}(a_0, \ldots, a_{i-2})$$

and the initial conditions $Q_0(\,) = 1$, $Q_1(a_0) = a_0$ imply that $\deg Q_{i+1}(a_0, \ldots, a_i) = \deg a_0 + \cdots + \deg a_i$. An argument similar to the proof of Theorem 6.2.3 shows that computing $Q_{n-1}(a_0, \ldots, a_{n-2})$ uses $O((\mathrm{dg}\, u)(\mathrm{dg}\, v))$ field operations; evidently this is also true of $Q_{n-2}(a_1, \ldots, a_{n-2})$. ∎

This proof gives bounds for the degrees of the multipliers expressing d as a linear combination of u and v. In fact, if u and v are nonzero, and not both constant polynomials, the polynomials a and b found by the extended Euclidean algorithm satisfy $\deg(a) < \deg(v)$, and $\deg(b) < \deg(u)$.

The greatest common divisor of two polynomials is only defined up to a nonzero constant multiple. It will be convenient to identify a canonical representative for the class of such polynomials. In the sequel, we will use the convention that $\gcd(u, v)$ is monic, unless both

u and v are zero (in which case it is zero). Theorems 6.2.3 and 6.2.4 remain true under this convention.

6.3 Continued Fractions

We let $k(X)$ denote the field of rational functions over k, and let $k((1/X))$ denote the set of all expressions of the form

$$f = \sum_{i \le d} c_i X^i = c_d X^d + \cdots + c_1 X + c_0 + c_{-1}/X + c_{-2}/X^2 + \cdots \tag{6.1}$$

where the c_i's are elements of k. It can be shown that $k((1/X))$ is a field, which contains a subfield isomorphic to $k(X)$ (see Exercise 8). In particular, then, any quotient of polynomials can be written as an element of $k((1/X))$; we illustrate with an example:

$$\frac{X^2+1}{X-1} = \left(X + \frac{1}{X}\right)\left(\frac{1}{1-1/X}\right) = X + 1 + 2/X + 2/X^2 + 2/X^3 + \cdots$$

(the last step uses the Taylor series for $(1-T)^{-1}$).

If f is expressed as in (6.1) above, we let

$$\lfloor f \rfloor = c_d X^d + \cdots + c_1 X + c_0$$

denote the *integral part* of f.

Any element f of $k((1/X))$ has a continued fraction expansion, whose quotients $a_0, a_1, a_2, \ldots$ can be found as follows:

$$\begin{aligned} f_0 &= f \\ a_0 &= \lfloor f_0 \rfloor \\ f_1 &= \frac{1}{f_0 - a_0} \\ a_1 &= \lfloor f_1 \rfloor \\ f_2 &= \frac{1}{f_1 - a_1} \\ &\vdots \end{aligned}$$

Continuing with our example, we compute the continued fraction of

$$f = \frac{X^2+1}{X-1} = X + 1 + 2/X + 2/X^2 + 2/X^3 + \cdots,$$

assuming k does not have characteristic 2. Since $f_0 = f$, the first quotient, a_0, equals $X+1$. Then

$$f_1 = (f - a_0)^{-1} = \frac{X}{2}\left(1 + \frac{1}{X} + \frac{1}{X^2} + \cdots\right)^{-1} = \frac{X}{2}\left(1 - \frac{1}{X}\right) = \frac{X-1}{2},$$

so that $a_1 = X/2 - 1/2$. This gives the expansion

$$\frac{X^2+1}{X-1} = a_0 + \frac{1}{a_1} = X + 1 + \frac{1}{X/2 - 1/2}.$$

We say that an element f of $k((1/X))$ is *rational* if it is a quotient of polynomials. Such elements are characterized by the following theorem.

THEOREM 6.3.1 An element f of $k((1/X))$ is rational iff its continued fraction expansion is finite.

Proof Imitate the proof of Theorem 4.5.1. ■

The quotients a_i produced by the continued fraction algorithm satisfy all the familiar formulas of Section 4.5. In particular, for polynomials u and v, the quotients of the continued fraction for u/v are the same as those produced when the Euclidean algorithm is applied to u and v.

In Chapter 4, we considered variants of the Euclidean algorithm in $\mathbb{Z}$ such as the least-remainder algorithm. For $k[X]$, such variants do not exist, as any function that satisfies "reasonable" axioms for a nearest-integer function must map f to $\lfloor f \rfloor$ (see Exercise 14).

One can also characterize rational functions by the behavior of their coefficient sequences. We say that a sequence $c_0, c_1, c_2, \ldots$ of elements in k satisfies a *linear recurrence* if it satisfies a linear relation with constant coefficients; that is, there is a positive number n and elements $a_0, \ldots, a_{n-1}$ of k such that for all $i \geq 0$,

$$c_{n+i} = a_{n-1}c_{n+i-1} + \cdots + a_0 c_i.$$

Such sequences are characterized by the following theorem.

THEOREM 6.3.2 An element $f(X) = \sum_{i \geq 0} c_i/X^i$ of $k((1/X))$ is rational iff $c_0, c_1, c_2, \ldots$ satisfies a linear recurrence.

Proof If f is rational, say $f = u/v$, then assume without loss of generality that v is monic. The expression for the coefficients of negative powers of X in $vf = u$ gives the required recurrence relation. Conversely, the coefficients and initial values in the recurrence relation for the coefficients of f give an expression for f as a rational function. ■

Sequences of this type are also called *shift-register sequences*, because they can be generated in the following way. Imagine a device with n *registers* $x_0, \ldots, x_{n-1}$, each of which can hold an element of k. Initially, x_i is set to c_i. Then, at each time step, the replacements

$$x_i \leftarrow \begin{cases} x_{i+1}, & \text{if } i < n; \\ \sum_{0 \le i < n} a_i x_i, & \text{otherwise.} \end{cases}$$

are done in parallel. This is particularly convenient when $k = \mathbb{F}_2$, a field for which hardware is easily built.

6.4 Computing in $k[X]/(f)$

In this section we discuss how algorithms and complexity bounds for the ring of integers $\mathbb{Z}$ generalize to the polynomial ring $k[X]$. We will take a "bootstrap" approach, where results are first proved on the bit complexity of operations in k, then used to estimate the running time of operations in $k[X]$.

If f is a polynomial in $k[X]$, with $|k| = p^m$, we define

$$\lg f = \begin{cases} 1, & \text{if } f = 0; \\ (1 + \deg f) \lg |k|, & \text{if } f \ne 0. \end{cases}$$

Roughly speaking, $\lg f$ is about $(\deg f)(\log |k|)$. Therefore f can be represented using about $\lg f$ bits, and if k is fixed then $\lg f$ is roughly proportional to the degree of f.

THEOREM 6.4.1 Let $k = \mathbb{F}_p[X]/(f)$, where f is a monic irreducible polynomial of degree n in $\mathbb{F}_p[X]$. In k, addition and subtraction can be done with $O(\lg |k|)$ bit operations, and multiplication and division can be done with $O((\lg |k|)^2)$ bit operations.

Proof Since $n \lg p = O(\lg |k|)$ (see Exercise 24), the bounds for the first three operations follow easily from Theorems 6.2.1 and 6.2.2. To analyze division, it suffices to show that if $v \ne 0$, then v^{-1} can be found in $O((\lg |k|)^2)$ bit operations. For this bound, we use the extended Euclidean algorithm to solve $av + bf = 1$ in $\mathbb{F}_p[X]$; this algorithm takes $O((\text{dg } f)^2)$ operations in $\mathbb{F}_p$, each of which has cost $O((\lg p)^2)$, so the result follows. ∎

COROLLARY 6.4.2 Let u and v be polynomials in $k[X]$. Then $u + v$ and $u - v$ can be computed with $O(\lg u + \lg v)$ bit operations, and uv can be computed using $O((\lg u)(\lg v))$ bit operations. If $v \ne 0$, we can find polynomials q and r with $u = qv + r$ and $\deg r < \deg v$ using $O((\lg q)(\lg v))$ bit operations. The gcd of u and v can be found using $O((\lg u)(\lg v))$ bit operations.

Proof Left to the reader. ∎

Given any polynomial f in $k[X]$, we can form the finite ring $R = k[X]/(f)$. Elements of this ring are represented by polynomials in $k[X]$ of degree less than $\deg f$; a ring element (represented by a polynomial g) is a unit precisely when g is relatively prime to f.

THEOREM 6.4.3 Let $f \in k[X]$ be a nonzero polynomial and let $R = k[X]/(f)$. We can add and subtract elements of R using $O(\lg f)$ bit operations, and we can multiply elements of R using $O((\lg f)^2)$ bit operations. If $g \in R^*$, we can compute g^{-1} using $O((\lg f)^2)$ bit operations.

Proof Similar to the proof of Theorem 6.4.1. ■

THEOREM 6.4.4 Let R be defined as in Theorem 6.4.3. Then for any a in R, a^e can be computed with $O((\lg e)(\lg f)^2)$ bit operations. In particular, computing a^e in $\mathbb{F}_q$ can be done with $O((\lg q)^3)$ bit operations when $e \le q$.

Proof Left to the reader. ■

6.5 Galois Theory

In the next two sections we go a little more deeply into the structure of finite fields and polynomial rings derived from them. In this section, we discuss properties of these fields and rings that follow easily from Galois theory. We will assume one standard result on finite fields, namely, that the multiplicative group of a finite field is cyclic. A proof of this fact, which can be read now if desired, appears in the next section (see Theorem 6.6.3). We will also assume the reader is familiar with Galois theory, or is willing to consult the references on this topic suggested in the Notes.

Let p be a prime and R a ring with $pR = 0$; in this case we say that R has *characteristic* p. Then the function $x \longmapsto x^p$ is a ring homomorphism τ, called the *Frobenius map*. R is also a vector space over $\mathbb{F}_p$, and τ is $\mathbb{F}_p$-linear.

The case $R = \mathbb{F}_{p^n}$ is especially important. In this ring, τ is an automorphism that fixes the base field $\mathbb{F}_p$. The *Galois group* of $\mathbb{F}_{p^n}$ is the group of automorphisms of $\mathbb{F}_{p^n}$ that fix $\mathbb{F}_p$; this group is cyclic of order n and is generated by τ. For example, taking $\mathbb{F}_5(\sqrt{2})$ as a model of $\mathbb{F}_{25}$,

$$\tau(1 + \sqrt{2}) = (1 + \sqrt{2})^5 = 1 - \sqrt{2},$$

that is, the Frobenius automorphism in $\mathbb{F}_{25}$ is just the ordinary conjugation $\sqrt{2} \rightarrow -\sqrt{2}$.

If q is a power of p, then similar considerations apply to $\mathbb{F}_q$-algebras, that is, rings containing $\mathbb{F}_q$ as a subring. Then the Frobenius map $\tau(x) = x^q$ is a ring homomorphism fixing $\mathbb{F}_q$. In the special case of $\mathbb{F}_{q^n}$, τ generates the *relative* Galois group of $\mathbb{F}_{q^n}/\mathbb{F}_q$, which is the group of automorphisms of $\mathbb{F}_{q^n}$ fixing $\mathbb{F}_q$.

Consider a finite field extension $\mathbb{F}_{q^n}/\mathbb{F}_q$. The *conjugates* of an element x of $\mathbb{F}_{q^n}$ are

$$x, x^q, x^{q^2}, \ldots, x^{q^{n-1}};$$

note that this list may contain duplications and is defined relative to a given field extension. The sum and product of conjugates are called *trace* and *norm* respectively:

$$\mathrm{Tr}(x) = x + x^q + \cdots + x^{q^{n-1}}$$
$$\mathrm{N}(x) = x^{1+q+\cdots+q^{n-1}}.$$

The trace is an $\mathbb{F}_q$-linear map from $\mathbb{F}_{q^n}$ onto $\mathbb{F}_q$. Similarly, the norm is a group homomorphism mapping $\mathbb{F}_{q^n}^*$ onto $\mathbb{F}_q^*$. We will show in the next section that both of these groups are cyclic; from this fact one may derive Hilbert's "theorem 90": an element of $\mathbb{F}_{q^n}$ has norm 1 iff it is of the form x^{q-1} for some nonzero x. (See Exercise 29.)

The subfields of $\mathbb{F}_{q^n}$ have an algebraic characterization: an element x belongs to $\mathbb{F}_{q^m}$ iff it satisfies $x^{q^m} = x$. This follows from the cyclic structure of $\mathbb{F}_{q^n}^*$, and the fact that a polynomial of degree n can have at most n zeroes. In particular, in any field containing $\mathbb{F}_q$,

$$x^q = x \iff x \in \mathbb{F}_q.$$

Consider a finite field extension $\mathbb{F}_{q^n}/\mathbb{F}_q$. Because the Frobenius map τ is an automorphism of order n, it follows from standard results of algebra that this extension is *separable* (see Exercise 10). This means that any polynomial irreducible over $\mathbb{F}_q$ must have distinct zeroes. Using this result, we can find the number of irreducible polynomials of degree n in $\mathbb{F}_q[X]$.

THEOREM 6.5.1 Let $I_q(n)$ denote the number of irreducible monic polynomials over $\mathbb{F}_q$ of degree n. Then

$$I_q(n) = \frac{1}{n}\sum_{d\,|\,n} \mu(d) q^{n/d},$$

where μ denotes the Möbius function. In particular, $(1 - 2q^{-n/2})/n \le I_q(n)/q^n \le 1/n$.

Proof We associate each element of $\mathbb{F}_{q^n}$ with the monic polynomial of least degree having it as a zero; the degree must divide n, as can be seen by considering field extensions. If we group the elements of $\mathbb{F}_{q^n}$ according to the degrees of these polynomials, we find

$$q^n = \sum_{d\,|\,n} \left|\{\alpha \in \mathbb{F}_{q^n} : \deg\alpha = d\}\right| = \sum_{d\,|\,n} dI_q(d).$$

The first result follows by Möbius inversion. To prove the upper bound, note that

$$\sum_{d \mid n} \mu(d) q^{n/d} = q^n \prod_{p \mid n} (1 - 1/q^p) \leq q^n.$$

The lower bound follows from the estimate

$$\left| nI_q(n) - q^n \right| \leq \sum_{\substack{d \mid n \\ d \neq n}} q^d \leq \frac{q^{n/2+1}}{q-1} \leq 2q^{n/2}. \qquad \blacksquare$$

Certain applications require an irreducible polynomial f for which X, a root of f, generates $\mathbb{F}_{q^n}^* = k[X]/(f)^*$; f is said to be *primitive* if this is the case. In general, when f is irreducible, the order e of X will be a divisor of $q^n - 1$. In this case e is said to be the *exponent* of f. (The situation for reducible f is more complicated; see Theorem 6.6.6.) For a number $e \geq 1$, we define $m(e)$ to be the smallest positive m such that $e \mid q^m - 1$.

THEOREM 6.5.2 Let $e \geq 1$. If f is a monic irreducible polynomial with exponent e, then f has degree $m(e)$. If φ denotes the Euler φ-function, then the number of such polynomials is $\varphi(e)/m(e)$; in particular, there are $\varphi(q^n - 1)/n$ monic primitive polynomials of degree n in $\mathbb{F}_q[X]$.

Proof Elements of order e will exist in $\mathbb{F}_{q^m}^*$ iff $e \mid q^m - 1$, so any polynomial with exponent e must have degree $m(e)$. Now to count the number of such polynomials, consider the cyclic group of order e contained in $\mathbb{F}_{q^m}^*$. It will have $\varphi(e)$ generators, and $m(e)$ of them form the zeroes of any such polynomial. To get the last result, take $e = q^n - 1$. $\blacksquare$

The last two theorems have probabilistic interpretations. We can state these as simply as possible if we assume the coefficient field k is fixed. First, if one picks a monic polynomial of degree n in $k[X]$ at random, it will be irreducible with probability about $1/n$; this is analogous to the prime number theorem, which implies that a random n-bit positive integer is a prime with probability roughly proportional to $1/n$.

Second, since $\varphi(n) = \Omega(n/\log\log n)$ (see Theorem 8.8.7), a random monic irreducible polynomial of degree n is primitive with probability $\Omega(1/\log n)$. In particular, then, X is a primitive root mod f, for infinitely many f. More generally, this has been shown to hold in $\mathbb{F}_q[X]$ when X is replaced by any polynomial of positive degree that is not a t-th power for any t, greater than 1, that divides $q - 1$. This is an analog of Artin's conjecture, which states that any a not equal to -1 or a square is a primitive root modulo p for infinitely many p.

Some applications require irreducible polynomials of a special form, for example, with a fixed constant term. There are many results dealing with the density of such polynomials; roughly, they state that polynomials with any "reasonable" property will be irreducible in the right proportion to their degree.

One can also ask about the density of polynomials in $k[X_1, \ldots, X_n]$ that are irreducible or absolutely irreducible, that is, do not factor in k or any extension field of k. For this question, the answer is simple: if $n > 1$, then "most" n-variable polynomials are absolutely irreducible over $\mathbb{F}_p$.

When k is a finite field, the polynomials in $k[X]$ satisfy an important reciprocity law.

THEOREM 6.5.3 (KÜHNE) Let f_1 and f_2 be monic irreducible polynomials in $k[X]$, of degrees n_1 and n_2 respectively. Let N_1 denote the norm from $k[X]/(f_1)$ to k, and N_2 the norm from $k[X]/(f_2)$ to k. Then

$$\mathrm{N}_1(f_2) = (-1)^{n_1 n_2} \mathrm{N}_2(f_1)$$

Proof Let $\alpha_1, \ldots, \alpha_{n_1}$ be the zeroes of f_1, and let $\beta_1, \ldots, \beta_{n_2}$ be the zeroes of f_2. Then

$$\begin{aligned} \mathrm{N}_1(f_2) &= \prod_{1 \le i \le n_1} f_2(\alpha_i) = \prod_{1 \le i \le n_1} \prod_{1 \le j \le n_2} (\alpha_i - \beta_j) \\ &= (-1)^{n_1 n_2} \prod_{1 \le j \le n_2} \prod_{1 \le i \le n_1} (\beta_j - \alpha_j) = (-1)^{n_1 n_2} \mathrm{N}_2(f_1). \end{aligned}$$ ∎

6.6 The Structure of $k[X]/(f)$

In this section we present algebraic structure theorems for finite fields and polynomial rings. The most important results in this regard are the Chinese remainder theorem for polynomials and the existence of generators for the multiplicative group of a finite field.

The additive structure can be summarized as follows. If f has degree n, then $R = k[X]/(f)$ is a n-dimensional vector space over k, with basis $\{1, X, X^2, \ldots, X^{n-1}\}$.

Several of the functions considered in this chapter can be viewed in terms of linear algebra. For example, to invert g in R, we can solve the linear system $af + bg = 1$ for polynomials a of degree $n-1$ and b of degree n. The linear transformation taking (a, b) to $af + bg$ has for its determinant the resultant of f and g, and this will be nonzero if f and g are relatively prime. For another example, suppose that f is irreducible. Then for any a in R, multiplication by a is a k-linear map, represented by a matrix M_a, and the trace and norm of a from $k[X]/(f)$ to k are equal to the trace and determinant of M_a, respectively.

The Frobenius map $a \mapsto a^q$ is an $\mathbb{F}_q$-linear transformation on R; we may also consider R as a vector space over $\mathbb{F}_p$, on which the absolute Frobenius map $a \mapsto a^p$ is an $\mathbb{F}_p$-linear transformation. Both functions may be represented as matrices relative to appropriate bases.

THEOREM 6.6.1 Let $f_1, \ldots, f_r$ be polynomials of positive degree in $k[X]$ that are pairwise relatively prime, and let f denote their product. Then

$$k[X]/(f) \cong k[X]/(f_1) \oplus \cdots \oplus k[X]/(f_r).$$

Proof The function $a \longmapsto (a \bmod f_1, \ldots, a \bmod f_r)$ is a homomorphism mapping the ring on the left to the ring on the right. Its kernel is 0, and since the two rings have the same size it is an isomorphism. ■

When considering this isomorphism, we will use a "component notation" and write $a = (a_1, \ldots, a_r)$ when $a_i = a \bmod f_i$, $i = 1, \ldots, r$. The next theorem gives bounds for the number of bit operations required to convert into and out of this representation.

THEOREM 6.6.2 Let f and $f_1, \ldots, f_r$ be as in Theorem 6.6.1. If $\deg a < \deg f$, then we can compute $a \bmod f_i$, $i = 1, \ldots, r$, in $O((\lg f)^2)$ bit operations. Conversely, given a_i, $i = 1, \ldots, r$, with $\deg a_i < \deg f_i$, we can find an a with $a \bmod f_i = a_i$, $i = 1, \ldots, r$, in $O((\lg f)^2)$ bit operations.

Proof Computing $a \bmod f_i$ requires $O((\mathrm{dg}\, f)(\mathrm{dg}\, f_i))$ operations in k, so the total is

$$(\mathrm{dg}\, f) \sum_{1 \le i \le r} \mathrm{dg}\, f_i \le 2(\mathrm{dg}\, f) \sum_{1 \le i \le r} \deg f_i \le 2(\mathrm{dg}\, f)^2.$$

Since each operation takes $O((\lg |k|)^2)$ bit operations, the first result follows.

For the converse, let $g_i = f/f_i$ and $e_i = g_i(g_i^{-1} \bmod f_i)$, $i = 1, \ldots, r$. We need $O((\mathrm{dg}\, f_i)(\mathrm{dg}\, f))$ operations to obtain e_i using these formulas, so all the e_i's can be found using $O((\mathrm{dg}\, f)^2)$ operations in k. Since $a = \sum e_i a_i$, we can find a using at most a constant times

$$\sum_{1 \le i \le r} (\mathrm{dg}\, a_i)\,(\mathrm{dg}\, f) = O((\mathrm{dg}\, f)^2)$$

such operations. From this the second assertion follows. ■

THEOREM 6.6.3 $\mathbb{F}_q^*$ is a cyclic group, of order $q - 1$.

Proof Let e denote the exponent of the group $\mathbb{F}_q^*$; that is, the least number e such that $x^e = 1$ for all nonzero x in $\mathbb{F}_q$. The polynomial $X^e - 1$ has at least $q - 1$ zeroes, namely the elements of $\mathbb{F}_q^*$. Because it can have at most e zeroes in all, $e \ge q - 1$. But since the order of an element in a group divides the order of that group, $e \mid q - 1$. Therefore $e = q - 1$ and $\mathbb{F}_q^*$ is cyclic. ■

We now discuss the group structure of $R = k[X]/(f)^*$ when f is arbitrary. By Lemma 5.6.1, we have

$$k[X]/(f)^* \cong k[X]/(f_1)^* \times \cdots \times k[X]/(f_r)^*.$$

and so it suffices to consider the case where f is a power of an irreducible polynomial. If f *is* irreducible, we are done by the above theorem; if not, the structure is more complicated but can be summarized as follows.

Let f be an irreducible polynomial, and let $R = k[X]/(f^e)$, $K = k[X]/(f)$. Then R^* is isomorphic to the direct product of K^* (a cyclic group of order $q^n - 1$) and $1 + Rf$ (an abelian group of p-power order). The precise structure of $1 + Rf$ is not easy to state, but roughly speaking, it is a direct product of many cyclic groups, mostly of order p. There may be cyclic factors of order p^i, $i > 1$, but their number goes down exponentially: about $1/p$ of the factors can have order p^2 or larger, about $1/p^2$ have order p^3 or more, and so on.

To state our results precisely we need some facts about abelian groups. Let A be an abelian group of p-power order. The fundamental theorem of abelian groups asserts that A is a product of cyclic groups, each of which also has p-power order. So,

$$A \cong \underbrace{C_p \times \cdots \times C_p}_{(e_1 \text{ copies})} \times \underbrace{C_{p^2} \times \cdots \times C_{p^2}}_{(e_2 \text{ copies})} \times \cdots \times \underbrace{C_{p^t} \times \cdots \times C_{p^t}}_{(e_t \text{ copies})},$$

where C_m denotes a cyclic group of order m. The structure of A is completely specified by the numbers e_i, or, equivalently, by the *ranks*

$$\begin{aligned} r_1 &= e_1 + e_2 + \cdots + e_t \\ r_2 &= e_2 + \cdots + e_t \\ &\vdots \\ r_t &= e_t. \end{aligned}$$

In terms of the ranks, A contains $r_i - r_{i+1}$ copies of C_{p^i} for $i < t$, and r_t copies of C_{p^t}. One often calls r_1 simply the *rank* of A; it is the number of elements required to generate A.

LEMMA 6.6.4 Let G be a finite abelian group with a subgroup A, for which $\gcd(|A|, |G/A|) = 1$. Then G has a subgroup B isomorphic to G/A, for which $G \cong A \times B$.

Proof Let m be the order of A and consider the homomorphism ϕ that takes an element of G to its m-th power. The kernel of ϕ is exactly A (it cannot be any larger, by Cauchy's theorem: if a prime p divides the order of a finite group, then the group has an element of order p). Hence $G^m \cong G/A$, and we may take B to be G^m; let n denote the order of this group. Since m and n are relatively prime, there exist integers a and b such that $am + bn = 1$, and if $x \in A \cap B$, $x^{am+bn} = x = 1$. Therefore $A \cap B = 1$, and since G is abelian, G is the direct product of A and B. ■

Now consider $G = R^*$, and its subgroup $A = 1 + Rf$. By counting coefficients, $|G| = q^{ne} - q^{n(e-1)} = q^{n(e-1)}(q^n - 1)$, and $|A| = q^{n(e-1)}$. By the lemma G has another subgroup B of order $q^n - 1$; since there is a homomorphism from R^* onto K^*, B must be cyclic.

Let $S = B \cup \{0\}$; since $|B|$ is relatively prime to q, it is also relatively prime to p, so $S^p = S$. The following lemma lets us use the elements of S as "digits" to represent elements of R.

LEMMA 6.6.5 Any element g in R can be written uniquely as

$$g = s_0 + s_1 f + s_2 f^2 + \cdots + s_{e-1} f^{e-1}$$

where $s_i \in S$, $i = 0, \ldots, e-1$.

Proof It will suffice to show that the elements of S mod f form a complete residue system, that is, every element of $k[X]/(f)$ is congruent to a unique element of S. Since $0 \in S$, we only need to show that the homomorphism from R^* to K^* that sends g to g mod f is 1-1 when restricted to B. But this is true, because the kernel of this homomorphism is A and $A \cap B = 1$. ■

Now we have done enough work to prove the following structure theorem.

THEOREM 6.6.6 Let f be an irreducible polynomial of degree n in $k[X]$, with $|k| = q = p^m$. Let $R = k[X]/(f^e)$. Then

$$R^* \cong K^* \times (1 + Rf)$$

where K^* is a cyclic group of order $q^n - 1$, isomorphic to the multiplicative group of $k[X]/(f)$, and $1 + Rf$ is a group of order $p^{mn(e-1)}$. The exponent of $1 + Rf$ is p^t, where $t = \lceil \log_p e \rceil$, and its ranks are given by $r_i = mn(\lceil e/p^{i-1} \rceil - \lceil e/p^i \rceil)$ for $i < t$, and $r_t = mn(\lceil e/p^{t-1} \rceil - 1)$.

Proof The first assertion of the theorem follows from the discussion above; it remains to compute the exponent and the ranks of $1 + Rf$. First, since $1 + Rf$ is a group of p-power order, its exponent must be a power of p. In this group $1 + f$ is an element of maximal order; $(1 + f)^{p^t} = 1 + f^{p^t} = 1$ whenever $p^t \geq e$, so the exponent must be $\max\{t : p^t \geq e\}$, as claimed.

Now let $A = 1 + Rf$ and consider the chain

$$A \supset A^p \supset A^{p^2} \supset \cdots \supset A^{p^t} = 1,$$

and observe that r_i is the rank of $V_i = A^{p^{i-1}}/A^{p^i}$, which is a direct product of cyclic groups of order p. For such groups, the rank is uniquely determined by the number of elements, so we have

$$r_i = \text{rank}(V_i) = \log_p |A^{p^{i-1}}| - \log_p |A^{p^i}|$$

Using Lemma 6.6.5, we can count the elements in A^{p^i} for $i < t$. An element of A^{p^i} is representable in the form

$$1 + s_1 f^{p^i} + s_2 f^{2p^i} + s_3 f^{3p^i} + \cdots$$

where $s_i \in S$ (because $S^p = S$). By omitting coefficients of f^e and higher powers, this representation can be made unique, implying that

$$\log_{|S|} \left| A^{p^i} \right| = \left| \{ j : p^i \le jp^i < e \} \right| = \left| \{ j : 1 \le j < e/p^i \} \right| = \left\lceil e/p^i - 1 \right\rceil .$$

Since $|S| = q^n = p^{mn}$, we find that when $0 \le i < t$,

$$\log_p \left| A^{p^i} \right| = mn(\left\lceil e/p^i \right\rceil - 1);$$

therefore $r_i = mn(\lceil e/p^{i-1} \rceil - \lceil e/p^i \rceil)$ as claimed. ■

As an example, let us take $k = \mathbb{F}_2$ and $R = k[X]/\left((X^2 + X + 1)^{11}\right)$. We have $m = 1$, $n = 2$, and $e = 11$. The 2-part of R^* has exponent 16, because $t = \lceil \log_2 11 \rceil = 4$. The ranks are

$$\begin{aligned} r_1 &= 2\left(11 - \lceil 11/2 \rceil\right) = 10, \\ r_2 &= 2\left(\lceil 11/2 \rceil - \lceil 11/4 \rceil\right) = 6, \\ r_3 &= 2\left(\lceil 11/4 \rceil - \lceil 11/8 \rceil\right) = 2, \\ r_3 &= 2\left(\lceil 11/8 \rceil - \lceil 11/16 \rceil\right) = 2. \end{aligned}$$

From these, we obtain $e_4 = 4$, $e_3 = 2 - 2 = 0$, $e_2 = 6 - 2 = 4$, and $e_1 = 10 - 6 = 4$. Therefore,

$$R^* \cong C_3 \times \underbrace{C_2 \times \cdots \times C_2}_{\text{(4 copies)}} \times \underbrace{C_4 \times \cdots \times C_4}_{\text{(4 copies)}} \times \underbrace{C_{16} \times C_{16}}_{\text{(2 copies)}}.$$

As a check, we verify that $|R^*| = 3 \cdot 2^{20} < |R| = 2^{22}$.

One application of Theorem 6.6.6 is in testing certain equations in $k[X]/(f)$ for solvability, without relying on the factorization of f. For example, there is a deterministic polynomial time algorithm to decide if a is a quadratic residue mod f (that is, if the equation $a = b^2$ can be solved in the group $k[X]/(f)^*$). (See Exercise 7.49.) We note that no such result is known for $\mathbb{Z}$.

We can also think of Theorem 6.6.6 as providing an analog to the Carmichael λ-function for polynomials in $k[X]$.

6.7 Characters

If G is a finite abelian group, then any homomorphism χ from G to $\mathbb{C}^*$ is called a *character*; since G is finite the image of χ is a finite cyclic group, composed of roots of unity. A character on G provides partial information about the elements of G; for example, if $G = \mathbb{F}_p^*$ and $\chi(x) = (\frac{x}{p})$, then $\chi(x) = 1$ precisely when x is a quadratic residue modulo p.

Now, assume that q is odd. Since $\mathbb{F}_q^*$ is cyclic, an element x of this group satisfies $x^{(q-1)/2} = 1$ iff x is a nonzero square; the function $\chi(x) = x^{(q-1)/2}$ provides a character that generalizes the Legendre symbol to any finite field of odd characteristic. This is computable in $O((\lg q)^3)$ bit operations, but using a reciprocity law, the running time can be improved. To do this we must generalize the Jacobi symbol to $k[X]$.

Let k be a finite field of odd characteristic, and let f and g be polynomials in $k[X]$, with f monic and of positive degree. If f is irreducible, then

$$\left(\frac{g}{f}\right) = \begin{cases} 1, & \text{if } f \text{ and } g \text{ are relatively prime and } g \text{ is a square mod } f; \\ -1, & \text{if } f \text{ and } g \text{ are relatively prime and } g \text{ is not a square mod } f; \\ 0, & \text{otherwise.} \end{cases}$$

If $f = f_1 \cdots f_r$ with each f_i monic and irreducible, then

$$\left(\frac{g}{f}\right) = \prod_{1 \le i \le r} \left(\frac{g}{f_i}\right).$$

By analogy with $\mathbb{Z}$, we call $\left(\frac{g}{f}\right)$ the *(generalized) Jacobi symbol.*

The above definition implies that if $x \in k$, $\left(\frac{x}{f}\right) = \chi(x)^{\deg f}$, where χ denotes the quadratic character on k^*. The analog of Gauss's quadratic reciprocity law is the following.

THEOREM 6.7.1 (DEDEKIND) Assume that $f, g \in k[X]$, with f, g monic and $\deg f$, $\deg g > 0$. Then

$$\left(\frac{g}{f}\right) = (-1)^{\frac{(q-1)}{2}(\deg g)(\deg f)} \left(\frac{f}{g}\right).$$

Proof Let m and n denote the degrees of f and g respectively. If f and g are irreducible, then by Theorem 6.5.3 we have

$$f^{(q^n-1)/(q-1)} \bmod g = (-1)^{mn} g^{(q^m-1)/(q-1)} \bmod f.$$

The result follows in this case from raising both sides of this equality to the $(q-1)/2$-th power. The general result follows from the multiplicativity of the symbol as a function of its upper and lower arguments. ■

This leads to the following algorithm.

POLY–JACOBI (u, v)
$u \leftarrow u \bmod v$
$c \leftarrow$ leading coefficient of u
$s \leftarrow \chi(c)^{\deg v}$ $\{ \chi$ is the quadratic character on $k \}$
if $\deg u = 0$ then return (s)
else
    $u \leftarrow u/c$
    if $(q-1)/2$, $\deg u$, $\deg v$ are all odd then $s \leftarrow -s$
    return $(s \cdot$ POLY–JACOBI $(v, u))$

THEOREM 6.7.2 Let $u, v \in k[X]$, and assume that v is monic and $\deg v > 0$. Then POLY–JACOBI(u, v) returns $\left(\frac{u}{v}\right)$, and uses $O(\mathrm{dg}\, v)$ evaluations of the quadratic character in k, and $O((\mathrm{dg}\, u)(\mathrm{dg}\, v))$ field operations. Consequently, the quadratic character in $\mathbb{F}_q$ can be evaluated with $O((\lg q)^2)$ bit operations.

Proof The correctness of the algorithm follows from Theorem 6.7.1. To analyze its running time, let

$$u_0 = a_0 u_1 + c_0 u_2$$
$$u_1 = a_1 u_2 + c_1 u_3$$
$$\vdots$$
$$u_{n-2} = a_{n-2} u_{n-1} + c_{n-2} u_n,$$

where $u_0 = u$, $u_1 = v$, v_i is monic for $i \geq 1$, $\deg u_i > \deg u_{i+1}$ for $i = 1, \ldots, n-1$, and u_n is the greatest common divisor of u and v; this is the Euclidean algorithm, modified to make each divisor monic. Since $n \leq \mathrm{dg}\, u_1 = \mathrm{dg}\, v$, and χ is evaluated at most n times, the first assertion holds. Counting field operations as in the proof of Theorem 6.2.3 proves the second.

To get the bit complexity bound, we assume that $\mathbb{F}_q = \mathbb{F}_p[X]/(f)$ where f is monic and has degree n. In $\mathbb{F}_p$, $\chi(x) = \left(\frac{x}{p}\right)$, and this can be evaluated in $O((\lg p)^2)$ bit operations, as shown in Section 5.9. Furthermore, we know that operations in $\mathbb{F}_p$ take $O((\lg p)^2)$ bit operations, so the total number of bit operations is a constant times

$$n^2(\lg p)^2 + n(\lg p)^2 = O(n^2 (\lg p)^2) = O((\lg q)^2). \qquad \blacksquare$$

The generalized Jacobi symbol gives a criterion for a polynomial to be a square in $k[X]/(f)$ when f is irreducible. Using Theorem 6.6.6, we see that when k has odd characteristic, a polynomial g in $k[X]/(f)^*$ is a square exactly when $(\frac{g}{f_i}) = 1$ for each irreducible monic divisor f_i of f. It would be of interest to find an algorithm with comparable running time when f is composite, but none seems to be known. (For a partial solution, see Exercise 7.49.)

It would also be of interest to generalize the result of this section to find a quadratic time algorithm to see if an element of $\mathbb{F}_q$ is a d-th power, when d is an arbitrary divisor of $q-1$. When $q = p$ (a prime), Euler's criterion, which uses $O(\lg q)$ multiplications in $\mathbb{F}_q$, is still the best known algorithm for this purpose. (If $d \mid p-1$, and q is not prime, the problem can be efficiently reduced to d-th power testing mod p. See the solution to Exercise 27.)

6.8 Exercises

1. Take any book on algebra, and prove all the theorems on finite fields without looking at the proofs given in that book.

2. Prove that a finite integral domain is a field.

3. Use the theory of splitting fields to show that any two finite fields of the same cardinality must be isomorphic.

4. Use the Euclidean algorithm to show that $k[X]$ is a principal ideal domain, and hence a unique factorization domain.

5. Generalize Exercise 4.16 to polynomials in $k[X]$.

6. (a) Let $f, g \in k[X]$ be polynomials of degree $< n$. Show how to compute the product fg using $O(n^{\log_2 3})$ field operations, chosen from $\{*, +, -\}$.

   (b) Suppose that g is monic. Show how to divide f by g (obtaining the quotient and remainder) within the same time bound.

   (c) Conclude that if f and g are two polynomials of the same degree in $\mathbb{F}_q[X]$, their product can be formed using can be done using $O((\lg f)^{\log_2 3})$ bit operations.

7. Use the quadratic formula and the identity $\mathrm{N}(x) = x\bar{x}$ to give another proof that inversion in $\mathbb{F}_q$ can be done in $O((\lg q)^2)$ bit operations when $q = p^2$.

8. Prove that $k((1/X))$ is a field.

9. Find necessary and sufficient conditions so that $f \in k((1/X))$ is a square.

10. Show that the Frobenius map on $\mathbb{F}_{p^n}$ is an automorphism of order n. Conclude that if $\mathbb{F}_{p^n} = \mathbb{F}_p(\alpha)$, then the conjugates of α are distinct, and the extension $\mathbb{F}_{p^n}/\mathbb{F}_p$ is separable.

11. Let A be an $n \times n$ matrix with coefficients in a finite field k. Then the ring of matrices $k[A]$ is isomorphic to $k[X]/(m)$, where m denotes the minimal polynomial of A. Use this to design an algorithm that computes A^e using $O((n + \lg e)n^2(\lg |k|)^2)$ bit operations. (Note that the power algorithm would use $O(n^3(\lg e)(\lg |k|)^2$.)

12. Define $I_q(n)$ as in Theorem 6.5.1.

    (a) Prove that for $n \geq 1$, we have $q^n/(2n) \leq I_q(n) \leq q^n/n$. Therefore, $\log I_q(n) = n \log q - \log n + O(1)$.

    (b) Conclude that $I_q(n)$ is not computable in polynomial time (if n is given in binary) but that it can be computed in time bounded by a polynomial in the output length $\lg I_q(n)$.

13. Show that a random degree n irreducible polynomial in $\mathbb{F}_q[X]$ is primitive with probability $\Omega((\log n + \log\log q)^{-1})$. (The implied constant is absolute.)

14. Let $F : k((1/X)) \to k[X]$ be such that $f - F(f) \in \frac{1}{X}k[[1/X]]$. Then $F(f) = \lfloor f \rfloor$.

15. Let $f, g \in k[X]$, with $n = \deg f$ and $m = \deg g$. (Here k can be any field.) Let f have zeroes $\alpha_1, \ldots, \alpha_n$ and leading coefficient a, and let g have zeroes $\beta_1, \ldots, \beta_m$, and nonzero leading coefficient b. The expression

$$R(f, g) = a^m b^n \prod_{1 \leq i \leq n} \prod_{1 \leq j \leq m} (\alpha_i - \beta_j)$$

    is called the *resultant* of f and g. It vanishes iff f and g have a common factor.

    (a) Show that $R(f, g) = (-1)^{mn} R(g, f)$.

    (b) Show that if $g \bmod f$ has degree d, then

$$R(f, g) = a^{m-d} R(f, g \bmod f).$$

    (c) Conclude that if $f, g \in \mathbb{F}_q[X]$, their resultant $R(f, g)$ can be computed using $O((\lg f)(\lg g))$ bit operations, and hence the norm of an element in $\mathbb{F}_q$ can be found using $O((\lg q)^2)$ bit operations.

16. An $n \times n$ matrix (a_{ij}) is called a *circulant* if a_{ij} depends only on $j - i \bmod n$. Show that the determinant of a circulant (with entries in $\mathbb{F}_q$) can be computed using $O(n^2(\lg q)^2)$ bit operations.

17. Show that the trace from $\mathbb{F}_q$ to $\mathbb{F}_p$ can be computed using $O((\lg q)^2)$ bit operations.

18. Show that the characters on a finite abelian group G form a group $\hat{G}$ under pointwise multiplication. Prove the orthogonality relation

$$\sum_{\chi \in \hat{G}} \chi(x)\bar{\chi}(x') = \begin{cases} |G|, & \text{if } x = x'; \\ 0, & \text{otherwise;} \end{cases}$$

and its dual

$$\sum_{x \in G} \chi(x)\bar{\chi}'(x) = \begin{cases} |G|, & \text{if } \chi = \chi'; \\ 0, & \text{otherwise.} \end{cases}$$

19. Extend Theorem 6.7.2 to show that over $\mathbb{F}_{2^n}$, any quadratic equation can be tested for solvability using $O(n^2)$ bit operations.

20. Study the bit complexity of operations in $\mathbb{F}_q$, when this field is implemented by iterated extensions.

21. For $i = 1, \ldots, k$, let α_i be an element of degree n_i over $\mathbb{F}_p$, satisfying the irreducible monic polynomial equation $f_i(\alpha_i) = 0$. Consider the ring

$$R = \mathbb{F}_p[X_1, \ldots, X_k]/(f_1(X_1), \ldots, f_k(X_k)).$$

(This is sometimes called the *tensor product*, and denoted by $\mathbb{F}_p(\alpha_1) \otimes \cdots \otimes \mathbb{F}_p(\alpha_k)$.)

(a) Show that if the degrees n_i are pairwise relatively prime, we have

$$R \cong \mathbb{F}_p(\alpha_1, \ldots, \alpha_k).$$

(b) Let $n = n_1 \cdots n_k$, where the n_i are as in (a), and $q = p^n$. Discuss the bit complexity of operations in $\mathbb{F}_q$, when this field is represented as the tensor product of the various $\mathbb{F}_{p^{n_i}}$.

22. Let $q = p^n$. Discuss the bit complexity of the basic operations in $\mathbb{F}_q$, when this field is implemented as an algebra over $\mathbb{F}_p$ (given by basis elements and a multiplication table).

23. (a) Let $q = p^n$. Show that there is a $\beta \in \mathbb{F}_q$ such that $\beta, \beta^p, \ldots, \beta^{p^{n-1}}$ are linearly independent over $\mathbb{F}_p$. (This is called a *normal basis*.)

(b) Suppose we are given a normal basis for $\mathbb{F}_q$, together with a multiplication table listing the products $\beta_i\beta_j$ relative to this basis. (Here we let $\beta_i = \beta^{p^i}$ for $i = 1, \ldots, n$.) Discuss the complexity of field operations, including exponentiation, in this representation.

(c) [Open] Define the *quality* of a normal basis to be the number of nonzero coefficients in the fundamental bilinear form for multiplication in that basis. Prove or disprove: $\mathbb{F}_{p^n}$ always has a normal basis of quality $O(n)$.

24. Our length function for polynomials is based on the assumption that an element of k can be represented using $\Theta(\lg |k|)$ bits. Justify this assumption, as follows. Let p and m be integers such that $p \geq 2$, $m \geq 1$, and let $q = p^m$. Show that

$$\frac{m \lg p}{2} \leq \lg q \leq m \lg p.$$

25. Prove the following result. Let f and g be monic irreducible polynomials in $k[X]$, with $\deg f = m$, $\deg g = n$. Let $d \mid q - 1$. Then

$$f^{\frac{q^n - 1}{d}} \bmod g = (-1)^{mn(q-1)/d} g^{\frac{q^m - 1}{d}} \bmod f.$$

This is a d-th power reciprocity law analogous to Theorem 6.7.1.

26. Show that for polynomials f and g in $k[X]$, the symbol $(\frac{f}{g})$ can depend on the field k.

27. Show that if K/k is an extension of finite fields, then x is a square in K iff $N(x)$ is a square in k. Use this to get an $O((\lg q)^2)$ algorithm for the quadratic character in $\mathbb{F}_q$, when $q = p^n$ and p is odd.

28. Let A denote the ring of integers modulo n, and let f be a monic polynomial in $A[X]$. For this exercise, you are to generalize some of the results of this chapter to the ring $R = A[X]/(f)$.

    (a) Show that if $g \in A[X]$, there are polynomials q and r for which $g = qf + r$, and $\deg(r) < \deg(f)$.

    (b) Show that elements of R can be added and subtracted using $O(\lg f)$ bit operations.

    (c) Show that two elements of R can be multiplied using $O((\lg f)^2)$ bit operations.

    (d) Show that $g \in R^*$ iff there exist $\alpha, \beta \in A[X]$ with $\deg \alpha < \deg f$, and $\deg \beta < \deg g$, such that $\alpha g + \beta f = 1$.

    (e) Use part (d) to design a polynomial time algorithm to find the multiplicative inverse of an element in R^*.

29. Show that a nonzero $a \in \mathbb{F}_{q^n}$ has norm 1 iff it has the form b^{q-1} for some $b \in \mathbb{F}_{q^n}^*$. (This is Hilbert's "theorem 90" for finite fields.)

30. Let $k = \mathbb{F}_q[X]/(f)$, where f is an irreducible monic polynomial of degree n. Let $a(X)$ (a polynomial of degree $< n$) be an element of k. Show that $m(X)$, the minimal polynomial of a, is given by

$$m(X) = \begin{cases} \prod_{1 \leq i < d} (X - a(X)^{q^i}) \bmod f(X) & \text{if } d < n, \\ \left(\prod_{1 \leq i < d} (X - a(X)^{q^i}) \bmod f(X)\right) + f(X) & \text{if } d = n. \end{cases}$$

(Here d is the least i such that $a^{q^i} = a$ in k.) Therefore the minimal polynomial may be computed without linear algebra, using $O(d \lg q)$ multiplications in k.

31. Let $X = \mathbb{F}_2^n$, and for $x, y \in X$, let $d(x, y)$ denote the number of nonzero bits in $x - y$. Show that X, with the distance function d, is a metric space. Show further, that the following problem is $\mathcal{NP}$-complete: given a linear subspace S of X (specified by a matrix that annihilates this subspace), does there exist an $x \in S$ of length k? (As usual, the length of x is $d(x, 0)$.)

32. Suppose we choose entries of an $n \times n$ matrix M independently and uniformly from $\mathbb{F}_q$. Show that the probability that M is nonsingular is $\prod_{1 \le i \le n}(1 - q^{-i})$.

33. Show that every group of order n is cyclic iff $\gcd(n, \varphi(n)) = 1$.

34. Let $\Phi_m(X)$ denote the monic irreducible polynomial in $\mathbb{Q}[X]$ whose zeroes are the primitive m-th roots of unity. This is called the *m-th cyclotomic polynomial*. Prove the following facts.

    (a) $\Phi_m(X)$ has integral coefficients.

    (b) $\Phi_m(X) = \prod_{d \mid m}(X^{m/d} - 1)^{\mu(d)}$.

    (c) Assume $\gcd(p, m) = 1$. In $\mathbb{F}_p[X]$, $\Phi_m(X)$ splits into distinct factors of degree k, where k is the order of p in $(\mathbb{Z}/(m))^*$.

    (d) Assume $p^r \,\|\, m$ for $r \ge 1$. In $\mathbb{F}_p[X]$, $\Phi_m(X) = (\Phi_{m/p^r}(X))^{(p-1)p^{r-1}}$.

35. (Continuation.) Assume that m is prime and congruent to 1 mod e. Let ζ be a primitive m-th root of unity. Consider the quantity

$$\eta = \sum_{x \in ((\mathbb{Z}/(m))^*)^e} \zeta^x.$$

(This is called a *Gaussian period*.) There is a unique monic irreducible polynomial in $\mathbb{Q}[X]$ that vanishes at η; call this polynomial $\Psi_e(X)$. (Note that this also depends on m.) Prove the following facts:

    (a) $\Psi_e(X)$ has degree e and integral coefficients.

    (b) For primes p not dividing the discriminant of $\Psi_e(X)$, the following are equal: (i) the order of p in the quotient group $(\mathbb{Z}/(m))^*/((\mathbb{Z}/(m))^*)^e$; (ii) the degree of any irreducible factor of $X^e - p$ mod m; (iii) the degree of any irreducible factor of $\Psi_e(X)$ mod p.

6.9 Notes on Chapter 6

6.1

Many books on algebra introduce finite fields, to exemplify the general theory of algebraic field extensions. Books in this spirit include Artin [1942]; Lang [1965]; and van der Waerden [1970]. For more detailed coverage of finite fields, see Albert [1961]; Ireland and Rosen [1990]; and the encyclopedic Lidl and Niederreiter [1983]. The books by Berlekamp [1968] and W. Peterson and Weldon [1972] give perspectives useful to a hardware designer.

Although Gauss discussed congruences modulo p in his Disquisitiones Arithmeticae (and apparently planned to devote an entire section to this topic), general finite fields were first studied by Galois [1830]. The theorem that all models of $\mathbb{F}_q$ are isomorphic is due to E. H. Moore [1896]. According to a famous theorem of MacLagan-Wedderburn [1905], any finite division ring must be commutative, that is, a field.

Our proof of the existence of finite fields follows Herstein [1987].

H. W. Lenstra [1991b] discussed various notions of an "explicit model" for a finite field. Naively, one might think an explicit model for $\mathbb{F}_q$ should be a set of computable functions to implement the various field operations. (This is used in mathematical logic; see, for example, Fröhlich and Shepherdson [1955] and Rabin [1960].) Lenstra proved, however, that the characteristic p cannot be computed using polynomially many operations in such a model unless integer factoring is easy. For this reason, we require that p be explicitly supplied as part of the model. (This requirement is related to the notion of "uniformity" that appears in circuit complexity.) It can be shown that various kinds of explicit data for $\mathbb{F}_q$, such as irreducible polynomials, normal bases (with multiplication tables), towers of extension fields, and tensor products, can be converted into models of $\mathbb{F}_q$ (in our sense) in deterministic polynomial time.

Because of their importance in coding theory and cryptography, finite fields have been extensively studied with an eye toward efficient implementations of arithmetic. A key early reference on this topic is the paper of Bartee and Schneider [1963]. For alternative presentations and more recent results, see Conway [1968]; D. Morrison [1968]; Laws and Rushworth [1971]; Redinbo [1979]; Willett [1980]; Chor [1982]; Yeh, Reed, and Truong [1984]; Davida and Litow [1985]; Beth [1986]; Cantor [1989]; Feng [1989]; Piontas [1989]; Diab [1991]; Morii and Takamatsu [1991]; Hasan and Bhargava [1992a, 1992b]; Shparlinski, Tsfasman, and Vladut [1992]; and Fenn, Benaissa, and Taylor [1993]. Most of these references assume "standard" models of finite fields (i.e., of the form $\mathbb{F}_p[X]/(f)$ with f irreducible). For a discussion of finite field implementations using normal bases, see the work cited in the solution to Exercise 23.

Thomas, Keller, and Larsen [1986] discussed arithmetic in $\mathbb{F}_p$, when p is a Mersenne prime.

Calmet [1985] and Lidl and Matthews [1988] discussed software packages for finite fields.

There is no essential difference between algorithms that manipulate elements of $\mathbb{F}_2$ and algorithms that use the standard Boolean operations $\{\vee, \wedge, \sim\}$. (Any operation in one system can be simulated by $O(1)$ operations in the other.) In what sense this remains true for larger finite fields is an interesting, and not completely solved, problem. For work in this area, see von zur Gathen and Seroussi [1991]; Boyar, Frandsen, and Sturtivant [1992]; and Sturtivant and Frandsen [1993].

6.2

S. Stevin [1585], was apparently the first to apply the Euclidean algorithm to polynomials. (See Netto and le Vavasseur [1907] for the relevant history.)

The running time bounds for the Euclidean algorithm are due to G. Collins [1968]. See G. Collins [1969] for related average-case results, and implementation studies.

The idea of using linear combinations of polynomials to extract information about their greatest common divisor is quite old. For example, the Euler-Bézout method for deciding if u and v have a root in common hinges on the following fact: u and v have a nontrivial gcd iff a and b can be chosen with $\deg a < \deg v$, $\deg b < \deg u$ so that $au + bv = 0$. (See Cauchy [1840].) By the middle of the 19th century, explicit determinantal formulas for polynomials a and b satisfying $au + bv = \gcd(u, v)$ were well known (see, for example, Cayley [1848]). The idea of using such formulas in mechanical computation occurred to Sylvester [1840]; after describing his method for obtaining the remainder sequence of two polynomials, he said:

> Through the well-known ingenuity and kindly proferred [sic] help of a distinguished friend, I trust to be able to get a machine made for working Sturm's theorem, and indeed all problems of derivation, after the method here expounded; on which subject I have a great deal more yet to say, than can be inferred from this or my preceding papers.

Although he accurately predicted his future prolixity, his machine seems never to have been built. However, the idea of solving linear equations to obtain remainders has been revived; Davida [1972] suggested that one invert elements of $\mathbb{F}_{2^n}$ in this way, and pointed out that the method was particularly suited to parallel computation. (See also Borodin, von zur Gathen, and Hopcroft [1982]; and Davida and Litow [1985].)

The binary gcd algorithm of Section 4.7 can be generalized to $k[X]$; for an analysis of this algorithm, see Norton [1985, 1989]. For other papers on computing the gcd of polynomials over finite fields, see Brent, Kung, and Luk [1983]; Brent and Kung [1983]; Strassen [1983]; Knopfmacher and Knopfmacher [1988, 1990a]; Norton [1989]; Ma and von zur Gathen [1990]; and Knopfmacher [1991].

6.3

The continued fraction algorithm for functions over finite fields can be found in the thesis of Artin [1924a].

Golomb [1967] devoted an entire book to shift-register sequences. For structural properties of these sequences, see Zierler [1959].

The proofs of Theorems 6.3.1 and 6.3.2 give an $O(n^3(\lg q)^2)$ algorithm to characterize a shift register of length n, given its output. A better method is that of Berlekamp [1968] and Massey [1969] which uses $O(n^2(\lg q)^2)$ bit operations. This is a variant of the continued fraction algorithm; the basic idea is to find a ratio of degree n polynomials that approximates about $2n$ terms of a rational power series. Many authors have discussed this connection, including Welch and Scholtz [1979]; Cheng [1984]; Dornsetter [1987]; Camion [1989]; and Dai and Zeng [1990]. We will discuss this algorithm, and its applications to cryptanalysis, more fully in Chapter 15.

Shift-register sequences are the coefficients of rational elements of $k((1/X))$. More generally, one may consider elements of $k((1/X))$ that are *algebraic*, that is, satisfy a polynomial equation with coefficients in $k[X]$. The coefficient sequences of such functions are called *automatic*; they may be characterized using finite automata and have many interesting properties. See Allouche [1987] for a discussion of this topic.

6.4

The bit complexity of polynomial arithmetic has not been studied as such. Rather, attention has focused on the number of field operations used. Many authors implicitly assume that the coefficient field is algebraically closed of characteristic zero, or at least contains enough roots of unity to support a Fast Fourier Transform (FFT). Readers who do not pay attention to these assumptions are likely to be confused by the running times reported in the literature. For example, Aho, Hopcroft, and Ullman [1974, p. 269] asserted that two degree n polynomials can be multiplied using $O(n \lg n)$ arithmetic operations. This is true over the complex numbers, but (as far as we know) it has never been proved for finite fields, in the sense that $O(n \lg n)$ operations in $\mathbb{F}_q$ suffice to multiply degree n polynomials in $\mathbb{F}_q[X]$.

For these reasons, it seems worthwhile to review the arithmetic complexity of polynomial arithmetic in some detail. We recall that

$$\mu(m, n) = \begin{cases} m(\lg n)(\lg\lg n), & \text{if } m \geq n; \\ n(\lg m)(\lg\lg m), & \text{otherwise.} \end{cases}$$

Let $u, v \in k[X]$, where k is a finite field. Then if u has degree m and v has degree n, the following bounds hold. We can add (or subtract) u and v using $O(m + n)$ operations in k, we can multiply u and v using $O(\mu(m, n))$ such operations, and we can divide u by v using $O(\mu(d, n))$ such operations, where d is the degree of the quotient. These results are due essentially to Schönhage [1977]. (See Bassalygo [1978] for weaker results along the same lines.) Moenck [1973] found an algorithm for the greatest common divisor that is asymptotically faster than the Euclidean algorithm; if we assume that $m \geq n$, then using his ideas, the continued fraction quotients of u/v, as well as the representation $\gcd(u, v) = au + bv$ ($\deg a < n$, $\deg b < m$), can be found using $O(m(\lg n)^2(\lg\lg n))$ operations in k. We note that the remainder sequence cannot be computed this quickly, since it has, in general, $\Theta(n^2)$ coefficients.

We may combine these results with the best known bounds for integer arithmetic and continued fractions to get estimates on bit complexity. In simplified form, we find the following. In $\mathbb{F}_q$, the four basic field operations (addition, subtraction, multiplication, and division) can be done using using $(\lg q)^{1+o(1)}$ bit operations. In $\mathbb{F}_q[X]$, the basic arithmetic operations (addition, subtraction, multiplication, division with remainder) on polynomials of binary length $\leq n$ can be done using $n^{1+o(1)}$ bit operations. The same holds for continued fraction quotients and the output of the extended Euclidean algorithm.

The practical utility of such "fast" polynomial arithmetic routines is not widely appreciated. Unlike the case of integer arithmetic, polynomial routines must explicitly manipulate each coefficient, and so we might expect fast routines to be useful for polynomials rather sooner than they are for integers.

As far as we know, only two authors have studied this matter carefully from an experimental point of view. P. Montgomery [1990] implemented polynomial multiplication in $\mathbb{F}_2[X]$ using a recursive algorithm (similar to integer multiplication à la Karatsuba and Ofman [1962]) and a more sophisticated algorithm of Cantor [1989] based on the FFT. He found the first algorithm was preferable for polynomials of degree less than 16,000. (His paper also gave many interesting ideas for implementing division, reciprocals, greatest common divisor, etc.) Shoup [1993b] gave corresponding data for classical multiplication and a FFT multiplication method in $\mathbb{F}_p[X]$, when p is a 100-bit prime. Linear interpolation of his timings suggest the crossover point is around degree 35. The rough conclusion one can draw from these experiments is that fast polynomial arithmetic becomes worthwhile when $\lg f \approx 1000$–10000.

For parallel computation, the situation is more complicated. It is fair to say that as far as $\mathcal{NC}$ algorithms are concerned, $\mathbb{F}_q$ behaves more like a ring than a field, unless its characteristic is small. That is, addition, subtraction, and multiplication are "easy" whereas inversion is "hard."

More precisely, we note the following results. Recall that the four fundamental arithmetic operations on integers (addition, subtraction, multiplication, division with remainder) can be done with $\mathcal{NC}$ algorithms. (See Exercise 3.27.) Therefore, the same is true for addition, subtraction, and multiplication in the prime field $\mathbb{F}_p$. (To get $xy \bmod p$, for example, we can compute $xy - p\lfloor (xy)/p \rfloor$.) There is no known $\mathcal{NC}$ algorithm for inversion mod p.

Now, consider polynomials in $\mathbb{F}_p[X]$. Addition, subtraction, and multiplication of polynomials have evident $\mathcal{NC}$ algorithms. If v is a monic polynomial, it is possible to compute approximations to $1/v$ by Newton's method, and this leads to an $\mathcal{NC}$ algorithm for division with remainder, provided the divisor is monic. (No such algorithm is known in the general case, as the example of $X \bmod aX - 1$ shows.)

Using the results for prime fields, it is not hard to see that all the assertions remain true for $\mathbb{F}_q$, provided that elements of this field are represented as residues modulo an irreducible monic polynomial in $\mathbb{F}_p[X]$. If f is monic, there are $\mathcal{NC}$ algorithms for addition, subtraction, and multiplication in $\mathbb{F}_q[X]/(f)$, but not for inversion.

No $\mathcal{NC}$ algorithms are known for computing the greatest common divisor in $\mathbb{F}_q[X]$, nor for modular exponentation. We note that inversion in $\mathbb{F}_q$ easily reduces to exponentiation (since $x^{-1} = x^{q-2}$), as does testing for units mod f (using Theorem 6.6.6 to obtain an exponent for $k[X]/(f)^*$).

Various authors have found parallel algorithms for these problems that are efficient in special circumstances. Litow and Davida [1988] showed that inversion of elements in $\mathbb{F}_{p^n}$ can be done using $n^{O(1)}$ operations in $\mathbb{F}_p$, arranged so as to have depth $O(\lg n)$. (See also von zur Gathen [1990].) Borodin, von zur Gathen, and Hopcroft [1982] found a parallel algorithm for the gcd in $k[X]$ that is in $\mathcal{NC}$ when the field k is fixed. (See also von zur Gathen [1984b].) Fich and Tompa [1985, 1988] and Golovanov and Solodovnikov [1987] gave $\mathcal{NC}$ algorithms for exponentiation in $k[X]/(f)$ for the case of small characteristic; von zur Gathen and Seroussi [1991] extended them to arbitrary finite algebras of small characteristic. Zeugmann [1990] gave an exponentiation algorithm useful when the modulus has only small irreducible factors. There are very efficient parallel algorithms for exponentiation when the characteristic is small and a normal basis is used. For these, see the references given for Exercise 23.

For surveys of the algebraic complexity of polynomial arithmetic, see Chapter 2 of Langemyr [1989]; and Eberly [1989].

For a VLSI implementation, see Eastman [1990].

For results in a different model (in which multiplication by fixed scalars is not counted), see Kaminski and Bshouty [1987, 1989]; Bshouty and Kaminski [1990]; Baum and Shokrollahi [1991]; Shokrollahi [1991, 1992a, 1992b]; Averbuch, Bshouty, and Kaminski [1992]; Bshouty [1992], and references therein.

6.5

The best introduction to Galois theory is Artin [1942]. Gaal [1971] is an introductory treatment, with many examples.

Hilbert's theorem 90 is from his "Zahlbericht" [1897]. Theorem 6.5.1 was first published by Dedekind [1857]; it was also known to Gauss and appears in his *Nachlass* [1889, §347]. The formula for the number of primitive polynomials is due to Pellet [1870].

It is easy to generate all primitive polynomials once one is known. If α has order $q-1$ in $\mathbb{F}_q^*$, then so does $\beta = \alpha^k$ whenever $\gcd(k, q-1) = 1$. Any efficient algorithm for minimal polynomials (e.g., Exercise 30) can be then used to find the primitive polynomial vanishing at β.

Bilharz [1937] reduced the analog of Artin's conjecture in $k[X]$ to the Riemann hypothesis for finite fields, which was later proved by Weil [1948a]. (See Hasse [1952].)

Kornblum [1919] proved the analog of Dirichlet's theorem on primes in progressions for $k[X]$. (See also Artin [1924a].) For generalizations, dealing with the density of irreducible polynomials of various types, see Hayes [1965]; Ree [1971]; and Leonard [1974]. S. Cohen [1968] showed that all but a negligible fraction of polynomials in $k[X_1, \ldots, X_n]$ are irreducible when $n \geq 2$; this was strengthened by Fredman [1972] to absolute irreducibility.

The concepts of norm and trace are due to Gauss [1832] and Dedekind [1882], respectively.

Theorem 6.5.3 is due to Kühne [1902]; the simple proof we give follows Ore [1934]. For other proofs, see F. Schmidt [1928]; Carlitz [1932]; Whiteman [1937]; and Hellegouarch [1986].

6.6

The running times for the Chinese remainder algorithm are from G. Collins [1968].

Asymptotically, the running time of Theorem 6.6.2 can be improved to $(\lg f)^{1+o(1)}$, where $f = \prod_{1 \leq i \leq n} f_i$ denotes the product of the moduli. This result appears to be folklore, although special cases can be found in Aho, Hopcroft, and Ullman [1974] and Borodin and Munro [1975].

The fundamental theorem of abelian groups is due to Frobenius and Stickelberger [1879].

Theorem 6.6.6 is apparently due to Claasen [1977]; see McDonald [1974] for general results on units in finite rings. Our proof relied on a certain "base f" representation appearing in Serre [1979, p. 32].

6.7

The quadratic reciprocity law in $k[X]$ is from Dedekind [1857]; Kühne [1902] was the first to generalize this to arbitrary finite fields. For other proofs, see Artin [1924a]; Vaidyanathaswamy [1927]; Ore [1934]; and references therein.

The running time estimates appear in Bach [1990b]. This paper also shows that asymptotically, the quadratic character in $\mathbb{F}_q$ can be computed using $(\lg q)^{1+o(1)}$ bit operations. For a different quadratic character algorithm, using $O(n(\lg q)^2)$ bit operations when $q = p^n$, see Itoh and Tsujii [1989a].

7 Solving Equations over Finite Fields

In this chapter, we give efficient methods for solving polynomial equations over finite fields. Because the integers modulo a prime p form a field, the methods we present can be used to find solutions to polynomial congruences modulo p. However, we will present them within the general framework of finite fields.

We discuss the following problems: computing square roots and solving binomial congruences over finite fields, factoring polynomials over finite fields, and construction of non-prime finite fields. Logically, the last topic should perhaps go first, as we have not yet presented any algorithm for constructing an extension field of the integers modulo p. However, the best methods for constructing such fields rely on polynomial factorization, and for this reason we introduce them after covering factorization. We also discuss methods to solve polynomial congruences in the finite ring $\mathbb{Z}/(p^e)$; strictly speaking, these methods are not finite field algorithms, but they are so closely related to finite fields that they belong here.

In contrast to the methods we have presented so far, many of the techniques of this chapter use randomization in an essential way. As of this writing, we do not know whether there is a deterministic polynomial-time algorithm for factoring polynomials over finite fields. We close the chapter with a section covering current knowledge on this question.

In this chapter we will adhere to the following notation: q is a prime power, with $q = p^n$. We denote the finite field of q elements by $\mathbb{F}_q$, or k.

7.1 Square Roots: Group-Theoretic Methods

Let q be a prime power. The results of Chapter 6 have the following implications. If q is odd, then half of the elements of $\mathbb{F}_q^*$ are squares; if q is even, all of the elements of $\mathbb{F}_q^*$ are squares. However, these results do not indicate any efficient methods for computing square roots. The purpose of this section is to present such methods. We start with a simple observation.

THEOREM 7.1.1 Let G be a group of odd order m, written multiplicatively. Let $a \in G$. Then the equation $x^2 = a$ has a unique solution in G, which is $x = a^{(m+1)/2}$.

Proof Because $a^m = 1$ for any element a of the group, we have $x^2 = a^{m+1} = a$. For the uniqueness, we note that the previous sentence shows that the map $x \rightarrow x^2$ is onto. Since G is finite, it must also be 1-1. ■

COROLLARY 7.1.2 If $q = 2^n$ or $q \equiv 3 \pmod 4$, square roots in $\mathbb{F}_q^*$ can be computed using $O((\lg q)^3)$ bit operations.

Proof We give explicit formulas. In $\mathbb{F}_{2^n}^*$, $\sqrt{a} = a^{2^{n-1}}$. If $q \equiv 3 \pmod 4$, $(\mathbb{F}_q^*)^2$ is a subgroup of odd order in $\mathbb{F}_q^*$, so $\sqrt{a} = a^{(q+1)/4}$ for any a in $(\mathbb{F}_q^*)^2$. ■

As nice as this solution is, however, it does not take care of all odd values of q. But it can be extended to the case $q \equiv 5 \bmod 8$. (See Exercise 1.)

To describe an algorithm that works for all odd q, we need to examine the structure of $\mathbb{F}_q^*$ more closely. Let $q - 1 = 2^s t$, where t is odd. Because $\mathbb{F}_q^*$ is cyclic (see Theorem 6.6.3), there is a unique subgroup H of order t contained in $\mathbb{F}_q^*$, and a chain of subgroups

$$H = G_0 \subset G_1 \subset \cdots \subset G_s = \mathbb{F}_q^*,$$

where G_i/G_{i-1} has order 2, $i = 1, \ldots, s$.

There is a homomorphism from $\mathbb{F}_q^*$ onto $\mathbb{F}_q^*/H$ that takes x to the coset xH; if g is not a square in $\mathbb{F}_q^*$, then the image of g under this homomorphism generates $\mathbb{F}_q^*/H$. This means that we can express a as $g^e h$, where $h \in H$. The idea of the algorithm is to find the square roots of g^e and h separately.

Computing the square root of h is easy; h belongs to a group of odd order, and we can use Theorem 7.1.1. The idea for finding e is the following. Suppose we know $e \bmod 2^{i-1}$. Then there are only two choices for $e \bmod 2^i$, and only the correct choice will satisfy $ag^{-e} \in G_{s-i}$. (This only depends on $e \bmod 2^i$.) Necessarily e is even, so once e is found, we know that a square root of g^e is $g^{e/2}$.

It remains to find the non-square g. We know of no efficient deterministic algorithm to do this, but we can use randomization and the following "subgroup principle." Let S be a proper subgroup of a group G. Then, because the order of a subgroup divides the order of a group, $|S| \leq |G|/2$. Taking S to be the subgroup of squares in $\mathbb{F}_q^*$, we see that a *random* element of $\mathbb{F}_q^*$ will be a non-square with probability at least $1/2$.

Here is the complete algorithm. The essential ideas are due to A. Tonelli, who published an algorithm to compute square roots modulo p in 1891.

TONELLI(a)
{ computes $b = \sqrt{a}$ in $\mathbb{F}_q^*$, q odd. }
choose $g \in \mathbb{F}_q^*$ at random.
if g is a square, fail.
let $q - 1 = 2^s t$, t odd.
$e \leftarrow 0$.
for $i \leftarrow 2$ to s do
        if $(ag^{-e})^{(q-1)/2^i} \neq 1$ then $e \leftarrow 2^{i-1} + e$.
$h \leftarrow ag^{-e}$.
$b \leftarrow g^{e/2} h^{(t+1)/2}$.
return (b).

We can use this method to compute square roots modulo primes. For example, suppose we want a square root of 10 modulo 13. Then if $q = 13$, we have $q - 1 = 2^2 \cdot 3$, so $t = 3$ and $s = 2$. Now 5 is a quadratic nonresidue modulo 13, and because $10^3 \equiv -1 \pmod{13}$, we have $e = 2$, and $h = 10 \cdot 5^{-2} \equiv 3$. So $b = 5 \cdot 3^2 \equiv 6$ is a square root of 10.

THEOREM 7.1.3 The randomized algorithm TONELLI has the following properties, when q is an odd prime power. It fails with probability $1/2$. If it does not fail, then it returns a square root of a, provided that a is a square in $\mathbb{F}_q^*$. Its expected running time is $O(\nu_2(q-1)(\lg q)^3)$ bit operations, which is $O((\lg q)^4)$.

Proof By Theorem 6.7.2, we can test if g is a square using $O((\lg q)^2)$ bit operations. A random choice of g will succeed exactly half of the time, because $\mathbb{F}_q^*$ is cyclic. Given that g is not a square, when the loop is entered for the i-th time, we have $ag^{-e} \in G_{s-i}$. This is enough to show that after the loop, $ag^{-e} \in H$; the correctness of the rest of the algorithm follows from Theorem 7.1.1. The running time bound holds because $s = \nu_2(q-1)$, and each power calculation in $\mathbb{F}_q$ takes $O((\lg q)^3)$ bit operations. ■

This algorithm gives us the power to quickly solve any quadratic equation over $\mathbb{F}_q$ when q is odd, by the usual quadratic formula. In complexity-theoretic language, we may therefore state that quadratic equations over $\mathbb{F}_q$ can be solved in random polynomial time by a Las Vegas algorithm.

7.2 Square Roots: Field-Theoretic Methods

We now present a different approach to computing square roots in $\mathbb{F}_q$, and give a probabilistic method, with an expected running time of $O((\lg q)^3)$ bit operations, that can be used for any odd q. This method is due to M. Cipolla; it is based on the theory of finite fields, and on the following observation in particular. Let a be an element of $\mathbb{F}_q$ whose square root we want. Suppose we find a quadratic extension of $\mathbb{F}_q$, and an element x of this extension field whose norm $\mathrm{N}(x)$ equals a. Then, because $\mathrm{N}(x) = x^{q+1}$, we may take $x^{(q+1)/2}$ as a square root of a. Here is the algorithm.

CIPOLLA(a)
choose $t \in \mathbb{F}_q$ at random.
if $t^2 - 4a$ is a square, fail.
$f \leftarrow X^2 - tX + a$.
$b \leftarrow X^{(q+1)/2} \bmod f$.
return (b).

To show this method works, we need two lemmas, in which it is assumed that q is odd, $a \neq 0$, and $\chi(x) = x^{(q-1)/2}$.

LEMMA 7.2.1 If t is chosen at random from $\mathbb{F}_q$, then $\chi(t^2 - 4a) = -1$ with probability $(q-1)/2q$.

Proof We count the polynomials $X^2 - tX + a$ that factor over $\mathbb{F}_q$. These are just the distinct polynomials $(X - \alpha)(X - \beta)$ with $\alpha, \beta \in \mathbb{F}_q$ and $\alpha\beta = a$. Once α is chosen, β is fixed; for every α except the two square roots of a, $\beta \neq \alpha$. Thus the number of polynomials that split is $(q-3)/2 + 2$, which is $(q+1)/2$. Therefore $(q-1)/2$ such polynomials don't split, from which the result follows. ■

LEMMA 7.2.2 If $\chi(t^2 - 4a) = -1$, then $\mathbb{F}_q[X]/(X^2 - tX + a)$ is a field of order q^2, in which X has norm a.

Proof Use the quadratic formula. ■

THEOREM 7.2.3 The randomized algorithm CIPOLLA has the following properties, assuming that q is an odd prime power and $a \neq 0$. It fails with probability $1/2 + 1/2q$. Otherwise, if a is a square in $\mathbb{F}_q^*$, it returns a square root of a. In bit operations, its expected running time is $O((\lg q)^3)$.

Proof We estimate the running time as follows. First, the quadratic character on $\mathbb{F}_q$ can be computed with $O((\lg q)^2)$ bit operations; this is negligible compared to the rest of the algorithm. The rest of the algorithm does a power calculation, and can be analyzed using Theorem 6.4.4.

For the correctness, we observe that $(x^{(q+1)/2})^2 = x^{q+1} = \mathrm{N}(x) = a$. Because a has only two square roots (in any field extension), the root found by the algorithm must lie in $\mathbb{F}_q$. ■

Now that we have two methods for computing square roots in $\mathbb{F}_q$, it is natural to wonder which is best. We discuss this briefly here.

Our first observation is that Cipolla's algorithm is oblivious to the structure of $\mathbb{F}_q^*$, whereas the running time of Tonelli's method is sensitive to this structure. For most q, 2 will not divide $q - 1$ a great number of times, and for such q, Tonelli's algorithm only requires a few power calculations in the simple structure $\mathbb{F}_q^*$. However, nothing prevents $q - 1$ from being highly divisible by 2, and for this reason, the $O((\lg q)^4)$ expected time bound of Tonelli's algorithm can be shown to be tight. (See Exercise 2.) For such "bad" values of q, Cipolla's method is preferable.

In using Tonelli's algorithm to compute many square roots in $\mathbb{F}_q$, the nonsquare g need only be found once; in contrast, Cipolla's method needs a different nonsquare each time it is used. However, this is not a great disadvantage. On the average, two trials should suffice to find a nonsquare, and by Theorem 6.7.2 any candidate nonsquare can be tested using $O((\lg q)^2)$ bit operations. For both methods, the time needed for this test is negligible compared to the total running time.

There are also other methods besides the two we have presented. For example, one may use a polynomial factorization algorithm to find the roots of $X^2 - a$ in $\mathbb{F}_q$. We will give such factorization methods in Section 7.4; for quadratic polynomials, their running times are comparable to that of Cipolla's algorithm. Finally, one can reduce square root computation in $\mathbb{F}_q$ to the solution of a quadratic equation in $\mathbb{F}_p$; if this is done, the time to compute square roots in $\mathbb{F}_q$ becomes $O((n + \lg p)(\lg p)^2)$. (See Exercise 34.)

The reader may wonder if randomization can be eliminated, perhaps by a more complicated procedure that still runs in polynomial time. We do not yet know if this is possible, although it would follow from the ERH. (See Theorem 7.8.4.) It can also be shown that the use of nonsquares is essential, in a certain sense. (See Exercise 5.) Finally, we note that R. Schoof has given a deterministic algorithm to compute the square root of a mod p that runs in polynomial time when a is fixed and p varies. (See Chapter 13.)

The proof of Corollary 7.1.2 gives an algorithm that finds square roots in $\mathbb{F}_q$, when $q = 2^n$. We now sketch another algorithm based on field theory, which may be preferable if several square roots are required. In any finite field of characteristic 2, the Frobenius map $\tau(x) = x^2$ is a linear transformation, whose matrix T relative to a basis of $\mathbb{F}_q/\mathbb{F}_2$ can be found and then inverted. This will take $O((\lg q)^3)$ bit operations. Once T^{-1} is known, multiplication by this matrix gives the square root of any element in $O((\lg q)^2)$ operations.

Here is an example. Take $\mathbb{F}_2[X]/(X^3 + X + 1)$ as a model of $\mathbb{F}_8$. Then $(aX^2 + bX + c)^2 = (a + b)X^2 + aX + c$. If we agree that the polynomial $aX^2 + bX + c$ is represented by the column vector $[a\, b\, c]^T$, then

$$T = \begin{bmatrix} 1 & 1 & 0 \\ 1 & 0 & 0 \\ 0 & 0 & 1 \end{bmatrix}; \qquad T^{-1} = \begin{bmatrix} 0 & 1 & 0 \\ 1 & 1 & 0 \\ 0 & 0 & 1 \end{bmatrix}.$$

Then if v represents $X + 1$, we have

$$T^{-1}v = \begin{bmatrix} 0 & 1 & 0 \\ 1 & 1 & 0 \\ 0 & 0 & 1 \end{bmatrix} \begin{bmatrix} 0 \\ 1 \\ 1 \end{bmatrix} = \begin{bmatrix} 1 \\ 1 \\ 1 \end{bmatrix},$$

so that $\sqrt{X + 1} = X^2 + X + 1$.

We note that a similar technique can be used to solve a quadratic equation over $\mathbb{F}_{2^n}$, in which case the usual quadratic formula involves a division by 2 and cannot be used. (See Exercise 10.)

7.3 Computing d-th Roots

The algorithms of the two previous sections can be extended to solve binomial equations $x^d = a$, although the methods are efficient only in certain cases. We indicate such extensions here.

To find d-th roots when d is relatively prime to $q-1$, we use an extension of Theorem 7.1.1.

THEOREM 7.3.1 Let G be a group of order m, and let d be relatively prime to m. Let $a \in G$. Then the equation $x^d = a$ has the unique solution $x = a^e$, where $de \equiv 1 \pmod{m}$. Consequently, if d is relatively prime to $q-1$, and satisfies $0 < d < q-1$, the equation $x^d = a$ can be solved in $\mathbb{F}_q$ using $O((\lg q)^3)$ bit operations.

Proof Left to the reader. ∎

Now let r be a prime divisor of $q-1$. L. Adleman, K. Manders, and G. Miller presented a generalization of Tonelli's algorithm to compute r-th roots in $\mathbb{F}_q$, which we give here. For variety's sake we present this using slightly different algebraic ideas than we used for Tonelli's algorithm.

If m is a divisor of $q-1$, let C_m denote the unique subgroup of order m contained in $\mathbb{F}_q^*$. (This exists because $\mathbb{F}_q^*$ is cyclic.) Let $q-1 = r^s t$, where r does not divide t. Because $\gcd(r^s, t) = 1$, we have the following isomorphism:

$$\mathbb{F}_q^* \cong C_{r^s} \times C_t.$$

Thus we may represent any element x of $\mathbb{F}_q^*$ as a pair (x_r, x_t), where $x_r \in C_{r^s}$ and $x_t \in C_t$. The transformations relating an element to its associated pair are given by $x \longmapsto (x^t, x^{r^s})$ and $(x_r, x_t) \longmapsto x_r^\alpha x_t^\beta$, where $\alpha t + \beta r^s = 1$.

The idea of the algorithm is to compute an r-th root by computing it in each direct factor separately. In C_t we use Theorem 7.3.1 and in C_{r^s} we use a process similar to Tonelli's algorithm, based on the chain of subgroups

$$1 = C_1 \subset C_r \subset C_{r^2} \subset \cdots \subset C_{r^s},$$

where C_i/C_{i-1} is a group of order r. This process requires an element of C_{r^s} that is not an r-th power, which we find with a randomized step.

AMM(a, r)
{ a is an r-th power in $\mathbb{F}_q^*$ }
choose $h \in \mathbb{F}_q^*$ at random.
if $h^{(q-1)/r} = 1$, fail.

let $q - 1 = r^s t$, where $r \not| t$.
$(a_r, a_t) \leftarrow (a^t, a^{r^s})$.
$g \leftarrow h^t$.
$e \leftarrow 0$.
for $i \leftarrow 0$ to $s - 1$ do
    select e_i, $0 \le e_i < r$, so that $(g^{e+e_i r^i} \cdot a_r)^{r^{s-i-1}} = 1$.
    $e \leftarrow e + e_i r^i$.
$r' \leftarrow r^{-1} \bmod t$.
$(b_r, b_t) \leftarrow (g^{-e/r}, a_t^{r'})$.
choose α, β so that $\alpha t + \beta r^s = 1$.
$b \leftarrow b_r^\alpha b_t^\beta$.
return (b)

THEOREM 7.3.2 With r a prime divisor of $q - 1$ and $a \in (\mathbb{F}_q^*)^r$, the randomized algorithm AMM has the following properties. It fails with probability $1/r$, and otherwise returns an r-th root of a. Its expected running time is $O(r\nu_r(q - 1)(\lg q)^3)$ bit operations, which is $O(r(\lg q)^4)$.

Proof Left to the reader. ∎

When $r \mid q - 1$, the equation $x^r = a$ will have not just one solution in $\mathbb{F}_q$, but several. If $r = 2$, there are two solutions, which differ by a sign. In general, however, the solutions are obtained by multiplying one solution by powers of a primitive r-th root of unity. It all solutions are desired, it is convenient to find this root of unity along with the extraction of r-th roots, as follows. If r is a prime dividing $q - 1$, and h is not an r-th power, as found by a successful run of the algorithm AMM, then $\zeta = g^{(q-1)/r}$ is a primitive r-th root of unity in $\mathbb{F}_q$. We may therefore output $\{b, \zeta b, \ldots, \zeta^{r-1} b\}$ as the complete solution set. The running time bound for finding all solutions is the same as in Theorem 7.3.2.

The time-consuming part of the algorithm AMM is solving an equation of the form $g^x = a$, where a belongs to a group of order r and g generates that group. This is an instance of the *discrete logarithm* problem, which we will discuss more fully in Chapter 12. Using specialized methods for this problem, the running time bounds in Theorem 7.3.2 can be reduced. For example, the "baby step-giant step" algorithm, due to D. Shanks, will solve the discrete logarithm problem in a group of order r using $O(\sqrt{r})$ group operations, and a comparable amount of storage. Thereby, the bound of Theorem 7.3.2 can be reduced to $O(\sqrt{r}\nu_r(q - 1)(\lg q)^3)$ bit operations. Another possibility is to find a generator g for $\mathbb{F}_q^*$, and find an integer y for which $g^y = a$. (This is a discrete logarithm problem in $\mathbb{F}_q^*$.) Then, a solution to $x^r = a$ is given by $x = g^z$, where $zr \equiv y \pmod{q - 1}$. If q is prime, there is a randomized algorithm for discrete logarithms in $\mathbb{F}_q^*$ using $e^{O(\sqrt{(\lg q)(\lg\lg q)})}$ bit operations. This leads to a randomized algorithm to solve $x^r = a$ with the same complexity.

Therefore, if q is prime, we should only use the algorithm AMM when r is small. As a rough guide, we need to have $\lg r = O(\sqrt{(\lg q)(\lg\lg q)})$. When r is really large, of course, trial and error is the best method. (See Exercise 11.)

Cipolla's algorithm can also be extended to compute r-th roots, but we leave the details of this to the reader. (See Exercise 9.)

The method of the last section to compute square roots in $\mathbb{F}_{2^n}$ can be extended to compute p-th roots in any finite field of characteristic p. For future reference we give the running time bound as a theorem.

THEOREM 7.3.3 Let $\mathbb{F}_q$ be a finite field of characteristic p, where $q = p^n$. There is an algorithm to compute the p-th root of any element in $\mathbb{F}_q$ with $O((n + \lg p)(\lg q)^2)$ bit operations, and any additional p-th root with $O((\lg q)^2)$ operations.

Proof Left to the reader. ■

We now explain how the ideas of this section can be combined to solve $x^d = a$ in $\mathbb{F}_q$, when d is arbitrary. First, we may assume that $0 < d < q - 1$, for if not, we reduce d modulo $q - 1$. Then, we find a factorization $d = rs$, where every prime factor of r divides $q - 1$, and $\gcd(s, q - 1) = 1$. (To do this, find a gcd-free basis for $\{d, q - 1\}$, as indicated in Chapter 4.) Let $r_1^{\mu_1} \cdots r_k^{\mu_k}$ be the prime factorization of r. Because s is prime to $q - 1$, we can solve $x_0^s = a$. Using a process similar to Tonelli's algorithm, we can then successively solve

$$\begin{aligned} x_1^{r_1^{\mu_1}} &= x_0 \\ x_2^{r_2^{\mu_2}} &= x_1 \\ &\vdots \\ x_k^{r_k^{\mu_k}} &= x_{k-1}. \end{aligned}$$

If $x = x_k$, then $x^d = a$ as desired. We note, however, that this method requires a partial factorization of $q - 1$; although easy to obtain when the r_i are small, this is not easy to get in general. (See Chapter 11 for a discussion of factorization.) It can be shown that this method uses $O(\gcd(d, q - 1)(\lg q)^3)$ bit operations. (See Exercise 12.)

Whether or not the methods of this section run in polynomial time depends on how the input is represented; we discuss the issue briefly here. In this book, we have adopted the convention that a polynomial is represented by explicitly listing its coefficients, and we could also agree that the above algorithm runs in polynomial time, because it finds a root of

$X^d - a$, in time bounded by a polynomial in d and $\lg q$. However, it is equally reasonable to argue that a binomial equation should be specified by its degree d and constant term a. Then the input is expressible with $O(\lg d + \lg q)$ bits, and we could not say that this algorithm runs in polynomial time, unless each r_i is small compared to q.

In computer algebra, the two representations mentioned above are called the *dense* and *sparse* representations of $X^d - a$, respectively. In discussing running times for polynomial root finding, one needs to specify whether the inputs are given in dense or sparse form. Our convention is to always assume dense representations, and the reader should bear this in mind in considering polynomial factoring, to which we now turn.

7.4 Polynomial Factoring Algorithms

All of the problems discussed so far in this chapter are special cases of the following one: given a finite field $\mathbb{F}_q$ and a polynomial $f \in \mathbb{F}_q[X]$, find a nontrivial factor of f. In this section we discuss general methods to do this.

We recall some algebraic results from Chapter 6. Let the factorization of f into irreducible polynomials be $f_1^{e_1} \cdots f_r^{e_r}$. Then

$$R = k[X]/(f) \cong k[X]/(f_1^{e_1}) \bigoplus \cdots \bigoplus k[X]/(f_r^{e_r}).$$

When $a \in R$, we will indicate the effect of this isomorphism by writing

$$a = (a_1, \ldots, a_r),$$

and call a_i, $i = 1, \ldots, r$, the *components* of a. We emphasize that the components of a are, in general, not available to an algorithm. Nevertheless, they give us an "invisible data structure" that allows clear expressions of proofs and algorithm descriptions.

The key idea in all polynomial factorization algorithms is to find a polynomial with both zero and non-zero components. For example, if f has two different irreducible factors f_1 and f_2 and $a = (0, 1)$, then $\gcd(a, f)$ is a nontrivial factor of f, because it is divisible by f_1 but not f_2.

A second important idea is the following one. Because R has characteristic p, the p-th power and q-th power maps are linear transformations, which may be represented by appropriate matrices relative to a basis of R. We will use this both to design algorithms and to speed up computations.

Two subrings of R play an important role in polynomial factoring algorithms; they are

$$\mathcal{A} = \{a \in R : a^p = a\}; \tag{7.1}$$

$$\mathcal{B} = \{a \in R : a^q = a\}.$$

Then $\mathcal{A} \subset \mathcal{B} \subset R$; we call $\mathcal{B}$ and $\mathcal{A}$ the *Berlekamp algebra* and *absolute Berlekamp algebra* of R respectively.

In most of the techniques we will describe below, one has a choice of using either $\mathcal{A}$ or $\mathcal{B}$. Our approach here is to present the conceptually simpler "relative" versions of the algorithms, which use $\mathcal{B}$, and leave the "absolute" methods to the exercises. If q is large, however, the absolute methods will be faster. (See Exercise 17.)

The following result characterizes the elements of $\mathcal{B}$.

THEOREM 7.4.1 If $a \in k[X]/(f)$, then $a \in \mathcal{B}$ iff $a_i \in k$, $i = 1, \dots, r$.

Proof If a has all its components in $\mathbb{F}_q$, then

$$a^q = (a_1, \dots, a_r)^q = (a_1^q, \dots, a_r^q) = (a_1, \dots, a_r) = a.$$

To prove the converse, it is enough to assume that $r = 1$, that is, g is irreducible and $f = g^e$. Let d be the degree of g, and choose i so that $q^i \geq e$. Then write

$$a = b + cg$$

where b is a polynomial of degree less than d. If $a^q = a$, then

$$a^q \equiv a \pmod{g},$$

which forces $b \in k$. Then,

$$a = a^{q^i} = (b + cg)^{q^i} = b,$$

so $a \in k$. ■

We now give the main ideas of a polynomial factoring algorithm due to E. Berlekamp. Let $\tau(x) = x^q$ be the Frobenius map on $k[X]/(f)$. Then, because τ leaves k fixed, τ is a k-linear transformation on R. The kernel of $\tau - 1$ is $\mathcal{B}$; we can compute a basis for this using linear algebra. Theorem 7.4.1 implies that if the kernel has dimension greater than 1, it must contain an element a that does not lie in k, with the component form $(a_1, \dots, a_r)$ and (say) $a_1 \neq a_2$. We do not know a_1, but subtracting some element of k from a will make the first component vanish but not the second. This element can then be used to split f.

BERLEKAMP(f)
find a linear transformation T for which $Ta = a^q$ for all $a \in R$.
choose a in the kernel of $T - I$, $a \notin k$.
for each $\alpha \in k$ do
    $g \leftarrow \gcd(a - \alpha, f)$.
    if $0 < \deg g < \deg f$ then return (g).

As an example of this method, we factor $f(X) = X^4 + 1$ over $\mathbb{F}_5$. First we compute the powers of X^5 modulo f, and find that $X^5 = X \cdot (X^4 + 1) - X \equiv -X$ modulo f, so that $X^{10} = (X^5)^2 \equiv X^2$ and $X^{15} \equiv -X^3$. Let T be the matrix for the Frobenius map relative to the basis $\{1, X, X^2, X^3\}$ of $\mathbb{F}_5[X]/(f)$; then we have

$$T = \begin{bmatrix} 1 & 0 & 0 & 0 \\ 0 & -1 & 0 & 0 \\ 0 & 0 & 1 & 0 \\ 0 & 0 & 0 & -1 \end{bmatrix}; \qquad T - I = \begin{bmatrix} 0 & 0 & 0 & 0 \\ 0 & -2 & 0 & 0 \\ 0 & 0 & 0 & 0 \\ 0 & 0 & 0 & -2 \end{bmatrix}.$$

The kernel of $T - I$ is all the vectors of the form $[t\ 0\ u\ 0]^T$, which represent polynomials of the form $t + uX^2$. Let $a = X^2$; then taking the elements of $\mathbb{F}_5$ in order, we find that X^2 and $X^2 + 1$ are relatively prime to $X^4 + 1$, but

$$\gcd(X^2 + 2, X^4 + 1) = X^2 + 2.$$

Therefore $X^4 + 1 = (X^2 + 2)(X^2 - 2)$; because 2 and -2 are quadratic nonresidues modulo 5, this factorization is complete.

To analyze the running time of Berlekamp's algorithm, we need some remarks on linear algebra.

THEOREM 7.4.2 Let $f \in k[X]$ be a monic polynomial of degree d. Then relative to the basis $\{1, X, \ldots, X^{d-1}\}$, a matrix for the Frobenius map $a \longmapsto a^q$ on R can be found using $O((d + \lg q)(\lg f)^2)$ bit operations.

Proof It is required to compute the q-th powers of the basis elements; if the polynomials are represented as column vectors, these powers form the columns of the matrix we seek. To find the powers, we compute $X^q \bmod f$, then $(X^q)^2, (X^q)^3, \ldots, (X^q)^{d-1}$. This requires one power calculation and $d - 1$ multiplications modulo f. The time bound follows from Theorems 6.4.3 and 6.4.4. ■

THEOREM 7.4.3 Let M be a $d \times d$ matrix of entries from $\mathbb{F}_q$, of rank r, and consider the linear transformation $v \longmapsto Mv$. A basis for the kernel of this transformation can be found using $O(rd^2(\lg q)^2)$ bit operations.

Proof Define an *elementary row operation* to be one of the following: a) replace a row by a nonzero constant multiple; b) swap two rows; c) add a multiple of one row to another. Each of these operations preserves the solution set of $Mv = 0$. We use Gaussian elimination and find a sequence of elementary row operations that transforms M into a matrix M' with the following properties: i) the first nonzero entry in any row is a 1; ii) a column containing the first nonzero entry of some row has all other entries zero (call such columns *basic*); iii) the rows are sorted according to the number of leading zeroes they contain.

Vectors v with $M'v = 0$ can be easily obtained in the following way. The values corresponding to non-basic columns may take arbitrary values; once these are determined, the other values are fixed. ■

For example, let

$$M = \begin{bmatrix} 1 & 2 & 3 \\ 4 & 5 & 6 \\ 7 & 8 & 9 \end{bmatrix}.$$

After elementary row operations we get

$$M' = \begin{bmatrix} 1 & 0 & -1 \\ 0 & 1 & 2 \\ 0 & 0 & 0 \end{bmatrix}.$$

The general solution to $Mv = 0$ is therefore obtained by solving

$$\begin{bmatrix} 1 & 0 & -1 \\ 0 & 1 & 2 \\ 0 & 0 & 0 \end{bmatrix} \begin{bmatrix} v_1 \\ v_2 \\ v_3 \end{bmatrix} = 0$$

by back-substitution; we get $v_1 - v_3 = v_2 + 2v_3 = 0$. The solution space of $Mv = 0$ is therefore generated by $v_1 = 1$, $v_2 = -2$, $v_3 = 1$.

THEOREM 7.4.4 Let $R = k[X]/(f)$, and let $\mathcal{B}$ denote the Berlekamp algebra of R, considered as a k-vector space. A basis for $\mathcal{B}$ can be found using $O((\mathrm{dg}\, f + \lg q)(\lg f)^2)$ bit operations.

Proof By Theorem 7.4.1, $\mathcal{B}$ is the kernel of $T - I$, where T is the matrix that represents the Frobenius map on $k[X]/(f)$. The time bound then follows from Theorems 7.4.2 and 7.4.3. ■

THEOREM 7.4.5 If f has at least two distinct monic irreducible factors, the algorithm BERLEKAMP outputs a nontrivial factor of f. Its uses $O((\mathrm{dg}\, f + q)(\lg f)^2)$ bit operations.

Proof Let $f_1^{e_1} \cdots f_r^{e_r}$ be the factorization of f. The algorithm will compute an a for which $\tau(a) = a$; by Theorem 7.4.1, $a = (a_1, \ldots, a_r)$, where $a_i \in k$, $r \geq 2$, and not all of the a_i's are the same. Choose the indices so that $a_1 \neq a_2$. Then for $\alpha = a_1$, the element $a - \alpha$ has the components

$$(a_1, a_2, \ldots, a_r) - (a_1, a_1, \ldots, a_1) = (0, a_2 - a_1, \ldots, a_r - a_1).$$

Then $a - \alpha \equiv 0 \pmod{f_1^{e_1}}$, and $a - \alpha \not\equiv 0 \pmod{f_2^{e_2}}$. Then $f_1^{e_1} \mid g$, $f_2^{e_2} \nmid g$, so that g is a nontrivial divisor of f.

The running time follows from Theorem 7.4.3. ■

As this bound uses q rather than $\lg q$, Berlekamp's algorithm does not run in polynomial time. However, we can assert that it is a polynomial time algorithm, if k is fixed and the input polynomial varies. This case is of practical importance. For example, applications in coding theory often use $k = \mathbb{F}_2$. Berlekamp's algorithm can also be modified so that it searches through $\mathbb{F}_p$ rather than k; this is indicated in Exercise 17.

To handle the case where the characteristic of k is large, we present a modification of Berlekamp's algorithm, due to D. Cantor and H. Zassenhaus. The ideas behind the modification are the following. Suppose we know a nonzero element a of the Berlekamp algebra $\mathcal{B}$, with the (unknown) components $(a_1, \ldots, a_r)$. We may as well assume that a is a unit mod f, for otherwise $\gcd(a, f)$ splits f. Then because each a_i is in k, $a^{(q-1)/2} = (\pm 1, \ldots, \pm 1)$. Intuitively, it seems likely that not all the signs will agree; if this is the case, $\gcd(a^{(q-1)/2} - 1, f)$ will split f.

Of course, it is possible that the signs will all be the same. But we can make this unlikely by choosing a *random* element of $\mathcal{B}$. We do this by taking a random linear combination of the elements that form a basis for $\mathcal{B}$. We give the details below.

CZ(f)
find a linear transformation T for which $Ta = a^q$ for all $a \in R$.
let $(b_1, \ldots, b_r)$ be a basis for the kernel of $T - I$.
choose $x_1, \ldots, x_r \in k$ at random.
$a \leftarrow \sum x_i b_i$.
$g \leftarrow \gcd(a, f)$.
if $0 < \deg g < \deg f$ then return (g).
$s \leftarrow a^{(q-1)/2}$.
$g \leftarrow \gcd(s - 1, f)$.
if $0 < \deg g < \deg f$ then return (g).

THEOREM 7.4.6 Let q be odd, and let $f \in \mathbb{F}_q[X]$ have r distinct monic irreducible factors. The randomized algorithm CZ fails to return a nontrivial factor of f with probability at most $1/2^{r-1}$. In bit operations, its expected running time is $O((\mathrm{dg}\, f + \lg q)(\lg f)^2)$.

Proof The element a constructed by the algorithm is either a unit in R or it is not. We show that in either case, the conditional probability of failure is at most $1/2^{r-1}$.

If a is a unit, then the algorithm fails precisely when s has components $(1, 1, \ldots, 1)$ or $(-1, -1, \ldots, -1)$. But given that a is a unit, s is a random element of the group

$(\pm 1, \pm 1, \ldots, \pm 1)$, which has 2^r elements. (See Exercise 18.) Therefore in this case the failure probability is $1/2^{r-1}$.

To handle the other case, we first note that because q is odd,

$$2^{r-1} + (q-1)^r \le (q-1)^{r-1} + (q-1)^r < q^r.$$

If a is not a unit, then only one choice of a is bad, namely $a = 0$. Therefore in this case the failure probability is

$$\frac{1}{q^r - (q-1)^r} < \frac{1}{2^{r-1}},$$

by the inequality above.

The running time follows from Theorem 7.4.3. ■

The Cantor-Zassenhaus algorithm will not work in fields of characteristic 2. It can be shown, however, that a similar procedure will factor polynomials in $k[X]$ when $k = \mathbb{F}_{2^n}$, with the same running time and error probability bounds. (See Exercise 19.)

It should be noted that neither of the algorithms presented in this section can factor an input of the form g^e, where g is irreducible. For this reason we would like to be able to quickly recognize polynomials of this form, or, even better, rearrange our polynomial factoring algorithms so that only legitimate inputs are presented to them. This can be done using the partial factorization procedures of the next section; these procedures are sufficiently fast that the running times given in Theorems 7.4.5 and 7.4.6 suffice to find a nontrivial factor of any polynomial that is not irreducible.

In complexity-theoretic terms, the extensions mentioned above show that there is a randomized polynomial-time procedure to factor polynomials over finite fields. Clearly, this also holds for finding the complete factorization of a polynomial, since there is a deterministic polynomial time test for irreducibility. (See Theorem 7.6.2.) We can therefore state that the complete factorization of a polynomial over a finite field can be found with a Las Vegas random polynomial time algorithm.

It is unknown if there is a deterministic polynomial-time procedure for this problem, even if the GRH is assumed. However, such results are known for special classes of polynomials and finite fields; we discuss these in Section 7.8.

7.5 Other Results on Polynomial Factoring

In this section we collect some other results on polynomial factoring, and give two useful algorithms to partially factor polynomials in $k[X]$. One of the procedures decomposes the input into products of distinct monic polynomials, and the other finds factors all of whose irreducible divisors have the same degree.

One may ensure good inputs for the polynomial factoring algorithms of the last section by finding divisors in which no irreducible polynomial appears more than once. We call such polynomials *squarefree*. The primary tool for decomposition into squarefree polynomials is the *derivative* of a polynomial

$$f = \sum_{0 \le i \le d} a_i X^i,$$

which is

$$f' = \sum_{0 \le i \le d} i a_i X^{i-1}. \tag{7.2}$$

Although this is defined by a familiar rule from calculus, the definition is purely algebraic; nothing resembling a limit process is intended. The reader may recall from calculus, however, that to remove double zeroes from f, one divides f by $\gcd(f, f')$. Something like this can be done in characteristic p, but the naive algorithm will not work without modifications.

The problem is that f' may be zero for a nonconstant polynomial f. However, inspection of (7.2) shows that this occurs only if the exponent of every power of X occurring in f with a nonzero coefficient is a multiple of p. In this case, we can easily factor f using the identity

$$a_0 + a_1 X^p + \cdots + a_d X^{pd} = (a_0^{1/p} + a_1^{1/p} X + \cdots + a_d^{1/p} X^d)^p$$

(which holds in characteristic p), and compute p-th roots of the coefficients a_i as indicated in the proof of Theorem 7.3.3. Also, if the factorization of f contains g^e with $e > 1$, and $p \not| e$, then $\gcd(f, f')$ is a nontrivial factor of f.

We give a polynomial-time algorithm for squarefree decomposition, using the ideas in the above paragraph in the SQUAREFREE algorithm given below.

SQUAREFREE(f)
{returns a list of pairs (f_i, e_i) with f_i squarefree, $\prod f_i^{e_i} = f$.}
$g \leftarrow \gcd(f, f')$
if $f' = 0$ then
  find h so that $h^p = f$
  $(h_1, e_1), \ldots, (h_s, e_s) \leftarrow$ SQUAREFREE (h)
  return ($(h_1, pe_1), \ldots, (h_s, pe_s)$)
else if $g = 1$ then
  return ($(f, 1)$)
else
  return (SQUAREFREE(g), SQUAREFREE(f/g))

THEOREM 7.5.1 If f is monic and $\deg f > 0$, the algorithm SQUAREFREE outputs a factorization $f = f_1^{e_1} \cdots f_r^{e_r}$ where each f_i is monic and squarefree.

Proof Because the degree of f' is less than that of f, exactly one of the following three conditions must hold: $f' = 0$, $\gcd(f, f') = 1$, or $\gcd(f, f')$ is a nontrivial divisor of f. If $f' = 0$, then f must be a pth power of some other polynomial h; we are done by induction. If f and f' are relatively prime, no nonconstant polynomial can divide f twice (a short computation shows this), so the output is correct. The result for the other case follows by induction, since $g \cdot (f/g) = f$. ■

THEOREM 7.5.2 Let $q = p^n$, and $k = \mathbb{F}_q$. Let f have degree $d \geq 1$. The algorithm SQUAREFREE uses $O((\mathrm{dg}\, f)(\lg f)^2)$ bit operations.

Proof Let $T(d)$ be the number of field operations (including p-th root computations) used for inputs of degree d. Computing g uses $O(d^2)$ field operations, and if h is required, it can be found with $O(d)$ operations. (Here the cost of dividing each exponent by p can be neglected.) Since $p \geq 2$, we have the estimate

$$T(d) \leq \max_{1 \leq i \leq d-1} \{T(d-i) + T(i)\} + O(d^2),$$

with $T(1) = 1$. By induction, it follows that $T(d) = O(d^3)$.

To get the result, it will suffice to show that p-th roots can be computed in k with $O((\lg q)^2)$ bit operations, after a precomputation taking time $O(d(\lg f)^2)$.

If it is necessary to take p-th roots at all, then $\lg p \leq p \leq d$. By applying Theorem 7.4.2 to the defining polynomial of k, we see that a matrix for the Frobenius automorphism on k can be found using $O((n + d)(\lg q)^2)$ bit operations, and the same bound suffices for inverting this matrix by Gaussian elimination. This is within the $O(d(\lg f)^2)$ time bound. Given a matrix for the inverse of the Frobenius map, p-th roots can be computed in k with $O((\lg q)^2)$ bit operations. (Compare Theorem 7.3.3.) ■

This algorithm can be improved, to run in nearly quadratic time. In particular, if k is fixed or $k = \mathbb{F}_p$, then the modified algorithm uses $O((\lg f)^2)$ bit operations. This is a result of D. Yun. (See Exercise 27.)

It is also possible to split a polynomial into the product of its linear factors, its degree 2 factors, and so on. This can be done in deterministic polynomial time, using the result that $\prod_{a \in \mathbb{F}_q}(X - a) = X^q - X$. (See Exercise 28.) The basic idea behind this algorithm is that if f is squarefree, then $\gcd(X^q - X, f(X))$ is the product of the degree 1 factors of f. If we divide f by this product, we may continue with $\gcd(X^{q^2} - X, f(X))$, $\gcd(X^{q^3} - X, f(X))$, and so on. To save time, we use the Frobenius map τ to compute q-th powers modulo f; we give the details below.

DISTINCT DEGREE(f)
$g \leftarrow X$
for $i \leftarrow 1$ to $\deg f$ do
    $g \leftarrow \tau(g) \bmod f$
    $f_i \leftarrow \gcd(g - X, f)$
    output(f_i)
    $f \leftarrow f/f_i$

THEOREM 7.5.3 Let f be a squarefree polynomial of degree d. Then the algorithm DISTINCT DEGREE outputs a list $f_1, \ldots, f_d$, where f_i is the product of all the monic degree i irreducible factors of f. This algorithm uses $O((\mathrm{dg}\, f + \lg q)(\lg f)^2)$ bit operations.

Proof By induction, $g = X^{q^i}$ at the i-th iteration of the loop; the correctness follows from this and the above remarks. The running time is dominated by the time needed to find a matrix for τ, which by Theorem 7.4.2 is $O((\mathrm{dg}\, f + \lg q)(\lg f)^2)$. ■

The distinct degree factorization of f is complete when f has at most one irreducible factor of each degree, that is, if $\deg f_i$ is 0 or i. For such polynomials, the algorithm gives a complete factorization rather quickly.

How numerous are these polynomials? Or, put another way, what is the chance that the algorithm DISTINCT DEGREE completely factors a random degree n monic polynomial in $\mathbb{F}_q[X]$? D. H. Lehmer proved that when n is fixed and $q \longrightarrow \infty$, the probability has a limit that is close to 56% for large n. (See Exercise 29.) When q is fixed and $n \longrightarrow \infty$, the probability tends to a limit depending on q. (For $q = 2$ the limit is about 40%.) This is a recent result of A. Knopfmacher and R. Warlimont; using their results, it is possible to show that DISTINCT DEGREE completely factors a fraction $\Omega(1)$ of the degree n polynomials in $\mathbb{F}_q[X]$.

7.6 Synthesis of Finite Fields

In Chapter 6, we proved that finite fields of every prime power order exist, without indicating any method for constructing them. In this section we give an algorithm to do this, or what is the same thing, to find an irreducible polynomial over $\mathbb{F}_p$ of specified degree. We also show how isomorphisms may be found between two different models of the same finite field. We first discuss irreducibility testing for polynomials, a topic of independent interest.

LEMMA 7.6.1 A monic polynomial f in $k[X]$ is squarefree iff $\gcd(f, f') = 1$.

Proof Left to the reader. ■

THEOREM 7.6.2 A monic polynomial f in $k[X]$ may be tested for irreducibility with $O((\mathrm{dg}\, f + \lg q)(\lg f)^2)$ bit operations.

Proof In light of the previous lemma, we may use the following criterion: f is irreducible over k iff $\gcd(f, f') = 1$ and the Berlekamp algebra of $k[X]/(f)$ has dimension 1 as a vector space over k. The gcd computation takes $O((\lg f)^2)$ steps, and the rank may be computed as in the proof of Theorem 7.4.4, in time $O((\mathrm{dg}\, f + \lg q)(\lg f)^2)$. ■

According to Theorem 6.5.1, approximately $1/d$ of all degree d monic polynomials over $\mathbb{F}_q$ are irreducible; here we will use the result that $\Omega(q^d/d)$ of them have this property. The following algorithm for finding an irreducible polynomial suggests itself: choose random monic polynomials and test them for irreducibility, and stop as soon as one is found that cannot be factored. Using the bound given in Theorem 7.6.2, we get the following theorem.

THEOREM 7.6.3 The expected number of bit operations to find an irreducible polynomial f of degree d in $k[X]$ is $O((d^2 + \lg f)(\lg f)^2)$.

Proof Let E denote the expected number of polynomials examined; by conditioning on whether the first polynomial is irreducible, we have $E = 1 + (1 - \Omega(1/d))E$. This implies that $E = O(d)$, and the result follows from Theorem 7.6.2. ■

COROLLARY 7.6.4 A model of $\mathbb{F}_q$ (specified by an irreducible monic polynomial in $\mathbb{F}_p[X]$) can be constructed in $O((\lg q)^4)$ expected time.

We now discuss the problem of finding isomorphisms between two finite fields of the same size. If k and k' are fields, then an *isomorphism* between k and k' is a pair of homomorphisms (φ, ψ), for which $\varphi : k \to k'$, $\psi : k' \to k$, and both $\varphi\psi$ and $\psi\varphi$ are identity maps. It follows from this that φ and ψ are 1-1 and onto.

We need a convention about how isomorphisms are represented. If the fields are given by standard models, say $k = \mathbb{F}_p(\alpha)$ and $k' = \mathbb{F}_p(\beta)$, then to specify φ it suffices to give $\varphi(\alpha)$ as a polynomial in β; a similar statement holds for ψ. It is convenient, however, to view k and k' as vector spaces over $\mathbb{F}_p$, and use matrices to represent the functions φ and ψ. With this convention, we have the following theorem.

THEOREM 7.6.5 Let $k = \mathbb{F}_p[X]/(f)$ and $k' = \mathbb{F}_p[Y]/(f')$ be two models of $\mathbb{F}_q$. Then an isomorphism between them can be found by arandomized algorithm that uses $O(n^2(\lg q)^3)$ expected bit operations. Once the isomorphism is found, it can be evaluated (in either direction) in time $O((\lg q)^2)$.

Proof Here is the algorithm. The polynomial f splits completely over k'; we find a linear factor, $X - g(Y)$, and let $\varphi(X) = g(Y)$. Then $\varphi(X^i) = g(Y)^i \bmod f'$ for $i = 0, \ldots, n-1$, which provides the matrix for φ. Inverting this matrix gives a matrix for ψ.

To analyze the running time, recall that if $g \in \mathbb{F}_q[X]$ is squarefree we can split it using $O((\deg g + \lg q)(\lg g)^2)$ expected bit operations. (This is Theorem 7.4.6.) To find a linear factor of f, we split f, then split the smaller degree factor, and continue until a degree 1 factor is found. The i-th factor found using this process will be a polynomial over k' of degree at most $n/2^i$, so the work for all factorizations will be at most a constant times

$$\sum_{i \geq 0} (n + \lg q) \left(\frac{n \lg q}{2^i} \right)^2 \leq 2n^2(n + \lg q)(\lg q)^2 = O(n^2 (\lg q)^3).$$

The other computations can be done within this time bound. ■

We note that for small values of n, the algorithm can be simplified somewhat. For example, there is a deterministic polynomial-time algorithm for this problem if $n = 2$. (See Exercise 33.)

7.7 Hensel's Lemma

Although the integers modulo p^e do not form a field, many of the results on polynomial factoring modulo p can be extended to polynomials over the ring $\mathbb{Z}/(p^e)$. The principal technique for this is an idea usually ascribed to Hensel.

We first explain how the technique works for finding zeroes of polynomials modulo p^e. Let $f(X)$ be a polynomial, of which we know a zero a_0 modulo p; we assume that a_0 is a simple zero, or, what is the same thing, that $f'(a_0) \not\equiv 0$ modulo p. Recall Taylor's formula for polynomials:

$$f(x + y) = f(x) + f'(x)y + f''(x)/2 \cdot y^2 + \cdots.$$

(In characteristic p we give it the following interpretation: any denominator is to be cancelled by a numerator arising from differentiation.) By this result, if $a_0 + a_1 p + a_2 p^2 + \cdots$ is a root of f modulo p^e, then

$$f(a_0 + a_1 p) \equiv f(a_0) + f'(a_0) a_1 p \equiv 0 \pmod{p^2}.$$

We can solve this for a_1, and find that

$$a_1 = -\frac{f(a_0)/p}{f'(a_0)} \bmod p;$$

the division in the numerator is legitimate because $f'(a_0) \not\equiv 0$ modulo p. This process can be repeated, to find a_2, a_3, and so on.

Nothing in the above algebraic development requires that p be prime; it is only necessary that $f'(a_0)$ be a unit modulo p. However, if $f'(a_0)$ is a unit mod p, then it is invertible modulo any power of p, and this will not change if we add a multiple of p to it. We can therefore use the same technique to go from a root of f modulo p^2 to a root modulo p^4, then to a root modulo p^8, and so on. We give the details below.

```
HENSEL-ZERO(f, a, p, n)
{ input: a simple zero of f modulo p }
{ output: a zero of f modulo p^n, n a power of 2 }
if n = 1 then return a
else
        x0 ← HENSEL-ZERO (f, a, p, n/2)
        x1 ← − ([f(x0)/p^(n/2)] mod p^n) / f'(x0)  mod p^(n/2)
        return (x0 + x1 p^(n/2))
```

THEOREM 7.7.1 The algorithm HENSEL-ZERO finds a zero of f modulo p^n, and uses $O((\operatorname{dg} f)n^2(\lg p)^2)$ bit operations.

Proof Since the correctness follows from the discussion above, we only need to discuss the running time. If we evaluate f and f' by Horner's rule, then the work at the top level is $O((\operatorname{dg} f)n^2(\lg p)^2)$. Thus the total cost is bounded by a constant times

$$(\operatorname{dg} f)(\lg p)^2\left[n^2+\left(\frac{n}{2}\right)^2+\left(\frac{n}{4}\right)^2+\left(\frac{n}{8}\right)^2+\cdots\right]\leq 2n^2(\operatorname{dg} f)(\lg p)^2. \qquad \blacksquare$$

In some situations the requirement that a_0 be a simple zero of f is overly restrictive. The algorithm will still work provided that $f(a_0)$ vanishes modulo a sufficiently high power of p (depending on f). For this reason, we can find a solution to $x^2 \equiv a \pmod{2^n}$ using $O(n^2)$ bit operations. (See Exercise 38.)

Hensel's method may also be used to find a factorization for a polynomial f mod p^n, assuming that we have information about the factorization mod p. When f is a monic polynomial in $\mathbb{Z}/(p^n)[X]$, it is sufficient to know coprime monic polynomials in $g, h \in \mathbb{F}_p[X]$ for which $f \equiv gh \pmod{p}$. We give an algorithm using these data below.

```
HENSEL-SPLIT(f, g, h, p, n)
{ input: monic polynomials f, g, h with g, h coprime and f ≡ gh (mod p) }
{ output: monic polynomials g, h such that f ≡ gh (mod p^n) }
find λ, μ ∈ F_p[X] such that
```

```
        λg + μh = 1, deg λ < deg h, deg μ < deg g
for i ← 2 to n do
{ find factorization mod p^i }
        q ← ((f − gh)/p^(i−1)) mod p
        u ← qμ mod g
        v ← qλ mod h
        g ← g + p^(i−1) u
        h ← h + p^(i−1) v
return(g,h)
```

THEOREM 7.7.2 The algorithm HENSEL-SPLIT factors f in $\mathbb{Z}/(p^n)[X]$, using $O(n(\lg f)^2)$ bit operations.

Proof We prove correctness by induction on the number of loop iterations. At the top of the loop, we can assume that $f \equiv gh \pmod{p^{i-1}}$ and that the leading coefficients of g and h are 1. Therefore, the division by p^{i-1} in the formation of q is legitimate, and, because the leading coefficient of $f - gh$ vanishes, $\deg q < \deg f$. In $\mathbb{F}_p[X]$, $uh + vg$ is a polynomial of degree less than $\deg f$, and congruent to q mod g and mod h. The polynomials g and h are relatively prime, so by the Chinese remainder theorem, $uh + vg = q$. Therefore,

$$
\begin{aligned}
(g + p^{i-1}u)(h + p^{i-1}v) &\equiv gh + p^{i-1}(uh + vg) \pmod{p^i} \\
&\equiv gh + p^{i-1}q \\
&\equiv f.
\end{aligned}
$$

Evidently these new factors are also monic.

We now analyze the running time. By Theorem 6.2.4, the polynomials λ and μ can be found using $O((\lg g)(\lg h))$ steps. The computations in step i of the loop involve polynomials of degree no more than $\deg f$, with coefficients from $\mathbb{Z}/(p^i)$ or $\mathbb{Z}/(p)$. Therefore, they can be done with $O((\mathrm{dg}\, f)^2 i^2 (\lg p)^2)$ bit operations. Summing over $i \leq n$ yields the result. ■

The above algorithm obtains the factorization of f modulo $p, p^2, p^3, p^4, \ldots, p^n$. This is in contrast to the root-finding algorithm HENSEL-ZERO, which squares the modulus at each stage. As $n \to \infty$, HENSEL-SPLIT requires $O(n^3)$ time to obtain a factorization mod p^n, whereas HENSEL-ROOT uses $O(n^2)$. (Of course, the implied constants depend on p and on the polynomial f.) It is possible to modify HENSEL-SPLIT so that it squares the modulus at each step, but we leave this refinement to the reader. (See Exercise 40.)

In solving a congruence modulo higher and higher powers of p, it is useful to think of the roots modulo prime powers as better and better approximations to some "true" solution.

The system of p-adic numbers allows one to express this intuition in a rigorous and useful fashion. We give a sketch of this system here, but caution the reader that a complete treatment is beyond the scope of this book.

The basic idea is to consider two numbers to be "close" if they agree modulo a high power of p, and extend the number system in the same way that the real numbers are an extension of the rationals. A *p-adic integer* is an infinite expression of the form

$$\cdots + a_3 p^3 + a_2 p^2 + a_1 p + a_0,$$

where $0 \leq a_i < p$; one defines *p-adic numbers* similarly by allowing negative powers of p. Because the normal arithmetic algorithms proceed from right to left, one may define addition, subtraction, and multiplication of such numbers in a straightforward manner. (Division is tricky; see Exercise 42.) In these terms, the algorithm HENSEL produces a sequence of solutions to $f(x) = 0$ that are valid modulo higher and higher powers of p, and converge to a p-adic solution. For example, 3 is a square root of 2 (mod 7), $10 = 7 + 3$ is a root (mod 49), $108 = 2 \cdot 7^2 + 1 \cdot 7 + 3$ is a root (mod 343), and so on. We may therefore assert that as 7-adic numbers,

$$\cdots + 6 \cdot 7^3 + 2 \cdot 7^2 + 1 \cdot 7 + 3 = \sqrt{2}.$$

From this viewpoint, the algorithm HENSEL-ZERO is really just Newton's method for finding better and better approximations to a root of f. For, if x_i is an approximate root of f, a better approximation is given by

$$x_{i+1} = x_i - \frac{f(x_i)}{f'(x_i)},$$

and the quotient is what the algorithm computes. Similarly, the algorithm to factor f modulo p^e may be thought of as computing approximate factors of f over the p-adic integers. All such methods are based on solving linear equations, which are derived from first-order Taylor expansions.

The principal technical advantage of p-adic numbers over congruences is that the p-adic numbers form a field of characteristic zero, with a topology that allows one to use continuity arguments. For this reason, it is sometimes convenient to prove results in this number system, even if one is primarily interested in congruences modulo p^e.

It is an interesting open question as to whether polynomials over the p-adic numbers can be factored in polynomial time. The principal difficulty is in getting started, that is, in finding factors that can be used as inputs for Hensel's lemma. In most situations, it suffices to factor the polynomial mod p, but there are cases where factorization modulo higher powers of p are required. (See Exercise 41.) The best known methods involve a search time that is polynomial in p, but not $\lg p$.

7.8 Complexity-Theoretic Results

Although polynomial factoring can be easily done in practice with randomized algorithms, it is still not known whether there are deterministic polynomial time algorithms for the problem. In this section we discuss results that bear on this problem.

The following theorem, which is of independent interest, shows that polynomial factoring can be reduced in deterministic polynomial time to factoring polynomials over the prime field, of a special type: those that split completely over the prime field.

THEOREM 7.8.1 Let q be a prime power p^n, $f \in \mathbb{F}_q[X]$, and assume that f has at least two irreducible factors. Then there are polynomials $g, h \in \mathbb{F}_q[X]$ with the following properties:

1. h splits completely in $\mathbb{F}_p[X]$,
2. for any zero a of h, $\gcd(g - a, f)$ is a nontrivial divisor of f,
3. $\deg g, \deg h \leq \deg f$.

These polynomials can be found using $O((\lg p + n + \operatorname{dg} f)(\lg f)^2))$ bit operations.

Proof We first give the construction, leaving the running time analysis for later. Let $\tau(x) = x^p$ be the absolute Frobenius map on $R = k[X]/(f)$. By our hypothesis on f, there is a nonconstant polynomial g in the kernel of $\tau - 1$. Then g has components $(a_1, \ldots, a_r)$, where $a_i \in \mathbb{F}_p$, $i = 1, \ldots, r$. Now let I denote the ideal in $\mathbb{F}_p[X]$ consisting of all polynomials F for which $F(g) \equiv 0 \bmod f$. The ideal I has a monic generator h, and by considering the components of h, we see that that $h = \prod_{a \in \{a_1, \ldots, a_r\}}(X - a)$. (We can find h by finding the least d such that $1, g, \ldots, g^d \bmod f$ are linearly dependent over $\mathbb{F}_p$.) By the Chinese remainder theorem, h has the properties asserted in the theorem.

It is now necessary to check the running time. Using ideas from the proof of Theorem 7.4.2, a matrix T for τ can be found in time $O((\lg p + n + \operatorname{dg} f)(\lg f)^2)$. (See Exercise 15.) We compute the kernel of $\tau - 1$, by applying Gaussian elimination to $T - I$, as indicated in the proof of Theorem 7.4.3. Because the rank of this matrix is at most d, elimination will take $O((\operatorname{dg} f)(\lg f)^2)$ steps. Once g is chosen, we need only consider powers of g up to $\deg f$, and this elimination step also takes $O((\operatorname{dg} f)(\lg f)^2)$ steps. From these three estimates the time bound follows. ■

Various authors have proved theorems suggesting that special classes of polynomials, or polynomials over special finite fields, can be quickly factored without randomization. In the remainder of this section, we discuss results of this type. We will present deterministic methods for solving quadratic and cubic equations in finite fields, and for finding certain roots of unity. We also give a deterministic method to factor polynomials over fields of characteristic p, when $p-1$ has small factors. These methods rely on efficient deterministic constructions for nonresidues and irreducible polynomials, which we will also present.

The algorithms used to prove such results are intricate and of theoretical interest only; the randomized algorithms given earlier suffice for any practical factorization problem. For this reason, we will omit some details from the algorithm descriptions, and not worry about obtaining the most efficient realizations of our algorithms.

The algorithms we will discuss below are only guaranteed to run in polynomial time if suitable generalizations of the Riemann hypothesis are true. These algorithms make use of auxiliary prime numbers with desired properties, and the extra hypotheses are needed to ensure that these primes can be found quickly, by searching through all primes up to a given bound. We will rely on the *extended Riemann hypothesis (ERH)*, which asserts that primes in arithmetic progressions are "well distributed," and the *generalized Riemann hypothesis (GRH)*, a corresponding assertion for algebraic numbers. The reader is referred to Chapter 8 for further discussion of these hypotheses and their consequences.

We observe that polynomials over any finite field of characteristic 2 can be factored in deterministic polynomial time, using Berlekamp's algorithm and Theorem 7.8.1. For this reason, in the remainder of this section, p denotes an odd prime.

THEOREM 7.8.2 Assume ERH. There is a deterministic algorithm to find a quadratic nonresidue mod p using $O((\lg p)^3)$ bit operations. If r is a prime divisor of $p-1$, an r-th power nonresidue mod p can be computed deterministically using $O((\lg p)^5)$ bit operations.

Proof We will use the following consequence of the ERH: if G is a nontrivial subgroup of $(\mathbb{Z}/(p))^*$, the smallest positive integer outside G is $O((\log p)^2)$. (See Theorem 8.5.3.)

To prove the first assertion, let G be the group of quadratic residues mod p. The least quadratic nonresidue must be prime (if not, there is some smaller prime that is also a nonresidue), and bounded by $O((\log p)^2)$. To find it, it suffices to compute the Legendre symbol for each of the $O((\log p)^2/(\lg\lg p))$ primes up to this bound. (Here we have used the prime number theorem.) The running time follows from Theorem 5.9.3.

For $r \neq 2$, the proof is similar, except that one must rely on Euler's criterion. ■

One way to look at this result is to say that certain types of irreducible polynomials in $\mathbb{F}_p[X]$ are easy to construct. A desirable generalization of this would be a deterministic polynomial time algorithm for constructing degree n irreducible polynomials in $\mathbb{F}_p[X]$. This is indeed possible if the GRH is true. We prove this next.

THEOREM 7.8.3 Assume GRH. There is a deterministic polynomial time algorithm to find an irreducible polynomial of degree n over $\mathbb{F}_p$. The algorithm uses $O(n^6(\lg p+\lg n)^4(\lg p)^2)$ bit operations.

Proof Let $\mathbb{F}_p(\alpha_1), \ldots, \mathbb{F}_p(\alpha_k)$ be extensions of $\mathbb{F}_p$, of degrees $n_1, \ldots, n_k$ respectively. If the n_i's are relatively prime, then $\mathbb{F}_p(\alpha_1, \ldots, \alpha_k)$ is generated as an extension of $\mathbb{F}_p$ by $\alpha_1 + \cdots + \alpha_k$. (To see this, use induction to reduce to the case $k = 2$. Then, observe that the degree $[\mathbb{F}_p(\alpha_1, \alpha_2) : \mathbb{F}_p(\alpha_1 + \alpha_2)]$ divides both n_1 and n_2, and is therefore 1.)

Using this result, we can reduce the problem to that of finding irreducible polynomials of prime-power degrees. Let $n = r_1^{e_1} \cdots r_k^{e_k}$ be the prime factorization of n, and f_i be an irreducible monic polynomial of degree $r_i^{e_i}$. Working in $\mathbb{F}_p[X_1, \ldots, X_k]/(f_1(X_1), \ldots, f_k(X_k))$, which is a model of $\mathbb{F}_{p^n}$, we use linear algebra to find the minimal polynomial of $\alpha = X_1 + \cdots + X_n$. Explicitly, this means that we must solve the linear equation

$$\alpha^n = \sum_{0 \le i < n} a_i \alpha^i$$

for $a_0, \ldots, a_{n-1}$. Then, $X^n - a_{n-1}X^{n-1} - \cdots - a_0$ is an irreducible monic polynomial of degree n with coefficients in $\mathbb{F}_p$.

There are now two cases to consider: either $r_i = p$, or $r_i \neq p$.

We construct irreducible polynomials of degree p^e using the following result: if the finite field $\mathbb{F}_q$ has characteristic p, the polynomial $X^p - X - \alpha$ is irreducible over $\mathbb{F}_q$ when $\mathrm{Tr}(\alpha) \neq 0$. (See Exercise 43.) This is done in stages; at the i-th stage, we construct a degree p^i irreducible monic polynomial $g_i \in \mathbb{F}_p[X]$. For the first stage, let $g_1 = X^p - X - 1$. At the i-th stage, $i \ge 2$, we will have a field of degree p^{i-1}, given by an irreducible polynomial g_{i-1} with zero β. Because the trace maps onto $\mathbb{F}_p$, there must be an j with $0 \le j < p^{i-1}$ for which $\mathrm{Tr}(\beta^j) \neq 0$. Choose such a j, and then, working in the field $\mathbb{F}_p(\beta)[X]/(X^p - X - \beta^j)$, find g_i, the minimal polynomial of X over $\mathbb{F}_p$. (Use linear algebra as before.) This will be irreducible, since X has degree p over $\mathbb{F}_p(\beta)$, and $[\mathbb{F}_p(\beta) : \mathbb{F}_p] = p^{i-1}$.

If $r^e \,||\, n$ for a prime $r \neq p$, we directly construct an irreducible polynomial h of degree r^e, using roots of unity, as follows. Choose the least prime $q \equiv 1 \pmod{r^e}$, different from p, with

$$p^{(q-1)/r} \not\equiv 1 \pmod{q}.$$

Assuming GRH, the least such q is $O(r^{2e}(\log p + \log r)^2)$. (Take $p = r$ and $n = p$ in Theorem 8.7.13.) Let ζ be a primitive q-th root of unity (in $\mathbb{C}$), and

$$\eta = \sum_{x \in ((\mathbb{Z}/(q))^*)^{r^e}} \zeta^x.$$

(This is a so-called *Gaussian period.*) Let $\Psi_{r^e}(X)$ be the irreducible monic polynomial in $\mathbb{Q}[X]$ vanishing at η. (This actually has integral coefficients, since the conjugates of η can be obtained by replacing ζ by ζ^p, ζ^{p^2}, and so on.) We choose $h = \Psi_{r^e} \bmod p$, which we can compute as follows. Let G be the subgroup of r^e-th power residues in $(\mathbb{Z}/(q))^*$, whose cosets are $C_0 = G, \ldots, C_{r^e-1}$. Use Euler's criterion to group the elements of $(\mathbb{Z}/(q))^*$ into cosets, and then (working in the ring $\mathbb{F}_p[X, Z]/(Z^{q-1} + \cdots + Z + 1)$) form

$$h(X) = \prod_{0 \le i < r^e} \left(X - \sum_{x \in C_i} Z^x \right).$$

This concludes the algorithm; we now discuss correctness. Except for $\Psi_{r^e}(X) \bmod p$, the polynomials generated by the algorithm are obtained as minimal polynomials of field elements, so they are irreducible. To handle the exceptional case, we argue as follows. Mod p, the q-th cyclotomic polynomial $Z^{q-1} + \cdots + Z + 1$ splits into distinct factors of degree m, where m is the order of p mod q. (See Exercise 6.34.) By our choice of q, we know that m is a multiple of r^e. Therefore, we have

$$A := \mathbb{F}_p[Z]/(Z^q - 1) \cong \mathbb{F}_p \oplus \mathbb{F}_{p^m} \oplus \cdots \oplus \mathbb{F}_{p^m}.$$

Let $\tau(x) = x^p$ denote the Frobenius map, and let

$$E = \sum_{x \in ((\mathbb{Z}/(q))^*)^{r^e}} Z^x.$$

Because multiplication by p cyclically permutes G's cosets, $E, \tau(E), \ldots, \tau^{r^e-1}(E)$ are distinct elements of A. Therefore, in at least one of the components $\mathbb{F}_{p^m}$, the homomorphic image $\bar{E}$ of E will satisfy

$$\bar{E} \ne \tau^{r^{e-1}} \bar{E}.$$

(This component cannot be $\mathbb{F}_p$, since τ is trivial here.) On the other hand, the degree $[\mathbb{F}_p(\bar{E}) : \mathbb{F}_p]$ has to be a divisor of r^e (since $E = \tau^{r^e} E$ in A, the same holds for $\bar{E}$). Therefore, $[\mathbb{F}_p(\bar{E}) : \mathbb{F}_p]$ equals r^e. Since

$$h(\bar{E}) = 0,$$

the polynomial h has to be the minimal polynomial for $\bar{E}$, and is therefore irreducible.

Finally, we must address the running time. We first note that operations in

$$\mathbb{F}_p[X_1, \ldots, X_k]/(f_1(X_1), \ldots, f_k(X_k))$$

can be done using $O(n^2(\lg p)^2))$ bit operations. This is proved in the answer to Exercise 6.21. (The weaker result in Exercise 6.20 would also suffice for our purposes.) We must compute n powers of $X_1 + \cdots + X_n$ and solve an $n \times n$ linear system with coefficients in $\mathbb{F}_p$. Thereby, we obtain f from $f_1, \ldots, f_k$ using $O(n^3(\lg p)^2)$ bit operations.

The i-th stage of the construction for degree p^e uses $O(p^{2i-1}(\lg p)^2)$ bit operations for the traces (by Exercise 6.17). Since there are p^i powers to compute (at a cost of $O((\lg p^i)^2)$

per power), and a $p^i \times p^i$ $\mathbb{F}_p$-linear system to solve, the work to get g_i is $O(p^{3i}(\lg p)^2)$. (Note that $i \leq p^i$.) Therefore, the total work for the p-th power degree construction is

$$\sum_{1 \leq i \leq e} O(p^{3i}(\lg p)^2) = O(p^{3e}(\lg p)^2) = O(n^3(\lg p)^2).$$

(Note that $p^3 + \cdots + p^{3e} \leq 2p^{3e}$.)

The construction for degree r^e ($r \neq p$) involves the following tasks: finding q, grouping $(\mathbb{Z}/(q))^*$ into cosets, and computing h. If we form h by successively combining partial results with linear polynomials, the work for the last task is $O(r^{2e}q^2(\lg p)^2)$. To find q, we use $O((\lg p)(\lg q') + (\lg q')^3)$ bit operations on candidates $q' \leq q$, so the work for this task is $O(q^2(\lg p)^2)$. Finally, grouping $(\mathbb{Z}/(q))^*$ into cosets uses $O(q(\lg q)^3) = O(q^2)$ bit operations. Thus the work for all tasks in the degree r^e construction is $O(r^{2e}q^2(\lg p)^2)$, which is $O(r^{6e}(\lg p + \lg n)^4(\lg p)^2)$ by the bound on q. Some calculus shows that

$$\sum_{r^e \,\|\, n} r^{6e} \leq n^6,$$

when we sum over the primes r dividing n. (See Exercise 44.) Therefore, the algorithm constructs all the degree r^e polynomials using $O(n^6(\lg p + \lg n)^4(\lg p)^2)$ bit operations. ■

We give an example of this construction, using it to find an irreducible polynomial of degree 6 over $\mathbb{F}_3$. Since $6 = 3 \cdot 2$, we must first find irreducible polynomials of degrees 3 and 2. The polynomial

$$g(X) = X^3 - X - 1$$

is irreducible over $\mathbb{F}_3$. (Although our proof used Exercise 43, the irreducibility can be verified directly by observing that h has no zero in $\mathbb{F}_3$.) To construct an irreducible polynomial of degree 2, we first find the auxiliary prime $q = 5$, satisfying $3^{(q-1)/2} \equiv -1 \pmod{q}$. The subgroup of squares in $(\mathbb{Z}/(5))^*$ is $\{\pm 1\}$, so we work with the periods

$$\eta_0 = \zeta + \zeta^4$$

and

$$\eta_1 = \zeta^2 + \zeta^3,$$

where ζ is a primitive 5-th root of unity. We compute $\eta_0 + \eta_1 = -1$, and $\eta_0\eta_1 = -1$, so an irreducible quadratic polynomial in $\mathbb{F}_3[Y]$ is given by

$$h(Y) = Y^2 + Y - 1.$$

(We can check this by computing its discriminant, 2, which is a quadratic nonresidue mod 3.) Now, we combine g and h to get an irreducible sextic polynomial. To do this, we look for a linear relation between powers of $X + Y$ in the ring $\mathbb{F}_3[X, Y]/(g(X), h(Y))$. This leads to the linear equation

$$\begin{bmatrix} 1 & 0 & 1 & 0 & 2 & 2 \\ 0 & 1 & 0 & 1 & 0 & 0 \\ 0 & 0 & 1 & 0 & 1 & 0 \\ 0 & 1 & 2 & 2 & 1 & 1 \\ 0 & 0 & 2 & 0 & 0 & 1 \\ 0 & 0 & 0 & 0 & 0 & 1 \end{bmatrix} \begin{bmatrix} a_0 \\ a_1 \\ a_2 \\ a_3 \\ a_4 \\ a_5 \end{bmatrix} = \begin{bmatrix} 1 \\ 0 \\ 1 \\ 2 \\ 1 \\ 0 \end{bmatrix},$$

in which the i-th column of coefficients expresses $(X + Y)^i$ relative to the basis $\{1, X, X^2, Y, XY, X^2Y\}$. A solution to this is $a_5 = 0$, $a_4 = a_3 = a_2 = 2$, $a_1 = a_0 = 1$. Therefore, the minimal polynomial of $X + Y$ is

$$f(X) = X^6 - \sum_{0 \le i < 6} a_i X^i = X^6 + X^4 + X^3 + X^2 + 2X + 2,$$

and this is irreducible.

The strength of Theorem 7.8.3 is in its generality and not in its running time, which is $n^{6+o(1)}$ for fixed p, and $O((\lg p)^6)$ for fixed n. For this reason, ad hoc methods are usually better for small n. When $n = 2$, for example, an irreducible quadratic polynomial f in $\mathbb{F}_p[X]$ can be found using $O((\lg p)^3)$ bit operations. (Use Theorem 7.8.2 to obtain a quadratic nonresidue a, and let $f = X^2 - a$.) Using an explicit formula for the defining equation of the Gaussian period, an irreducible cubic in $\mathbb{F}_p[X]$ can be found within the same time bound. (See Exercise 46.)

We now discuss deterministic algorithms for solving quadratic and cubic equations in finite fields. These methods combine classic results of algebra with the deterministic techniques we have presented above.

THEOREM 7.8.4 Assume ERH. There is a deterministic algorithm to solve quadratic equations over $\mathbb{F}_q$, using $O((\lg q)^4)$ bit operations.

Proof By Theorem 7.8.1, any quadratic equation over $\mathbb{F}_q$ may be deterministically reduced to a quadratic equation over $\mathbb{F}_p$. This can be solved by the quadratic formula, using the algorithm TONELLI. This requires a quadratic nonresidue mod p; by Theorem 8.5.3, the least one is $O((\lg p)^2)$.

Combining Theorems 6.2.3,7.1.3, and 7.8.1, we see that $O((\lg p + n)(\lg q)^2 + (\lg p)^4)$ bit operations suffice; this is within the stated time bound. ■

THEOREM 7.8.5 Assume GRH. There is a deterministic polynomial time algorithm to compute cube roots in $\mathbb{F}_q$, using $O((\lg q)^4)$ bit operations.

Proof We may as well assume $3 \mid q - 1$, for otherwise we can use Theorem 7.3.1. (Take $G = \mathbb{F}_q^*, m = q-1$, and $d = 3$.) We can also assume $p > 3$. (For $p = 3$, use Theorem 7.3.3.)

Let $a \in (\mathbb{F}_q^*)^3$. We first describe an algorithm for finding a cube root of a.

First, we find a primitive cube root of unity $\zeta \in \mathbb{F}_q$. This is a zero of $X^2 + X + 1$, which can be found by solving a quadratic equation. Let $\mathbb{F}_{p^m}$ be the smallest extension of $\mathbb{F}_p$ containing ζ; we have $m \le 2$, and $m = 1$ iff $3 \mid p - 1$.

Let $R = \mathbb{F}_q[X]/(X^3 - a)$. The ring R has the automorphism σ, which sends X to ζX. If $x \in R$, then we may arrange the components x_i of x so that

$$(x_1, x_2, x_3)^\sigma = (x_2, x_3, x_1).$$

Then σ is an $\mathbb{F}_p$-linear transformation on R.

Find a nonzero solution in R to the simultaneous $\mathbb{F}_p$-linear equations $u^{p^m} = u$ and $u^\sigma = \zeta u$.

A solution to these equations must have components $(t, \zeta t, \zeta^2 t)$, for some nonzero $t \in \mathbb{F}_{p^m}$. Let $b = u^3$; all components of b are equal, so $b \in \mathbb{F}_{p^m}$. If we can factor $T^3 - b$ over this smaller field, then the problem is solved, for a root β of this polynomial must be a component of u, and then $\gcd(u - \beta, X^3 - a)$ splits $X^3 - a$, and provides a cube root of a. The problem is therefore reduced to that of finding cube roots in $\mathbb{F}_{p^m}$, which can be done with the algorithm AMM.

This algorithm requires an element of $\mathbb{F}_{p^m} - (\mathbb{F}_{p^m})^3$.

If $m = 1$, we construct a cubic nonresidue as follows. First, find an irreducible polynomial $f \in \mathbb{F}_p[X]$ of degree 3, and let $K = \mathbb{F}_p(\alpha)$, where α is a zero of f. Consider

$$\gamma = \alpha + \zeta\alpha^p + \zeta^2\alpha^{p^2}.$$

By a direct computation, $\gamma^p = \zeta^{-1}\gamma$, so that $\gamma \notin \mathbb{F}_p$, unless it vanishes. In this case replace α by α^2. On the other hand, $\gamma^3 = \gamma(\zeta\gamma)(\zeta^2\gamma)$, so that

$$(\gamma^3)^p = (\gamma^p)(\zeta\gamma^p)(\zeta^2\gamma^p) = (\zeta^2\gamma^p)(\gamma^p)(\zeta\gamma^p) = \gamma^3.$$

Therefore, $\theta = \gamma^3$ is a cubic nonresidue in $\mathbb{F}_p$. (Compare Theorem 7.8.2.)

If $m = 2$, the construction is similar, except that we must work with an extension of degree 6. This is most easily handled in the following way. Let f be an irreducible cubic as before, with root β. Working in $\mathbb{F}_p(\zeta, \beta)$ (which we can realize as $\mathbb{F}_p[X, Y]/(X^2 + X + 1, f(Y))$, we form $\alpha = \beta + \zeta$ and define

$$\gamma = \alpha + \zeta\alpha^{p^2} + \zeta^2\alpha^{p^4}.$$

As observed in the proof of Theorem 7.8.3, $\mathbb{F}_p(\alpha)$ is a degree 6 extension of $\mathbb{F}_p$. Arguing as before (but this time using the p^2-th power map), $\theta = \gamma^3$, or the corresponding element formed using α^2, is an element of $\mathbb{F}_{p^m} - (\mathbb{F}_{p^m})^3$.

We obtain the running time bound as follows. We can find ζ by solving a quadratic equation, which costs $O((\lg q)^4)$, by Theorem 7.8.4. To handle the linear equations, observe that the solutions in R to $u^\sigma = \zeta u$ are exactly the polynomials of the form bX, with $b \in \mathbb{F}_q$. Furthermore,

$$X^{p^m-1} = (X^3)^{(p^m-1)/3} = a^{(p^m-1)/3}$$

Putting $c = a^{(p^m-1)/3}$, we have $(bX)^{p^m} = bX$ iff b satisfies the $\mathbb{F}_p$-linear equation $b^{p^m} = c^{-1}b$. By Theorem 7.4.2, this can be set up and solved using $O((\lg q)^3)$ bit operations. According to Exercise 46, the cost to find an irreducible cubic is $O((\lg p)^3) = O((\lg q)^3)$. The work for computing θ and applying AMM is within this time bound. ■

We illustrate this with an example. Take $\mathbb{F}_7(\sqrt{3})$ as a model of $\mathbb{F}_{49}$, in which we wish to compute a cube root of $a = 3 - \sqrt{3}$. Let $R = \mathbb{F}_{49}[X]/(X^3 - a)$; The solutions to $v^7 = v$ are the polynomials of the form

$$v = v_1 + v_2(1 - \sqrt{3})X + v_3(5 + \sqrt{3})X^2,$$

with $v_i \in \mathbb{F}_7$. Taking $\zeta = 2$ (a cube root of unity in $\mathbb{F}_7$), we find that if v also satisfies $v^\sigma = \zeta v$, then $v_1 = v_3 = 0$. So let $u = (1 - \sqrt{3})X$; then $u^3 = 6$, and 3 is a cube root of 6 (mod 7). We have

$$u - 3 = (1 - \sqrt{3})X - 3 = (1 - \sqrt{3})(X - (2 + 2\sqrt{3})),$$

and $X - (2 + 2\sqrt{3})$ divides $X^3 - a$, so that $(2 + 2\sqrt{3})^3 = a$.

The next theorem uses the solution by radicals for cubic equations, traditionally ascribed to G. Cardano.

THEOREM 7.8.6 Assume GRH. There is a deterministic polynomial time algorithm to solve cubic equations in $\mathbb{F}_q$, using $O((\lg q)^4)$ bit operations.

Proof Let $f(X) = 0$ be a reducible cubic equation with coefficients in $\mathbb{F}_q$. Using Theorem 7.8.1, we compute an auxiliary polynomial $h(X)$ of degree 2 or 3 that splits completely in $\mathbb{F}_p[X]$ and use its zeroes to split f.

If h has degree 2, we find its zeroes using Theorem 7.8.4, and then split f into linear and irreducible quadratic factors. This yields the unique root of $f(X) = 0$ in $\mathbb{F}_q$.

Therefore, assume that $h(X) = X^3 + aX^2 + bX + c$ has degree 3, with three zeroes in $\mathbb{F}_p$ to be found. We may as well assume $p \neq 3$. (Otherwise, $p = 3$, and the only possible h is $X^3 - X$, with zeroes $\{0, \pm 1\}$.) Making the substitution $X = Y - a/3$, we can bring the equation $h = 0$ into the form

$$\ell(Y) = Y^3 + BY + C = 0.$$

We can assume $B \neq 0$, for otherwise we can use Theorem 7.8.5. Let $\mathbb{F}_{p^m}$ with $m \leq 2$ be the smallest extension of $\mathbb{F}_p$ containing a primitive cube root of unity ζ. We find ζ by solving $\zeta^2 + \zeta + 1 = 0$, when $3 \mid p - 1$, and otherwise work in the extension field $\mathbb{F}_p[Z]/(Z^2 + Z + 1)$.

Choose a root $u \in \mathbb{F}_{p^m}$ of $U^2 + CU - B^3/27 = 0$, then let $v_1^3 = u$, and $v_2 = -B/(3v_1)$. (It is immaterial which cube root of u is used.) The roots y_1, y_2, y_3 of $\ell(Y) = 0$ are then obtained by solving the simultaneous $\mathbb{F}_{p^m}$-linear equations

$$\begin{aligned} y_1 + \quad y_2 + \quad y_3 &= 0, \\ y_1 + \ \zeta y_2 + \zeta^2 y_3 &= 3v_1, \\ y_1 + \zeta^2 y_2 + \ \zeta y_3 &= 3v_2. \end{aligned}$$

We leave the correctness of this to the reader as Exercise 47. The running time bound is a consequence of Theorem 7.8.1, Theorem 7.8.4, and Theorem 7.8.5. ■

We illustrate this result with an example. Let us find the roots of $h(X) = X^3+X^2+20X+2$ mod 31. Substituting $X = Y + 10$ yields $\ell(Y) = Y^3 - Y$, which evidently has distinct zeroes mod 31. To find a cube root of unity, we search for a cubic nonresidue. We cannot use 2, because $2^{10} \equiv 1 \pmod{31}$, but $3^{10} \equiv 25 \pmod{31}$. Therefore, $\zeta = 5$ is a primitive cube root of unity. We have $B = -1$ and $C = 0$, so the auxiliary quadratic equation is

$$U^2 + CU - B^3/27 = U^2 - 8 = 0.$$

We can take $u = 2\sqrt{2} = 16$, $v_1 = 8$, and from this obtain $v_2 = -B/(3v_1) = 22$. Therefore, the zeroes of $\ell(Y)$ are given by

$$\begin{aligned} y_1 + \quad y_2 + \quad y_3 &= 0, \\ y_1 + \ 5y_2 + 25y_3 &= 24, \\ y_1 + 25y_2 + \ 5y_3 &= 4. \end{aligned}$$

These equations have the unique solution $y_1 = -1$, $y_2 = 0$, $y_3 = 1$. Therefore, the zeroes of the original equation $h = 0$ are $\{9, 10, 11\}$.

Continuing this line of investigation, it is possible to obtain deterministic algorithms for solving various other types of equations. We will limit ourselves to one more result of this type, which states that the n-th cyclotomic polynomial $\Phi_n(X)$ (the defining polynomial for primitive n-th roots of unity) can be factored over $\mathbb{F}_p$ in polynomial time, if $p \equiv 1 \pmod{n}$ and the ERH is true.

THEOREM 7.8.7 Assume ERH. Let $p \equiv 1 \pmod{n}$, and let $\Phi_n(X)$ denote the n-th cyclotomic polynomial. There is a deterministic algorithm to find a primitive n-th root of unity in $\mathbb{F}_p^*$, using $O(n(\lg p)^5)$ bit operations. Given such a root of unity, the n-th cyclotomic polynomial $\Phi_n(X)$ can be completely factored in $\mathbb{F}_p[X]$ using $O(n(\lg p)^2)$ bit operations.

Proof We first show how to construct primitive n-th roots of unity. There are two cases: $n = q^e$ (a prime power), and general n. Since $n \leq p$, a complete factorization of n can be found within our time bound, and we will assume this is available.

If $n = q^e$, we use Theorem 7.8.2 to find a q-th power nonresidue a belonging to $\mathbb{F}_p^*$. Then

$$\zeta_q = a^{(p-1)/q^e} \bmod p$$

is a primitive q^e-th root of unity. (Its order must divide q^e, and $\zeta_q^{q^{e-1}} \neq 1$ by Euler's criterion.) This uses $O((\lg p)^5)$ steps.

If n is not a prime power, let its prime factorization be $n = q_1^{e_1} \cdots q_r^{e_r}$. We use the construction above to to get ζ_{q_i}, a root of unity of order q^i, for $i = 1, \ldots, r$. Then

$$\zeta = \prod_{i \leq i \leq r} \zeta_{q_i}$$

is a root of unity of order n. The total work to get ζ is

$$O(n) + r \cdot O((\lg p)^5) = O(n(\lg p)^5).$$

It remains to find the factorization of $\Phi_n(X)$. Since ζ is a primitive n-th root of unity in $\mathbb{F}_p$, we have

$$\Phi_n(X) = \prod_{i \in (\mathbb{Z}/(n))^*} (X - \zeta^i) .$$

(Every such ζ^i is a zero of $\Phi_n(X)$, and there cannot be any more zeroes, since $\Phi_n(X)$ has degree $\varphi(n)$. Within our time bound, we can compute all ζ^i for $1 \leq i < n$ (using $O(n(\lg p)^2)$ bit operations), and output $X - \zeta^i$ whenever $\gcd(i, n) = 1$ (using $O(n(\lg n)^2) = O(n(\lg p)^2)$ operations for the gcd's). ∎

We note that the time to construct one root of unity could be improved by using a sharper time bound for factorization.

It is possible to use similar methods to factor $\Phi_n(X)$ over $\mathbb{F}_p$, even when $p \not\equiv 1 \pmod{n}$. The main idea is to find an extension field $\mathbb{F}_q$ for which $q \equiv 1 \pmod{n}$, and compute a primitive n-th root of unity ζ_n in this extension field. The irreducible factors of Φ_n mod p are products of the conjugates of $X - \zeta_n$. (If $p^e \,\|\, n$, for $e > 0$, we can easily reduce the problem to one of factoring Φ_{n/p^e}; see Exercise 6.34.) The same technique used to prove Theorem 7.8.7 will work, provided we can resolve the obstruction of finding elements of $\mathbb{F}_q^*$ that are not r-th powers. Unfortunately, we know of no elementary way to do this, and we refer the reader to the Notes for a further discussion of this problem.

This concludes our presentation of deterministic algorithms to factor special types of polynomials. We now give a theorem of the opposite type, which shows that polynomials over certain large finite fields may be quickly factored, if the ERH holds.

THEOREM 7.8.8 Assume ERH. For a prime power $q = p^n$, let $S(p)$ denote the largest prime factor of $p - 1$. There is a deterministic algorithm that splits any $f \in \mathbb{F}_q[X]$ using $O((S(p) + \lg p)(\lg f)^4)$ bit operations.

Proof Using Theorem 7.8.1, we may as well assume that $q = p$, and that f is squarefree, with (at least two) distinct roots in $\mathbb{F}_p$. (This reduction can be accomplished using $O((\lg f)^3)$ bit operations.) Within our time bound, we can also obtain the complete factorization of $p - 1$, using, say, trial division. Consider the following algorithm.

Let $R = \mathbb{F}_p[X]/(f)$, and choose a nonconstant $a \in R$ satisfying $a^p = a$. We may as well assume that $a \in R^*$, for otherwise $\gcd(a, f)$ splits f. Find a prime divisor s of $p - 1$ for which $a^{(p-1)/s^\nu} \notin \mathbb{F}_p$, where $\nu = \nu_s(p-1)$. There exist $i < \nu$ for which $b = a^{(p-1)/s^i} \in \mathbb{F}_p$ (for example, $i = 0$); choose the largest such i. Then $c = a^{(p-1)/s^{i+1}}$ is an s-th root of b in R. Consider the isomorphism $R \cong \mathbb{F}_p \oplus \cdots \oplus \mathbb{F}_p$. In component notation, we have

$$c = (c_1, c_2, \ldots, c_r)$$

where $c_1, \ldots, c_r$ are s-th roots of b in $\mathbb{F}_p$; they are not all equal. Let h be the least element of $\mathbb{F}_p^*$ with $h^{(p-1)/s} \not\equiv 1 \pmod p$. Use this, and the algorithm AMM, to find the set T of s-th roots of b in $\mathbb{F}_p^*$. We have

$$T = \{\sqrt[s]{b}, \zeta\sqrt[s]{b}, \ldots, \zeta^{s-1}\sqrt[s]{b}\},$$

where $\zeta = h^{(p-1)/s}$. For at least one such root, say, $\hat{c}$, $\gcd(c - \hat{c}, f)$ splits f.

To show correctness, we must check that a suitable prime divisor s exists. In component notation, we have

$$a = (a_1, a_2, \ldots, a_r),$$

choosing indices so that $a_1 \neq a_2$. If $a^{(p-1)/s^\nu} \in \mathbb{F}_p$, then $(a_1/a_2)^{(p-1)/s^\nu} = 1$, so the order of a_1/a_2 in $\mathbb{F}_p^*$ is relatively prime to s. If this happens for all s, the order must be 1. Then $a_1/a_2 = 1$, that is, $a_1 = a_2$. This contradiction shows that at least one s must work.

We now analyze the running time. Finding a uses $O((\lg f)^3)$ steps, by Theorem 7.4.4. Testing a value of s uses $O((\lg p)(\lg f)^2)$ steps, by Theorem 6.4.3. Since $p - 1$ has $O(\lg p)$ prime divisors, the work to find a suitable s is $O((\lg p)^2(\lg f)^2)$. By Theorem 7.8.2, h can be found within $O((\lg p)^5)$ steps, assuming ERH. Finally, by Theorem 7.3.2, the set T can be found within $O(s(\lg p)^4)$ steps. We get the bound of the theorem by observing that $s \leq S(p)$ and $\lg p \leq \lg f$. ■

We give an example of this method. Let $p = 5$, and $f(X) = X^4 + 1$. The element $a = X^2$ is a nonconstant element of the Berlekamp algebra of $\mathbb{F}_5[X]/(X^4 + 1)$, as can readily be verified. Since $p - 1 = 2^2$, our only choice for s is 2. We find that $a^2 = X^4 \equiv -1 \pmod{f}$, so we take $c = X^2$. The square roots of -1 in $\mathbb{F}_5$ are ± 2, and

$$\gcd(c - 2, f) = \gcd(X^2 - 2, X^4 + 1) = X^2 - 2$$

splits f.

With a little more care in the analysis, the running time bound in Theorem 7.8.8 can be improved, to

$$O((\mathrm{dg}\, f + n + (\lg p)^2)(\lg f)^2 + (S(p) + \lg p)(\lg p)^4).$$

When q is a fixed prime p, we thus obtain an unconditional deterministic algorithm using $O((\mathrm{dg}\, f)^3)$ bit operations. (The ERH is not necessary, since p is fixed.) This is comparable to the running time of Berlekamp's algorithm. We also note that situations where many polynomials need to be factored over the same field, it may be useful to precompute the factorization of $p - 1$ and a set of nonresidues.

7.9 Exercises

1. Show that if $q \equiv 5 \pmod 8$, then in $\mathbb{F}_q$,

$$\sqrt{a} = \begin{cases} a^{(q+3)/8} \\ 2^{-1}(4a)^{(q+3)/8} \end{cases}$$

accordingly as $a^{(q-1)/4} \equiv 1$ or -1.

2. Use Linnik's theorem (see Section 8.5) to show that in Theorem 7.1.3, $(\lg q)^4$ cannot be replaced by any smaller function of q.

3. Here is the original version of Tonelli's algorithm. Prove that it finds a square root of a in $\mathbb{F}_q^*$, given a $b \notin (\mathbb{F}_q^*)^2$.

   Set $e_1 = (q - 1)/2$ and $e_2 = q - 1$. As long as e_1 remains even, divide e_1 and e_2 by 2, adding $(q - 1)/2$ to e_2 whenever $a^{e_1} b^{e_2} = -1$. At the end of the process, $a^{(e_1+1)/2} b^{e_2/2}$ is a square root of a.

4. Let $p \equiv 3 \pmod 4$. Note that a model of $\mathbb{F}_{p^2}$ can be given as $\mathbb{F}_p(i)$, where i is a root of $X^2 + 1 = 0$. Show how to quickly find a number x, $1 \le x < p - 1$, with $(\frac{x}{p}) = +1$ and $(\frac{x+1}{p}) = -1$. Using x, construct a non-square in $\mathbb{F}_{p^2}$. Conclude that square roots in $\mathbb{F}_{p^2}$ can be computed in deterministic polynomial time.

5. Show that the problem of finding an element of $\mathbb{F}_q$ that is not a square can be reduced to finding square roots in $\mathbb{F}_q$, in deterministic polynomial time.

6. Show that if a is not a square, the algorithm CIPOLLA either fails or returns an element of $\mathbb{F}_{q^2} - \mathbb{F}_q$.

7. Suppose we implemented the algorithm CIPOLLA using the sparse polynomial $g(Y) = Y^2 - \Delta$. By how much could we expect to speed up the calculation?

8. Let $q \equiv 1 \pmod 4$, and write $q - 1 = 2^e d$, with d odd. Consider the following randomized algorithm to compute a square root of $a \in \mathbb{F}_q^*$: Choose $t \in \mathbb{F}_q$ at random, and find the least i such that $(X + t)^{2^i d}$ has no constant term, when evaluated mod $X^2 + a$. Then if $(X + t)^{2^{i-1}d} \equiv rX + s$, a square root of a is given by s/r. Fill out the details, and prove that the resulting algorithm uses $O((\lg q)^3)$ bit operations, and has failure probability about $1/2^{e-1}$.

9. Show that a randomly selected monic polynomial in $\mathbb{F}_q[X]$ of degree k with nonzero constant term a is irreducible with probability $\Omega(1/k)$. Using this result, extend Cipolla's algorithm to a method to solve $x^k = a$ when $k \mid q - 1$.

10. Show that quadratic equations over $\mathbb{F}_{2^n}$ can be solved using $O(n^3)$ bit operations.

11. Let $r \mid q - 1$. What is the expected cost of finding a solution to $x^r = a$ in $\mathbb{F}_q^*$ by random search?

12. (a) Let r be a prime divisor of $q - 1$. Show that if $r^\mu \le q - 1$, then a solution to $x^{r^\mu} = a$ in $\mathbb{F}_q$ can be found using $O(r\nu_r(q - 1)(\lg q)^3)$ bit operations.

    (b) Assume $1 \le d \le q - 1$. By considering the algorithm given at the end of Section 7.3, show that a solution to $x^d = a$ in $\mathbb{F}_q$ can be found using $O(\gcd(d, q-1)(\lg q)^3)$ bit operations.

13. When using Berlekamp's algorithm, are you better off extracting all possible factors using nonconstant elements of $\mathcal{B}$, or running the algorithm recursively on each factor you find?

14. Let $\mathcal{A}$ denote the absolute Berlekamp algebra of $k[X]/(f)$. Prove that if $a = (a_1, \ldots, a_r)$, then $a \in \mathcal{A}$ iff $a_i \in \mathbb{F}_p$, $i = 1, \ldots, r$.

15. Let $f \in \mathbb{F}_q[X]$ be a polynomial of degree d. Show that a matrix for the absolute Frobenius map $a \longmapsto a^p \bmod f$ can be computed using $O((d + n + \lg p)(\lg f)^2)$ bit operations.

16. Let f be defined as in Exercise 15. Show how to compute a basis for the absolute Berlekamp algebra $\mathcal{A}$ of $k[X]/(f)$ using $O((n + d + \lg p)(\lg f)^2)$ bit operations.

17. Design and analyze an "absolute" version of Berlekamp's algorithm.

18. Let $\phi : G \to H$ be a homomorphism of finite groups. Show that if ϕ is onto, and x is a randomly chosen element of G, then $\phi(x)$ is a random element of H.

19. If $k = \mathbb{F}_{2^n}$, then $\mathrm{Tr}(x) = x + x^2 + \cdots + x^{2^{n-1}}$ is an $\mathbb{F}_2$-linear map from k onto $\mathbb{F}_2$. Consider the following method to factor a polynomial $f \in k[X]$: choose a random element a of the Berlekamp algebra $\mathcal{B}$ of $k[X]/(f)$, let $b = a + a^2 + \cdots + a^{2^{n-1}}$, and compute $\gcd(b, f)$. Show that it splits f with probability $1 - 1/2^{r-1}$, if f has r distinct irreducible factors. What is its running time?

20. Consider the following algorithm to factor f: first reduce to the case where all factors have degree m, then try to split f with $\gcd(r^{(q^m-1)/2} - 1, f)$, where r is a random element of $k[X]/(f)$. Show that this method splits f with high probability, and uses $O((\lg f)^3)$ bit operations.

21. Consider the following "absolute" version of the Cantor-Zassenhaus algorithm: To factor f, find a basis for the absolute Berlekamp algebra $\mathcal{A}$ of $k[X]/(f)$. Then choose a random $r \in \mathcal{A}$, and try to split f with $\gcd(r^{(p-1)/2} - 1, f)$. Show that this algorithm will split f with high probability, using $O((n + \mathrm{dg}\, f + \lg p)(\lg f)^2)$ bit operations. For which values of n, d, and p would this be preferable to the algorithm CZ?

22. Analyze the following algorithms to find a root of f in k.

    (a) If q is odd, choose $a \in k$ at random. Then try to split f using $\gcd((X + a)^{(q-1)/2} - 1, f)$.

    (b) If $q = 2^n$, let $T(g) = g + g^2 + \cdots + g^{2^{n-1}}$, a linear map from $\mathcal{B}$ onto $\mathcal{A}$. Try to split f using $\gcd(T(aX), f)$, where $a \in k$ is chosen at random.

23. (Niederreiter.) Assume k has characteristic 2. Consider a monic squarefree polynomial in $k[X]$ with the irreducible factorization $f = g_1 \cdots g_m$. Prove the following results.

    (a) The mapping $T(h) = (fh)' + h^2$ is an $\mathbb{F}_2$-linear transformation on $k[X]$.

    (b) The kernel of T comprises the polynomials of the form $h_b = (f/b)b'$, where b is a monic factor of f. (Thus it has dimension m.)

    (c) If $b \neq 1, f$, then $\gcd(h_b, f)$ splits f.

    Use these to give an alternative proof that polynomials in $k[X]$ can be factored in deterministic polynomial time.

24. Give a deterministic polynomial time algorithm to construct a normal basis of $\mathbb{F}_q$. Assume that $q = p^n$, and that a degree n irreducible monic polynomial in $\mathbb{F}_p[X]$ is provided as input.

25. Let f be a monic polynomial of degree n in $\mathbb{F}_q[X]$ with discriminant D and ν irreducible factors, where q is odd. Show that if f has no repeated factors, then $n \equiv \nu \bmod 2$ iff D is a square in $\mathbb{F}_q$. Conclude that the analog of the Möbius function in $k[X]$ can be computed in $O((\lg f)^2)$ steps, if k has odd characteristic.

26. (Continuation.) Let r and p be odd primes. Apply the previous exercise to the factorization of $(X^r - 1)/(X - 1) \bmod p$, and deduce the law of quadratic reciprocity.

27. Let $k = \mathbb{F}_q$, where $q = p^n$. Let $f \in k[X]$ have degree $d \geq 1$. Give a modification of the algorithm SQUAREFREE using $O((\lg f)^2)$ bit operations, assuming a matrix is provided for the linear map $a \longmapsto a^{1/p}$ on $\mathbb{F}_q$. Conclude that a squarefree decomposition of f may be found using $O((1 + n/d^2)(\lg f)^2)$ bit operations.

28. Show that $\prod_{a \in \mathbb{F}_q}(X - a) = X^q - X$.

29. Let D_n be the probability that a random monic polynomial in $\mathbb{F}_q$ of degree n has factors of distinct degrees; let P_n be the probability that such a polynomial is irreducible. Show that

$$1 + D_1 X + D_2 X^2 + \cdots = \prod_{n \geq 1}(1 + P_n X^n)$$

For large q, we will have $P_n \approx 1/n$, so it is reasonable to replace this generating function by

$$\prod_{n \geq 1}\left(1 + \frac{X^n}{n}\right) = 1 + X + \frac{1}{2}X^2 + \frac{5}{6}X^3 + \frac{7}{12}X^4 + \cdots = \sum_{n \geq 0} \tilde{D}_n X^n.$$

Show that $\lim_{n \to \infty} \tilde{D}_n = e^{-\gamma}$.

30. Show that a random monic polynomial in $\mathbb{F}_q[X]$ of degree $d > 1$ is squarefree with probability $1 - 1/q$.

31. Consider the following test to see if f, a polynomial of degree d, is irreducible. Check that $X^{q^d} \equiv X \pmod{f}$, then verify that $\gcd(X^{q^{d/l}} - X, f) = 1$ for each prime divisor l of d. Show that this uses $O((\lg \lg \lg d)(\lg f)^3)$ bit operations.

32. We can use distinct-degree factorization as an irreducibility test, as follows. A degree d polynomial $f \in \mathbb{F}_q[X]$ is irreducible iff

$$\gcd(f, X^{q^i} - X) = 1$$

for $1 \le i \le n/2$. We can find an irreducible polynomial of degree d by choosing random polynomials, and applying the above test. Show that the expected number of bit operations used by this method is $O(d(\lg d)(\lg q)(\lg f)^2)$.

33. Show that an isomorphism between two models of $\mathbb{F}_{p^2}$ can be found using $O((\lg p)^4)$ bit operations. (No randomness is required.)

34. Design a randomized algorithm that computes square roots in $\mathbb{F}_q$, $q = p^n$, using $O((n + \lg p)(\lg p)^2)$ bit operations, on the average.

35. Give a "relative" version of Theorem 7.8.1 that reduces factorization of polynomials in $k[X]$ to root finding over k.

36. Find examples of the following pathologies in $\mathbb{Z}/(p^n)[X]$:

    (a) It is not always true that $\deg fg = \deg f + \deg g$.

    (b) $\mathbb{Z}/(p^n)[X]$ does not have unique factorization.

    (c) It need not be the case that $\gcd(f, g)$ is a linear combination of f and g.

37. Let $f \in \mathbb{Z}[X]$, $a_0 \in \mathbb{Z}$, and assume that $n = \nu_p(f(a_0)) < \infty$. Show that the algorithm HENSEL-ZERO is still correct provided that $n \ge 2\nu_p(f'(a_0)) + 1$.

38. (a) Let $n \ge 3$. Prove that an odd integer a is a square mod 2^n iff $a \equiv 1 \pmod 8$.

    (b) If $x^2 \equiv a \pmod{2^n}$, the square roots of a are $\pm x, \pm x + 2^{n-1}$.

    (c) Show that a square root of $a \pmod{2^n}$ can be computed using $O(n^2)$ bit operations.

39. Let $q = p^n$ with $n \ge 2$, and p odd. Assume a is a p-th power in $(\mathbb{Z}/(p^n))^*$. Show how to solve the congruence $x^p \equiv a \pmod{p^n}$ using $O((n + \lg p)(\lg q)^2)$ bit operations.

40. Modify the algorithm HENSEL-SPLIT so that it obtains factors of f mod $p, p^2, p^4, \ldots$. Conclude that a nontrivial factor of $f \in \mathbb{Z}/(p^n)[X]$ can be found using $O((\lg f)^2)$ bit operations, using the same inputs as HENSEL-SPLIT.

41. Let p, q be odd primes with $(\frac{q}{p}) = +1$, and let

$$f(X) = X^2 - p^2 q.$$

Show that in $\mathbb{F}_p[X]$, f is the square of a linear polynomial, but f has two distinct factors modulo all sufficiently high powers of p.

42. Give algorithms to add, subtract, and multiply p-adic numbers. Show that a p-adic integer x has an inverse iff $x \not\equiv 0 \bmod p$, and use this to design a division algorithm.

43. Assume $\mathbb{F}_q$ has characteristic p, and prove the following result. If $\alpha \in \mathbb{F}_q$, the polynomial

$$X^p - X - \alpha$$

is irreducible in $\mathbb{F}_q[X]$ when $\mathrm{Tr}(\alpha) \neq 0$, and splits into distinct linear factors otherwise.

44. Let $m, n \geq 2$ be integers. Assume that the prime factorization of n is $r_1^{e_1} \cdots r_k^{e_k}$. Show that

$$\sum_{1 \leq i \leq k} r_i^{me_i} \leq n^m.$$

(No prime number theory is required.)

45. Let q be a prime with $q \equiv 1 \pmod{e}$. Let ζ be a primitive q-th root of unity. Let

$$\eta = \sum_{x \in ((\mathbb{Z}/(q))^*)^e} \zeta^x,$$

a Gaussian period. Let $\Psi_e(X)$ be the irreducible monic polynomial in $\mathbb{Q}[X]$ vanishing at η. Prove the following results.

(a) For $e = 2$,

$$\Psi_2(X) = X^2 + X + \frac{1 - (\frac{-1}{q})q}{4}.$$

Thus, Ψ_2 can be computed using $O(\lg q)$ bit operations.

(b) When $e = 3$,

$$\Psi_3(X) = X^3 + X^2 - \frac{q-1}{3}X - \frac{q(L+3)-1}{27},$$

where the integers L and $|M|$ are determined by the requirements $L \equiv 1 \pmod{3}$ and $4q = L^2 + 27M^2$. Via this formula, Ψ_3 can be computed using $O(\sqrt{q}(\lg q)^2)$ bit operations.

46. (Continuation.) Assume GRH. Show that for $n = 2, 3, 6$, an irreducible polynomial of degree n in $\mathbb{F}_p[X]$ can be found using $O((\lg p)^3)$ bit operations.

47. Verify that the solution given to $h(X) = 0$ in the proof of Theorem 7.8.6 is correct.

48. Let

$$f(X) = X^4 + a_3X^3 + a_2X^2 + a_1X + a_0$$

be a monic degree 4 polynomial in $\mathbb{F}_q[X]$. Show that the roots in $\mathbb{F}_q$ of $f(X) = 0$ can be found using $O((\lg q)^4)$ bit operations, assuming GRH.

49. Give a deterministic polynomial time algorithm to decide if a is a square mod f, when $a, f \in \mathbb{F}_q[X]$.

50. Let f denote a monic polynomial in $\mathbb{F}_q[X]$. Show that the following functions can all be computed in deterministic polynomial time:

    (a) $\varphi(f)$, the size of $\mathbb{F}_q[X]/(f)^*$;

    (b) $\omega(f)$, the number of distinct monic irreducible divisors of f;

    (c) $\omega'(f)$, the number of monic irreducible divisors of f;

    (d) $d(f)$, the number of monic polynomials dividing f;

    (e) $\sigma(f)$, the sum of $q^{\deg d}$ over all monic polynomials d dividing f;

    (f) $\lambda(f)$, the exponent of $(\mathbb{F}_q[X]/(f))^*$.

51. [Open] Assume GRH. Give a deterministic algorithm for factoring polynomials in $\mathbb{F}_q[X]$, using $(\lg f)^{O(1)}$ bit operations.

7.10 Notes on Chapter 7

Shparlinski [1992] devoted an entire book to the algorithmic aspects of finite fields. For recent survey articles emphasizing open problems, see H. W. Lenstra [1990]; Lidl [1991]; Niederreiter [1991, 1993b, 1993c]; Shparlinski [1991, 1992].

7.1

Lagrange [1769, §54] gave the explicit formula for square roots modulo p when $p \equiv 3 \pmod 4$.

Gauss [1801] gave several "exclusion" methods to compute square roots modulo p. Although these methods all use $\Omega(\sqrt{p})$ trials, 19th century mathematicians apparently considered them quite practical and even commented on their complexity. (See, for example, H. J. S. Smith [1860, §68].)

Although he did not state his results in our language, Tonelli [1891] gave what is apparently the first random polynomial-time algorithm to compute square roots modulo p. (See also Shanks [1972]; H. Williams [1972]; Adleman, Manders, and Miller [1977]; and Huang [1985, 1991a]).

Lindhurst [1995] showed that Tonelli's algorithm in $\mathbb{F}_p$ uses $O((\lg p)^3)$ bit operations on average, when p and a are chosen at random.

7.2

Our algorithm CIPOLLA is a streamlined version of an algorithm from Cipolla [1903], which computes square roots modulo p. (See also D. H. Lehmer [1969] and Mignotte [1980a].) As an alternative to finite fields, one can implement the algorithm CIPOLLA using Lucas sequences. In our language, such an implementation amounts to using $X^2 - \Delta$ as the defining polynomial of $\mathbb{F}_{q^2}$. (See Exercise 7.) Choosing such a sparse polynomial will speed up the algorithm, but only by a constant factor.

Although this algorithm is still the best we know of—no known method for computing square roots in $\mathbb{F}_p$ uses $o(\lg p)$ field operations—its efficacy was not immediately appreciated. According to Cipolla's editors, his method represented only "a step toward effective calculation," because of the high power it requires. Throughout the 20th century, new exponential time algorithms for the problem were published. See Tamarkine and Friedmann [1906]; Cipolla [1907a]; Cunningham [1908]; Vandiver [1929]; Chang [1960].

The proof of Lemma 7.2.1 was pointed out to us by B. Plumstead.

The idea that one could use the quadratic character to predict success in a randomized square root algorithm appears in Berlekamp [1967]. (See also Itoh [1987, 1989]; Bach [1990b].)

Using elliptic curves, Schoof [1985] proved that the square root of an integer a mod p, should it exist, can be computed using $|a|^{1/2+o(1)}(\lg p)^9$ bit operations. This is polynomial time for fixed a. (The exponent of $\lg p$ can be reduced to $6 + o(1)$ using fast arithmetic.)

Although theoretically convenient, the inverse matrix used to find square roots in $\mathbb{F}_{2^n}$ should probably be replaced in practice by LU factorization. For a discussion of the operation counts motivating this choice, see Strang [1980, §1.5].

7.3

Legendre [1785] gave a general treatment of binomial congruences modulo p, which included the method of Theorem 7.3.1 for the group $\mathbb{F}_p^*$.

Schwarz [1949] gave an explicit formula (which can be evaluated in polynomial time) for the number of degree k irreducible factors of the binomial $X^n - a \in \mathbb{F}_p[X]$, when $\gcd(n, p) = 1$. (See also Schwarz [1946]; Rédei [1950].)

Our algorithm for r-th roots was sketched by Adleman, Manders, and Miller [1977]. See also Shanks [1972]; Elia [1993]. (The latter paper assumes a primitive root is given, when in fact only a nonresidue is necessary.) H. Williams [1972] presented a different algorithm, which does not use the composition series of C_{r^s} and can take exponential time on certain inputs. Improvements to this algorithm (but still not reaching polynomial time) were given by K. Williams and Hardy [1993].

The "baby-step giant-step" algorithm is due to Shanks [1969]. For rigorous subexponential time algorithms for discrete logarithms, see Pomerance [1986].

Bach [1987, 1990b, 1991b] gave some algorithms for computing roots in finite fields that make efficient use of randomness.

7.4

The first general discussion of polynomial factoring over a finite field seems to be that of Legendre [1785]. In this paper, Legendre outlined a method whereby one could find zeroes of a polynomial in

$\mathbb{F}_p[X]$, using two ideas that are key to modern factorization algorithms: using $\gcd(X^{(p-1)/2} \pm 1, f(X))$ to separate the roots of $f = 0$ according to their quadratic character, and replacing $f(X)$ by $f(X + a)$ when this failed to separate the roots. (He also suggested using auxiliary binomial equations to find roots, an idea that seems less useful today.) Unfortunately, Legendre's ideas did not take hold, and were not common knowledge until their rediscovery by modern researchers interested in randomized algorithms.

For the early history of methods to solve binomial and other equations over finite fields, see Dickson [1919]. Many useful techniques were already known to Arwin [1920], but his paper did not have great impact. For earlier reviews of the state of the art, see Zassenhaus [1968]; Berlekamp [1972]; and Mignotte [1976a].

Our proof of Theorem 7.4.1 is due essentially to Zassenhaus [1969].

The deterministic factorization algorithm for $\mathbb{F}_q$ is from Berlekamp [1967]. According to Slisenko [1981] and Grigoriev [1984], similar algorithms were found independently by Skopin and Faddeev in the 1950's. (These seem not to have been published.)

The idea that one should use the matrix of the Frobenius map on $k[X]/(f)$ to obtain information about the factorization of f goes back at least to Petr [1937]. See also Schwarz [1939, 1940, 1946, 1956]; Butler [1954]; Tu [1974]; J. M. Chen and Li [1977]; and Gunji and Arnon [1981].

Beard [1974] proposed factoring f by searching for zero divisors in $k[X]/(f)$ (in general this takes exponential time). Lloyd and Remmers [1967] advocated trial division, aided by factor tables. For (exponentially large) explicit formulas for the roots of f, see Cipolla [1930]; Mann [1974]; Prešić [1970].

Shoup [1990a] showed that the average time required to factor polynomials $f \in \mathbb{F}_p[X]$ (assuming the input is chosen from the uniform distribution on all degree n monic polynomials) is polynomial in n and $\lg p$. Shparlinski [1993a] gave corresponding results for bivariate polynomials in $\mathbb{F}_p[X, Y]$.

For equations of low degree over small finite fields, it is feasible to design methods utilizing table lookup. For such approaches, see Blokh [1964]; Polkinghorn [1966]; Chien, Cunningham, and Oldham [1969].

Our randomized factorization algorithm comes from Cantor and Zassenhaus [1981]; the same idea was presented by Camion [1980], but without the sharp error probability bound. Berlekamp [1970] gave a random polynomial-time algorithm to find the roots of a polynomial modulo p. All of these methods use Legendre's idea of separating roots according to quadratic character. For other variants of this method, see Rabin [1980a]; Ben-Or [1981]; Camion [1982, 1983a, 1983b]; Lazard [1982]; van Oorschot and Vanstone [1990a, 1990b]; Rothstein and Zassenhaus [1994].

For an "additive" version of the algorithm (inspired by the theory of block designs), see Menezes, van Oorschot and Vanstone [1988]; van Oorschot [1988]; van Oorschot and Vanstone [1989]. (This is mainly of theoretical interest, as other methods have the same or lower complexity.)

Huber [1992] gave a deterministic algorithm to find roots of polynomials in $\mathbb{F}_{p^n}[X]$ based on the following idea. Consider the group of transformations on $\mathbb{F}_{p^n}$, generated by the Frobenius automorphism and two other maps (inversion and a variant of the discrete logarithm), then use the orbits of this group to separate roots. Apparently the algorithm uses polynomial time when $p = 2$ (a complete analysis has not yet been done).

The sharpest known estimate on the asymptotic complexity of polynomial factoring is due to Kaltofen and Shoup [1995]. With their algorithm, the expected number of $\mathbb{F}_q$-operations used to factor a degree n polynomial in $\mathbb{F}_q[X]$ is $O(n^{1.815} \lg q)$. This improved a weaker result of von zur Gathen and Shoup [1992a, 1992b].

Trevisan and Wang [1991] compared the performance of several factorization algorithms for polynomials over $\mathbb{F}_p$. They recommended combining the algorithms of Berlekamp [1967], Cantor and Zassenhaus [1981], and Shoup [1990a], using easily computed properties of the input to choose between these methods. For other performance studies of algorithms to factor polynomials over finite fields, see Poli and Gennero [1988]; Menezes, van Oorschot and Vanstone [1988].

Von zur Gathen [1984b] showed that polynomials in $\mathbb{F}_p[X]$ can be factored with a randomized $\mathcal{NC}$ algorithm (here p is fixed).

There are random polynomial-time algorithms to factor polynomials in $\mathbb{F}_q[X_1, \ldots, X_n]$. (Any deterministic algorithm, could, of course, be used when $n = 1$.) For such results, see A. Lenstra [1983]; Grigoriev [1984]; Chistov [1984b]; von zur Gathen and Kaltofen [1985]; Wan [1990]; and Viry [1993].

The *skew-polynomial ring* $\mathbb{F}_q[X, \sigma]$ is a noncommutative polynomial ring in which X is an indeterminate, σ is an automorphism of $\mathbb{F}_q$, and $Xa = a^\sigma X$ for $a \in \mathbb{F}_q$. Giesbrecht [1992a] presented efficient factorization algorithms for this ring.

More references on polynomial factorization algorithms can be found in the answers given for Exercises 19–23.

7.5

It has long been clear that when solving a polynomial equation over a finite field, one should first remove double roots by taking the gcd with its derivative. For example, Galois [1830] stated that "one can always prepare the given congruence $f(X) = 0$ so as to have no equal roots, or, in other words, so as to have no common factor with $f'(X) = 0$, and the way to do this is evidently the same as for ordinary equations."

The distinct degree factorization technique is also quite old; hints of it are contained in Galois [1830], and a full treatment appears in Serret [1885]. See also Gauss [1889, §370–371]; and Golomb, Welch, and Hales [1959].

Agou [1976] gave a distinct-degree factorization algorithm based on symmetric functions, but did not analyze its running time.

Let $D(n, q)$ be the probability that a random degree n polynomial in $\mathbb{F}_q[X]$ has distinct degree factors. D.H. Lehmer [1972] showed that $D(n, q)$ has a limit D_n as $q \to \infty$, and that $\lim_{n\to\infty} D_n = e^{-\gamma}$. Knopfmacher and Warlimont [1995a] showed that for fixed q, $D(n, q)$ has a limit $L(q)$ as $n \to \infty$, and proved that

$$D(n, q) = L(q) + O\left(\frac{1}{n}\right).$$

(The constant implied in this result is absolute.) It is known that $L(2) \doteq 0.3967$, and that $L(q)$ increases monotonically with q, to the limit $e^{-\gamma} \doteq 0.5615$. See also Knopfmacher and Warlimont [1995b].

It is interesting to consider the analogous question for $\mathbb{Z}$. We note that an integer n has prime factors of different sizes iff it is squarefree. According to a result of Gegenbauer [1885], the squarefree integers have density $6/\pi^2$. (See Exercise 9.36.) Therefore, the fraction of k-bit integers with prime factors of different sizes is about $3/\pi^2 \doteq 0.3040$.

Bach and Shoup [1990] showed how to factor polynomials in $\mathbb{F}_q[X]$ using a "small" amount of randomness.

Mignotte [1975, 1976a, 1976b] showed how to determine the "splitting degree" of $f \in \mathbb{F}_q[X]$, that is, the smallest r such that all roots of f lie in $\mathbb{F}_{q^r}$. Because r can be large, his algorithm is exponential. We note that r can be computed in polynomial time, as the least common multiple of the degrees of the irreducible factors of f. (These numbers are easily obtained from the algorithm DISTINCT DEGREE.) An algorithm based on symmetric functions (which can be evaluated quickly by computing powers of the companion matrix of f) appears in Feit and Rees [1978]. (See also Schwarz [1940].)

For results about the distribution of the number of irreducible factors of polynomials over finite fields, see Mignotte [1980c]; Knopfmacher and Knopfmacher [1993]; Knopfmacher, Knopfmacher, and Warlimont [1994]; and references therein.

7.6

The irreducibility criterion used to prove Theorem 7.6.2 is due to Butler [1954]. Earlier, Petr [1937] gave an irreducibility criterion for $k[X]$ that can be quickly checked, but he did not describe it in algorithmic terms. Hellegouarch [1986] gave some finite field analogs of the randomized primality tests of Chapter 9; although theoretically interesting, they are not as efficient as Theorem 7.6.2.

Galois [1830] claimed that irreducible polynomials could be easily selected "par le tâtonnement," that is, by trial and error. Theorem 7.6.3, which analyzes such an algorithm, is from Calmet and Loos [1980]. (See also Rabin [1980a] and Ben-Or [1981].) Shoup [1993a, 1994] found a probabilistic algorithm for generating a degree n irreducible polynomial in $\mathbb{F}_q[X]$ using an expected number of $O((n^2 + n \lg q)(\lg n)(\lg \lg n))$ operations in $\mathbb{F}_q$. (Asymptotically, this is the fastest known algorithm.)

A proof that irreducible polynomials over $\mathbb{F}_q$ can be generated in deterministic polynomial time when the characteristic of q is fixed was given by Chistov [1984]. For proofs, see Semaev [1988]; Shoup [1988, 1990c]. The last paper gives the following complexity bound: a degree n irreducible polynomial in $\mathbb{F}_p[X]$ can be found using $p^{1/2+o(1)}n^{4+o(1)}$ bit operations. For an $\mathcal{NC}$ algorithm to generate irreducible polynomials over $\mathbb{F}_p$ (p fixed), see Frandsen [1991a]. Rifà and Borrell [1991] discussed efficient hardware implementations.

There are many results leading to the construction of irreducible polynomials of special degrees. The history is too involved to review here, and we refer the interested reader to the books of Lidl and

Niederreiter [1983, Chapter 3] and Shparlinski [1992, Section 2.1]. Recent papers in this area include S. Cohen [1992]; and Shparlinski [1993b].

Additional references on constructing irreducible polynomials can be found in the Notes to Section 7.8.

Zierler [1974] gave a deterministic algorithm to find an isomorphism between two models of $\mathbb{F}_{2^n}$; it uses $O(n^5)$ bit operations. H. W. Lenstra [1991a] showed that an isomorphism between any two models of $\mathbb{F}_q$ (given by, say, irreducible polynomials of the same degree in $\mathbb{F}_p[X]$) can be found in deterministic polynomial time. This improved an earlier result of Evdokimov [1989], who solved the problem assuming GRH. See Pinch [1992] for heuristic methods (which do not appear to run in polynomial time).

7.7

Although the algorithms of this section are usually called "Hensel methods," most of the basic ideas predate Hensel.

Special cases of the root-finding algorithm appear in Lagrange [1769, §55] (a linearly convergent iteration for solving $x^2 \equiv a \pmod{p^n}$) and Legendre [1798, §360] (a quadratically convergent iteration for $x^d \equiv a \pmod{p^n}$). Cauchy [1847] showed how to lift roots of congruences modulo p to congruences modulo p^n, and Gauss's *Nachlass* [1889] contains an algorithm to lift factorizations of relatively prime polynomials. See also Demeczky [1891].

Hensel [1904] used these ideas to show that there is an algorithm to compute the factorization of a polynomial over the p-adic numbers. (See also Hensel [1908].)

The algorithms of Sections 7.1 and 7.2 can be extended to find square roots modulo p^e; see Tonelli [1892, 1893] and Cipolla [1904a].

Hensel [1904] invented p-adic numbers. For a delightful introduction to this number system, see the book of Koblitz [1977]. Lewis [1969] surveyed their applications to the study of diophantine equations. For applications of p-adic numbers to algebraic number theory, see e.g., Lang [1986].

In the p-adic number system, carries propagate in the direction of least significance. For this reason, roundoff errors do not accumulate. This is in striking contrast to the behavior of ordinary arithmetic, and several authors have proposed that p-adic numbers be used to obtain accurate results in numerical computations. For this approach, see Hehner and Horspool [1979]; Gregory [1980]; Gregory and Krishnamurthy [1984]; J. Morrison [1988].

Zeugmann [1989] discussed parallel p-adic arithmetic.

Chistov [1987, 1991] gave a deterministic algorithm to find the p-adic factors of a polynomial in $f \in \mathbb{Z}[X]$. His algorithm takes time bounded by a polynomial in the length of f (degree times the length of the largest coefficient) and p. It is unknown if the dependence on p can be reduced to polynomial in $\lg p$, even if randomization or the ERH is used. The difficult case is when p divides the discriminant of f, for otherwise, f is squarefree mod p, and the algorithms of this chapter suffice to find its p-adic factors to any desired precision.

7.8

A deterministic polynomial-time reduction of polynomial factorization to root finding was given by Zassenhaus [1969]; we have given his method in an "absolute" version. For another reduction, see Berlekamp [1970].

Adleman and H. W. Lenstra [1986] and Evdokimov [1989] proved, independently, that irreducible polynomials of degree n over $\mathbb{F}_p$ can be generated in polynomial time under GRH. (Neither paper gave any running time estimate more precise than $(n \lg p)^{O(1)}$.) The idea of using Gaussian periods to get irreducible polynomials appeared already in the factoring algorithm of Bach and Shallit [1985, 1989]. For some heuristic ideas on deterministic irreducible polynomial generation, see von zur Gathen [1986a]. For deterministic polynomial time algorithms to construct irreducible polynomials in $\mathbb{F}_p[X]$ of degree close to n, see Adleman and H. W. Lenstra [1986]; Shparlinski [1993b].

In forming the defining equation of the Gaussian periods, it is possible to avoid roots of unity (which may have high degree) and work with the periods directly. See Bach and Shallit [1985, 1989]. Mathews [1892, §178] suggested a different way to obtain the defining polynomial of a Gaussian period η, which amounts to the following: find the characteristic polynomial of the matrix that represents multiplication by η relative to the basis consisting of η's conjugates.

Shoup [1988, 1989a, 1990c] showed that construction of irreducible polynomials in $\mathbb{F}_p[X]$ reduces in deterministic polynomial time to factorization in $\mathbb{F}_p[X]$.

For the checkered history of Cardano's formula (which was published but not invented by him), we refer to Kramer [1981, p. 96]. Cauchy [1829] used this formula to find the roots of cubic equations that split completely over $\mathbb{F}_p$. He gave a similar algorithm for quartic polynomials. (See Exercise 48.)

Huang [1984a, 1991b] proved Theorem 7.8.7, but did not estimate the running time more precisely than $(n \lg p)^{O(1)}$. Our use of Euler's criterion follows von zur Gathen [1987a].

Huang [1985, 1991a] gave an algorithm which, when given primes r and p, constructs a model of $\mathbb{F}_{p^m}$, where m is order of r mod p, together with an element outside $(\mathbb{F}_{p^m}^*)^r$. The algorithm uses algebraic number theory, and runs in polynomial time under GRH. (See also Evdokimov [1989].) Buchmann and Shoup [1991] showed how to find r-th nonresidues in $\mathbb{F}_{p^m}$, when $r \mid p^m - 1$, using $(\lg p)^{O(m)}$ steps (also on GRH).

For q prime, Gauss [1801] solved the cyclotomic equation $\Phi_q(X) = 0$ by radicals, and Poinsot observed that this reduces finding a q-th root of unity mod p, when $q \mid p - 1$, to the solution of various binomial equations. He believed that one could find all needed radicals in $\mathbb{F}_p$, but this is false, as shown by H. J. S. Smith [1860, §66]. Thus, it may be necessary to solve binomial equations in extensions of $\mathbb{F}_p$. In any case, the resulting method is probably inferior to the use of Euler's criterion when q is large.

Pila [1990] showed that for each n, a primitive n-th root of unity in $\mathbb{F}_p^*$ (should it exist) can be constructed using $(\lg p)^{O(1)}$ bit operations. (The implied constant depends severely upon n.)

Various authors have given results suggesting that special classes of irreducible polynomials in $\mathbb{Z}[X]$ can be factored mod p in deterministic polynomial time. Huang [1985, 1991a] proved this for polynomials with an abelian Galois group, which includes the important special case of cyclotomic polynomials. Rónyai [1989b, 1992] generalized this result to normal polynomials (those splitting completely over the field generated by a root), and Evdokimov [1989] proved this for polynomials whose Galois group is solvable. All of these results assume GRH, and rely on techniques from constructive Galois theory, which we cannot go into here. Taking a different approach, Rónyai [1987] showed that for every m, polynomials of degree $\leq m$ in $\mathbb{F}_p[X]$ can be factored using $(\lg p)^{O(1)}$ bit operations. This result also assumes GRH; the implied constant depends on m but not on p.

Evdokimov [1994] gave a deterministic algorithm for factoring polynomials in $\mathbb{F}_q[X]$ using $(n^{\lg n} \lg q)^{O(1)}$ steps, on GRH.

As far as we know, the time required to factor polynomials in $\mathbb{F}_p[X]$ is exponential in $\lg p$. The best current estimate of the dependence on p is due to Shparlinski [1992, p. 10]: a degree n polynomial in $\mathbb{F}_p[X]$ can be factored using $p^{1/2}(\lg p)n^{O(1)}$ operations in $\mathbb{F}_p$. (See also Shoup [1991b].) This improved similar estimates of Shoup [1989a, 1990a, 1991a] and H. W. Lenstra (cited in Shoup [1989a]). Several authors have stated that the exponent of p can be reduced to $1/4 + o(1)$, but this has never been proved. (Beliefs in such a statement seem to come from misreading the character sum bounds of Burgess [1962a].) Shparlinski [1992, p. 13] observed that there is a set S of primes, whose density is positive, with the following property. There is an algorithm that factors any degree n polynomial in $\mathbb{F}_p[X]$, when $p \in S$, using $p^{0.153}n^{O(1)}$ bit operations.

The algorithm used to prove Theorem 7.8.8 was given, independently, by von zur Gathen [1987a] and Mignotte and Schnorr [1988]. (Von zur Gathen gave an asymptotic running time bound for the algorithm, too complicated to quote here.) Kempfert [1969] employed similar ideas but did not analyze the running time of his algorithm. See also Moenck [1977] (for an algorithm based on partial factorization of $p - 1$), and Menezes, van Oorschot, and Vanstone [1992] (for a good overview of the method). Sharper versions of this theorem were given by Rónyai [1989a] and Shoup [1991a]. The result remains true if $p - 1$ is replaced by any cyclotomic polynomial in p. See Bach and von zur Gathen [1988]; Bach, von zur Gathen, and Lenstra [1994]. An analogous result based on elliptic curves appeared in Rónyai [1992].

It is curious that polynomials over $\mathbb{Z}$ can be factored in deterministic polynomial time, even though no such algorithm is known for $\mathbb{F}_p$. See A. K. Lenstra, Lenstra, and Lovász [1982].

8 Prime Numbers: Facts and Heuristics

This chapter provides a survey of prime number theory, focusing on topics that are useful in the design and analysis of algorithms. Here we concentrate on questions related to the distribution of primes with various properties. Algorithmic problems involving primes, such as primality testing, are discussed in Chapters 9 and 13.

Most of the results of this chapter are proved by means of classical analysis. This is in contrast to the algebraic and combinatorial techniques that pervade the rest of algorithmic number theory.

One feature of this subject that might not be immediately apparent is its strong connection with probability theory. Because many of the important questions involve the "average" behavior of number-theoretic functions, probability theory is a useful language for expressing results.

Even more important, perhaps, is the use of probability as a guide to the intuition. Many number-theoretic functions exhibit such irregular behavior that they appear to have been chosen at random. In such situations, one can often guess an answer even when a rigorous proof is not available. Such glorified guesses, often called heuristic arguments, are an indispensable tool of the algorithm designer.

Analytic number theory is a large area, and we make no attempt to examine it all. In particular, most of the results we will discuss belong to multiplicative number theory. This, roughly speaking, studies the factorization of numbers and the properties of multiplicative structures such as unit groups. In this chapter, we concentrate on results that are useful in primality testing and in the design of algorithms for finite fields. Results that are used in factorization algorithms will be covered in a later chapter. We pay particular attention to inequalities and growth rate estimates, because of their evident connection to the analysis of algorithms.

In contrast to the earlier sections of this book, some parts of this chapter will require more mathematical sophistication on the part of the reader. In particular, we will assume that the reader has had the equivalent of a first course in complex analysis, and (for Section 8.7) some exposure to algebraic number theory. Our main goal, however, is to build the reader's intuition about how the primes "behave," so we will not try to prove every statement we use. For a first reading, we suggest Sections 8.1–8.2, in which it is proved that the prime number theorem holds up to a constant factor, using only real-variable arguments. (This suffices for most of the "big-O" results in theoretical computer science that rely on the prime number theorem.) Next, the reader may wish to review the theorems and conjectures in the chapter, almost all of which can be stated within elementary number theory. For readers who desire a more systematic treatment of analytic number theory, excellent texts are available, and we have listed several in the Notes.

8.1 Some History

The concept of primality dates from early times, and has fascinated mathematicians ever since. The central African Ishango bone, dating from c. 20,000 B.C., contains three rows of notches, one of which displays 11, 13, 17, and 19 notches. Some have interpreted this as an indication that these people of the Pleistocene knew of prime numbers.

Certainly the Pythagoreans (c. 500 B.C.) were aware of the difference between prime and composite numbers, and Euclid's famous geometry book included a proof that the primes are infinite in number. Somewhat later, Eratosthenes described his sieve, whereby one could quickly enumerate the primes.

Besides having an intuitive appeal, the idea of primality is now central to mathematics. Doubtless this results from the extreme importance of divisibility in modern algebra. If one thinks of the primes as the "building blocks" of mathematical structure, then it is natural to ask how common they are. Below we survey the history of attempts to answer this question precisely.

It is intuitively clear that large primes are distributed more sparsely than small ones. The first precise descriptions of this phenomenon were made by Gauss and Legendre around the turn of the 18th century. According to a letter Gauss wrote late in life, he started his investigations into the density of primes around 1792, with the observation that $\pi(x)$, the number of primes less than x, was about $\int dx/\log x$. Legendre independently came to a similar conclusion; in his text on number theory, he stated that $\pi(x)$ was very well approximated by a function of the form $x/(A\log x + B)$, and provided the values $A = 1$ and $B = -1.08366$. Either assertion implies that

$$\pi(x) \sim \frac{x}{\log x},$$

which is nowadays called the *prime number theorem*.

The first rigorous results along these lines are due to Chebyshev; in our notation, he proved that

$$\pi(x) = \Theta(x/\log x),$$

that is, the ratio $\big(\pi(x)\log x\big)/x$ is bounded both above and below. Although a breakthrough for its time, this result is not too difficult to prove and we shall do so in the next section.

Chebyshev published his theorem in 1852. Already, it was known that the behavior of primes was connected to the *zeta function*, defined by

$$\zeta(s) = \sum_{n\geq 1} \frac{1}{n^s}$$

for real numbers $s > 1$. Indeed, Euler had argued that from properties of the zeta function, one can obtain Euclid's result, as follows. Start with the identity

$$\zeta(s) = \prod_p \left(1 - \frac{1}{p^s}\right)^{-1},$$

take logarithms, and substitute $s = 1$ (!) to obtain

$$\log(\zeta(1)) = \sum_p \frac{1}{p} + \cdots,$$

where ... denotes an infinite sum with a finite value. Because the left hand side is infinite, the sum on the right hand side must diverge, so there are infinitely many primes. (For a rigorous version of this proof, see Exercise 2.)

The preeminence of the zeta function in prime number theory dates from an 1859 paper by Riemann. In this short note, Riemann extended $\zeta(s)$ to a meromorphic function on the complex plane, and used its function-theoretic properties to derive a new asymptotic formula for $\pi(x)$, seemingly more accurate than those given by Gauss and Legendre. The complex zeroes of the zeta function figured explicitly in his formula, and for these Riemann made his now-famous conjecture that all such zeroes have real part 1/2.

Although Riemann's paper presented a wealth of ideas, it did not provide a proof of the prime number theorem. The next step in this direction was taken by F. Mertens, who established an asymptotic formula for the fraction of numbers not divisible by the primes less than x. Finally, in 1896, J. Hadamard and C.-J. de la Vallée Poussin, expanding on Riemann's ideas, independently proved the prime number theorem.

Prime number theory can be extended in various directions, in particular to the study of prime numbers in arithmetic progressions. Although special cases of this theorem were known before, Dirichlet was the first to show that every arithmetic progression that does not obviously exclude primes contains an infinite number of them. He did this by introducing a function similar to the zeta function—nowadays called an *L-function*—and considering its behavior near 1. Just as ordinary prime number theory is controlled by the zeta function, the theory of primes in progressions is reflects the behavior of L-functions.

One can also consider generalizations of the zeta function to algebraic number theory; this was done in the early 20th century by R. Dedekind, E. Landau, E. Hecke, E. Artin, and others. One benefit of this work was Artin's formulation and proof of a far-reaching generalization of quadratic reciprocity. Although algebraic proofs of this theorem are now known, the first proof used properties of these generalized zeta functions.

The "extended" or "generalized" Riemann hypothesis, in its most comprehensive form, is the conjecture that none of these zeta-like functions vanish to the right of the line

$\Re(s) = 1/2$. If true, this hypothesis would have a number of interesting algorithmic consequences, including the existence of a deterministic polynomial-time algorithm for primality. No one at present has any idea how to prove this hypothesis, but it is supported by empirical evidence and can be thought of as a convenient heuristic assumption that "explains" many interesting and useful facets of number-theoretic behavior.

Another important aspect of modern number theory is the refinement of elementary methods, that is, methods not relying on the theory of functions. A high point of this tradition is P. Erdős and A. Selberg's proof of the prime number theorem using only real variable techniques. Another is V. Brun's proof that the twin primes are sparse, in which sieves played a crucial role. Although we will not discuss sieves in detail in this chapter, the reader should be aware that many important estimates cannot presently be attained any other way.

8.2 The Density of Primes

Our first task is to understand *intuitively* why the fraction of primes in $\{1, \ldots, x\}$ should be about $1/\log x$. To do this, we observe that if 1 out of every $\log n$ numbers near n were prime, we would be able to make the following approximation:

$$\vartheta(x) = \sum_{p \le x} \log p \sim \sum_{n \le x} \frac{\log n}{\log n} \sim x.$$

Conversely, such a result should allow us to count the primes less than or equal to x. We reason as follows.

$$\pi(x) = \sum_{p \le x} 1 = \sum_{n \le x} \frac{\vartheta(n) - \vartheta(n-1)}{\log n}.$$

If $\vartheta(n)$ is about n on average, we should have

$$\pi(x) \sim \sum_{n \le x} \frac{1}{\log n} \sim \frac{x}{\log x}.$$

Instead of proving that $\vartheta(x) \sim x$, it turns out to be easier to show the same for

$$\psi(x) = \sum_{p^k \le x} \log p.$$

The difference between these two functions is negligible; that is, nontrivial prime powers are so much rarer than primes that this replacement has no effect asymptotically.

We now show why it is plausible that $\psi(x) \sim x$. Recall that if x is a positive integer and p is a prime,

$$\nu_p(x!) = \sum_{k \geq 1} \left\lfloor \frac{x}{p^k} \right\rfloor .$$

(See Exercise 2.15.) By considering the prime factorization of x, we see that

$$\begin{aligned}
\log(x!) &= \sum_{p^k} \left\lfloor \frac{x}{p^k} \right\rfloor \log p \\
&= \sum_{p^k} \sum_{dp^k \leq x} \log p \\
&= \sum_{d} \sum_{p^k \leq x/d} \log p \\
&= \sum_{d \leq x} \psi(x/d).
\end{aligned}$$

On the other hand, we know that $\log x! \sim x \log x$ by Stirling's approximation, and

$$1 + \frac{1}{2} + \cdots + \frac{1}{x} \sim \log x$$

by Euler's summation formula (Corollary 2.5.2). This strongly suggests that we should have $\psi(x) \sim x$; indeed, the only possible limit for the ratio is 1. (See Exercise 13.)

For many purposes it is enough to know that $\pi(x)$, the number of primes less than or equal to x, is $\Theta(x/\log x)$. This is easy to prove and we do so below.

LEMMA 8.2.1 We have

$$0 \leq \psi(x) - \vartheta(x) \leq \sqrt{x} \log^2 x.$$

Proof Left to the reader as Exercise 9. ■

LEMMA 8.2.2 We have $\vartheta(x) = O(x)$.

Proof Let m be a power of 2, $m \geq 2$, and consider

$$\binom{m}{m/2} = \frac{m!}{(m/2)!^2}$$

No prime larger than $m/2$ appears in the denominator of the fraction above, so

$$\prod_{m/2<p\leq m} p \leq \binom{m}{m/2} \leq 2^m.$$

Taking logarithms, we find that

$$\vartheta(m) - \vartheta(m/2) \leq m \log 2,$$

which gives $\vartheta(m) \leq 2m \log 2$ for powers of 2. This implies $\vartheta(x) \leq 4x \log 2$ for all $x > 0$. ■

LEMMA 8.2.3 We have $\psi(x) = \Omega(x)$.

Proof Consider

$$I = \int_0^1 x^n(1-x)^n dx.$$

If we expand $(1-x)^n$ by the binomial theorem and interchange summation and integration, we introduce no denominator larger than $2n+1$. Therefore $I \cdot \operatorname{lcm}(1, \ldots, 2n+1)$ is an integer. Since $0 < x(1-x) \leq 1/4$ in the range of integration, we have $0 < I < 4^{-n}$, so

$$4^{-n} \operatorname{lcm}\{1, \ldots, 2n+1\} \geq 1.$$

Therefore

$$\psi(2n+1) = \log(\operatorname{lcm}\{1, \ldots, 2n+1\}) \geq n \log 4.$$

This implies that $\psi(x) \geq \lfloor \frac{x-1}{2} \rfloor \log 4$ for all $x > 0$. ■

THEOREM 8.2.4 We have

$$\psi(x) = \Theta(x),$$

and the same estimate holds for $\vartheta(x)$.

Proof Combine the estimates in Lemmas 8.2.1–8.2.3. ■

THEOREM 8.2.5 (CHEBYSHEV) We have

$$\pi(x) = \Theta(x/\log x).$$

Proof Write $\pi(x)$ as a Stieltjes integral and integrate by parts:

$$\pi(x) = \int_{2^-}^{x} \frac{d\vartheta(t)}{\log t} = \frac{\vartheta(x)}{\log x} + \int_{2^-}^{x} \frac{\vartheta(t)dt}{t \log^2 t}.$$

Using Theorem 2.6.1, the last integral is $O(x/\log^2 x)$, so

$$\pi(x) = \Theta\left(\frac{x}{\log x}\right).$$ ■

Once we know that $\pi(x) = \Theta(x/\log x)$, several useful order-of-magnitude estimates follow. These are given in Exercise 4.

The prime number theorem can be given an appealing probabilistic interpretation: the "chance" that n is prime is about $1/\log n$. A rigorous version of this statement is that a random integer, drawn uniformly from $\{1, \ldots, n\}$, will be prime with probability asymptotic to $1/(\log n)$. It follows, then, that the expected number of draws needed to obtain a prime is asymptotic to $\log n$.

To express the prime number theorem precisely, we introduce the *logarithmic integral*, defined by

$$\mathrm{li}(x) = \int_2^x \frac{dt}{\log t}.$$

Because li is a continuous increasing function it has an inverse function, which we denote by li^{-1}. We also let $\lambda(x) = (\log x)^{3/5}(\log\log x)^{-1/5}$. The following results, which we give without proof, give the sharpest known forms of the prime number theorem.

THEOREM 8.2.6 There is a $c > 0$ such that

$$\pi(x) = \mathrm{li}(x) + O(xe^{-c\lambda(x)}) \sim \frac{x}{\log x}.$$

COROLLARY 8.2.7 We also have

$$\psi(x) = x + O(xe^{-c\lambda(x)}) \sim x,$$

and the same bound holds with $\psi(x)$ replaced by $\vartheta(x)$.

COROLLARY 8.2.8 Let p_n denote the n-th prime. Then

$$p_n = \mathrm{li}^{-1}(n) + O(ne^{-c\lambda(n)}) \sim n\log n.$$

The above three results are similar, in the sense that each one easily implies the others. A fourth result, which we state below, was proved by Mertens, using only Chebyshev's theorem. Because Chebyshev's theorem does not specify an exact constant, Mertens's theorem is in some intuitive sense a weaker version of the prime number theorem. We state it below with the best known error term.

COROLLARY 8.2.9 We have

$$\prod_{p\le x}\left(1-\frac{1}{p}\right)=\frac{e^{-\gamma}}{\log x}+O(e^{-c\lambda(x)})\sim\frac{e^{-\gamma}}{\log x}.$$

For most purposes it is enough to know that $e^{\lambda(x)}$ grows more rapidly than any power of $\log x$, but more slowly than any positive power of x. Thus, for example, we have $\pi(x)=x/\log x+O(x/\log^2 x)$.

In the remainder of this section, we examine how complex analysis can be used to estimate sums over primes. In particular, we will sketch a proof of the prime number theorem.

We summarize the properties of the zeta function that we will need. First, $\zeta(s)$ can be extended to a function on the entire complex plane; it is single-valued and analytic everywhere except for a simple pole at $s=1$, with residue 1. The zeta function vanishes at negative even integers and at infinitely many places in the *critical strip* $0<\Re(s)<1$. There are no other zeroes; in particular, the zeta function does not vanish on the line $\Re(s)=1$. There are various ways to express this last statement quantitatively; we will use the bound $|\frac{\zeta'}{\zeta}(\sigma+it)|=(\log t)^{O(1)}$, valid uniformly when $\sigma\ge 1$, as $|t|\longrightarrow\infty$.

We now sketch a proof that $\psi(x)\sim x$, but warn the reader that the argument below makes no pretensions to completeness.

Recall von Mangoldt's lambda function, defined by

$$\Lambda(n)=\begin{cases}\log p, & \text{if } n=p^k \text{ for some prime } p;\\ 0, & \text{otherwise.}\end{cases}$$

It is not hard to see that

$$\psi_1(x):=\int_0^x \psi(t)\,dt=\sum_{n\le x}(x-n)\Lambda(n).$$

(See Exercise 7.) Roughly speaking, the prime number theorem is equivalent to the idea that in such sums, $\Lambda(n)$ should be ignored; we therefore expect that

$$\psi_1(x)\sim\sum_{n\le x}(x-n)\sim\frac{x^2}{2}.$$

Conversely, asymptotic differentiation is permissible in this case, and the above relation implies $\psi(x)\sim x$. (See Exercise 8.)

It therefore suffices to show $\psi_1(x)\sim x$; to do this, we first express $\psi_1(x)$ as an integral involving the zeta function:

$$\psi_1(x)=\frac{-1}{2\pi i}\int_{2-i\infty}^{2+i\infty}\frac{x^{s+1}}{s(s+1)}\frac{\zeta'}{\zeta}(s)\,ds.$$

If we subtract off the pole at $s = 1$, move the line of integration to the left, and divide by x^2 we get

$$\frac{\psi_1(x)}{x^2} - \frac{1}{2}\left(1 - \frac{1}{x}\right)^2 = \frac{-1}{2\pi i}\int_{1-i\infty}^{1+i\infty} \frac{x^{s-1}}{s(s+1)}\left(\frac{\zeta'}{\zeta}(s) + \frac{1}{s-1}\right) ds.$$

Away from the pole at 1, we have $|\frac{\zeta'}{\zeta}(1+it)| = (\log t)^{O(1)}$, so if we substitute $s = 1 + it$ into the integral, it becomes

$$\frac{1}{2\pi}\int_{-\infty}^{\infty} h(t)e^{it\log x}\, dt,$$

where $\int_{-\infty}^{\infty} |h| < \infty$. By the Riemann-Lebesgue theorem, the integral tends to 0 as $x \longrightarrow \infty$, so $\psi_1(x) \sim x^2/2$.

8.3 Sharp Estimates and the Riemann Hypothesis

It is natural to ask how accurate the estimate of Theorem 8.2.6 is, and a quick check of the numerical data suggests that improvements should be possible. For example, we have

$$\pi(10^{10}) = 455\,052\,511$$

and, to the nearest integer,

$$\mathrm{li}(10^{10}) = 455\,055\,614.$$

Here, then, about half of the digits of $\pi(x)$ are given correctly by the approximation $\mathrm{li}(x)$. This pattern persists throughout the range where $\pi(x)$ is known exactly. This suggests that the true error in the prime number theorem is not that given by Theorem 8.2.6, but rather something closer to $\sqrt{x}$.

Unfortunately, nothing resembling this has been proved, although there is plenty of evidence to suggest it is true. Indeed, it is one of the most notorious open problems in mathematics. This conjecture, or more precisely an equivalent statement about the complex zeroes of the zeta function, was first made by Riemann, and has been known ever since as the Riemann hypothesis. We state it precisely below.

CONJECTURE 8.3.1 (RIEMANN HYPOTHESIS) For any $\epsilon > 0$, we have

$$\pi(x) = \mathrm{li}(x) + O(x^{1/2+\epsilon}).$$

This conjecture holds iff the zeta function does not vanish in the half-plane $\Re(s) > 1/2$, and is equivalent to other estimates for prime number sums. For example, the Riemann hypothesis holds iff $\psi(x) = x + O(x^{1/2+\epsilon})$. (See Exercise 18.)

We now explain why the Riemann hypothesis is plausible on probabilistic grounds. We will take the following lemma from probability theory, which is a quantitative version of the law of large numbers, without proof.

LEMMA 8.3.2 Let $X_1, X_2, X_3, \ldots$ be a sequence of independent random variables with mean 0. Let $b_1, b_2, b_3, \ldots$ be an increasing sequence of positive numbers with $\sum E(X_k^2)/b_k^2 < \infty$. If $S_n = \sum_{1 \le k \le n} X_k$, then $S_n = o(b_n)$ with probability 1.

Our probabilistic model for the distribution of primes is that each integer $k \geq 2$ is labelled "prime" with probability $1/\log k$, and that different labellings are independent events. Then, for any $x > 0$, we have the random variable

$$\Pi(x) = \left|\{k \le x \ : \ k \text{ is labelled "prime"}\}\right|.$$

By Euler's summation formula (Corollary 2.5.2)

$$\mathrm{E}(\Pi(x)) = \mathrm{li}(x) + O(1).$$

We can use Lemma 8.3.2 to estimate the deviation from this expected value, as follows. Let T_k be 1 or 0 according to whether the integer k was labelled "prime" or not, and set $X_k = T_k - 1/\log k$. Then, choosing $b_k = k^{1/2+\epsilon}$, we obtain

$$\Pi(x) = \mathrm{li}(x) + O(x^{1/2+\epsilon}),$$

almost surely.

This suggests that the Riemann hypothesis might be true, but for no good reason. It may simply be a consequence of the irregular way in which the primes are distributed.

It is interesting that if $\pi(x) = \mathrm{li}(x) + O(x^{1/2+\epsilon})$ holds for arbitrary positive ϵ, then a stronger result is actually true. More precisely, we have the following theorem, due to H. von Koch.

THEOREM 8.3.3 Assume the Riemann hypothesis. Then the following bounds hold:

$$\pi(x) = \mathrm{li}(x) + O(\sqrt{x}\log x),$$

$$\psi(x) = x + O(\sqrt{x}(\log x)^2),$$

$$p_n = \mathrm{li}^{-1}(n) + O(\sqrt{n}(\log n)^{5/2}),$$

$$\prod_{p \le x}\left(1 - \frac{1}{p}\right) = \frac{e^{-\gamma}}{\log x} + O(x^{-1/2}).$$

The Riemann hypothesis is supported by some striking empirical evidence. Many researchers have computed the zeroes of the Riemann zeta function in the critical strip, without finding a single zero whose real part is different from 1/2. The most thorough investigations are those of J. van de Lune, H. te Riele, and D. Winter, who found that the first 1.5 *billion* zeroes satisfy the Riemann hypothesis.

We emphasize that these results provide a proof—once all roundoff error has been accounted for—that all the zeroes in a given rectangle have real part 1/2. It is interesting to note how this is done. First, one finds a function $\xi(s)$ that vanishes in the critical strip exactly when the zeta function does, and is real on the line $1/2 + it$. It follows that one can detect zeroes on the line by finding the places where $\xi(1/2 + it)$ changes sign. One also needs to know when one has found all the zeroes. To do this, one uses the argument principle, according to which the zeroes of an analytic function in a region can be counted by integrating its logarithmic derivative around the boundary of that region.

We now delve more deeply into the connection between the zeroes of the zeta function and error terms in the prime number theorem. We take the viewpoint that it is useful to estimate prime number sums, and ask how knowledge about the zeta function can help in such estimates. The answer involves *explicit formulas*, which express a given sum in terms of the zeroes and poles of the zeta function.

Roughly speaking, these formulas give information about prime number sums in the same way that Laplace transforms are used to obtain information about the solutions to linear differential equations. Readers familiar with the use of these transforms in applied mathematics may wish to think of the zeta function as taking its arguments from a kind of complex "frequency domain." As in engineering, the real parts of the poles are what count, in determining the behavior of a sum.

As an example, we discuss the sum

$$\psi_1(x) = \sum_{n \leq x} (x - n)\Lambda(n),$$

which figured in the proof of the prime number theorem. As before, we start with the integral formula

$$\psi_1(x) = \frac{-1}{2\pi i} \int_{2-i\infty}^{2+i\infty} \frac{x^{s+1}}{s(s+1)} \frac{\zeta'}{\zeta}(s)\, ds.$$

This can be integrated as the sum of its residues in the half-plane to the left of the line of integration. We argue formally, but claim that the calculation can be fully justified. Each singularity of the integrand produces a term; arranging them in order by growth rate we obtain

$$\psi_1(x) = \frac{x^2}{2} - \sum_{\rho} \frac{x^{\rho+1}}{\rho(\rho+1)} + O(x).$$

Here the leading term comes from the pole of ζ'/ζ at 1, and the sum is over the zeroes ρ of the zeta function in the critical strip. This is our desired explicit formula.

Because the zeta function is real on the real axis, its zeroes come in conjugate pairs. Let the n-th such pair be $\beta_n \pm i\gamma_n$, where $0 < \gamma_1 \leq \gamma_2 \leq \gamma_3 \leq \cdots$. It is known that $\gamma_n \sim 2\pi n/\log n$, and this implies that $\sum_\rho |\rho(\rho+1)|^{-1}$ converges. If we knew that the zeroes all occurred on the line $\Re(s) = 1/2$, then we could combine this fact with $|x^{\rho+1}| = x^{3/2}$ to obtain the estimate

$$\psi_1(x) = \frac{x^2}{2} + O(x^{3/2}).$$

The exponent 3/2 cannot be replaced by anything smaller, because the complex zeroes of the zeta function are also known to lie symmetrically about the line $\Re(s) = 1/2$.

Arguments of this type provide a direct connection between the horizontal location of the zeroes of the zeta function and the accuracy of asymptotic estimates for prime number sums. We can sharply estimate the error by combining an explicit formula with an estimate of the first zero, if any, that lies off the line.

We illustrate with an estimate of $\psi_1(x)$, giving enough of the argument to convince the reader that providing numerical bounds would not be a difficult task. The heart of the matter is to find a good estimate for

$$\sum_\rho \frac{x^{\rho+1}}{\rho(\rho+1)}.$$

To do this we break the sum into a sum over "good" zeroes (those with real part $1/2$) and a sum over "bad" zeroes, and estimate each separately. The sum over the good zeroes is $O(x^{3/2})$, because $\sum_\rho |\rho|^{-2} < \infty$. A good value for the constant in this estimate can be obtained by observing that for any good zero, $|\rho|^2 = 1/\rho + 1/\bar{\rho}$, and using the known formula

$$\sum_\rho \left(\frac{1}{\rho} + \frac{1}{\bar{\rho}}\right) = \gamma + 2 - \log(4\pi).$$

Let $N(t)$ denote the number of zeroes of the zeta function whose imaginary part lies between 0 and t. If B is a bound on the height (i.e., imaginary part) of the first bad zero, the sum over the bad zeroes is no more than twice

$$x^2 \int_B^\infty \frac{dN(t)}{t^2} = x^2 \left(-\frac{N(B)}{B^2} + 2\int_B^\infty \frac{N(t)dt}{t^3}\right).$$

(Here we have used the fact that the bad zeroes are symmetric around the critical line.) This is easy to estimate once we know an explicit bound for $N(t)$ and a value for B. The known relation

$$N(t) = \frac{t}{2\pi} \log \frac{t}{2\pi} + \frac{t}{2\pi} + O(\log t)$$

together with a short calculation, shows that the sum over the bad zeroes is at most

$$x^2\left(\frac{\log B}{\pi B}+O\left(\frac{1}{B}\right)\right).$$

We may summarize the above paragraph by stating that

$$\psi_1(x)=\left(\frac{1}{2}+\frac{\log B}{\pi B}+O\left(\frac{1}{B}\right)\right)x^2+O(x^{3/2}).$$

It is known that $B \geq 5\times 10^8$. Roughly speaking, then, the relative error incurred in our estimate of $\psi_1(x)$ by ignoring bad zeroes of the zeta function is about 1 part in 10 million. In this case, therefore, the Riemann hypothesis is "practically true."

8.4 Primes in Arithmetic Progressions and the ERH

Examination of the sequence of primes shows that they are almost evenly distributed among the various residue classes mod n. For example, among the 1229 primes less than 10 000, there are 611 congruent to 1 mod 3 and 617 congruent to 2 mod 3. The first precise result along these lines was conjectured by Legendre and proved by Dirichlet.

THEOREM 8.4.1 (DIRICHLET) If $\gcd(a, n) = 1$, then there are infinitely many primes congruent to a modulo n.

If the primes are equidistributed mod n, then a sum over the primes in a congruence class should equal, as a general rule, the corresponding sum over the primes divided by $\varphi(n)$, the number of allowable congruence classes. In particular, $\pi(x, n, a)$, the number of primes less than or equal to x and congruent to a mod n, should be about $x/(\varphi(n)\log x)$. This was proved by de la Vallée Poussin in 1896, and published shortly after his proof of the prime number theorem. We give a modern version with the best known error term.

THEOREM 8.4.2 Let $\lambda(x) = (\log x)^{3/5}(\log\log x)^{-1/5}$. For each n there is a $c > 0$ such that if $\gcd(a, n) = 1$,

$$\pi(x,n,a)=\frac{1}{\varphi(n)}\mathrm{li}(x)+O(xe^{-c\lambda(x)})\sim\frac{x}{\varphi(n)\log x}.$$

It is often useful to know that the primes are equidistributed relative to subgroups of the multiplicative group of integers mod n. For this purpose we state the following result.

COROLLARY 8.4.3 Let G be a subgroup of $(\mathbb{Z}/(n))^*$ of index d. Let C be a coset of G, and let $\pi(x, C)$ denote the number of primes less than or equal to x whose residue classes mod

n lie in C. Then

$$\pi(x, C) = \frac{1}{d}\mathrm{li}(x) + O(xe^{-c\lambda(x)}) \sim \frac{x}{d \log x}.$$

In these theorems, n is fixed, and $x \to \infty$. The situation when n and x both vary is more complicated, and we refer interested readers to the Notes.

Just as with the ordinary prime number theorem, one can use probability theory to speculate about the error terms in the above estimates. Again we assume that the chance of labelling an integer k congruent to a modulo n "prime" is inversely proportional to $\log k$. Now, however, we must take this probability to be $n/(\varphi(n) \log k)$; this can be thought of as the conditional probability that k is prime, given that it is relatively prime to n. This leads to the following conjecture, which can be derived in a fashion similar to Conjecture 8.3.1. (See Exercise 28.)

CONJECTURE 8.4.4 (EXTENDED RIEMANN HYPOTHESIS) Let n and a be relatively prime integers. Then for any $\epsilon > 0$, we have

$$\pi(x, n, a) = \frac{\mathrm{li}(x)}{\varphi(n)} + O(x^{1/2+\epsilon}).$$

This conjecture is a form of the *Extended Riemann Hypothesis (ERH)*. We will give an equivalent statement in function-theoretic terms, but some definitions are required.

Let χ be a character on the multiplicative group $(\mathbb{Z}/(n))^*$; that is, a homomorphism from $(\mathbb{Z}/(n))^*$ into the unit circle. Abusing notation slightly, we will also use χ to denote a complex-valued function on the integers, given by

$$\chi(m) = \begin{cases} \chi(m \bmod n), & \text{if } \gcd(m, n) = 1; \\ 0, & \text{otherwise.} \end{cases}$$

Such functions are known as *Dirichlet characters*; if χ never takes a value different from 0 or 1 it is called *principal*. It is possible that χ is induced by a character χ' of smaller modulus, in the sense that they both agree whenever χ is nonzero. If this is not the case, χ is called *primitive*. Every Dirichlet character is induced by a unique primitive one.

The *Dirichlet L-function* corresponding to χ is defined by the formula

$$L(s, \chi) = \sum_{n \geq 1} \frac{\chi(n)}{n^s}$$

when $\Re(s) > 1$, and analytic continuation otherwise. If χ is not principal, then $L(s, \chi)$ is an entire function. The function $L(s, \chi)$ has an infinite number of zeroes in the critical strip $0 < \Re(s) < 1$. The ERH is the statement that all such zeroes have real part 1/2. This is equivalent to Conjecture 8.4.4; we leave this for the analytically inclined reader to prove. (See Exercise 19.)

For many applications, it is useful to know that if the ERH holds, then the error term in Conjecture 8.4.4 depends only moderately on n. A theorem exhibiting this dependence was proved (essentially) by E. Titchmarsh in 1930.

THEOREM 8.4.5 Assume ERH. Let $\gcd(a, n) = 1$. Then

$$\pi(x, n, a) = \frac{\mathrm{li}(x)}{\varphi(n)} + O(x^{1/2}(\log x + \log n)),$$

and the constant implied by the "O" term is absolute.

There is also a version of this theorem for primes lying in cosets of subgroups of $(\mathbb{Z}/(n))^*$. Using the notation of Corollary 8.4.3, we have

THEOREM 8.4.6 Assume ERH. Then

$$\pi(x, C) = \frac{\mathrm{li}(x)}{d} + O(x^{1/2}(\log x + \log n)),$$

where the constant implied by the "O" term is absolute.

8.5 Applications of the ERH

Perhaps the most useful implication of the ERH is that numbers with desired multiplicative properties can often be quickly found by direct search. In this section we discuss a number of results of this type.

The first problem we will discuss is that of finding a quadratic nonresidue modulo a prime p. Such numbers are useful in computation; for example, Tonelli's algorithm to compute square roots modulo p runs in deterministic polynomial time when provided with a number a satisfying $\left(\frac{a}{p}\right) = -1$. (See Chapter 7.) This leads to the problem of estimating the least quadratic nonresidue modulo p; any such estimate bounds the amount of searching that must be done to find a nonresidue.

It has long been observed that quadratic residues and nonresidues are distributed irregularly, in a pseudo-random fashion. This suggests that the least quadratic nonresidue modulo p should be small, and that if one needs a nonresidue, the best strategy is simply to try $2, 3, 5, \ldots$ in order until one is found.

We first investigate the consequences of the heuristic assumption that for $q = 2, 3, 5, \ldots$, the quadratic character $\left(\frac{q}{p}\right)$ is chosen at random. This is similar to flipping coins until a head occurs; the expected number of trials is 2. We can also ask the following question.

Suppose that this experiment is repeated for each prime p independently. There will be occasional "bad primes" for which the least nonresidue is unusually large. How large are these nonresidues likely to be? Using probability theory, one can argue heuristically that the least quadratic nonresidue mod p is $O((\log p)(\log\log p))$. (See Exercise 37.)

Perhaps motivated by such probabilistic considerations, various authors have conjectured that the least quadratic nonresidue mod p is very small. For example, I. Vinogradov conjectured that it is $O(p^\epsilon)$ for every $\epsilon > 0$. Even Vinogradov's conjecture is not strong enough to imply that a nonresidue can be found in deterministic polynomial time. However, a bound that is polynomial in $\log p$ can be proved from the ERH, as was done by N. Ankeny in 1952.

This is not difficult to see, in the following way. Let $\pi_+(x)$ denote the number of primes less than or equal to x that are quadratic residues of p. From Theorems 8.4.6 and 8.3.3 we have

$$\pi(x) = \mathrm{li}(x) + O(x^{1/2}\log x)$$

and

$$\pi_+(x) = \frac{1}{2}\mathrm{li}(x) + O(x^{1/2}(\log x + \log p)).$$

If all positive numbers less than or equal to x are quadratic residues, we must have $\pi(x) = \pi_+(x)$, and therefore

$$\frac{1}{2}\mathrm{li}(x) \sim \frac{x}{2\log x} = O(x^{1/2}(\log x + \log p)).$$

This implies that x cannot be too large; in particular, $x = O((\log p)^2(\log\log p)^2)$.

Both the generality and the accuracy of this theorem can be improved, if we are willing to work with more complicated prime number sums. In particular, we can obtain better results if we replace the prime-counting functions (such as $\pi(x)$) by suitably weighted averages. We will need the following technical lemma, which we quote without proof.

LEMMA 8.5.1 Assume ERH. Let χ be a Dirichlet character, defined modulo $n \geq 2$. Let $N_\chi(t)$ be the number of zeroes of $L(s,\chi)$ with real part between 0 and 1 and imaginary part $\leq t$ in absolute value. Then

$$N_\chi(t) = O(t\log(nt))$$

for $t \geq 2$ and

$$\frac{L'}{L}(1/4 + it) = O(\log(n(|t| + 2)))$$

for all t. (The implied constants do not depend on n.)

We now define 1_χ to be 1 if χ is principal and 0 otherwise.

Roughly speaking, the first part of Lemma 8.5.1 tells us that the zeroes of $L(s, \chi)$ in the critical strip have density proportional to $\log n$ for large n. Using this idea, we can guess an approximation for the averaged sum $\sum_{p \le x} \Lambda(p)\chi(p)(x-p)$. Since most prime powers are really primes, this should not differ too much from the corresponding sum over all numbers, so we can use the explicit formula

$$\sum_{m \le x} \Lambda(m)\chi(m)(x-m) = 1_\chi \frac{x^2}{2} - \sum_\rho \frac{x^{\rho+1}}{\rho(\rho+1)} + \cdots,$$

similar to that used in the prime number theorem. Since the corresponding sum over roots of the zeta function converges absolutely, we expect that

$$\left| \sum_\rho \frac{x^{\rho+1}}{\rho(\rho+1)} \right| \le x^{3/2} \times (\text{density of } L\text{'s zeroes}) \times O(1),$$

so that

$$\sum_{p \le x} \Lambda(p)\chi(p)(x-p) = 1_\chi x^2/2 + O(x^{3/2} \log n).$$

The next lemma proves this, assuming ERH.

LEMMA 8.5.2 Assume ERH. Let χ be as in Lemma 8.5.1, with $1_\chi = 1$ when χ is principal and 0 otherwise. Then

$$\sum_{p \le x} \Lambda(p)\chi(p)(x-p) = 1_\chi \frac{x^2}{2} + O(x^{3/2} \log n),$$

where the constant in the "O" symbol does not depend on n.

Proof By interchanging summation and integration, we find

$$\sum_{m \le x} \Lambda(m)\chi(m)(x-m) = \frac{-1}{2\pi i} \int_{2-i\infty}^{2+i\infty} \frac{x^{s+1}}{s(s+1)} \frac{L'}{L}(s, \chi)\, ds.$$

We may move the integration line to the left and obtain

$$\sum_{m \le x} \Lambda(m)\chi(m)(x-m) = 1_\chi \frac{x^2}{2} - \sum_\rho \frac{x^{\rho+1}}{\rho(\rho+1)} - \frac{1}{2\pi i} \int_{1/4-i\infty}^{1/4+i\infty} \frac{x^{s+1}}{s(s+1)} \frac{L'}{L}(s, \chi)\, ds,$$

where the sum runs over the zeroes of $L(s, \chi)$ in the critical strip. (This is similar to the proof of the prime number theorem, as done in Exercise 16.) Using the ERH and applying the first part of Lemma 8.5.1 we get

$$\left|\sum_\rho \frac{x^{\rho+1}}{\rho(\rho+1)}\right| \le x^{3/2}\left(\frac{4}{3}N(2) + \int_2^\infty \frac{dN_\chi(t)}{t^2}\right) = O(x^{3/2}\log n)$$

after integration by parts. Also,

$$\left|\int_{1/4-i\infty}^{1/4+i\infty} \frac{x^{s+1}}{s(s+1)}\frac{L'}{L}(s,\chi)\,ds\right| \le 2x^{5/4}\int_0^\infty \frac{O(\log(n(t+2)))}{(t+1/4)^2}dt = O(x^{5/4}\log n).$$

Therefore,

$$\sum_{m\le x} \Lambda(m)\chi(m)(x-m) = 1_\chi\frac{x^2}{2} + O(x^{3/2}\log n).$$

This is not quite the result we need, because prime powers are included in this sum. However,

$$\sum_{\substack{p^k\le x\\ k\ge 2}} \Lambda(p^k)(x-p^k) \le x\left(\sum_{2\le k\le \log_2 x} \vartheta(x^{1/k})\right)$$

$$\le x\left(\vartheta(x^{1/2}) + (\log_2 x)\vartheta(x^{1/3})\right) = O(x^{3/2}).$$

Subtracting the two sums proves the theorem. ■

We now introduce the function

$$G(n) = \min\{x : (\mathbb{Z}/(n))^* \text{ is generated by primes } \le x\}.$$

Note that $G(n)$ is defined for all n, by the unique factorization theorem. On probabilistic grounds, we might guess that $G(n) = O((\log n)(\log\log n))$. (See Exercise 37.) The ERH implies a growth rate somewhat larger than this.

THEOREM 8.5.3 (ANKENY) Assume ERH. Then $G(n) = O((\log n)^2)$. In particular, every nontrivial subgroup of $(\mathbb{Z}/(n))^*$ omits a positive number that is $O((\log n)^2)$.

Proof We note that if x is positive,

$$\sum_{\substack{p\le x\\ p\mid n}} \Lambda(p)(x-p) \le x\log n.$$

Now, assume that the primes less than or equal to x do not generate $(\mathbb{Z}/(n))^*$. Then they generate a proper subgroup G of $(\mathbb{Z}/(n))^*$, and we can choose a nonprincipal Dirichlet character χ such that $\chi(n) \in \{0, 1\}$ for all $n < x$. Since χ is not principal, we have

$$\sum_{p \leq x} \Lambda(p)\chi(p)(x - p) = O(x^{3/2} \log n).$$

On the other hand, our assumption on x implies that

$$\sum_{p \leq x} \Lambda(p)\chi(p)(x - p) = \sum_{p \leq x} (x - p) \log p - \sum_{\substack{p \leq x \\ p \mid n}} \Lambda(p)(x - p) = \Omega(x^2 - x \log n).$$

Thus $x^2 - x \log n = O(x^{3/2} \log n)$. This implies $\sqrt{x} = O(\log n)$, that is, $x = O((\log n)^2)$. ■

Ankeny's theorem is extremely important in algorithmic number theory, for the following reason. In many situations, we need a number that lies outside a certain proper subgroup of $(\mathbb{Z}/(n))^*$. Such numbers are easy to find by guessing. The bound given by Ankeny's theorem allows this randomization to be replaced by a direct search. This has applications to finding roots of polynomial equations over finite fields (see Section 7.8) and primality testing (see Section 9.5).

It is also useful to know that primes tend to have very small primitive roots. Again, this is an empirical observation that has so far eluded proof. It can be given some justification by probabilistic arguments: if one pretends that each prime has an equal chance of being a primitive root mod p, this chance being the fraction of primitive roots modulo p, then one obtains an almost sure bound of $O((\log p)(\log \log p)^2)$ for the least primitive root. (See Exercise 37.)

If the ERH is true, then the least primitive root mod p is bounded by a polynomial in $\log p$. V. Shoup has recently proved the following theorem.

THEOREM 8.5.4 Assume ERH. Let $g(p)$ denote the least positive primitive root modulo p, and let ω denote the number of primes dividing $p - 1$. Then

$$g(p) = O(\omega^4 (\log \omega)^4 (\log p)^2) = O((\log p)^6).$$

In some sense, the notion of primitive root is a generalization of quadratic nonresidue: a nonresidue lies outside the group of squares, but a primitive root lies outside *every* proper subgroup. Now, quadratic reciprocity and Dirichlet's theorem imply that every prime is a quadratic nonresidue modulo infinitely many other primes. One would naturally ask, then, if something similar holds for primitive roots.

A conjecture of this type was made by Artin, supporting the the notion that primes have small primitive roots. More precisely, *Artin's conjecture* is the following statement: if a is neither -1 nor a square, then a is a primitive root for infinitely many primes.

Artin gave a simple probabilistic argument for the density of such primes, which we reproduce here. For a to be a primitive root mod q, it is necessary and sufficient that for each prime p, we have $a^{(q-1)/p} \not\equiv 1 \pmod{p}$ whenever $q \equiv 1 \pmod{p}$. By Theorem 8.4.2, about 1 out of every $p-1$ primes is congruent to 1 (mod p), and if q is such a prime, we expect $a^{(q-1)/p}$ to be congruent to 1 about $1/p$ of the time. If the requirements for the various p were independent, then the density of primes for which a is a primitive root would equal

$$A = \prod_p \left(1 - \frac{1}{p(p-1)}\right) \doteq 0.373956.$$

This value is nowadays called *Artin's constant.*

As appealing as this argument is, it contains a subtle error. The problem is that the requirements for various p are not always independent. We illustrate this for $a = 5$. We need to exclude the possibility that a is a perfect fifth power for primes of the form $5m+1$. For such a prime p, we have

$$\left(\frac{5}{p}\right) = \left(\frac{5m+1}{5}\right) = \left(\frac{1}{5}\right) = +1$$

by quadratic reciprocity. Therefore, if 5 is not a fifth power mod p, it has to be a square mod p.

In general, one needs to multiply A by a correction factor, computable from the factorization of a, to obtain a density that agrees with numerical data. The dependence of this factor on a is complicated, and we refer interested readers to the Notes for the full story.

In 1967, C. Hooley proved that Artin's conjecture follows from the generalized Riemann hypothesis (GRH). (This is a generalization of the ERH to algebraic numbers, which we will discuss in Section 8.7.) We give a simplified version of his theorem below.

THEOREM 8.5.5 Assume GRH. Let $a \ge 2$ be a squarefree integer, with $a \not\equiv 1 \pmod{4}$. Let $\pi_a(x)$ denote the number of primes less than or equal to x for which a is a primitive root. Then as $x \longrightarrow \infty$,

$$\pi_a(x) = A\frac{x}{\log x} + O\left(\frac{x(\log a + \log\log x)}{(\log x)^2}\right).$$

(The constant implied by the "O" symbol is absolute.)

It is now known that Artin's conjecture holds for infinitely many bases a, although we still cannot specify a particular prime for which it is true. This follows from a recent theorem of R. Heath-Brown: if a, b, and c are distinct primes, then the number of $p \le x$ for which at least one of these is a primitive root is $\Omega(x/(\log x)^2)$.

We know from Dirichlet's theorem that every arithmetic progression prime to n must contain a prime. How large is the least one? Numerical data suggests that the minimal prime in a progression mod n is bounded by a moderate function of n. For example, if $n \leq 10^4$, every residue class in $(\mathbb{Z}/(n))^*$ is represented by a prime $\leq 1.23 \times 10^6$. In this regard, Yu. Linnik proved the following remarkable theorem:

THEOREM 8.5.6 (LINNIK) Let $P(n, a)$ denote the least prime congruent to a mod n, when $\gcd(n, a) = 1$, and $n \geq 2$. Then $P(n, a) = n^{O(1)}$. (The constant implied by the "O" symbol is absolute.)

We emphasize that Linnik's theorem does *not* assume the ERH. The best estimate of the exponent in Linnik's theorem is also due to Heath-Brown.

THEOREM 8.5.7 We have $P(n, a) = O(n^{11/2})$, where the implied constant is absolute.

For many purposes, we would like the exponent to be as small as possible, and it is not difficult to show that if the ERH holds, it can be chosen arbitrarily close to 2. To see this, assume that no prime up to x is congruent to a mod n. Then Theorem 8.4.5 implies that

$$\pi(x, n, a) = \frac{\mathrm{li}(x)}{\varphi(n)} + O(\sqrt{x}(\log x + \log n)) = 0.$$

Since $\mathrm{li}(x) \sim x/(\log x)$, this implies a bound on x of $O(\varphi(n)^2(\log n)^4)$. Using averaging as in the proof of Ankeny's theorem, we can reduce this slightly.

THEOREM 8.5.8 Assume ERH. Let $\gcd(a, n) = 1$. Then the smallest prime congruent to a mod n is $O(\varphi(n)^2(\log n)^2)$, where the constant implied by the "O" symbol does not depend on n.

Proof Recall Lemma 8.5.2, which we write as

$$1_\chi \frac{x^2}{2} - \sum_{p \leq x} \Lambda(p)\chi(p)(x - p) = O(x^{3/2} \log n).$$

Suppose that all primes less than x differ from a mod n. Taking the sum over all characters of $(\mathbb{Z}/(n))^*$, the above formula implies

$$\sum_\chi \bar{\chi}(a) \left(1_\chi \frac{x^2}{2} - \sum_{p \leq x} \Lambda(p)\chi(p)(x - p) \right) = \frac{x^2}{2} = O(x^{3/2}\varphi(n) \log n).$$

(Here we have used Exercise 6.18.) This gives a bound on x; dividing by $x^{3/2}$ and squaring, we obtain the theorem. ∎

We also have the following result.

THEOREM 8.5.9 Assume ERH. Let G be a nontrivial subgroup of $(\mathbb{Z}/(n))^*$ of index d, and let C be a coset of G. The least prime whose residue belongs to C is $O(d^2(\log n)^2)$.

Proof Similar to the proof of Theorem 8.5.8; the only modification necessary is to sum over characters of $(\mathbb{Z}/(n))^*/G$. ∎

Let us consider briefly what these results imply for problem of actually *finding* a prime in an arithmetic progression. Suppose we wanted to find a prime congruent to a modulo n, where a and n are relatively prime. An obvious strategy is to test the numbers $a, a + n, a + 2n, \ldots$ for primality. Now, even if deterministic primality testing can be done quickly—and this is still an open question—it is not obvious that the search will terminate quickly. If the ERH holds, Theorem 8.5.8 implies that $a + kn$ is prime for some $k = O(\varphi(n)(\log n)^2)$, but this bound is not polynomial in $\lg n$. (Using randomization, we can solve this problem efficiently, if we are willing to settle for a somewhat larger prime. See Exercise 30.)

Probabilistic considerations suggest that a much shorter search ought to be sufficient. To see this, consider the coupon collector's problem: we choose samples (with replacement) from a set of size m until each element has been drawn at least once. The expected time until the process finishes is asymptotic to $m \log m$. (See Exercise 36.) We now let our set be the $\varphi(n)$ elements of $(\mathbb{Z}/(n))^*$. If the primes were distributed at random among the residue classes, we would expect to use $\varphi(n)\log(\varphi(n))$ primes before each residue class contains a prime. Now, we have $\log(\varphi(n)) \sim \log n$, and $p_k \sim k \log k$. Using a different probabilistic argument, S. Wagstaff made the following conjecture in 1979.

CONJECTURE 8.5.10 (WAGSTAFF) Let a and n be positive integers with $\gcd(a, n) = 1$. Then the least prime congruent to a mod n is $O(\varphi(n)(\log n)^2)$. (The constant implied by the "O" symbol is absolute.)

One implication of this conjecture is that searching for a prime of the form $a + kn$ will never require us to examine more than $O((\log n)^2)$ values of k. We note that this bound is polynomial in $\lg n$.

8.6 Other Conjectures about Primes

Armed with the belief that the primes behave randomly, one can make up conjectures and hypotheses ad infinitum. We mention a few of the more famous ones in this section.

The prime number theorem implies that the distance between consecutive primes can be arbitrarily large, and that the average distance between the primes less than x is about $\log x$. Gaps much larger than this seem to be rare.

The simplest way to make this observation quantitative is to state that $p_n - p_{n-1} \leq n^{\alpha+o(1)}$, where α is a small positive constant. Applying the prime number theorem and Bertrand's postulate, we see that we can take $\alpha = 1$. (See Exercise 11.) This is not the best possible value for α, however. In 1930, G. Hoheisel proved the following result.

THEOREM 8.6.1 Let p_n be the n-th prime. We have $p_n - p_{n-1} \leq n^{\alpha+o(1)}$, with $0 < \alpha < 1$.

The best current bound for the exponent is due to R. Baker and G. Harman, who showed that it suffices to take $\alpha = 107/200 = .535$. If the Riemann hypothesis holds, we can use $\alpha = 1/2$. An examination of the numerical data on prime gaps suggests that even these estimates are too large. More precisely, there is evidence to suggest that $p_n - p_{n-1}$ is never much larger than $(\log n)^2$. For example, the largest known value of $p_n - p_{n-1}$ is 778, occurring at $p_n = 42\,842\,283\,926\,129$. For this value of p_n, $(\log n)^2 = 783$ to the nearest integer.

A simple probabilistic argument supports this as well. Consider $k - 1$ random samples from the uniform distribution on $(0, 1)$. This produces k intervals whose total length is 1. The maximum interval, that is, the maximum gap between the random numbers, has expected value H_k/k. (See Exercise 38.) Now, let us imagine that the prime numbers in the interval $(0, x)$ represent independent random choices. (We ignore the fact that small primes are packed together more densely than large ones, but that has little effect on the result.) We guess that the largest gap between primes should be about

$$\frac{x \log(\pi(x))}{\pi(x)} \sim (\log x)^2.$$

This suggests the following conjecture, which the probabilist H. Cramér derived from a different heuristic assumption. (See Exercise 39.)

CONJECTURE 8.6.2 (CRAMÉR) Let p_n denote the n-th prime. Then $p_n - p_{n-1} = O((\log p_n)^2)$.

This conjecture has the implication that a search for the smallest prime $\geq n$ will terminate very quickly, after testing $O((\lg n)^2)$ numbers for primality. It also has the implication that there is a Las Vegas (probably fast, always correct) random polynomial-time algorithm for testing primality. We will discuss this in Chapter 13.

It is not known whether there are an infinite number of *twin primes*, that is, whether n and $n + 2$ are both prime for infinitely many n. One can, however, derive a conjecture about their density using a simple probabilistic argument.

Suppose that p is prime, and m, m', n are large random numbers, roughly equal. Then

$$\frac{\Pr[n, n+2 \not\equiv 0 \bmod p]}{\Pr[m, m' \not\equiv 0 \bmod p]} \approx \begin{cases} 2, & \text{if } p = 2; \\ (1 - 2/p)/(1 - 1/p)^2, & \text{if } p \text{ is odd.} \end{cases}$$

Assuming that the divisibility conditions for the various p are independent, we guess that

$$\Pr[n, n+2 \text{ both prime}] = \frac{\Pr[n, n+2 \text{ both prime}]}{\Pr[m, m' \text{ both prime}]} \cdot \Pr[m, m' \text{ both prime}]$$

$$\approx 2 \prod_{p>2} \frac{p(p-2)}{(p-1)^2} \cdot \frac{1}{(\log n)^2}.$$

Therefore, the number of twin primes less than x should be asymptotic to

$$2 \prod_{p>2} \frac{p(p-2)}{(p-1)^2} \int_2^x \frac{dt}{(\log t)^2}.$$

The coefficient of the integral in this expression is often called the *twin prime constant*; we have $2 \prod_{p>2} p(p-2)/(p-1)^2 \doteq 1.32032$.

This conjecture is supported by numerical data. For example, the above formula predicts that there should be 1249 pairs of twin primes $\leq 10^5$. There are, in fact, 1224.

A natural generalization of this conjecture is the following. Let $a_1, a_2, \ldots, a_k$ be a sequence of distinct integers. We are interested in the values n for which $n + a_1, n + a_2, \ldots, n + a_k$ are simultaneously prime. For this to be possible for infinitely many n, it is clearly necessary that

$$\nu(p) = |\{a_1 \bmod p, \ldots, a_k \bmod p\}|$$

be less than p, for each prime p. Call the sequence *admissible* if that is the case. The following is called the *strong prime tuples conjecture*:

CONJECTURE 8.6.3 Let $a_1, \ldots, a_k$ be an admissible sequence. There are infinitely many n such that $n + a_1, n + a_2, \ldots, n + a_k$ are all prime. More precisely, if

$$C(a_1, \ldots, a_k) = \prod_p \frac{(1 - \nu(p)/p)}{(1 - 1/p)^k},$$

then the number of $n \leq x$ for which $n + a_1, \ldots, n + a_k$ are all prime is asymptotic to

$$C(a_1, \ldots, a_k) \int_2^x \frac{dt}{(\log t)^k}.$$

We note that the infinite product converges, because $\nu(p) = k$ for sufficiently large p. This conjecture can be justified heuristically as was the density of twin primes. (See Exercise 42.)

The prime tuples conjecture has a number of interesting consequences, among them Artin's conjecture. It also implies a lower bound on the density of Carmichael numbers, which would have implications for primality testing. (See Chapter 9.)

One can extend the prime tuples conjecture from linear polynomials to polynomials of higher degree. Roughly speaking, the extended conjecture states that any set of polynomials not prevented in some obvious fashion from representing primes will do so simultaneously, infinitely often. We refer readers to the Notes for a discussion of this generalization.

8.7 Extensions to Algebraic Numbers

Although this book is not concerned with algebraic numbers per se, there are many problems in elementary number theory that can be readily handled with the aid of algebraic numbers. In this section we discuss how the analytic theory of algebraic numbers can be used to obtain algorithmically useful estimates. We assume some familiarity with algebraic number theory; readers are also encouraged to review the material on groups, rings, and fields in Section 2.8.

Let K be an extension field of the rationals of finite degree. We will call such fields *number fields*. A number belonging to K is called an *algebraic integer* if it satisfies a monic polynomial equation with integral coefficients. The algebraic integers form a subring of K, which we denote by A.

As with finite fields, a number field can be presented in a number of equivalent ways. We will assume that K is represented as $\mathbb{Q}[X]/(f)$, where f is an irreducible monic polynomial in $\mathbb{Z}[X]$; thus, our data for a number field always includes one specific algebraic integer α, together with its minimal polynomial.

The "complexity" of a number field K is measured by two important integer invariants, the *degree* and the *discriminant*. The first is simply the dimension of K as a vector space over $\mathbb{Q}$, and can be read off from the presentation of the field. The second is related to the size of the coefficients in a presentation. Because of its appearance in analytic theorems, we treat it in detail, after some preliminary definitions.

According to Galois theory, a number field of degree n has n distinct embeddings $\sigma_1, \ldots, \sigma_n$ into $\mathbb{C}$; these are determined by the factorization of f over $\mathbb{C}$. As with finite fields, $\mathrm{N}(x) = \prod_{1 \le i \le n} \sigma_i(x)$, and $\mathrm{Tr}(x) = \sum_{1 \le i \le n} \sigma_i(x)$ define the *norm* and *trace* of an element x. It is useful to know that both the norm and the trace can be computed efficiently, without knowing the σ_i's. (See Exercise 47.)

The ring of integers A always has an *integral basis*, that is, every element of A is a unique integer linear combination of n numbers $\beta_1, \ldots, \beta_n$. We define Δ_K, the *discriminant* of K, to be $\det(\mathrm{Tr}(\beta_i \beta_j))$.

Once we have an integral basis, we can use the above formula to compute the discriminant. However, there is no known polynomial-time algorithm for finding an integral basis, and no way to compute discriminants that does not, at least implicitly, involve factorization. (See Exercise 55.) In practice we have to settle for estimates; the following theorem is useful in this regard.

THEOREM 8.7.1 Let $f(X)$ be an irreducible monic polynomial with integer coefficients. Let $K = \mathbb{Q}(\alpha)$ be an algebraic number field, where α is a root of $f(X) = 0$. Then the discriminant of K divides the resultant of f and f'.

We leave the proof of this and several other facts about discriminants to the reader. (See Exercises 48–49.)

A good example to keep in mind is the *cyclotomic field* $K = \mathbb{Q}(\zeta_m)$, where ζ_m is a primitive m-th root of unity. In this case, it can be shown that K has degree $\varphi(m)$, and discriminant dividing $m^{\varphi(m)}$. Furthermore, the set $\{1, \zeta_m, \ldots, \zeta_m^{\varphi(m)-1}\}$ forms an integral basis.

It can be shown that if $\mathbf{I}$ is a nonzero ideal of A, then $|A/\mathbf{I}|$ is finite. This is called the *norm* of $\mathbf{I}$, and denoted $\mathrm{N}\mathbf{I}$. A nonzero ideal $\mathbf{P}$ of A is called *prime* if $A/\mathbf{P}$ is an integral domain. (For brevity's sake, we will call any such ideal a prime of K.) Because any finite integral domain is a field, $\mathrm{N}\mathbf{P} = p^f$ for some some ordinary prime number p. We call f the *degree* of $\mathbf{P}$.

Much of the theory of prime numbers can be generalized to prime ideals in a number field. For example, any nonzero ideal of A can be written as an essentially unique product of prime ideals. (Unique factorization will not hold in general for numbers.) The analog of the prime number theorem is the *prime ideal theorem*, proved by E. Landau in 1903. The theorem below presents the best error term known to date.

THEOREM 8.7.2 Let K be an algebraic number field of degree n. Let $\pi_K(x)$ denote the number of prime ideals whose norm is $\le x$. Let $\lambda(x) = (\log x)^{3/5}(\log\log x)^{-1/5}$. There is a $c > 0$ (depending on K) such that

$$\pi_K(x) = \mathrm{li}(x) + O(xe^{-c\lambda(x)}) \sim \frac{x}{\log x}.$$

By comparing this with the ordinary prime number theorem (Theorem 8.2.6), we see that in some sense "most" prime ideals have degree 1. (See Exercise 46.)

Every number field has an associated complex function called its *Dedekind zeta function*. This is a complex function, defined for $\Re(s) > 1$ by the formula

$$\zeta_K(s) = \sum_{\mathbf{A}} \frac{1}{\mathrm{N}\mathbf{A}^s},$$

and analytic continuation otherwise. This is a generalization of the Riemann zeta function and has similar properties. In particular, it has a simple pole at $s = 1$ and infinitely many zeroes in the critical strip $0 < \Re(s) < 1$. It is conjectured that all such zeroes have real part $1/2$; this is equivalent to the following estimate for the error term in the prime ideal theorem.

CONJECTURE 8.7.3 (GENERALIZED RIEMANN HYPOTHESIS) If K is an algebraic number field, then for any $\epsilon > 0$, we have

$$\pi_K(x) = \mathrm{li}(x) + O(x^{1/2+\epsilon}).$$

In contrast to the ordinary Riemann hypothesis and the ERH, we will not present any probabilistic arguments for this conjecture. (One can get an idea of the complications by noting that for quadratic fields, prime ideals of norm p occur in pairs. Hence the crude idea that p "chooses" to be the norm of a prime ideal with probability $1/\log p$ is wrong.) Rather, the GRH is supported mainly on formal grounds, and by numerical calculation.

In some cases, the Dedekind zeta function can be expressed using the Riemann zeta function and Dirichlet L-functions. For example, if K is the n-th cyclotomic field $\mathbb{Q}(\zeta_n)$, we have

$$\zeta_K(s) = \prod_{\chi} L(s, \chi'),$$

where the product runs over the $\varphi(n)$ primitive characters inducing the characters of $(\mathbb{Z}/(n))^*$. Thus we see that the Riemann hypothesis for Dedekind zeta functions includes the ERH as presented earlier.

In applications of the GRH, it is useful to know how the error term depends on K; we state this below.

THEOREM 8.7.4 Assume GRH. Let K be a number field of degree n and discriminant Δ. For $x \geq 2$, we have

$$|\pi_K(x) - \mathrm{li}(x)| = O\big(\sqrt{x}(n \log x + \log|\Delta|)\big)$$

(Here the constant implied by the "O"-symbol is absolute.)

For any ordinary prime number p, the ideal pA can be written as a product of prime ideals. In most cases, this factorization can be explicitly computed, using algorithms from Chapter 7 and the following theorem of Dedekind. This important result links the factorization of a polynomial over a finite field to an ideal factorization in a number field.

THEOREM 8.7.5 Let E and K be number fields. Assume that $K = E(\alpha)$, where α is a zero of monic irreducible polynomial f, with coefficients in E's ring of integers A. Let $\mathbf{p}$ be a prime of E, relatively prime to the discriminant of K, with norm q. If $f(X) = f_1(X) \cdots f_r(X)$ is the factorization of f in $\mathbb{F}_q[X]$, then $\mathbf{p}A = \mathbf{P}_1 \cdots \mathbf{P}_r$, where $\mathbf{P}_i = \mathbf{p}A + f_i(\alpha)A$. Furthermore, $\mathrm{N}\mathbf{P}_i = q^{\deg(f_i)}$, so each $\mathbf{P}_i$ is prime.

If the ideal $\mathbf{p}A$ is the product of distinct prime ideals of norm q, we say that $\mathbf{p}$ *splits*. It is useful to have a symbol to describe the splitting behavior of a prime ideal. Suppose that K/E is an extension of number fields; we assume that any conjugate of an element of K, relative to E, actually belongs to K. (Such extensions are called *normal*.) Let G denote the Galois group of K/E. If G is abelian, it can be shown that if $\mathbf{p}$ is a prime of E, relatively prime to the discriminant of K, there is a unique $\sigma \in G$, with the property that for every algebraic integer $x \in K$, and prime divisor $\mathbf{P}$ of $\mathbf{p}$,

$$\sigma(x) \equiv x^{\mathrm{N}\mathbf{p}} \pmod{\mathbf{P}},$$

This automorphism is called the *Artin symbol*, and denoted by $\left(\frac{K/E}{\mathbf{p}}\right)$. If K/E is not abelian, then the automorphism is only defined up to conjugation, and we use the symbol to refer to the conjugacy class.

At this point, it is helpful to consider an example. Consider the field $K = \mathbb{Q}(\sqrt{-3})$, which is the same as the cyclotomic field $\mathbb{Q}(\zeta_3)$. The minimal polynomial for ζ_3 is X^2+X+1, so $\zeta_3 = (-1+\sqrt{-3})/2$. By quadratic reciprocity, $\left(\frac{-3}{p}\right) = \left(\frac{p}{3}\right)$, so an ordinary prime p splits iff it is congruent to 1 mod 3. Let σ denote the automorphism of K that changes the sign of $\sqrt{-3}$. We therefore have

$$\left(\frac{K/\mathbb{Q}}{p}\right) = \begin{cases} 1, & \text{if } p \equiv 1 \pmod 3; \\ \sigma, & \text{if } p \equiv 2 \pmod 3. \end{cases}$$

It will be observed that in this case, the order of $\left(\frac{K/\mathbb{Q}}{p}\right)$ equals the degree of an irreducible factor of $X^2 + X + 1$ mod p. This is true in general, for normal extensions. For this reason, one does not usually care about the precise value of the Artin symbol, but only its order.

Another interesting fact is that the splitting behavior of primes in abelian extensions is determined by congruence conditions. (Note that this occurs in the above example.) The following theorem was proved by T. Takagi in 1920.

THEOREM 8.7.6 Let K/E be an abelian extension of number fields. Then there is an ideal $\mathbf{f}$ (of E's ring of integers), with the following property. If $\mathbf{p} = (\pi)$ is a principal prime ideal, all real conjugates of π are positive, and $\pi \equiv 1 \bmod \mathbf{f}$, then $\mathbf{p}$ splits in K/E.

Any such ideal is called a *conductor* of the extension K/E. There is always a conductor $\mathbf{f}$ for which $\mathrm{N}\mathbf{f}$ divides the discriminant of K. (See the Notes for better estimates.)

In the example above, we found that a prime ideal (p) in $\mathbb{Z}$ splits in $K = \mathbb{Q}(\sqrt{-3})$ whenever $p \equiv 1 \pmod 3$. This shows that 3 is a conductor for $K/\mathbb{Q}$. We note that -3 is the discriminant of K, as well.

The analog of Ankeny's theorem for number fields is the following.

THEOREM 8.7.7 Assume GRH. Let K/E be a proper abelian extension of number fields, with a conductor $\mathbf{f}$. Let Δ denote the discriminant of E. Then there is a degree 1 prime ideal $\mathbf{p}$ of E that does not split in K, relatively prime to Δ and $\mathbf{f}$, satisfying $\mathrm{N}\mathbf{p} = O((\log|\Delta| + \log \mathrm{N}\mathbf{f})^2)$. The constant implied by the "O" symbol is absolute.

This theorem gives a bound on the least prime ideal whose Artin symbol is nontrivial. There is a related theorem, which can be thought of as a generalization of Linnik's theorem, which bounds the least prime ideal with a given Artin symbol.

THEOREM 8.7.8 Assume GRH. Let K/E be normal extension of number fields, with $E \neq \mathbb{Q}$. Let σ be an automorphism of K/E, lying in the conjugacy class $\langle\sigma\rangle$. If Δ is the discriminant of K, there is a degree 1 prime ideal $\mathbf{p}$ of E with $\left(\frac{K/E}{\mathbf{p}}\right) = \langle\sigma\rangle$, relatively prime to the discriminant of K, satisfying $\mathrm{N}\mathbf{p} = O((\log|\Delta|)^2)$. The constant implied by the "O" symbol is absolute.

Finally, we give a generalization of the prime number theorem for arithmetic progressions. A weaker version of this was proved by N. Chebotarev in 1927; it is usually called the *density theorem*. We give an unconditional version, and a version that assumes GRH, with a sharper error term.

THEOREM 8.7.9 Let K/E be normal extension of number fields, of degree $m > 1$. Let σ be an automorphism of K/E, belonging to a conjugacy class of size c. Let $\pi_\sigma(x)$ denote the number of degree 1 prime ideals $\mathbf{p}$ of E, with norm $\leq x$, for which $\left(\frac{K/E}{\mathbf{p}}\right)$ is conjugate to σ. Then

$$\pi_\sigma(x) \sim \frac{c}{m}\mathrm{li}(x).$$

THEOREM 8.7.10 Assume GRH. Using the notation of Theorem 8.7.9, let Δ be the discriminant of K, and let n be its degree. Then

$$\pi_\sigma(x) = \frac{c}{m}\mathrm{li}(x) + O\left(\frac{c\sqrt{x}}{m}(n\log x + \log|\Delta|)\right).$$

The constant implied by the "O" symbol is absolute.

We now give some useful applications of the above theorems. They are all based on the following idea: if we want an ordinary prime number with certain properties, this can sometimes be achieved by first finding a suitable prime ideal dividing that prime number.

Recall that when p and q are primes with $p \mid q-1$, Theorem 8.5.3 gives a bound on the least p-th power nonresidue mod q. We now give results applicable to the dual problem, when we have fixed n and p, and want a prime q for which n is a p-th power nonresidue mod q. We first need a lemma that is purely field-theoretic in nature.

LEMMA 8.7.11 Let p be a prime, and $e \geq 1$. Let $E = \mathbb{Q}(\zeta)$ be the cyclotomic field generated by a primitive p^e-th root of unity ζ. If $n > 1$ is an integer that is not a p-th power in $\mathbb{Z}$, and $p \nmid n$, the polynomial $X^p - n$ is irreducible over E.

Proof We consider the cases $p = 2$ and $p > 2$ separately. If $p = 2$, we observe that the discriminant of E is a power of 2, whereas the discriminant of $\mathbb{Q}(\sqrt{n})$ contains some prime $q \neq 2$. (See Exercises 51 and 54.) Therefore, $\sqrt{n} \notin E$, and the degree $[E(\sqrt{n}) : E] = 2$.

If $p > 2$, let $\omega = \zeta^{p^{e-1}}$ and $F = \mathbb{Q}(\omega)$. By our choice of n, there is a prime $q \neq p$ with $\nu_q(n) \not\equiv 0 \bmod p$. Considering the norm of n, we see that $\alpha = \sqrt[p]{n} \notin F$. We can choose $\sigma \neq 1$ in the Galois group of $F(\alpha)/F$ such that $\sigma(\alpha) = \omega\alpha$. Then the order of σ is at least p, so $[F(\alpha) : F] = p$.

We now observe that $\mathbb{Q}(\omega, \alpha)$ is not abelian over $\mathbb{Q}$. (If τ is an automorphism sending ω to its complex conjugate, then $\sigma\tau(\alpha) \neq \tau\sigma(\alpha)$.) Neither, then, is the larger field $K = \mathbb{Q}(\zeta, \alpha)$. Therefore, $\alpha \notin E$, and as before, $[E(\alpha) : E] = p$. ■

Having a proper field extension in hand, we can now apply the density theorem.

THEOREM 8.7.12 Let p^e be a prime power with p prime. Let $n > 1$ be an integer that is neither divisible by p nor a p-th power. There are infinitely many primes $q \equiv 1 \pmod{p^e}$ with $n^{(q-1)/p} \not\equiv 0, 1 \pmod{q}$.

Proof Let ζ be a primitive p^e-th root of unity. We apply Theorem 8.7.9 to the case where $E = \mathbb{Q}(\zeta)$ and $K = E(\sqrt[p]{n})$. By Lemma 8.7.11, K has an automorphism $\sigma \neq 1$ fixing E. (Note that the extension K/E is normal, since E contains a primitive p-th root of unity.) Therefore, by Theorem 8.7.9, there are infinitely many degree 1 primes $\mathbf{q}$ of E, not dividing the discriminant of K, for which $\left(\frac{K/E}{\mathbf{q}}\right) = \sigma$. Let $\mathbf{q}$ be such a prime, and let $\mathbf{Q}$ be a prime of K containing $\mathbf{q}$. Then $\mathrm{N}\mathbf{q} = q$, an ordinary prime congruent to 1 mod p^e, and (by the uniqueness of Artin symbols) there is some algebraic integer $x \in K$ with $x^q \not\equiv x \pmod{\mathbf{Q}}$. Applying Theorem 8.7.5, we see that the polynomial $X^p - n$ has no linear factors mod $\mathbf{q}$, and therefore no linear factors mod q. This means that n is not a p-th power in the finite field $\mathbb{F}_q$, so $n^{(q-1)/p} \not\equiv 0, 1 \pmod{q}$, by Euler's criterion. ■

THEOREM 8.7.13 Assume GRH. Under the assumptions of Theorem 8.7.12, there is a prime $q \equiv 1 \pmod{p^e}$ with $n^{(q-1)/p} \not\equiv 0, 1 \pmod{q}$, for which $q = O(p^{2e}(\log n + \log p)^2)$.

Proof Define ζ, E, and K as in the proof of Theorem 8.7.12. We make a similar argument, except that Theorem 8.7.7 is used to get a bound on q. We first observe that Δ, the discriminant of E, divides p^{p^e}. (See Exercise 51.) Furthermore, we can take $\mathbf{f} = (1 - \zeta)^{p^e} n$ as a conductor of K/E. (See Exercise 52.) From these estimates, we obtain $|\Delta|\mathrm{N}\mathbf{f} \leq p^{2p^e} n^{p^e}$.

By Theorem 8.7.7, there is a degree 1 prime ideal $\mathbf{q}$ of E, satisfying $\mathbf{Nq} = O(p^{2e}(\log n + \log p)^2)$, that is relatively prime to np, and does not split in K/E. We observe that $\left(\frac{K/E}{\mathbf{q}}\right) \neq 1$, and finish up with an argument identical to the proof of Theorem 8.7.12. ■

By considering other number fields, it is possible to obtain many other estimates of an elementary nature. We have given some of these in Exercises 56–58.

8.8 Some Useful Explicit Estimates

In this section, we give many explicit estimates for functions related to prime numbers, without proof. These estimates are extremely useful in the analysis of number-theoretic algorithms.

THEOREM 8.8.1 We have

$$\pi(x) > \frac{x}{\log x}, \quad \text{for } x \geq 17;$$

$$\pi(x) < 1.25506\frac{x}{\log x}, \quad \text{for } x > 1;$$

$$\frac{x}{\log x + 2} < \pi(x) < \frac{x}{\log x - 4}, \quad \text{for } x \geq 55.$$

THEOREM 8.8.2 We have

$$\psi(x) \geq \vartheta(x), \quad \text{for } x \geq 0;$$

$$\psi(x) < 1.03883x, \quad \text{for } x > 0.$$

THEOREM 8.8.3 We have

$$\vartheta(x) > .84x, \quad \text{for } x \geq 101;$$

$$\vartheta(x) < 1.000081x, \quad \text{for } x > 0.$$

THEOREM 8.8.4 We have

$$p_n > n \log n, \quad \text{for } n \geq 1;$$

$$p_n < n(\log n + \log\log n), \quad \text{for } n \geq 6.$$

In the next theorem, we let

$$B = \gamma + \sum_p \left(\frac{1}{p} + \log\frac{p-1}{p}\right) \doteq .261497.$$

We call this the *prime-reciprocal constant*.

THEOREM 8.8.5 We have

$$\sum_{p\le x}\frac{1}{p} > \log\log x, \quad \text{for } x > 1;$$

$$\sum_{p\le x}\frac{1}{p} < \log\log x + B + \frac{1}{(\log x)^2}, \quad \text{for } x > 1.$$

THEOREM 8.8.6 We have

$$\prod_{p\le x}\left(1-\frac{1}{p}\right) < \frac{e^{-\gamma}}{\log x}\left(1+\frac{1}{2(\log x)^2}\right), \quad \text{for } x > 1;$$

$$\prod_{p\le x}\left(1-\frac{1}{p}\right) > \frac{e^{-\gamma}}{\log x}\left(1-\frac{1}{(\log x)^2}\right), \quad \text{for } x > 1.$$

THEOREM 8.8.7 We have

$$\varphi(n) > \frac{n}{e^{\gamma}\log\log n + \frac{3}{\log\log n}}, \quad \text{for } n \ge 3;$$

$$\varphi(n) < n, \quad \text{for } n > 1.$$

THEOREM 8.8.8 We have

$$\sigma(n) > \frac{6}{\pi^2}\frac{n^2}{\varphi(n)}, \quad \text{for } n \ge 1;$$

$$\sigma(n) < \frac{n^2}{\varphi(n)}, \quad \text{for } n > 1.$$

THEOREM 8.8.9 We have

$$\log_2 d(n) < 1.5379\frac{\log n}{\log\log n}, \quad \text{for } n \ge 3.$$

THEOREM 8.8.10 For $n \ge 3$, we have

$$\omega(n) \le \frac{\log n}{\log\log n} + 1.45743\frac{\log n}{(\log\log n)^2}.$$

THEOREM 8.8.11 For $x \ge 4$ and $2 \le y \le x$ we have $\Psi(x, y) > x^{1-(\log\log x)/(\log y)}$.

THEOREM 8.8.12 Let χ be a nonprincipal Dirichlet character mod n. Then for all k, ℓ,

$$\left|\sum_{k\le x<k+\ell}\chi(x)\right| \le 2\sqrt{n}\log n.$$

In the next theorem, we let

$$\psi(x, n, a) = \sum_{\substack{m \le x \\ m \equiv a \pmod{n}}} \Lambda(m),$$

and define $\vartheta(x, n, a)$ to be the corresponding sum restricted to primes.

THEOREM 8.8.13 For $x \ge 43$ and $a \in \{1, 3\}$ we have

$$0.40x \le \psi(x, 4, a) \le 0.59x;$$

also for $x \ge 151$, and $a \in \{1, 2\}$ we have

$$0.41x \le \vartheta(x, 3, a) \le 0.51x.$$

THEOREM 8.8.14 We have

$$\pi(x, n, a) < \frac{2x}{\varphi(n)\log(x/n)}, \quad \text{if } x > n.$$

THEOREM 8.8.15 Let $N(T)$ denote the number of zeroes of the zeta function with $0 < \Im(\rho) < T$. For all $T \ge 2$, we have

$$\left| N(T) - \frac{T}{2\pi}\log\frac{T}{2\pi} - \frac{T}{2\pi} + \frac{7}{8} \right| \le 0.137\log T + 0.443\log\log T + 1.588.$$

THEOREM 8.8.16 Let $\rho = \beta + i\gamma$ be a zero of the zeta function in the critical strip. If $\beta \ne 1/2$, then $|\gamma| \ge 545\,439\,823$ and $\beta < 1 - (R\log|\gamma/17|)^{-1}$, where $R = 9.64591$.

THEOREM 8.8.17 Assume ERH. Let G be a nontrivial subgroup of $(\mathbb{Z}/(n))^*$. Then there is an $m > 0$, not belonging to G, with $m \le 2(\log n)^2$. The least $m > 0$ in $(\mathbb{Z}/(n))^* - G$ satisfies $m \le 3(\log n)^2$.

THEOREM 8.8.18 Assume ERH. Then for $x \ge 2$, we have

$$\left| \pi(x, n, a) - \frac{\mathrm{li}(x)}{\varphi(n)} \right| \le \sqrt{x}(\log x + 2\log n).$$

THEOREM 8.8.19 Assume ERH. If $n > 2$ and $a \in (\mathbb{Z}/(n))^*$, there is a prime $p \equiv a \pmod{n}$ satisfying $p < 2(n\log n)^2$.

THEOREM 8.8.20 Assume GRH. Let K/E be a proper abelian extension of number fields, with a conductor $\mathbf{f}$. Let Δ denote the discriminant of E. There is a degree 1 prime ideal $\mathbf{p}$ of E, not splitting in K, relatively prime to $\Delta\mathbf{f}$, with $\mathrm{N}\mathbf{p} \le 18\log(|\Delta|^2\mathrm{N}\mathbf{f})^2$.

THEOREM 8.8.21 Assume GRH. Let K/E be a normal extension of number fields, with $K \ne \mathbb{Q}$. Let n and Δ be the degree and discriminant of K. If σ is an element of the Galois group of K/E, there is a degree 1 prime $\mathbf{p}$ of E, not dividing the discriminant of E, such that $\left(\frac{K/E}{\mathbf{p}}\right) = \sigma$, with $\mathrm{N}\mathbf{p} \le (4 \log|\Delta| + 2.5n + 5)^2$.

THEOREM 8.8.22 Assume GRH and use the notation of Theorem 8.8.21. Let $\pi_\sigma(x)$ be the number of primes $\mathbf{p}$ of E, not dividing the discriminant of K/E, for which $\left(\frac{K/E}{\mathbf{p}}\right)$ is conjugate to σ. Then

$$\left|\pi_\sigma(x) - \frac{c}{m}\mathrm{li}(x)\right| \le \frac{c}{m}\sqrt{x}\left(2\log|\Delta| + n\log x\right).$$

(Here m denotes the index $[K : E]$ and c is the size of σ's conjugacy class.)

8.9 Exercises

1. Program the sieve of Eratosthenes (see Chapter 9), and count the primes less than x for various values of x. Compare your results with the following approximations:

$$\pi(x) \approx \frac{x}{\log x - 1}, \qquad \text{[Chebyshev]};$$

$$\pi(x) \approx \int_2^x \frac{dt}{\log t}, \qquad \text{[Gauss]};$$

$$\pi(x) \approx \sum_{n\ge 1} \frac{\mu(n)}{n}\mathrm{li}(x^{1/n}). \qquad \text{[Riemann]}.$$

Exercises 2–15 are "elementary" in the sense that no complex analysis is required.

2. Prove that if $s > 1$, then

$$\sum_{n\ge 1} \frac{1}{n^s} = \prod_p \left(1 - \frac{1}{p^s}\right)^{-1}.$$

   Show also that $\lim_{s\to 1} \zeta(s) = \infty$, and conclude from this that there are infinitely many primes.

3. This exercise gives a tantalizing, but incorrect, heuristic argument for the prime number theorem. Using the sieve, we guess that the fraction of primes in $[1, x]$ should be about

$$\Delta(x) = \prod_{p\le\sqrt{x}} (1 - 1/p).$$

   Taking logarithms, replacing the sum by an integral, using the Taylor series for $\log(1 - 1/t)$, and ignoring second-order terms, we obtain $\log \Delta(x) = -\int_2^{\sqrt{x}} \Delta(t)/t\, dt$. Therefore

$$\frac{\Delta'}{\Delta}(x) = -\frac{\Delta(\sqrt{x})}{2x}.$$

Verify that $\Delta(x) = 1/(\log x)$ is a solution to the above equation. How do you reconcile this with Corollary 8.2.9?

4. Observe that when x is integral, $\log x! = \sum_{n \le x} \lfloor x/n \rfloor \Lambda(n)$. Now use Chebyshev's result (Theorem 8.2.5) to show the following:

   (a) $\sum_{p \le x} \frac{\log p}{p} = \log p + O(1)$;

   (b) There is a constant B such that $\sum_{p \le x} \frac{1}{p} = \log \log x + B + O(\frac{1}{\log x})$;

   (c) There is a constant C such that $\prod_{p \le x}(1 - \frac{1}{p}) \sim \frac{C}{\log x}$;

   (d) $p_n = \Theta(n \log n)$;

   (e) $\omega(n) = O(\log n / \log \log n)$;

   (f) $d(n) = 2^{O(\log n / \log \log n)}$;

   (g) $\varphi(n) = \Omega(n / \log \log n)$.

5. For this exercise you are to evaluate the constants B and C above, thereby proving Mertens's theorem.

   (a) Let $\alpha > 0$. Show that $\int_0^\infty e^{-\alpha x} \log x dx = -(\gamma + \log \alpha)/\alpha$, where γ is Euler's constant.

   (b) Show that $\zeta(s) \sim 1/(s-1)$ as $s \longrightarrow 1^+$.

   (c) Let $S(x)$ denote the sum in Exercise 4, part b), and assume $s > 1$. Integrate $\int_{2^-}^\infty x^{1-s} dS(x)$ by parts and use the result to show that $\sum_p p^{-s} \sim -\log(s-1) - \gamma + B + o(1)$, as $s \longrightarrow 1^+$.

   (d) Compare this to $\sum_p \log(1 - p^{-s})$, and conclude that

   $$B = \gamma + \sum_p \sum_{k \ge 2} \frac{1}{kp^k}$$

   and

   $$C = e^{-\gamma}.$$

6. Prove the following results, which lend credence to the idea that most prime powers are actually primes:

   (a) $\sum_{p^k \le x} 1 = \Theta(x / \log x)$;

   (b) $\sum_{p^k \le x} 1/p^k = \log \log x + O(1)$.

7. Let $a_1, a_2, a_3, \ldots$ be a sequence of complex numbers, and let $S(x) = \sum_{n \le x} a_n$. Prove that

$$\int_0^x S(t)\,dt = \sum_{n\le x}(x-n)a_n.$$

8. Let ψ be a nondecreasing function, for which $\int_0^x \psi(t)\,dt \sim x^2/2$. Show that

$$\limsup_{x\longrightarrow\infty}\frac{\psi(x)}{x} = \liminf_{x\longrightarrow\infty}\frac{\psi(x)}{x} = 1,$$

and conclude that $\psi(x) \sim x$.

9. Show that $\psi(x) - \vartheta(x) = O(\sqrt{x}(\log x)^2)$, without appealing to Chebyshev's theorem (Theorem 8.2.5).

10. (Cesàro.) Derive the asymptotic series

$$\mathrm{li}(x) \sim \sum_{n\ge 1}\frac{(n-1)!\,x}{(\log x)^n} = \frac{x}{\log x} + \frac{x}{(\log x)^2} + \cdots$$

and invert this to obtain an asymptotic series for $\mathrm{li}^{-1}(y)$. Apply Corollary 8.2.8 to obtain

$$\frac{p_n}{n} = \log n + \log\log n - 1 + \frac{\log\log n - 2}{\log n} - \frac{(\log\log n)^2/2 - 3\log n + 11/2}{(\log n)^2}$$
$$+ O\left(\left(\frac{\log\log n}{\log n}\right)^3\right).$$

11. Find a $c > 0$ with the following property: there is always a prime between x and cx. (For $c = 2$, this is called "Bertrand's postulate.")

12. Let $n > 1$ be an integer. Show that the least k such that n is not a k-th power is $O(\log\log n)$.

13. By considering $\log x! = \sum_{d\le x}\psi(x/d)$, show that the only possible limit for $\psi(x)/x$ is 1.

14. Show that $\sigma(n)\varphi(n) = \Theta(n^2)$. We have, in fact,

$$\frac{6}{\pi^2} \le \frac{\sigma(n)\varphi(n)}{n^2} \le 1.$$

15. (E. Landau.) Let $g(n) = \max\{\mathrm{lcm}(a_1, a_2, \ldots, a_k)\}$, where the maximum is taken over all tuples of positive integers summing to n.

    (a) Show that $g(n)$ is the largest possible order of any permutation on n letters.

    (b) Use the prime number theorem to show $\log g(n) \sim \sqrt{n\log n}$.

Exercises 16–19 assume the reader has some acquaintance with analytic number theory.

16. Fill in the details of the proof of the prime number theorem, as sketched in Section 8.2. The major steps are the following.

    (a) Let $f(X) = \prod_{1\le i\le r}(X - a_i)$ be a polynomial with simple zeroes, all to the left of $\Re(s) = c$. Show that if $r \ge 2$,

    $$\frac{1}{2\pi i}\int_{c-i\infty}^{c+i\infty}\frac{x^s}{f(s)}\,ds = \begin{cases}\sum_{1\le i\le r}\frac{x^{a_i}}{\prod_{j\ne i}(a_i-a_j)}, & \text{if } x \ge 1;\\ 0, & \text{if } 0 < x < 1.\end{cases}$$

    (b) When $\Re(s) > 1$,

    $$\frac{\zeta'}{\zeta}(s) = -\sum_{n\ge 1}\Lambda(n)n^{-s}.$$

    (c) For $x \ge 1$, we have

    $$\psi_1(x) - \frac{1}{2}(x-1)^2 = \frac{-1}{2\pi i}\int_{2-i\infty}^{2+i\infty}\frac{x^{s+1}}{s(s+1)}\left(\frac{\zeta'}{\zeta}(s) + \frac{1}{s-1}\right)ds.$$

    (d) The integrand in (c) is an analytic function for $1 \le \Re(s) \le 2$.

    (e) For t large, we have $\zeta'/\zeta(1 \pm it) = (\log t)^{O(1)}$.

    (f) The formula of (c) holds if the line $\Re(s) = 2$ is replaced by $\Re(s) = 1$.

    (g) If $f(t)$ is integrable over the real line, then $\int_{-\infty}^{\infty} x^{it} f(t)\,dt \longrightarrow 0$ as $x \longrightarrow \infty$.

17. Prove the following fact. Suppose that for some $\sigma > 0$, we have $\sum_{1\le i\le x} a_i = O(x^\sigma)$, as $x \longrightarrow \infty$. Then $\sum_{n\ge1} a_n n^{-s}$ represents an analytic function in the half plane $\Re(s) > \sigma$.

18. Let $1/2 \le \sigma < 1$. Show that the following statements are equivalent:

    (a) The Riemann zeta function has no zeroes with real part greater than σ;

    (b) $\psi(x) = x + O(x^{\sigma+\epsilon})$ for every $\epsilon > 0$;

    (c) $\vartheta(x) = x + O(x^{\sigma+\epsilon})$ for every $\epsilon > 0$;

    (d) $\pi(x) = \mathrm{li}(x) + O(x^{\sigma+\epsilon})$ for every $\epsilon > 0$.

19. (Continuation.) Generalize this to primes in arithmetic progressions. In particular, show the following are equivalent:

    (a) No Dirichlet L-function has a zero with real part greater than σ;

    (b) $\pi(x, n, a) = \mathrm{li}(x)/\varphi(n) + O(x^{\sigma+\epsilon})$, for every $\epsilon > 0$.

20. (Backlund.) The zeta function has the following asymptotic expansion:

$$\zeta(s) \sim \sum_{1\le n<N} n^{-s} + \frac{N^{1-s}}{s-1} + \frac{N^{-s}}{2} + \sum_{k\ge 1} \frac{B_{2k}}{(2k)!} s^{\overline{2k-1}} N^{-s-2k+1}.$$

Here $s^{\overline{m}}$ denotes the m-th rising power $s(s+1)\cdots(s+m-1)$, and B_{2k}, for $k \ge 1$, are the Bernoulli numbers. (We have $B_2 = 1/6$, $B_4 = -1/30$, and so on.) Show that the truncated series, up to but not including the B_{2k} term, approximates $\zeta(s)$ with an error no more than

$$\left| \frac{s+2k-1}{\Re(s)+2k-1} \right|$$

times the absolute value of the B_{2k} term.

21. (Stieltjes.) For the logarithm of the gamma function, we have the following asymptotic expansion:

$$\log\Gamma(s) \sim \left(s - \frac{1}{2}\right)\log s - s + \frac{1}{2}\log(2\pi) + \sum_{k\ge 1} \frac{B_{2k}}{2k(2k-1)s^{2k-1}}.$$

Let $s = re^{i\theta}$ with $r > 0$ real and $-\pi < \theta < \pi$. Show that the truncated series, up to but not including the B_{2k} term, approximates $\log\Gamma(s)$ with an error no more than $(\cos(\theta/2))^{-2k}$ times the absolute value of the B_{2k} term.

22. (Riemann.) Prove the functional equation

$$\zeta(1-s) = 2(2\pi)^{-s}\Gamma(s)\cos\left(\frac{\pi s}{2}\right)\zeta(s).$$

Together with the results in Exercises 20–21, this gives a way to compute the zeta function in the left half plane.

23. It can be shown that

$$\xi(s) = \Gamma(s/2-1)(s-1)\pi^{-s/2}\zeta(s)$$

is real on the line $s = 1/2 + it$. Use this formula to compute the first few complex zeroes of $\zeta(s)$.

24. Although the behavior of the zeta function on the critical line is not well understood, it does have an explicit Fourier transform there. Show that

$$\frac{\zeta(1/2+it)}{1/2+it} = \int_{-\infty}^{\infty} (e^{-x/2}\lfloor e^x \rfloor - e^{x/2})e^{-ixt}\,dx.$$

25. The purpose of the following exercise is to demonstrate that the zeta function appears in the real world.

    (a) Prove that if $s > 1$,

    $$\int_0^\infty \frac{x^{s-1}\,dx}{e^x - 1} = \Gamma(s)\zeta(s).$$

    (b) Planck's radiation law states that the energy density of the blackbody spectrum (per unit frequency ν) is

    $$\frac{8\pi h}{c^3}\frac{\nu^3\,d\nu}{e^{h\nu/kT} - 1}.$$

    (Here T is the temperature, and c, h, and k are physical constants.) Integrate this over all frequencies, to show that the total energy varies as the fourth power of the temperature.

26. Let $P(n) = \max\{P(n,a) : a \in (\mathbb{Z}/(n))^*\}$, where $P(n,a)$ is defined as in Theorem 8.5.6. Show that $P(n) = \Omega_\infty(n \log n)$.

27. (Continuation.) Assuming the truth of Conjecture 8.5.10, show how to compute $P(n)$ using $n^{1+o(1)}$ bit operations. (It is not known how to do this in polynomial time.)

28. Extend Cramér's argument for the Riemann hypothesis to a heuristic argument that $\pi(x, n, a) = \mathrm{li}(x)/\varphi(n) + x^{1/2+o(1)}$.

29. (Shapiro.) This exercise requires some familiarity with mathematical logic. Find a decideable predicate, say $\phi(x_1, \ldots, x_r)$, with the property that the ERH holds iff $\forall x_1 \cdots x_r \phi(x_1, \ldots, x_r)$ is a true statement about the natural numbers.

30. Assume ERH. Give a randomized algorithm for finding a prime p congruent to a mod n. Your algorithm should produce a prime $p = n^{O(1)}$, and run in expected polynomial time.

31. Let C denote an acceptable value of Linnik's constant (i.e., any number greater than 2, assuming ERH). Show that there is an infinite sequence of primes $q_1, q_2, \ldots$ for which the least quadratic nonresidue mod q_i is $\geq (1/C + o(1)) \log q_i$. Show further that this holds for the least primitive root.

32. Assume ERH. Show that the k-th smallest quadratic nonresidue mod p is $O((\log p + k)^2)$.

33. [Open] Assume ERH. Let $a \in \mathbb{F}_p^*$. Show that the least x with $\left(\frac{x^2-a}{p}\right) = -1$ is $(\log p)^{O(1)}$.

34. (H. W. Lenstra.) Show that if χ is a character of $(\mathbb{Z}/(n))^*$, with a kernel at most $\sqrt{n}$ in size, then there is a number a, with $a = O((\log n)^2)$, for which $\chi(a) \neq 0, 1$.

35. Let $k > 1$. Use Euler's criterion and Linnik's theorem to design an algorithm for testing whether an integer n is a k-th power, that runs quickly on average.

36. (a) Consider the following process: choose random samples from $\{1, \ldots, m\}$ until every element of this set has been observed. Show that the expected number of samples is mH_m, where H_m is the m-th harmonic number. (By Theorem 2.5.3, this is asymptotic to $m \log m$.)

    (b) Suppose we do the experiment of (a) for each $m \geq 1$. Let $W(m)$ be the number of samples obtained at the m-th trial. Show that with probability 1, $W(m) \leq (2 + \epsilon)m \log m$.

    (c) (McCurley.) Define $P(n)$ as in Exercise 26. Give a heuristic argument that $P(n) \leq (2 + o(1))\varphi(n)(\log n)^2$.

37. Let $n > 1$ be an integer. Consider the following random experiment. For each $p = 2, 3, 5, \ldots$ choose a random element of $(\mathbb{Z}/(n))^*$ (with the uniform distribution). Prove the following results:

    (a) If n is prime, the least p at which a quadratic nonresidue is obtained is almost surely $O((\log n)(\log \log n))$;

    (b) The least p at which a primitive root is obtained is almost surely

    $$O((\log n)(\log \log n)^2);$$

    (c) In general, the least p at which the samples generate $(\mathbb{Z}/(n))^*$ is

    $$O((\log n)(\log \log n)).$$

38. Consider the following experiment. Choose $n - 1$ independent random samples from the uniform distribution on $[0, 1]$, and arrange them in order, to get random variables $X_1 < X_2 < \cdots < X_{n-1}$. Further, let $X_0 = 0$ and $X_n = 1$.

    (a) Show that the largest gap (i.e. $X_i - X_{i-1}$) has expected value H_n/n, where H_n is the n-th harmonic number.

    (b) If the experiment is done independently for each $n \geq 1$, the largest gap in the n-th experiment is almost surely $\leq (2 + o(1))(\log n)/n$.

39. (Cramér.) Give a heuristic argument that $p_n - p_{n-1} = O((\log n)^2)$, based on the idea that n is prime with probability $1/\log n$.

40. (Backlund.) Show how to construct a sequence of m successive composite integers, using $O(m^2)$ bit operations.

41. (Lander and Parkin.) The following method can be used to search for "record" values of $p_n - p_{n-1}$, using a small amount of space. Suppose that p is a prime, below which all gaps between primes are $< \Delta$. Starting from $q = p + \Delta$, search backward until a prime p' is found. If $p' = p$, a new record value is computed by finding the first prime $\geq q$. If $p' < p$, the value of p' is replaced by p and the search begins anew. Analyze this algorithm.

42. Give a heuristic argument for Conjecture 8.6.3.

43. (Riesel.) Show that a sequence $a_1, \ldots, a_k$ can be checked for admissibility (as in Conjecture 8.6.3) in polynomial time.

44. Recall that a Mersenne prime has the form $2^p - 1$ (p is necessarily prime), and a Fermat prime has the form $2^{2^k} + 1$. Give heuristic arguments to support the conjecture that there are infinitely many Mersenne primes, and only finitely many Fermat primes.

45. (Feynman.) Let $N = m^n$, a perfect n-th power. Using a Taylor approximation, we have $(m+1)^n - m^n \approx nN^{1-1/n}$. This suggests we should assign a "probability" of $1/(nz^{1-1/n})$ to the event that z is an n-th power, and that the "expected number" of integer solutions to $x^n + y^n = z^n$ with $x, y \geq B$ is about

$$\int_B^\infty \int_B^\infty \frac{dx\,dy}{n(x^n + y^n)^{1-1/n}}.$$

Estimate this integral. What does your estimate suggest about Fermat's last theorem?

Exercises 46–58 assume some knowledge of algebraic number theory.

46. Let $\mathbf{p}$ stand for a prime of the number field K. Show that

$$|\{\mathbf{p} : \mathbf{Np} \leq x \text{ and } \deg \mathbf{p} = 1\}| \sim \mathrm{li}(x),$$

as $x \longrightarrow \infty$.

47. Let $K = \mathbb{Q}(\alpha)$ be a number field, of degree n. Consider K as a vector space over $\mathbb{Q}$, and let $A = (a_{ij})$ be the linear operator representing multiplication by α.

    (a) Show that $\mathrm{Tr}(\alpha) = a_{11} + \cdots + a_{nn}$.

    (b) Show that $\mathrm{N}(\alpha) = \det A$.

    (c) Give deterministic polynomial time algorithms for computing traces and norms.

48. (Continuation.) This exercise gives several other definitions (and a non-definition!) of the discriminant.

    (a) Let $\beta_1, \ldots, \beta_n \in K$. Prove that $\det(\mathrm{Tr}(\beta_i \beta_j)) = \det(\sigma_i(\beta_j))^2$. (This gives another way to calculate Δ_K, if you know an integral basis and its conjugates.)

    (b) The *discriminant* of an element $\gamma \in K$ is $\mathrm{disc}(\gamma) = \det(\mathrm{Tr}(\gamma^{i+j}))$. (This agrees with the field discriminant if $\{1, \gamma, \ldots, \gamma^{n-1}\}$ forms an integral basis.) Show that if $g(X)$, the minimal polynomial for γ, has degree n, then

    $$\mathrm{disc}(\gamma) = (-1)^{\binom{n}{2}} \mathrm{N}(g'(\gamma))$$

    (c) Under the same assumptions, show that

    $$\mathrm{disc}(\gamma) = (-1)^{\binom{n}{2}} R(g, g').$$

    (Here R is the resultant, defined as in Exercise 15.)

    (d) (Dedekind.) Consider the number field K defined by a root of $X^3 - X^2 - 2X - 8$. Let A be the ring of algebraic integers in K. Show that the discriminant of K does not equal $\gcd_{\alpha \in A} \mathrm{disc}(\alpha)$.

49. (Continuation.) Let α be an algebraic integer in K, which is a zero of a monic polynomial $f(X) \in \mathbb{Z}[X]$. (This need not be the minimal polynomial.) Prove the following results, which are useful in estimating discriminants.

    (a) Δ_K, the discriminant of K, divides $\mathrm{disc}(\alpha)$.

    (b) $\mathrm{disc}(\alpha)$ divides the norm of $f'(\alpha)$.

    (c) $\mathrm{disc}(\alpha)$ divides the resultant of f and f'.

50. Let $f(X) \in \mathbb{Z}[X]$ be a monic polynomial of degree n, whose coefficients are bounded by B in absolute value. Show how to compute the resultant $R(f, f')$ in time bounded by a polynomial in n and $\log B$.

51. Prove the following bounds.

    (a) If d is a nonzero integer, the discriminant of $\mathbb{Q}(\sqrt{d})$ divides $4d$.

    (b) More generally, the discriminant of $\mathbb{Q}(\sqrt[n]{a})$ divides $n^n a^{n-1}$, when a is a nonzero integer.

    (c) The discriminant of $\mathbb{Q}(\zeta_n)$ divides $n^{\varphi(n)}$.

52. Define ζ, K, and E as in the proof of Theorem 8.7.13. Prove that $(1 - \zeta)^{p^e} n$ is a conductor for K/E.

53. In the text, we gave generalizations of Ankeny's theorem, Linnik's theorem, and the prime number theorem for arithmetic progessions to algebraic numbers. Verify that these results imply the corresponding theorems for $\mathbb{Z}$.

54. Let d be a squarefree integer. Prove the following facts.

   (a) Verify that the ring of integers in $\mathbb{Q}(\sqrt{d})$ has the basis $\{1, \beta\}$, where $\beta = (1 + \sqrt{d})/2$ (if $d \equiv 1 \pmod 4$) or $\sqrt{d}$ (otherwise).

   (b) The discriminant is d (if $d \equiv 1 \pmod 4$) or $4d$ (otherwise).

55. If $n > 1$ is an integer, it has a unique representation as $n = m^2 n'$, where m and n' are positive integers and n' is squarefree. We call n' the *squarefree part* of n.

   (Chistov.) Let D be a function with the following properties. When $f \in \mathbb{Z}[X]$ is irreducible, $D(f)$ is the discriminant of the number field generated by a root of $f(X) = 0$. (We don't care what value D has when its input is not irreducible.) Suppose we charge 1 step for the evaluation of D. Show how to compute the squarefree part of any $n > 1$ in polynomial time.

56. Assume GRH. Show that the least quadratic nonresidue mod p is $\Omega_\infty((\log p)(\log\log p))$.

57. Assume GRH. Let p and q be distinct prime numbers, and $k > 1$. Show that there is a prime $n \equiv 1 \pmod k$, such that p generates $(\mathbb{Z}/(n))^*/((\mathbb{Z}/(n))^*)^k$ and $q \in ((\mathbb{Z}/(n))^*)^k$, with $n = O(k^6(\log(pqk))^2)$.

58. Assume GRH. Assume $n > 1$. Let $a, d \in (\mathbb{Z}/(n))^*$ (so that $|a|, |d| \leq n$). Show that there is a prime p, such that $\left(\frac{d}{p}\right) = +1$ and $p \equiv a \pmod n$, with $p = O(n^2(\log n)^2)$.

8.10 Notes on Chapter 8

8.1

For the Ishango bone, see de Heinzelin [1962], who determined a date of c. 6500 B.C. for this artifact. Recent scholarship (see, e.g., Brooks and Smith [1987]; Brooks et al. [1995]) suggests that this artifact probably dates to c. 20,000 B.C.

For the Pythagoreans and prime numbers, see T. Heath [1921, pp. 72–74].

Edwards [1974] is the best reference on the early history of analytic number theory. Our abbreviated history is based on the following sources: Euler [1744]; Gauss [1849]; Legendre [1798, 1830]; Chebyshev [1852]; Riemann [1860]; and Mertens [1874a]. Dirichlet was apparently familiar with the prime number theorem; see Dirichlet [1889, p. 372]. The prime number theorem was proved independently by Hadamard [1896] and de la Vallée Poussin [1896a].

For some entertaining treatments of prime number lore, see Shanks [1978, 1985]; Zagier [1977]; and Ribenboim [1988b, 1994].

Several notions of complexity theory have their orgins in the analytic number theory tradition. In particular, the emphasis on the asymptotic growth rates of functions and the consequent de-emphasis on constant factors is probably due to the influence of E. Landau [1909, 1927a]. Bachmann [1894] invented big-O notation, which is now a staple of computer science.

Some introductory texts on analytic number theory include Ingham [1932]; Apostol [1976]; and H. Davenport [1980]. More advanced texts include Landau [1927a]; Prachar [1957]; Ivić [1985]; and Titchmarsh [1986]. For historical information, the reader should consult Landau [1909]; Bohr and Cramér [1923]; and Edwards [1974].

The article by Diamond [1982] provides a review of elementary methods in prime number theory, including references for elementary proofs of the prime number theorem. For sieves, see the survey article of Deshouillers [1979] and the books by Halberstam and Richert [1974] and Hooley [1976].

8.2

Our plausibility argument for $\psi(x) \sim x$ was given, in a slightly different form, by G. Hardy [1940b]. (See also Huxley [1972].) It can be made rigorous using so-called Tauberian theorems; see Ingham [1945].

Lemma 8.2.1 is not best possible, and can be improved to $\psi(x) - \vartheta(x) = \Theta(\sqrt{x})$. (See the solution to Exercise 9.)

Theorem 8.2.5 is due to Chebyshev [1851, 1852]. Our proof follows Erdős [1932] for the upper bound, and Nair [1982] for the lower bound. (The idea of Nair's proof also appears in Gel´fond [1946].)

Many authors use li(x) to denote the function

$$\mathrm{li}_0(x) = \lim_{t \to 0^+} \int_0^{1-t} \frac{dt}{\log t} + \int_{1+t}^{x} \frac{dt}{\log t}.$$

(This limit is a so-called *Cauchy principal value*.) This differs from our logarithmic integral by the constant $\mathrm{li}_0(2) \doteq 1.045164$. This makes no difference to the statement of the prime number theorem, unless one is interested in explicit bounds.

De la Vallée Poussin [1896a] showed that Theorem 8.2.6 and Corollary 8.2.7 hold with $\lambda(x) = \sqrt{\log x}$. This estimate was improved by various authors, who found successively larger zero-free regions for the zeta function. The error term in Theorem 8.2.6 derives from work of I. Vinogradov and Korobov, but the final details are due to A. Walfisz and H.-E. Richert. (See Walfisz [1963].) Vinogradov and Korobov claimed that one could take $\lambda(x) = (\log x)^{3/5}$, but this was apparently unjustified. (Unfortunately, this latter bound is often quoted.)

Ivić [1985] and Karatsuba [1990] gave full proofs of Theorem 8.2.6, and reviewed efforts to improve the prime number theorem.

The error term in Theorem 8.2.8 follows easily from the other forms of the prime number theorem, but we have not found it in the literature. Theorem 8.2.9 is a corrected version of a result from A. Vinogradov [1963].

Theorem 8.2.9 gives the asymptotic density of the numbers divisible by no prime $\leq x$. In applications, it is often useful to estimate the fraction of such numbers in a finite interval. Jurkat and Richert [1965] proved that

$$|\{n \leq y : \text{ no } p \leq x \text{ divides } n\}| = y \prod_{p \leq x} \left(1 - \frac{1}{p}\right)\left(1 + O\left(\frac{1}{\log y}\right)\right),$$

for $\log x \leq (\log y)/(2 \log\log 3y)$. (See also Warlimont [1970].) For the analogous problem concerning numbers relatively prime to a given modulus, see Suryanarayana [1974] and references therein.

References for our proof that $\psi(x) \sim x$ can be found in the solution to Exercise 16. For other short proofs of the prime number theorem, see Littlewood [1971] and Newman [1980].

Many authors have studied the distribution of integers that are, in some sense, "approximately prime." An example of this sort of result is the theorem of Gegenbauer [1885], which states that the fraction of squarefree numbers $\leq x$ is asymptotic to $6/\pi^2$. As another example, E. Landau [1900] proved that for each $k \geq 1$, the number of positive integers $\leq x$ that are products of k primes is asymptotic to

$$\frac{1}{(k-1)!} \frac{x(\log\log x)^{k-1}}{\log x}.$$

(See also E. Landau [1911].) Wintner [1941] gave this result a probabilistic interpretation.

8.3

For references to tables comparing $\pi(x)$ to various approximations, see the solution given for Exercise 1.

The proof of Lemma 8.3.2 can be found in Feller [1971, p. 238].

Riemann [1860] called Conjecture 8.3.1 (or, more exactly, the equivalent assertion about the zeta function) "very likely." The heuristic model we have used to argue for the Riemann hypothesis is due to Cramér [1935, 1937]. He showed that

$$\limsup_{x \to \infty} \frac{|\Pi(x) - \mathrm{li}(x)|}{\sqrt{(x \log\log x)/(\log x)}} = \sqrt{2},$$

almost surely. (See also Ducray [1965]; one presumes the argument there is heuristic.) If this held for $\pi(x)$, it would improve Theorem 8.3.3. On the other hand, Littlewood [1914] proved

$$|\pi(x) - \mathrm{li}(x)| = \Omega_\infty\left(\frac{\sqrt{x}\log\log\log x}{\log x}\right).$$

Among other things, this invalidates the idea that Riemann's formula, $\mathrm{li}(x) + \sum_{k>1} \mu(k)\mathrm{li}(x^{1/k})/k$, is always a better approximation to $\pi(x)$ than $\mathrm{li}(x)$ is.

Denjoy [1931] gave another "probability" argument for the Riemann hypothesis. Briefly, the idea is the following. Consider the sum

$$M(x) = \sum_{n<x} \mu(n);$$

if n is a large squarefree number, there is no particular reason to suppose it should have an odd or an even number of prime factors. Therefore we could regard $\mu(n)$ as a bounded random variable whose distribution is symmetric around 0. If the various values of $\mu(n)$ were independent, we would have $M(x) = O(x^{1/2+\epsilon})$ for every $\epsilon > 0$, which implies the Riemann hypothesis. (We would also have $M(x) \neq O(\sqrt{x})$, Mertens's conjecture to the contrary.) Good and Churchhouse [1968] showed that the hypothesis of independence agrees fairly well with the numerical data. Similar probabilistic arguments for the Riemann hypothesis were given by Wintner [1944] and again by Denjoy [1964]. See also P. Davis [1981]; and Wilf [1987].

Clearly, no one can mistake these probabilistic arguments for rigorous mathematics and remain in a state of grace. Nevertheless, they are useful in making educated guesses as to how number-theoretic functions should "behave." It is interesting to read Cramér's thoughts on this topic:

> If we are interested in the distribution of a given sequence S of integers, we then consider S as a member of an infinite class C of sequences, which may be concretely interpreted as the possible realizations of some game of chance. It is then in many cases possible to prove that *with a probability* = 1, a certain relation R holds in C, i.e. that in a definite mathematical sense "almost all" sequences of C satisfy R. Of course we cannot in general conclude that R holds for the particular sequence S, but results suggested in this way may sometimes afterwards be rigorously proved by other methods.

Others were deeply, and rightly, suspicious of such methods. Although they themselves contributed several heuristic estimates to the literature, G. Hardy and Littlewood [1922] wrote a scathing critique of probabilistic sieve arguments for the density of primes, and summed up their opinion with the following footnote: "*Probability* is not a notion of pure mathematics, but of philosophy or physics."

Littlewood [1965] believed the Riemann hypothesis was false.

The history of number theory is rife with plausible conjectures that have turned out to be false. A classic example is the inequality $\pi(x) \leq \mathrm{li}(x)$, which holds for every $x < 10^8$ but must eventually fail, by a theorem of Littlewood [1914]. In other cases, reasonable conjectures are inconsistent. For example, because the average density of primes decreases, we might guess that $\pi(n+x) - \pi(n) \leq \pi(x)$, whenever $x \geq 2$. (This conjecture is from Hardy and Littlewood [1922, p. 54].) However, Hensley and Richards [1974] showed that this contradicts the prime tuples conjecture.

Heuristic arguments such as we have given must be distinguished from *probabilistic number theory*, which provides rigorous proofs of number-theoretic facts using probability theory. A prototypical result in this area is the theorem of Erdős and Kac [1940], which states that if we choose a random number $n \leq x$, then $\omega(n)$ has an approximate normal distribution, with mean and variance asymptotic to $\log\log x$. (See Rényi and Turán [1958] for an estimate of the error in this approximation.) To be sure, this is suggested by the heuristic idea that the residues $n \bmod p$ are independent; the point is to obtain a rigorous proof. Expositions of this theorem have been given by Billingsley [1969, 1973]. For more on probabilistic number theory, see Kac [1959]; and Elliott [1979, 1980].

For a dictionary between analytic number theory and probability theory, see Elliott [1972]. Shlesinger [1986] used a fractal random walk to argue against the Riemann hypothesis. Julia [1990] discussed the connections between prime number theory and statistical mechanics. Lapidus and Maier [1991] related the Riemann hypothesis to fractal geometry.

Various authors constructed other models for the primes, supposedly more realistic than Cramér's. One appealing model is the *random sieve,* due to Hawkins [1958]. Starting with the number 2, one repeatedly performs the following operation: choose the smallest number m that has not been labelled, label it "prime," and with probability $1/m$ cross off each higher number in the list. Many theorems and conjectures of analytic number theory can be proved for the random sieve, including the Riemann hypothesis. For this work, see Hawkins [1974]; Wunderlich [1974, 1976]; Neudecker [1975]; and Heyde [1978]. We note that the random sieve distributes primes fairly among arithmetic progressions, but the analog of the ERH does not seem to have been proved. For other deterministic prime number simulacra, see Gardiner et al. [1955]; Erdős and Jabotinsky [1958], P. Stein [1974], and references therein.

The bounds of Theorem 8.3.3 are due to von Koch [1901], except for the estimate for p_n, which appeared in Robin [1983]. We give explicit versions of these results below.

$$|\pi(x) - \mathrm{li}(x)| \le \frac{\sqrt{x}\log x}{8\pi} \qquad \text{for } x \ge 2657\,.$$

$$|\psi(x) - x| \le \frac{\sqrt{x}(\log x)^2}{8\pi} \qquad \text{for } x \ge 73\,.$$

$$|p_n - \mathrm{li}_0^{-1}(n)| \le \frac{\sqrt{n}(\log n)^{5/2}}{2\pi} \qquad \text{for } x \ge 109\,.$$

$$\left|\prod_{p<x}\left(1 - \frac{1}{p}\right) - \frac{e^{-\gamma}}{\log x}\right| \le \frac{e^{-\gamma}}{8\pi}\left(\frac{3 + 5/\log x}{\sqrt{x}}\right) \qquad \text{for } x \ge 8\,.$$

The first two and the fourth are from Schoenfeld [1976]. The third can be proved by combining (6.3) of this paper with Theorem 9 of Robin [1983]; here $\mathrm{li}_0(x) = \int_0^x dt/\log t$.

The Riemann hypothesis is supported by various theorems stating that the zeta function has "many" zeroes with real part 1/2. The best result along these lines is due to Conrey [1989], who showed that the density of such zeroes is at least 2/5.

Experimental investigation of the Riemann hypothesis has always been a challenging computational problem. (See A. Cohen [1988] for a review.) Riemann computed several zeroes of the zeta function, finding the first to be about $1/2 + 14i$ (see Figure 3 of Hejhal [1987]). Similar calculations were done by Stieltjes, and also by Gram (see Gram [1895]). We summarize work since then in the table below; a number T in the second column indicates that all zeroes with $|\gamma| < T$ have real part 1/2. (In two cases we estimated this with Theorem 8.8.15, the largest zero not having been reported.)

Riemann apparently did his computations by hand (see the carries in column additions in Edwards [1974, p. 5]). Gram and Backlund may have used mechanical calculators; Hutchinson and Titchmarsh certainly did. All the later computations used electronic computers.

Table 8.1
Systematic Checks on the Riemann Hypothesis

Date	Height	Reference	Comments
1902	50	Gram [1902]	Euler-Maclaurin expansion
1918	200	Backlund [1918]	
1925	300	Hutchinson [1925]	
1935	390	Titchmarsh [1935]	Riemann-Siegel formula
1936	1468	Titchmarsh [1936]	Zero count off by 1!
1950	1540	Turing [1953]	Examined $24938 \leq t \leq 25735$
1956	9878	D. H. Lehmer [1956a]	
1956	21943	D. H. Lehmer [1956b]	
1958	29750	Meller [1958]	Started at height 14036
1966	170571	Lehman [1966]	
1968	> 1 894 434	Rosser et al. [1969]	To number 3 502 494; see Musser et al. [1969]
1979	> 35 018 259	Brent [1979]	To number 81 000 001
1982	81 702 130	Brent et al. [1982]	
1983	119 590 809	van de Lune et al. [1983]	
1986	545 439 823	van de Lune et al. [1986]	

The largest known zeroes of the zeta function have been found by Odlyzko [1987a]. This paper reports that all zeroes numbered $10^{12} + 1$ (approximately 26 76533 95649) to $10^{12} + 10^5$ (approximately 26 76534 21321) satisfy the Riemann hypothesis. Since then, Odlyzko (personal communication) has computed 175 million consecutive zeroes near number 10^{20} (about 1.5×10^{19}) and found all of them to have real part $1/2$.

Shanks [1962] proposed checking the Riemann hypothesis by examining the Padé table of $\zeta(s/(s+1))$. For a computational check of a necessary condition for the Riemann hypothesis, see Matiyasevich [1982].

The simplest way to compute the zeta function in the critical strip is to use the asymptotic expansion given in Exercise 20. (Although Euler [1741] had evaluated the zeta function at small integers in this way, the extension to complex arguments was apparently first used by Gram [1902].) Roughly speaking, one needs $O(t)$ terms of this expansion to evaluate $\zeta(\sigma + it)$ in the critical strip. All modern checks of the Riemann hypothesis have used the so-called *Riemann-Siegel formula* (actually from Siegel [1932]), which requires $O(\sqrt{t})$ terms. Odlyzko and Schönhage [1988] gave an algorithm to evaluate $\zeta(s)$ in the critical strip with good amortized behavior. For more about these methods, see Odlyzko [1987b] and references therein.

Various authors have made conjectures about the vertical distribution of the zeroes of the zeta function in the critical strip. H. Montgomery [1973, 1975, 1980] derived a limiting distribution for $\pi(x) - \mathrm{li}(x)$ from a "pair correlation hypothesis," suggested by the idea that the zeroes come from the eigenvalues of some (as yet unknown) Hermitian operator. Odlyzko [1987a] and Hejhal [1987] compared the zero distribution to the limiting distribution coming from random Hermitian matrices, and found excellent numerical agreement.

The explicit formula for ψ_1 can be found in Ingham [1932, p. 73]. For the evaluation of $\sum_\rho \left(1/\rho + 1/\bar{\rho}\right)$, see Edwards [1974, p. 67]. (This was apparently known to Riemann.) The asymptotic formula for $N(t)$ is due to von Mangoldt [1905]; for a proof, see Edwards [1974, pp. 132 ff.].

8.4

Legendre reduced the law of quadratic reciprocity to a then-unproved assumption, that every arithmetic progression contains a prime. Dirichlet [1837a, 1837b] showed that there are infinitely many primes congruent to $a \pmod{n}$, when a and n are relatively prime. The crucial step in his proof is showing that $L(1, \chi) \neq 0$. The prime number theorem for arithmetic progressions was proved by de la Vallée Poussin [1896b].

The error term in Theorem 8.4.2 is from Walfisz [1963]. It is a difficult problem to obtain accurate estimates in which the dependence on n is made explicit. Roughly speaking, there is a tradeoff between the size of the error term and the number of moduli to which the estimate applies. The book of H. Davenport [1980] provides a good review of this problem.

Walfisz [1936] showed that if $k > 0$, there is a $c > 0$ for which

$$\pi(x, n, a) - \frac{\mathrm{li}(x)}{\varphi(n)} = O\left(xe^{-c\sqrt{\log x}}\right),$$

uniformly for $n \leq (\log x)^k$. However, this relies on a theorem of Siegel [1935], which was proved nonconstructively. If $k > 1$, there is no known algorithm to compute the constant implied by the "O" symbol. According to Hinz [1980, Korollar 1.4], it is possible to replace $\sqrt{\log x}$ by $\lambda(x)$ in Walfisz's theorem. This range of n can also be enlarged if one is willing to throw out the multiples of a few "bad" moduli. See, e.g., Alford, Granville, and Pomerance [1994a, §3].

One can also improve the estimate if one knows that no $L(s, \chi)$ arising from a character of $(\mathbb{Z}/(n))^*$ has a real zero close to 1 (in a sense we will not make precise). Such zeroes, called *Siegel zeroes*, can only occur when χ is a real character; the modulus n is called *exceptional* if one exists.

There are many "equivalent forms" of the prime number theorem for arithmetic progressions, similar to Corollaries 8.2.7 and 8.2.8. Usually, one obtains these by simply dividing by $\varphi(n)$, to compensate for the fact that only one in $\varphi(n)$ primes belong to the residue class. Thus, for example,

$$\sum_{\substack{p<x \\ p\equiv a \pmod{n}}} \Lambda(p) \sim \frac{x}{\varphi(n)}.$$

For such results, see Landau [1909] and Shapiro [1946]. A generalization of Mertens's theorem to primes in progressions, too complicated to quote here, appears in K. Williams [1974].

Apparently, Piltz [1884] first conjectured that all the complex zeroes of Dirichlet L-functions are on the critical line.

A theorem of Bombieri [1965] shows that the ERH is, roughly speaking, "true on average." In particular, the maximum error in the approximation $\pi(x, n, a) \approx \mathrm{li}(x)/\varphi(n)$ is bounded by $x^{1/2+o(1)}$, if one averages over all moduli $\leq N$ and takes $x = N^{2+o(1)}$. For a bound on the "variance" in this estimate, see Gallagher [1968].

Systematic computations of zeroes of Dirichlet L-functions, showing that the ERH holds in certain rectangles, were performed by Davies and Haselgrove [1961], Davies [1965], Spira [1969], and Rumely [1993]. This last computation verified the ERH for all moduli ≤ 13 up to height 10 000, all moduli ≤ 72 up to height 2500, and miscellaneous other moduli to various heights. Zeroes of L-functions were also computed, but not systematically, by Weinberger [1975] and Hejhal [1987]. Shanks [1973] compared $L(1, \chi)$ with asymptotic estimates of Littlewood [1928] and found no evidence against the ERH.

Rosser [1949, 1950] proved that $L(s, \chi)$ has no positive real zeroes, when χ is defined mod $n \leq 227$. Low [1968] and G. Purdy, Terras, Terras, and Williams [1979] verified that L-functions derived from imaginary quadratic fields of discriminant $d > -115147$ have no positive real zeroes, but did not consider real fields. Thus the best bound is still Rosser's: any exceptional modulus must exceed 227.

Although we have made an attempt to standardize the terminology, the reader should be aware that the "extended" or "generalized" Riemann hypothesis is interpreted differently by different authors. Current practice has been summed up by Narkiewicz [1990, §VIII]:

> We use ERH... to denote the statement that every conceivable zeta-function which should not have zeroes in the critical strip indeed does not have them.

Theorems 8.4.5 and 8.4.6 follow easily from results in Titchmarsh [1930]. We sketch how this can be done. Although the dependence on n was not stated explicitly, Titchmarsh proved

$$\psi(x, \chi) := \sum_{m \leq x} \Lambda(m)\chi(m) = 1_\chi x + O\Big(\sqrt{x}(\log x)(\log x + \log n)\Big)$$

whenever χ is a primitive character mod n. Since

$$\sum_{\substack{m \leq x \\ \gcd(m,n)>1}} \Lambda(m) \leq (\log_2 n)(\log x)$$

(see H. Davenport [1980, p. 121]), the restriction to primitive characters is unnecessary. Furthermore, since $\psi(x) - \vartheta(x) = O(\sqrt{x}(\log x)^2)$ (see Exercise 9), we also have

$$\vartheta(x, \chi) := \sum_{p \leq x} \Lambda(p)\chi(p) = 1_\chi x + O(\sqrt{x}(\log x)(\log x + \log n)).$$

By an argument similar to the proof of Theorem 2.7.1, we get

$$\pi(x, \chi) := \sum_{p \leq x} \chi(p) = \int_{2^-}^{x} \frac{d\vartheta(x, \chi)}{\log t} = 1_\chi \mathrm{li}(x) + O\Big(\sqrt{x}(\log x + \log n)\Big).$$

Multiplying this by $\bar{\chi}(a)$, summing over the characters of $(\mathbb{Z}/(n))^*$, and using Exercise 6.18, we get Theorem 8.4.5. Theorem 8.4.6 is proved similarly, using the characters of $(\mathbb{Z}/(n))^*/G$.

8.5

Cobham [1966] was apparently the first to explicitly use the ERH in the analysis of an algorithm.

The best reference on small nonresidues and primitive roots is Narkiewicz [1986].

Lemma 8.5.1 is from Prachar [1957]. The first estimate appears on p. 221. The second can be easily verified by combining Satz 3.3 (p. 220) with Satz 4.1 (p. 225).

Theorem 8.5.3 is due in essence to Ankeny [1952]. He proved this bound for the least quadratic nonresidue (when n is prime) and remarked that it also held for the least r-th power nonresidue; Y. Wang [1964] gave a proof of this. The first complete proof (i.e. valid for all n) appeared in H. Montgomery [1971]. Our group-theoretic interpretation is due to Vélu [1978]. Hahn [1992] gave the bound for p-th power nonresidues modulo p^2. We note that $G(n) \le (\log n)^{O(1)}$ would follow if we knew that all L-functions were zero-free in a strip $\Re(s) > 1 - \epsilon$. (This is a weak version of the ERH.)

Bach and Huelsbergen [1993] give heuristic arguments and numerical data to support the idea that $G(n)$ is never much larger than $(\log n)(\log\log n)/\log 2$.

Burthe [1995] showed that $G(n) = O(n^{3/(8\sqrt{e})+\epsilon})$, where the constant in the big-O depends on ϵ. Note that $3/8\sqrt{e} \doteq .22745$.

The first estimate for small nonresidues was by Gauss [1801, §129]: if $p \equiv 1 \pmod 8$, then some number less than $2\sqrt{p} + 1$ is a quadratic nonresidue. The best asymptotic bound for $N_2(p)$, the least quadratic nonresidue mod p, is due to Burgess [1957, 1963a]:

$$N_2(p) \le p^{1/(4\sqrt{e})+o(1)}.$$

Sharper bounds of this type (i.e. with smaller exponents on p) are known for k-th power nonresidues. For this work, see Y. Wang [1964], Norton [1971, 1973], Kolesnik and Straus [1978], and references therein.

For lower bounds on k-th power nonresidues, see the solutions to Exercises 31 and 56, and work cited therein.

Many authors have noted that the residues and nonresidues of powers mod p are distributed "randomly" in the interval $[0, p-1]$. For results of this type, see Moroz [1961] and Peralta [1992]. Damgård [1990] proposed using the successive values of $\left(\frac{x}{p}\right)$ as pseudo-random numbers.

Let $N_k(p)$ denote the least k-th power nonresidue, when $k \mid p-1$. Elliott [1967, 1970] proved there a constant c_k such that

$$\frac{1}{\pi(x,k,1)} \sum_{\substack{p<x \\ p\equiv 1 \pmod k}} N_k(p) \sim c_k,$$

generalizing a result of Erdős [1961] for $k = 2$. If k is prime, we have

$$c_k = (k-1)\sum_{n\ge 1} \frac{p_n}{k^n},$$

the result predicted by a naive probabilistic argument. It is interesting that this does not hold in general; for example, $c_8 \doteq 2.3224$, about 7% higher than we would guess. Baillie and Wagstaff [1980] showed that $L(n)$, the least positive m with $\left(\frac{m}{n}\right) \ne 1$, satisfies

$$\frac{2}{x}\sum_{\substack{n<x \\ n \text{ odd}}} L(n) \sim 1 + \sum_{j\ge 2} \frac{p_j+1}{2^{j-1}} \prod_{1\le i<j} \left(1 - \frac{1}{p_i}\right) \doteq 3.147755,$$

which is in agreement with probability theory.

Bach and Huelsbergen [1993] proved that the average value of $G(n)$ for $n < x$ is at least $(1 + o(1))(\log\log x)(\log\log\log x)$. Numerical evidence and heuristic arguments suggest that this is also an upper bound.

Theorem 8.5.4 is from Shoup [1990b, 1992]. This theorem shows that there is a polynomial-time algorithm to construct a "small" set that is guaranteed to contain a primitive root mod p, assuming ERH. For another such construction, not relying on estimates for primitive roots, see Bach [1995].

The best unconditional estimate for the least primitive root was proved by Y. Wang [1959] and Burgess [1962a]: if $g(p)$ is the least primitive root mod p, we have

$$g(p) \le p^{1/4+o(1)}.$$

Landau [1927a] proved $g(p) \le d(p-1)\sqrt{p}\log p$, improving a similar bound of I. Vinogradov [1918]. For another explicit bound, see Grosswald [1981]. S. Cohen, Odoni, and Strothers [1974] generalized this result to primitive roots mod p^2.

The data of Western and Miller [1968] lend credence to the idea that $g(p) = (\log p)^{O(1)}$. This is known to be true, on average. Burgess and Elliott [1968] showed

$$\frac{1}{\pi(x)}\sum_{p<x} g(p) = O\Big((\log x)^2(\log\log x)^4\Big).$$

Assuming ERH, Murata [1991a,1991b] improved this to $O((\log x)(\log\log x)^7)$.

Theorem 8.5.5 is from Hooley [1967]. The dependence on a was analyzed by Murata [1991c]. For expositions of this theorem, see Narkiewicz [1986] (this is delightfully short); and Hooley [1976]. The exact hypothesis required is the following. Let ζ denote a primitive k-th root of unity, and let $K = \mathbb{Q}(\zeta, a^{1/k'})$, where k' divides k. Then the Dedekind zeta function of K satisfies the Riemann hypothesis.

It is interesting that the error in Artin's original conjecture was discovered experimentally. D. H. and E. Lehmer [1962] counted the primes for which a is a primitive root, for various a, and found discrepancies from Artin's original calculation. The correct version was given by Lang and Tate, (see their preface to Artin [1965]) and in an elementary form by Heilbronn (see Western and Miller [1968]) as follows. Let $a \ne 0, \pm 1$ be not a square, and write $a = a_1 a_2^2$ with a_1 squarefree. Let h be the largest integer for which a is an h-th power. Let

$$A_0 = \prod_{q\,|\,h}\left(1 - \frac{1}{q-1}\right)\prod_{q\,\nmid\,h}\left(1 - \frac{1}{q(q-1)}\right).$$

Then Theorem 8.5.5 holds, with $A = A_0$ if $a \not\equiv 1 \pmod 4$, and

$$A = A_0\left[1 - \mu(|a_1|)\prod_{q\,|\,\gcd(a_1,h)}\left(\frac{1}{q-2}\right)\prod_{\substack{q\,|\,a_1\\ q\,\nmid\,h}}\left(\frac{1}{q^2-q-1}\right)\right]$$

otherwise.

H. W. Lenstra [1977a] generalized Artin's conjecture to primes in arithmetic progressions. The presumed density of primes p for which a has index n in $(\mathbb{Z}/(p))^*$ has also been computed. See H. W. Lenstra [1977b]; Wagstaff [1982]; and Murata [1991c].

Schinzel and Sierpiński [1958] proved that the prime tuples conjecture implies Artin's conjecture.

Heath-Brown [1986] proved the following theorem: Let a, b, c be three nonzero integers that are multiplicatively independent (that is, their logarithms are linearly independent over $\mathbb{Q}$). If no element in the set $\{a, b, c, -3ab, -3ac, -3bc, abc\}$ is a square, then the number of primes $p \leq x$ for which at least one of a, b, c is a primitive root is $\Omega(x/(\log x)^2)$. A similar, but more complicated, result was obtained by Gupta and Ram Murty [1984]. (See Ram Murty [1988] for an overview.)

Artin's constant was computed to 45 places by Wrench [1961]. Bach [1994] proved that the first t bits of Artin's constant can be computed using $O(t^3 \lg t)$ bit operations.

Linnik [1944a] showed the existence of a $C > 0$ for which $P(n, a) \leq O(n^C)$, and various authors followed with estimates for Linnik's "constant" C. The best current value, $C = 11/2$, is due to Heath-Brown [1992]. (This paper also reviews previous calculations of C.) It would be useful to have an explicit version of Linnik's theorem (i.e., one in which the constant implied by the "O" symbol is computed), but none seems to be known.

Theorem 8.5.8 is due to Y. Wang, Hsieh, and Yu [1965].

Wagstaff [1979] gave a different heuristic argument that $P(n, a) = O\left(\varphi(n)(\log n)(\log \varphi(n))\right)$. His heuristic model is similar to Cramér's: each integer x in an arithmetic progression chooses to be prime with probability $n/(\varphi(n) \log x)$. Pursuing this further, McCurley [1986] argued that that the constant in the above estimate can be taken arbitrarily close to 2.

For more about small primes in arithmetic progressions, see work cited in the solution to Exercise 26 and the papers of Granville [1989a]; and Heath-Brown [1992].

8.6

For an excellent survey article on prime gaps, see Heath-Brown [1988].

Piltz [1884] conjectured that $p_n - p_{n-1} = O(p^\alpha)$ for every $\alpha > 0$, in his inaugural dissertation. After computing numerical data, Gram [1884] stated that the largest difference between prime powers $\leq n$ grew rather like $(\log n)^2$.

There are several stochastic models for the primes, all more or less equivalent. The formulation we used in arguing for Conjecture 8.6.2 is due to Cadwell [1971]. Based on the idea that each number chooses to be prime or composite at random, Cramér [1937] conjectured that $\limsup_{n\to\infty}(p_n - p_{n-1})/(\log n)^2 = 1$, and Shanks [1964] suggested that the maximum gap between primes below p_n is asymptotic to $(\log n)^2$. Some authors have stated that the distribution of primes simulates a Poisson process, thus suggesting that the number of primes in a small interval should obey a Poisson distribution. (See, e.g., Kosambi [1963]; Stein and Ulam [1967].) Polvani [1933] applied statistical tests (based on the central limit theorem) to the empirical counts of primes in intervals.

The work of Hildebrand and Maier [1989] suggests that these models need to be revised, to account for anomalous prime counts in small intervals. For example, let $X(m)$ denote the number of primes in

$[m, m + (\log m)^3]$. Under the Cramér model, $X(m)$ has an approximate normal distribution with mean and variance about $(\log m)^2$, so $X(m) - (\log m)^2 = O((\log m)^{3/2})$, with probability 1. This contradicts a theorem of Hildebrand and Maier, according to which $X(m) - (\log m)^2 \geq (\log m)^{2-O(1/\log\log m)}$ infinitely often. (See also Maier [1981, 1985].)

For a critique of Cramér-style models, and possible replacements, see Granville [1994].

Assuming the Riemann hypothesis, Selberg [1943] showed that if $f(x) \to \infty$, then for almost all $x \geq 0$, $[x, x + f(x)\log^2 x]$ contains a prime. ("Almost all" means that the set of real numbers x for which the interval fails to contain a prime has density 0.)

Numerous authors showed $p_n - p_{n-1} = O(p_n^{\alpha})$, for various exponents in the range $1/2 < \alpha < 1$. Theorem 8.6.1 was proved by Hoheisel [1930], who showed one could take any $\alpha > 1 - 1/33000$. A long sequence of improvements followed, which have been reviewed by Ivić [1985]. Currently, the best result seems to be that of Baker and Harman (cited in Granville [1994]), who proved $\alpha \geq 1/2 + 7/200$.

It is an old conjecture that $p_n - p_{n-1} = O(\sqrt{p_n})$; see, e.g. Legendre [1830, p. 78]. Theorem 8.3.3 implies a bound of $O(\sqrt{p_n}(\log p_n)^2)$, assuming the Riemann hypothesis, and Cramér [1921] improved this to $O(\sqrt{p_n}\log p_n)$. If one adds a pair correlation hypothesis, the bound can be further reduced to $o(\sqrt{p_n \log p_n})$; see Heath-Brown [1982] and Heath-Brown and Goldston [1984].

Heath-Brown [1978a] showed the second moment of $p_n - p_{n-1}$ for primes below x is $\leq x^{1/3}(\log x)^{O(1)}$. (The constant subsumed by the "O" is not trivial.) Selberg [1943] gave a result implying a sharper bound of $O((\log x)^4)$, assuming the Riemann hypothesis.

As soon as one knows that $\pi(x) = o(x)$, one can conclude that the difference $p_n - p_{n-1}$ is not bounded. The first result showing the existence of non-trivial large gaps (i.e., larger than the mean value) was apparently that of Backlund [1929]: $p_n - p_{n-1}$ exceeds $(2 - \epsilon)\log p_n$ infinitely often. (His proof was rediscovered by S. Johnson [1965].) Many researchers successively improved this bound; we refer to Erdős [1935a] and Maier and Pomerance [1990] for the history. This last paper gives the best known result:

$$\liminf_{n\to\infty} \frac{(p_n - p_{n-1})(\log\log\log p_n)^2}{(\log p_n)(\log\log p_n)(\log\log\log\log p_n)} \geq 2.337765\ldots.$$

Many investigators have searched for large gaps between consecutive primes, though not always systematically.

For studies indicating the first appearance of gaps of given sizes, see Western [1934]; Lander and Parkin [1967b], Brent [1973]; and Brent [1980a]. Young and Potler [1989] examined the gaps between primes $\leq 7.263 \times 10^{13}$, and found the largest to be 778.

For other empirical data on prime gaps, see Glaisher [1878b]; M. Stein and Ulam [1967]; M. Jones, Lal, and Blundon [1967]; Cadwell [1971]; and Weintraub [1981, 1982, 1991, 1993].

Tables comparing large prime gaps to various growth rates appear in Western [1934]; Brent [1973]; and Riesel [1985a, p. 85].

For computational study of a conjecture related to Cramér's, see Golomb [1981].

A weakened form of Cramér's conjecture follows from work of Iwaniec [1978]. Thinking of the elements of $(\mathbb{Z}/(n))^*$ as "approximately prime," we let $J(n)$ denote the maximum distance between integers relatively prime to n. (This is sometimes called *Jacobsthal's function.*) Then $J(n) = O((\log n)^2)$.

For overviews on twin primes and related groupings of prime numbers, see Riesel [1985a]; and Ribenboim [1988b].

We do not know who first conjectured that there are infinitely many twin primes. (The term is apparently due to Stäckel [1916].) De Polignac [1849] stated that every even number occurs as the difference of consecutive primes infinitely often. Piltz [1884, pp. 46–47] remarked that the groups of two primes near x should have average density proportional to $1/(\log x)^2$, groups of three $1/(\log x)^3$, and so on.

Using sieves, Brun [1920] proved that there are $O(x/(\log x)^2)$ twin primes $\leq x$. Hence $\sum 1/p$, where p ranges over all twin primes, converges. Brent [1975] called this sum *Brun's constant*. The only known way to compute it is to enumerate the primes up to a limit and estimate the remainder. Even if one uses Conjecture 8.6.3 for this estimate, the cost to compute the k-th bit is doubly exponential in k. This gives an example of a naturally occurring constant that seems hard to compute.

Many authors have given formulas for the putative density of twin primes. The definitive treatment is probably by Hardy and Littlewood [1922], who proved several earlier conjectures to be incorrect. See also Sutton [1937] (beware of historical errors and omissions); Lévy [1949]; Pólya [1959]; and Rubinstein [1993]. Brent [1975] empirically analyzed the error in Hardy and Littlewood's approximation.

Our heuristic argument for the density of twin primes follows Cherwell and Wright [1960].

Heath-Brown [1983] showed that if some L-function has a Siegel zero, then there are an infinite number of twin primes.

Crandall and Penk [1979] discussed some techniques useful in searching for large twin primes. The largest known twin primes (approximately 10^{3388}) were found by Parady, Smith, and Zarantonello [1990].

Empirical data on prime twins can be found in the references above, as well as Glaisher [1878c]; Cherwell [1946]; Früchtl [1950]; Sexton [1954] (which reviews earlier studies); Selfridge [1959]; Shanks [1965]; M. Jones, Lal, and Blundon [1967]; and Bohman [1973].

Dickson [1904] conjectured that, unless obviously prevented from doing so, any k linear polynomials are simultaneously prime infinitely often. (He gave no density estimate.) G. Hardy and Littlewood [1922] derived the density estimate of Conjecture 8.6.3, in a more sophisticated fashion than we have done here. (See also Cherwell [1946].)

Using sieve methods, it can be shown that the number of $n \leq x$ at which k linear polynomials are simultaneously prime is $O(x/(\log x)^k)$. (One has to exclude special cases, such as identical polynomials.) For theorems of this sort, see Prachar [1957, pp. 45–52]; and Halberstam and Richert [1974, p. 172].

H. F. Smith [1956] gave an algorithm for the following problem: given k, minimize $a_k - a_1$ over all admissible sequences of length k. His algorithm is exponential; it is not clear if this can be done in polynomial time.

One can generalize the definition of prime twins to triplets, quadruplets, etc. For empirical data of these, see the papers on twins cited above, as well as Sexton [1955]; Golubew [1972a].

The strong prime tuples conjecture implies that there are arbitrarily long arithmetic progressions of primes; the challenge of finding them has attracted many researchers. For historical reviews, see Karst [1970]; and Pritchard [1985]. Grosswald and Hagis [1979] gave a heuristic theory predicting at which point a progression of length k should appear. (See Riesel [1970] for a related conjecture.) Pritchard [1983b] discussed algorithms to search for such progressions; the main devices employed are sieves and a theorem of Waring that the difference between the primes in a progression of length k must be divisible by the first $k-1$ primes.

For empirical data on primes in arithmetic progressions, see the papers cited above, as well as Karst [1969]; Karst and Root [1973]; Root and Karst [1973]; Weintraub [1977a, 1977b]; and Pritchard [1983c]. The longest known progression of primes has length 22, and was found by Pritchard, Moran, and Thyssen [1995].

The strong prime tuples conjecture implies that if a and b are relatively prime, with $a > 1$, there are arbitrarily long sequences of primes of the form $p, ap+b, a(ap+b)+b, \ldots$. (These are sometimes called *Cunningham chains.*) For information on these, see Löh [1989] and references therein.

Bateman and Horn [1962] generalized the strong prime tuples conjecture as follows. Let $f_1(X), \ldots, f_k(X)$ be distinct irreducible polynomials in $\mathbb{Z}[X]$ with positive leading coefficients. Assume $d_i = \deg(f_i) > 0$, and let $\nu(p)$ denote the number of zeroes in $\mathbb{F}_p$ of the polynomial $f_1 \cdots f_k$. Then $f_1, \ldots, f_k$ should represent primes infinitely often, and the density

$$N(x) := |\{n \le x : f_i(n) \text{ is prime}, i = 1, \ldots, k\}|$$

should satisfy

$$N(x) \sim \frac{\prod_p (1-\nu(p)/p)(1-1/p)^{-k}}{\prod_{1\le i\le k} d_i} \int_2^x \frac{dt}{(\log t)^k}.$$

This generalizes and extends conjectures made by various other authors. For these, see Bouniakowsky [1857]; G. Hardy and Littlewood [1922]; Schinzel and Sierpiński [1958]; and Schinzel [1961].

Using sieve methods, Bateman and Stemmler [1962] proved that $N(x) = O(x/(\log x)^k)$; the implied constant is worse than the conjectured one by a factor $k!2^k$.

Schinzel [1963] extended the Bateman-Horn conjecture to include an additive requirement that $n - g(n)$ is also prime, where $g(X)$ is another polynomial. This includes Goldbach's conjecture and several related conjectures due to G. Hardy and Littlewood [1922] as special cases.

Brent [1974] computed the presumed density for a restricted type of prime tuple, where the intermediate numbers are required to be composite.

Empirical data on the number of prime values taken by polynomials may be found in Western [1922]; Poletti [1929]; Shanks [1960, 1961, 1963]; Bateman and Horn [1962]; Bateman and Stemmler [1962]; Lal [1967]; and Golobew (the indefatigable!) [1972b].

We will not discuss additive number theory here. Readers interested in Goldbach's conjecture and related problems can consult the original papers collected by Y. Wang [1984]. The books by Guy [1994] and Ribenboim [1985] are good sources for computational aspects of these problems.

8.7

The book of Marcus [1977] is a good introduction to algebraic number theory. For historical information, consult Hilbert [1897]; Dickson et al. [1923]; and Vandiver and Wahlin [1928].

The standard reference on analytic aspects of algebraic number theory is Narkiewicz [1990]. Other texts include Landau [1927b]; Lang [1986]; and L. Goldstein [1971]. For a short survey see Stark [1975].

The computational aspects of algebraic number theory form a subject in itself. We will not go into this here, except to note that the main problems are to construct the Galois group, find an integral basis, and give algorithms for computing in the class group and the group of units. For survey articles, see Buchmann [1990]; and H. W. Lenstra [1992]. Books on the subject include Zimmer [1972]; Pohst and Zassenhaus [1989]; and H. Cohen [1993].

Most of the theory of discriminants is due to Dedekind [1879, 1882].

When considering relative extensions K/E, the discriminant definition given in the text will not work, because relative integral bases need not exist. Hilbert [1897] gave a definition equivalent to the following: Let n be the degree of K/E, and choose a set $\beta_1, \ldots, \beta_r$ of integers in K that generates A as a module over E's ring of integers. (Steinitz [1912] proved we can take $r \le n+1$.) Then the *relative discriminant* $\mathbf{\Delta}_{K/E}$ is the ideal generated by the $n \times n$ subdeterminants of $(\mathrm{Tr}(\beta_i\beta_j))$. Equivalently, the numbers $f'(\alpha)$, where f is the minimal polynomial of an integral element α, generate an ideal called the *different*; the discriminant is then the relative norm of the different. As before, this allows one to estimate $\mathbf{\Delta}_{K/E}$ by choosing one α.

The following two results are useful for estimating discriminants. First, for $i = 1, 2$, let K_i be a number field of degree n_i and discriminant Δ_i. Then the discriminant of the composed field K_1K_2 always divides $\Delta_1^{n_2}\Delta_2^{n_1}$. (This follows from the theorem of Tôyama [1955].) Second, if $\mathbb{Q} \subset E \subset K$ is a tower of extensions, then the discriminant of K equals

$$\Delta_E^{[K:E]} \mathrm{N}\mathbf{\Delta}_{K/E}.$$

(This is Theorem 39 of Hilbert [1897].) Both of these results are true for relative extensions as well.

For the discriminants of polynomials of degree ≤ 5, see Dickson [1926]. Swan [1962] computed the discriminant of the trinomial $X^n + aX^m + b$. For a review of discriminant computations for particular fields, see Narkiewicz [1990, pp. 78–82].

If one knows the primes appearing in the discriminant, one can use p-adic methods due to Hensel [1908]. The following estimate follows from this work. Suppose K is a number field of degree n and discriminant Δ. Let $\mathbf{p}$ be a prime of K, of degree $f_{\mathbf{p}}$, with $\mathbf{p}^{e_{\mathbf{p}}} \,||\, p$. Then

$$\nu_p(\Delta) = \sum_{\mathbf{p}|p} f_{\mathbf{p}}(\bar{e}_{\mathbf{p}} - 1),$$

where $\bar{e}_{\mathbf{p}} = e_{\mathbf{p}}$ if this is a unit mod p, and $e_{\mathbf{p}} + 1 \le \bar{e}_{\mathbf{p}} \le (\nu_p(e_{\mathbf{p}}) + 1)e_{\mathbf{p}}$ otherwise. To compute $\bar{e}_{\mathbf{p}}$ precisely, one can analyze the action of the Galois group mod $\mathbf{p}^{\nu}$ (see, e.g. H. Cohn [1978]). From this we have the bound $\nu_p(\Delta) = O(n \log_p n)$; see Ore [1926] for an explicit bound showing this is best possible.

The discriminant of a field K grows exponentially with the degree n of the field. Although not best possible, the following bound of Minkowski [1891] is explicit:

$$\Delta_K \geq \frac{n^{2n}}{n!^2} \geq \frac{(\pi e^2/4)^n}{2\pi n}.$$

(Here we have used Stirling's approximation.) Sharper results may be found in the survey of Odlyzko [1990].

Washington [1982] gave a comprehensive treatment of cyclotomic fields. Kronecker [1854] proved the irreducibility of the m-th cyclotomic polynomial. Its discriminant (which is the discriminant of $\mathbb{Q}(\zeta_m)$) is

$$\pm \frac{m^{\varphi(m)}}{\prod_{p \mid m} p^{\varphi(m)/(p-1)}},$$

a formula apparently due to Dedekind [1882]. (See Rados [1906] for an elementary proof.)

Dedekind [1879] first defined the zeta function of a number field. For generalized L-functions (which we will not go into here), see, e.g., L. Goldstein [1971].

The prime ideal theorem is due to E. Landau [1903b]. The error term we have given in Theorem 8.7.2 is due to Mitsui [1968] and Sokolovskiĭ [1968], independently. Neither author made the dependence on the field K explicit; the complications involved are roughly the same as for Dirichlet's theorem, involving possible real zeroes of the zeta function. An error term making the dependence explicit can be obtained from work of Lagarias and Odlyzko [1977]. See also Serre [1981].

The prime ideal theorem can be expressed in several equivalent forms. Because the density of prime ideals with norm near x is about 1 in $\log x$, we can usually obtain a correct result if we take a prime number sum and replace p everywhere by the norm $\mathbf{Np}$ of a prime ideal. Thus, for example, we have

$$\sum_{\mathbf{Np} \leq x} \log \mathbf{Np} \sim \sum_{p \leq x} \log p \sim x.$$

For this and similar formulas, see Landau [1903b, 1906].

Chebyshev-style bounds for $\pi_K(x)$ are known. Friedlander [1980] proved that for any $\epsilon > 0$,

$$\pi_K(x) = \Omega\left(\frac{x}{(\sqrt{|\Delta_K|}\log(2x))^{1+\epsilon}}\right).$$

(His proof does not allow one to compute the implicit constant from ϵ.) We also have

$$\pi_K(x) \leq n\pi(x),$$

where n is the degree of the field K.

Theorem 8.7.4 is from Lang [1971]. This paper also gives a useful heuristic for converting analytic estimates over $\mathbb{Z}$ to algebraic numbers. Lang's principle boils down to the following observation: the density of the zeroes of $\zeta_K(s)$ at height t in the critical strip is $\Theta(n \log t + \log |\Delta|)$.

The GRH is supported mainly by formal arguments, and the numerical computations of Dirichlet L-functions cited earlier. Davies [1961] and Lagarias and Odlyzko [1979] computed zeroes of more general zeta and L-functions, and found none with real part different from 1/2.

Theorem 8.7.5 was first proved by Dedekind [1878], for the case $E = \mathbb{Q}$. (Zolotarev [1874] gave a similar theorem, but it does not apply to all number fields.) The condition that $\mathbf{p}$ not divide Δ_K is overly restrictive; it suffices that the ring of $\mathbf{p}$-integral elements be generated by α. Thus there are no exceptions if K's ring of integers has a relative integral basis composed of powers of α.

In general, $\mathbf{p}A = \prod_{1\le i\le r} \mathbf{P}_i^{e_i}$, where $\sum_{1\le i\le r} e_i f_i = [K : E]$. If K/E is normal, all the $\mathbf{P}_i$'s are conjugate, so the values of e_i and f_i do not depend on i. Dedekind [1882] proved that the primes with $e_i > 1$, the so-called *ramified primes*, are exactly those appearing in the discriminant.

Ore [1927] proved the following theorem. Let $K = \mathbb{Q}(\alpha)$, where $f(X)$ is α's minimal polynomial. If $f = f_1 \cdots f_r$ is a factorization into irreducible polynomials over the p-adic field, then in K's ring of integers, $p = \mathbf{P}_1^{e_1} \cdots \mathbf{P}_r^{e_r}$, with $\mathrm{N}\mathbf{P}_i^{e_i} = p^{\deg f_i}$. (If p^α exactly divides the discriminant of f, it suffices to do the factorization modulo $p^{\alpha+1}$.) He further showed how to compute the e_i's, by factoring the f_i's over suitable extensions of the p-adic numbers.

The Artin symbol was first employed by Artin [1924b]; the name and the symbolism $\left(\frac{K/E}{\mathbf{p}}\right)$ are due to Hasse [1926, Part II].

Because there are connections between class groups of number fields and abelian extensions, the theory of abelian extensions is called *class field theory*. See Wyman [1972] for a short introduction. For full expositions, see Hasse [1926, 1967]; and Cassels and Fröhlich [1967]. For short proofs of the main theorems (requiring some specialized machinery), see Neukirch [1986]. The books of H. Cohn [1978, 1985] are notable for their many examples.

Theorem 8.7.6 is from T. Takagi [1920]. The other main theorem of class field theory is the reciprocity law of Artin [1927], which gives an explicit isomorphism between the Galois group of K/E and a "generalized class group" of E. We will not go into this here.

Readers familiar with class field theory will note that we have defined "conductor" in a non-standard way. (Technically, we ignore infinite primes.) This is done primarily to make results such as Theorem 8.7.7, which depend only on the norm of a conductor, more accessible.

To use Theorem 8.7.7 in any practical situation, it is necessary to estimate a conductor. The main tool for this is the following product formula of Hasse [1926]. Suppose K/E is an abelian extension, with Galois group G. Then, for any character χ on G, the kernel of χ corresponds (via Galois theory) to a subfield K_χ; we define $\mathbf{f}_\chi$ to be the minimal conductor of K_χ/E. Hasse's formula states that

$$\mathbf{\Delta}_{K/E} = \prod_\chi \mathbf{f}_\chi.$$

For cyclic extensions of degree m, we have the bound $\mathbf{f}^{\varphi(m)} \mid \mathbf{\Delta}_{K/E}$. (This is exact when m is prime, as shown by T. Takagi [1920].)

Hasse [1967, p. 140] proved the following theorem: Suppose K/E is abelian, with a conductor $\mathbf{f}$. Let L be any finite extension of E. Then KL/L is also an abelian extension, with a conductor dividing $\mathbf{f}$.

Hasse [1967, p. 232] gave the following bound for a conductor $\mathbf{f}$ of the extension

$$K\left(\sqrt[n]{a_1}, \ldots, \sqrt[n]{a_r}\right)/K.$$

(It is assumed that K contains a primitive nth root of unity.) If $\mathbf{p}$ is a prime (of K) not dividing n, but dividing some a_i, then at most one copy of $\mathbf{p}$ appears in $\mathbf{f}$. If $\mathbf{p}^\nu \,||\, n$ and $\mathbf{p}^s \,||\, p$ (where $\nu \ge 1$ and $p \in \mathbb{Z}$), then at most $s/(p-1) + \nu s + 1$ copies of $\mathbf{p}$ appear in $\mathbf{f}$. No other primes divide $\mathbf{f}$.

It is known that if K is a finite abelian extension of $\mathbb{Q}$, then $K \subset \mathbb{Q}(\zeta_m)$ for some m, and one can take m as a conductor of $K/\mathbb{Q}$. This is usually called the *Kronecker-Weber theorem*. We refer the reader to Washington [1982, pp. 319 ff.] for the history and a proof.

Theorems 8.7.7 and 8.7.8 are from Lagarias, Montgomery, and Odlyzko [1979]. Chebotarev [1926] proved Theorem 8.7.9 for the weaker notion of Dirichlet density, generalizing an earlier theorem of Frobenius. Theorem 8.7.10 is from Lagarias and Odlyzko [1977].

It is possible to state results such as Theorem 8.7.7 in an "internal" way, without reference to field extensions. (We did this, for example, with Theorem 8.5.3.) To do this properly requires a generalization of arithmetic progressions, based on class groups. We will not go into this here, except to state the following consequence of Theorem 8.7.7: the ideal class group of a number field with discriminant Δ is generated by primes of norm $\leq (\log |\Delta|)^2$.

For generalizations of the prime ideal theorem to arithmetic progressions, see L. Goldstein [1970a]; Hinz [1980]; and references therein. Generalizations of Linnik's theorem to algebraic numbers can be found in Fogels [1961]; and Weiss [1983]. For a generalization of Artin's conjecture to number fields, see L. Goldstein [1970b].

8.8

Two useful compendia of analytic estimates are Spearman and Williams [1975] and Mitrinović and Popadić [1978].

The results in this section come from the following sources.

Theorems 8.8.1–8.8.7: Rosser and Schoenfeld [1962] except for the lower bound in Theorem 8.8.4 (which was first proved by Rosser [1939]), and the upper bound in Theorem 8.8.3 (from Schoenfeld [1976, p. 360]). Also see Rosser [1941, 1963]; Hanson [1972]; Rosser and Schoenfeld [1975]; Grimson and Hanson [1977]; Costa Pereira [1985]. For results assuming the Riemann hypothesis, see the Notes for Section 8.3.

Theorem 8.8.8: G. Hardy and Wright [1985].

Theorem 8.8.9: Nicolas and Robin [1983]. We note that a bound of $5(\log n)/(\log \log n)$ can be easily obtained from the proof of Theorem 317 in G. Hardy and Wright [1985], by setting $\epsilon = 4$.

Theorem 8.8.10: Robin [1983].

Theorem 8.8.11: Konyagin and Pomerance [1994].

Theorem 8.8.12: H. Davenport [1980, Chapter 23]. This is an explicit version of the Pólya-Vinogradov inequality.

Theorem 8.8.13: McCurley [1984b]; and Livingston [1986]. For other moduli, see McCurley [1984a]; and Ramaré and Rumely [1994].

Theorem 8.8.14: H. Montgomery and Vaughan [1973]. This is an explicit form of the Brun-Titchmarsh theorem (for which see Titchmarsh [1930]).

Theorem 8.8.15: Rosser [1941, Theorem 19]; Rosser and Schoenfeld [1975]; van de Lune, te Riele, and Winter [1986]. For analogous results on Dirichlet L-functions, see McCurley [1984a]; and Ramaré and Rumely [1994].

Theorem 8.8.16: van de Lune, te Riele, and Winter [1986]; Rosser and Schoenfeld [1975]. For explicit zero-free regions for Dirichlet L-functions, see McCurley [1984b, 1984c]; and Ramaré and Rumely [1994].

Theorem 8.8.17: Bach [1985, 1990a].

Theorem 8.8.18: Oesterlé [1979]. (Take $E = \mathbb{Q}$, $K = \mathbb{Q}(\zeta_n)$, and $\sigma(\zeta_n) = \zeta_n^a$ in Theorem 8.8.22.)

Theorem 8.8.19: Bach and Sorenson [1994].

Theorem 8.8.20: Bach [1990].

Theorem 8.8.21: Bach and Sorenson [1994].

Theorem 8.8.22: Oesterlé [1979].

The prime-reciprocal constant (B in Theorem 8.8.5) was evaluated to 23 places by Terrill and Sweeny [1946].

9 Prime Numbers: Basic Algorithms

In this chapter, we discuss algorithms for testing primality, for finding the n-th prime, for computing the number of primes $\leq x$, and for creating a table of the primes from 1 to n. The methods used in this chapter are primarily based on the concepts introduced in Chapters 5–6. More advanced methods for primality testing, based on algebraic geometry, will be discussed in Chapter 13.

These problems have attracted the attention of great mathematicians throughout the ages. Eratosthenes (c. 250 B.C.) gave a method for determining the all primes below a given limit. In the early 13th century, Fibonacci pointed out that to determine if a number n is prime or not, it suffices to trial divide by the integers $\leq \sqrt{n}$. Euler proved that if a number $n \equiv 1 \pmod 4$ can be written as the sum of two squares in essentially one way, and if further those squares are relatively prime, then n is prime; using this, he showed that $1000^2 + 3^2$ is prime. As we have already mentioned in Chapter 1, Gauss emphasized the importance of the problem of distinguishing prime numbers from composite numbers, and he gave methods for abbreviating the work of trial division.

Until the work of E. Lucas in 1876, however, the problems of primality testing and factoring were not considered separately. Although Fermat's theorem had been known since 1640, it seems that Lucas in 1876 was the first to recognize that it could be useful in determining whether a number is composite. Lucas showed that, at least in some cases, it was much *easier* to prove that a number was prime or composite than to discover its factors, if any. Several investigators, particularly M. Kraitchik and D. H. Lehmer in the 1920's, refined Lucas' work and tested the primality of many large numbers, particularly those of the form $2^n \pm 1$.

More recently, primality testing has attracted the attention of computer scientists and cryptographers. Cryptosystems such as RSA (see Chapter 15) depend on the fact that it is relatively easy to produce 100-digit primes, while it is apparently hard to factor the product of two such primes.

In his text *Théorie des Nombres*, published posthumously in 1891, Lucas listed five problems for which no satisfactory solution existed:

1. Find a prime number larger than a given prime.
2. Find a [nontrivial] function that takes on only prime values.
3. Find the prime which follows a given prime.
4. Compute $\pi(n)$, the number of primes less than n.
5. Compute p_n, the n-th prime.

To these five problems, we might add

6. Given an integer $n \geq 2$, decide whether it is prime or composite.
7. Find a "random" prime in a given interval $[x, y]$.

Even today, we do not have a truly satisfactory solution to any of these problems. For example, there is no known deterministic polynomial time algorithm to test primality. Recent results of Adleman and Huang (see Chapter 13) show that the set of primes (written in binary) is in the complexity class $\mathcal{ZPP}$, but their algorithm appears to be useless in practice. There are some excellent primality testing algorithms based on algebraic number theory (see section 9.6) or the theory of elliptic curves (see Chapter 13). Although these algorithms do not run in polynomial time or cannot yet be proved to do so, they still allow the primality of numbers of up to 1000 digits to be proved in a reasonable length of time.

The situation for problems 4 and 5 above is even worse. The best current algorithm for either problem uses $O(n^{1/2+\epsilon})$ bit operations, and a polynomial-time algorithm is not presently in sight.

9.1 Primality Proofs and Fermat's Theorem

Recall Fermat's theorem: if p is a prime and $\gcd(a, p) = 1$, then $a^{p-1} \equiv 1 \pmod{p}$; see Theorem 2.1.3. Using this theorem, we can often prove that a composite number n is indeed composite—all we have to do is produce an $a \not\equiv 0 \pmod{n}$, such that $a^{n-1} \not\equiv 1 \pmod{n}$. For most composite n, these numbers a (called *Fermat witnesses*, or just *witnesses*) abound, and are easily found by random choice.

If the converse to Fermat's theorem were true, we would have a simple, fast way of proving that a number is prime.

Unfortunately, the converse to Fermat's theorem is false: for example, $4^{14} \equiv 1 \pmod{15}$, but 15 is not a prime. There also exist infinitely many composite numbers n for which the converse of Fermat's theorem is "as false as possible"—for these numbers we have $a^{n-1} \equiv 1 \pmod{n}$ for *every* a with $\gcd(a, n) = 1$. Such numbers are called *Carmichael numbers*; the smallest example is 561.

Thus it is clear that we cannot use the simple converse to Fermat's theorem to prove primality. Three separate approaches have been devised to handle this problem. Each approach gives up exactly *one* of the following desirable properties: speed (we'd like the algorithm to run in polynomial time, if possible); generality (we want the algorithm to work for all prime numbers, not just those of certain types), or correctness (we want a *proof* of primality, if possible).

In the first approach, we give up speed, but retain generality and correctness. This type of algorithm requires as input extra information that may not be quickly obtainable for all numbers. The result is a test that is *guaranteed* to produce a proof of primality, if the input is prime. In general, however, these tests require factorization, for which no fast algorithm is known. Although often useful in practice, no known test of this type runs in polynomial time.

In the second approach, we restrict our attention to numbers of specific forms, such as $2^p - 1$ (Mersenne numbers), or $2^{2^k} + 1$ (Fermat numbers). The result is a test that answers "prime" or "composite" and provides a proof of its answer; furthermore, the test runs in polynomial time. This approach has been extremely successful, and has led to the discovery of the largest known primes, but it lacks generality.

In the third approach, we give up being able to actually produce a *proof* of primality. Here there are two distinct kinds of tests. The first kind, the so-called "probabilistic" test, uses a source of random numbers. It has the property that the answer may be incorrect, but the error probability is small no matter what number is being tested. The most popular versions of this test have the additional property that if the number is prime, the test never mistakenly asserts that it is composite. This kind of test is extremely fast, but the probabilistic nature of the results may leave some uncomfortable about its reliability. The second test uses an algorithm based on an unproved (but widely believed) hypothesis from number theory, the Extended Riemann Hypothesis (ERH), discussed in Chapter 8. This method gives correct results *if* ERH is true.

In this section, we focus on the first of these approaches. The other two approaches are covered in later sections.

The principal idea behind proofs by the converse of Fermat's theorem is the following: a number n is a prime if one can find an element of large enough order in the appropriate multiplicative group.

For example, n is a prime iff $(\mathbb{Z}/(n))^*$ contains an element of order $n - 1$. The proof is simple: if n is a prime, then by Theorem 5.6.3, we know n has a primitive root a, which is of order $n - 1$. On the other hand, if $(\mathbb{Z}/(n))^*$ has an element of order $n - 1$, then every k, $1 \le k \le n - 1$, must be relatively prime to n, and hence n is prime.

This lays the groundwork for the following theorem, first enunciated by M. Kraitchik and D. H. Lehmer, although weaker versions had previously been formulated by E. Lucas and F. Proth.

THEOREM 9.1.1 Let n be an integer. Then n is prime if and only if there exists a such that $a^{n-1} \equiv 1 \pmod{n}$ and $a^{(n-1)/q} \not\equiv 1 \pmod{n}$ for all primes $q \mid n - 1$.

Proof Let n be a prime, and let a be a generator for $(\mathbb{Z}/(n))^*$. Then $a^{n-1} \equiv 1 \pmod{n}$ and $a^{(n-1)/q} \not\equiv 1 \pmod{n}$ for all primes q dividing $n - 1$.

On the other hand, suppose there exists a such that $a^{n-1} \equiv 1 \pmod{n}$ and $a^{(n-1)/q} \not\equiv 1 \pmod{n}$ for all $q \mid n - 1$. Let $m = \operatorname{ord}_n a$; we will show that $m = n - 1$. Since $a^{n-1} \equiv 1 \pmod{n}$, we know from Exercise 5.1 that $m \mid n - 1$. Suppose $m < n - 1$. Then there exists a prime p that divides $(n - 1)/m$, so $(n - 1)/m = kp$. Then

$$a^{(n-1)/p} \equiv a^{mk} \equiv 1 \pmod{n},$$

a contradiction. Hence $m = n - 1$ and n is prime. ■

COROLLARY 9.1.2 If the prime factorization of $n-1$ is given, and n is prime, we can produce a proof of that fact in random polynomial time. The expected number of bit operations used is $O((\lg n)^4)$.

Proof We simply choose $a \in (\mathbb{Z}/(n))^*$ at random, and check if a satisfies the two conditions (i) $a^{n-1} \equiv 1 \pmod{n}$ and (ii) $a^{(n-1)/p} \not\equiv 1 \pmod{n}$ for all primes p dividing $n-1$. If both of these are satisfied, then as above, n is prime.

The primality proof will succeed if a is a primitive root, mod n. Since n is prime, by Exercise 5.23 it has $\varphi(n-1)$ primitive roots. By Theorem 8.8.7 we have $\varphi(n-1) = \Omega(n/(\log\log n))$. We must compute $a^{(n-1)/q}$ for all primes $q \mid n-1$; by Exercise 8.4, there are at most $O((\log n)/(\log\log n))$ such primes. Each exponentiation can be done using $O((\lg n)^3)$ bit operations. Putting all these estimates together shows that the expected number of bit operations is $O((\lg n)^4)$. ■

We point out that if n is composite, the algorithm suggested in the previous corollary will diverge (run forever). This unpleasant feature can be avoided by running one of the tests in section 9.4 in parallel with the test above; the resulting algorithm gives an errorless ($\mathcal{ZPP}$) algorithm for determining whether n is prime or composite in $O((\lg n)^4)$ bit operations—provided that one knows the factorization of $n-1$.

In fact, Fellows and Koblitz recently showed that one can determine if a number n is prime or composite in *deterministic* polynomial time, given the complete factorization of $n-1$. This can be done with the following algorithm:

FELLOWS-KOBLITZ($n, q_1, e_1, q_2, e_2, \ldots, q_k, e_k$)
{ the complete factorization of $n-1 = q_1^{e_1} \cdots q_k^{e_k}$}
(1) for $a \leftarrow 2$ to $\lfloor (\log n)^2 \rfloor$ do
    (2) if $a^{n-1} \not\equiv 1 \pmod{n}$ then
        return('composite')
    (3) compute $\mathrm{ord}_n\, a$ using the algorithm in Exercise 5.8
    (4) for each prime $q \mid \mathrm{ord}_n\, a$ do
        (5) if $\gcd(a^{(\mathrm{ord}_n\, a)/q} - 1, n) > 1$ then
            return('composite')
(6) compute $h = \mathrm{lcm}_{2 \le a \le (\log n)^2}\, \mathrm{ord}_n\, a$
(7) if $h < \sqrt{n}$ then
    return('composite')
else
    return('prime')

THEOREM 9.1.3 Given that n is odd and the complete factorization of $n-1$ is $q_1^{e_1}\cdots q_k^{e_k}$, the algorithm FELLOWS-KOBLITZ correctly determines whether n is prime or composite and uses $O((\lg n)^6/(\lg\lg n))$ bit operations.

Proof First, let us prove correctness. Note that if $a^{n-1}\not\equiv 1 \pmod n$, then n is composite, and so the algorithm terminates at step (2). Otherwise, $a^{n-1}\equiv 1 \pmod n$, and so we can determine $\operatorname{ord}_n a$ using the algorithm of Exercise 5.8. (Note that the exercise is stated only for primes p, but the proof goes through without change if $a^{n-1}\equiv 1 \pmod n$.) If $\gcd(a^{(\operatorname{ord}_n a)/q}-1, n) > 1$ in step (5), then we have found a nontrivial divisor of n, and hence n is composite.

Now assume that step (6) is reached. We claim that $\operatorname{ord}_n a$ is actually equal to $\operatorname{ord}_p a$ for each prime p dividing n. For by Exercise 5.1, $\operatorname{ord}_p a \mid \operatorname{ord}_n a$, and if $\operatorname{ord}_p a < \operatorname{ord}_n a$, we would have $a^{(\operatorname{ord}_n a)/q}\equiv 1 \pmod p$ for some q dividing $\operatorname{ord}_n a$. But this would have been discovered in step (5).

Let p be the least prime dividing n; then $p\le\sqrt{n}$. But $h\mid p-1$. If $h\ge\sqrt{n}$ in step (7), then $p-1\ge\sqrt{n}$, a contradiction.

Now assume $h<\sqrt{n}$ in step (7). Suppose n is a prime, and consider the subgroup G of $(\mathbb{Z}/(n))^*$ generated by the elements $2, 3, \ldots, \lfloor(\log n)^2\rfloor$. By Exercise 5.28, we have $h=|G|$. But certainly G contains all the $(\log n)^2$—smooth positive integers $\le n$. It follows that $h\ge\Psi(n, (\log n)^2)$. By Theorem 8.8.11, we know $\Psi(n, (\log n)^2) > \sqrt{n}$, a contradiction.

It now remains to bound the running time of this algorithm. The time for the loop is clearly dominated by steps (3) and (4); for each a, these steps can be done in $O((\lg n)^4/(\lg\lg n))$ bit operations. Since the loop is performed $O((\lg n)^2)$ times, the total time is $O((\lg n)^6/(\lg\lg n))$. ■

Of course, the real drawback to these methods is the need for the factorization of $n-1$. For numbers of special forms, however, the method can be of great value. For example, let us prove that the 17-digit number $n=1+118^8$ is prime. The only primes dividing $n-1$ are 2 and 59. We find that $2^{n-1}\equiv 1 \pmod n$, while $2^{(n-1)/2}\equiv 18794296013353089 \pmod n$, and $2^{(n-1)/59}\equiv 36951497246592513 \pmod n$. Hence n is prime.

Theorem 9.1.1 also has a surprising complexity-theoretic application. Recall from Chapter 3 that `PRIMES` (respectively `COMPOSITES`) denotes the set of prime numbers (respectively composite numbers), written in binary. It is clear that `COMPOSITES` $\in\mathcal{NP}$, as we can guess a nontrivial factorization and verify it in polynomial time. Although not immediately obvious, the same result holds for the set `PRIMES`:

THEOREM 9.1.4 (PRATT) `PRIMES` $\in\mathcal{NP}$.

Proof Given a prime n, we can "guess" an element a of order $n-1$ and the factorization of $n-1$, and then, using Theorem 9.1.1, check in polynomial time that a is indeed of order

$n - 1$. One problem remains, however: how do we know that the complete factorization of $n - 1$ is indeed what we have guessed? (It could contain composite numbers masquerading as primes.)

To handle this problem, we *recursively* carry out primality proofs for all the primes dividing $n-1$. The recursion ends when we reach the prime 2. For example, the computation tree proving the primality of $n = 617$ might look like Figure 9.1.

Note that the children of each node labeled m are the primes dividing $m - 1$. The parenthesized numbers are elements of order $m - 1$ in $(\mathbb{Z}/(m))^*$.

Such a tree proves the primality of a prime number, and we cannot construct such a tree for a composite number.

Let $T(n)$ be the number of nodes in a computation tree for the prime n; note that $T(2) = 1$. Then for $n \geq 3$, we have $T(n) = 1 + \sum_{p \mid n-1} T(p)$. How big can $T(n)$ be?

We will prove by induction that $T(n) < 2 \log_2 n$. This is true for $n = 2$, as its tree just consists of a single node labeled 2. Now assume the bound is true for all primes $p < n$, and let $T(n)$ denote the total number of nodes in the tree with root labeled n. Let $n - 1 = p_1 p_2 p_3 \cdots p_k$, where $p_1 = 2$, and the p_i are (not necessarily distinct) primes. Then the tree with root n has (i) a node for n; (ii) a node for $p_1 = 2$; and (iii) subtrees with roots labeled with the other primes dividing $n - 1$. Thus, for $n \geq 3$,

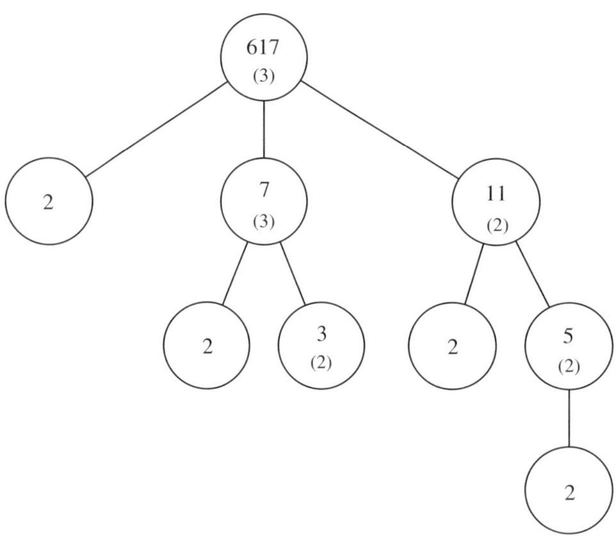

Figure 9.1
A primality proof for 617

$$
\begin{aligned}
T(n) &\le 1 + T(2) + \sum_{2\le i\le k} T(p_i) \\
&\le 2 + \sum_{2\le i\le k} 2\log_2 p_i \\
&= 2 + 2\log_2((n-1)/2) < 2\log_2 n.
\end{aligned}
$$

Each node labeled m of the tree has $O(\lg n)$ bits of information associated with it (m, the exponent to which m appears in its parent node, a generator for $(\mathbb{Z}/(m))^*$, and $O(\lg\lg n)$ bits for pointers), so the entire tree can be represented using $O((\lg n)^2)$ bits.

Now, how long does it take us to verify the computations in the tree with root labeled n? Let m be a node with parent r. With the node labeled m we associate the cost of (i) multiplying together the pieces of the factorization of $m-1$ given by its children, and verifying that their product is indeed $m-1$; (ii) checking that $a^{m-1} \equiv 1 \pmod{m}$; and (iii) verifying that $a^{(r-1)/m} \not\equiv 1 \pmod{r}$ for the parent node r. We can do (i) in $O((\lg n)^2)$ bit operations by Exercise 3.3, and (ii) and (iii) can be done in $O((\lg n)^3)$ steps. Since there are $O(\log n)$ nodes, the total cost is $O((\lg n)^4)$ bit operations, and the proof that PRIMES $\in \mathcal{NP}$ is complete. ■

We next provide another way to prove primality, which does not depend on the *complete* factorization of $n-1$, but only a partial factorization:

THEOREM 9.1.5 (POCKLINGTON) Suppose $n-1 = q^k r$, where q is a prime and $q \nmid r$. If there exists a such that $a^{n-1} \equiv 1 \pmod{n}$ and $\gcd(a^{(n-1)/q} - 1, n) = 1$, then for every prime $p \mid n$ we have $p \equiv 1 \pmod{q^k}$.

Proof Suppose a satisfies the hypotheses of the theorem. Let p be a prime dividing n, and let $t = \operatorname{ord}_p a$. Since $a^{n-1} \equiv 1 \pmod{n}$, we know that $a^{n-1} \equiv 1 \pmod{p}$, and so $t \mid n-1 = q^k r$. Since $\gcd(a^{(n-1)/q} - 1, n) = 1$, we have $a^{(n-1)/q} \not\equiv 1 \pmod{p}$, so we see that $t \nmid (n-1)/q = q^{k-1}r$. Hence $q^k \mid t$. However, we know that $t \mid p-1$. Thus $q^k \mid p-1$, as desired. ■

We also have

COROLLARY 9.1.6 Suppose $n-1 = fr$, where $f > \sqrt{n} - 1$ and $\gcd(f, r) = 1$. If there exists an a for which $a^{n-1} \equiv 1 \pmod{n}$ but $\gcd(a^{(n-1)/q} - 1, n) = 1$ for every prime $q \mid f$, then n is prime.

Proof Assume n is composite, and let p be the smallest prime factor of n. Clearly $p \le \sqrt{n}$. From Theorem 9.1.5, $p \equiv 1 \pmod{q^k}$ for all $q^k \,\|\, f$. Thus, by the Chinese remainder theorem, $p \equiv 1 \pmod{f}$. Hence $p \ge f + 1$, and so $p > \sqrt{n}$, a contradiction. ■

One can also design primality proofs that depend on knowing the factorization of $n + 1$. Since $|\mathbb{F}_{n^2}^*| = (n - 1)(n + 1)$, it seems reasonable to carry out the proof using not $\mathbb{Z}/(n)$, but rather a finite ring which will turn out to be isomorphic to $\mathbb{F}_{n^2}$ if indeed n is prime.

THEOREM 9.1.7 Suppose $c \in (\mathbb{Z}/(n))^*$ is such that $\left(\frac{c}{n}\right) = -1$. Define $f(X) = X^2 - c$. Let $e, f \in (\mathbb{Z}/(n))^*$ satisfy $\gcd(e, f, n) = 1$, and define $a = (f + eX)/(f - eX)$. (All arithmetic will be done in the polynomial ring $\mathbb{Z}[X]/(f(X), n)$.) Then n is prime if and only if there is some choice of e, f above such that (i) $a^{n+1} = 1$; and (ii) for all primes $q \mid n + 1$ we have $a^{(n+1)/q} = gX + h$, where $\gcd(n, g, h - 1) = 1$.

Proof If n is prime, then $f(X)$ is irreducible (mod n), and hence the ring $\mathbb{Z}[X]/(f(X), n)$ is isomorphic to $\mathbb{F}_{n^2}$. Also $a = (f - eX)^{n-1}$; if $f - eX$ generates $\mathbb{F}_{n^2}$, then $\operatorname{ord} a = n + 1$.

On the other hand, since $\left(\frac{c}{n}\right) = -1$, there must be a prime $p \mid n$ such that $\left(\frac{c}{p}\right) = -1$. Then $f(X)$ is irreducible (mod p), and conditions (i) and (ii) imply that a is an element of order $n + 1$. But then $n + 1 \mid p^2 - 1$. The conditions $p \mid n$ and $n + 1 \mid p^2 - 1$ together imply that $p = n$ (see Exercise 26); thus n itself must be prime. ■

COROLLARY 9.1.8 If the prime factorization of $n + 1$ is given, and n is prime, we can produce a proof of that fact in random polynomial time. The expected number of bit operations used is $O((\lg n)^4)$.

9.2 Primality Tests for Numbers of Special Forms

In this section, we give primality tests for numbers of special forms, such as $M_p = 2^p - 1$ (Mersenne numbers) and $F_k = 2^{2^k} + 1$ (Fermat numbers). These tests are fast, simple, and provide rigorous proofs that their results are correct.

We start with a simple test due to Pepin, for the primality of the Fermat numbers $F_k = 2^{2^k} + 1$:

THEOREM 9.2.1 (PEPIN'S TEST) For $k \geq 2$, F_k is prime if and only if $5^{(F_k-1)/2} \equiv -1 \pmod{F_k}$.

Proof Let F_k be a prime. We first show that 5 is a quadratic nonresidue of F_k for $k \geq 2$. Since $2^4 \equiv 1 \pmod 5$, it is easy to prove by induction that $2^{2^k} \equiv 1 \pmod 5$ for all $k \geq 2$. Hence $F_k \equiv 2 \pmod 5$ for $k \geq 2$. Thus

$$\left(\frac{5}{F_k}\right) = \left(\frac{F_k}{5}\right) = \left(\frac{2}{5}\right) = -1,$$

where we have used quadratic reciprocity. Thus 5 is a nonresidue. By Euler's criterion (Theorem 5.7.3), we have $5^{(F_k-1)/2} \equiv -1 \pmod{F_k}$.

To prove the converse, we use Theorem 9.1.1, with $a = 5$. The only prime that divides $F_k - 1$ is 2, and by hypothesis $a^{(F_k-1)/2} \not\equiv 1 \pmod{F_k}$. On the other hand, since $a^{(F_k-1)/2} \equiv -1 \pmod{F_k}$, it follows that $a^{F_k-1} \equiv 1 \pmod{F_k}$. Thus the hypotheses of Theorem 9.1.1 are satisfied, and F_k is prime. ■

COROLLARY 9.2.2 We can decide whether the Fermat number $n = 2^{2^k} + 1$ is prime or composite in deterministic polynomial time. Pepin's test uses $O((\lg n)^3)$ bit operations.

With the aid of this test, and other methods, it is now known that F_k is prime for $0 \le k \le 4$ and composite for $5 \le k \le 23$. The character of F_{24} is presently unknown.

Proth gave a sufficient condition for the primality of numbers of the form $f \cdot 2^s + 1$:

THEOREM 9.2.3 (PROTH'S TEST) Let $n = f \cdot 2^s + 1$, where $1 \le f < 2^s$. Then n is prime iff $a^{(n-1)/2} \equiv -1 \pmod{n}$ for some a.

Proof Use Corollary 9.1.6. ■

The last and most important test of this section is the Lucas-Lehmer test for the primality of Mersenne numbers. As mentioned in Section 1.2, prime numbers of the form $2^p - 1$, now known as Mersenne primes, have attracted particular attention because of their connection with *perfect numbers*. A number n is said to be perfect if $\sigma(n) = 2n$. Such a number (examples include 6 and 28) is the sum of its "aliquot parts"—that is, the sum of its divisors excepting itself. Many writers, including St. Augustine (c. 400 A.D.), read mystical significance into the fact that for such numbers, the "whole" was equal to the sum of its "parts". Euclid proved that if $2^p - 1$ is a prime, then $(2^p - 1)2^{p-1}$ is perfect; Euler proved that every *even* perfect number is of this form. There are currently no odd perfect numbers known, but the smallest, if it exists, must exceed 10^{300}. Thus, the best way to produce perfect numbers is to test numbers of the form $2^p - 1$ for primality.

THEOREM 9.2.4 (LUCAS-LEHMER TEST) Let p be a prime > 2, and define $n = 2^p - 1$. Also define $S_1 = 4$ and $S_{k+1} = S_k^2 - 2$ for $k \ge 1$. Then n is prime if and only if $S_{p-1} \equiv 0 \pmod{n}$.

Proof The definition of S_n may appear unmotivated; we need to see why it appears in the test.

Consider the polynomial

$$f(X) = X^2 - 2^{(p+1)/2}X - 1.$$

Let q be any prime dividing n, and let $\alpha, \beta \in \mathbb{F}_{q^2}$ be the roots of $f(X) = 0$ over $\mathbb{F}_q$. Note that $\alpha + \beta = 2^{(p+1)/2}$ and $\alpha\beta = -1$.

Write $V(k) = \alpha^k + \beta^k$, and let $\overline{S}_k$ denote the value of $S_k \pmod{q}$. We first show by induction on k that $\overline{S}_k = V(2^k)$.

It is true for $k = 1$, since

$$V(2) = \alpha^2 + \beta^2 = (\alpha + \beta)^2 - 2\alpha\beta = (2^{(p+1)/2})^2 - 2(-1) = 2^{p+1} + 2 = 4.$$

Now assume $\overline{S}_k = V(2^k) = \alpha^{2^k} + \beta^{2^k}$. Then

$$\overline{S}_{k+1} = (\alpha^{2^k} + \beta^{2^k})^2 - 2 = \alpha^{2^{k+1}} + \beta^{2^{k+1}} + 2\alpha^{2^k}\beta^{2^k} - 2 = \alpha^{2^{k+1}} + \beta^{2^{k+1}} = V(2^{k+1}).$$

Now assume $n = 2^p - 1$ is prime. We show that $S_{p-1} \equiv 0 \pmod{n}$. First, we note that $f(X)$ is irreducible over $\mathbb{F}_n$. For by the quadratic formula, $f(X) = aX^2 + bX + c$ is irreducible iff the discriminant $\Delta = b^2 - 4ac$ is not a square. Now

$$\Delta = (2^{(p+1)/2})^2 - 4(-1) \equiv 6 \pmod{n}.$$

Thus $\left(\frac{\Delta}{n}\right) = \left(\frac{2}{n}\right)\left(\frac{3}{n}\right)$. But $\left(\frac{2}{n}\right) = +1$, since $(2^{(p+1)/2})^2 \equiv 2 \pmod{n}$. It is also easy to see that $\left(\frac{3}{n}\right) = -\left(\frac{n}{3}\right) = -\left(\frac{1}{3}\right) = -1$, where we have used quadratic reciprocity. Thus $\left(\frac{\Delta}{n}\right) = -1$, and so $f(x)$ is irreducible. Since $\alpha^{n+1} = \mathrm{N}(\alpha)$ (see section 6.6.5),

$$\alpha^{n+1} = \beta^{n+1} = \alpha\beta = -1,$$

and so $\alpha^{n+1} + \beta^{n+1} = \overline{S}_p = -2$. Since $S_p = S_{p-1}^2 - 2$, we must have $S_{p-1} \equiv 0 \pmod{n}$.

Now assume that $S_{p-1} \equiv 0 \pmod{n}$, but n isn't prime. Then n must have a prime divisor q, with $q^2 \le n$.

Since $S_{p-1} \equiv 0 \pmod{n}$, we have $S_{p-1} \equiv 0 \pmod{q}$. As above, consider $f(x)$ over $\mathbb{F}_q$, and let α and β be the roots of $f(x) = 0$. Then $\alpha^{2^{p-1}} + \beta^{2^{p-1}} = 0$, so $\alpha^{2^p} + (\alpha\beta)^{2^{p-1}} = 0$. Hence $\alpha^{2^p} = -1$.

Thus the order of α in $\mathbb{F}_{q^2}$ is 2^{p+1}. But $\alpha \in \mathbb{F}_{q^2}$, so $\operatorname{ord}(\alpha) \mid q^2 - 1$. But then $2^{p+1} \mid q^2 - 1$, which is impossible, since $2^{p+1} > n$, and $q^2 \le n$. This gives us a contradiction, and so n must be prime. ■

COROLLARY 9.2.5 We can decide whether the Mersenne number $n = 2^p - 1$ is prime or composite in deterministic polynomial time. The test can be done using $O((\lg n)^3)$ bit operations.

Proof The theorem above uses $O(p)$ squarings and $O(p)$ subtractions modulo a number with p bits. ■

Theorem 9.2.4 gives a very practical test for the primality of Mersenne numbers. For example, let us test $M_7 = 127$. Calculating modulo 127, we find $S_1 \equiv 4$, $S_2 \equiv 14$, $S_3 \equiv 67$, $S_4 \equiv 42$, $S_5 \equiv 111$, and $S_6 \equiv 0$. Hence 127 is a prime.

We now know 33 Mersenne primes, corresponding to the exponents $p = 2, 3, 5, 7, 13, 17, 19, 31, 61, 89, 107, 127, 521, 607, 1279, 2203, 2281, 3217, 4253, 4423, 9689, 9941, 11213, 19937, 21701, 23209, 44497, 86243, 110503, 132049, 216091, 756839$, and

859433. All of the numbers greater than $2^{31} - 1$ were tested by the Lucas-Lehmer test or one of its variants.

9.3 Pseudoprimes and Carmichael Numbers

Let n be composite. If there is an a such that $a^{n-1} \equiv 1 \pmod{n}$, then n "looks like a prime" for the base a. This gives rise to the following definition:

DEFINITION 9.3.1 An odd composite number n such that $a^{n-1} \equiv 1 \pmod{n}$ is called a *pseudoprime* to the base a. The set of all such odd composite n is denoted by psp(a).

In the literature, these numbers are sometimes called *ordinary* or *Fermat* pseudoprimes.

The set psp(2) in particular has been examined by many authors. For example, it is known that there are only 5597 such numbers below 10^9; the smallest is 341.

Instead of holding a fixed and looking at pseudoprimes to the base a, we can also look at those bases relatively prime to n for which n is a prime or pseudoprime:

DEFINITION 9.3.2 The *group of Fermat liars* $F(n)$ is defined be

$$F(n) = \{a \in (\mathbb{Z}/(n))^* \ : \ a^{n-1} \equiv 1 \pmod{n}\}.$$

As mentioned before, there exist composite numbers n such that $F(n) = (\mathbb{Z}/(n))^*$; that is, n is a pseudoprime to all bases relatively prime to it. Such a number can never be *proved* composite by using Fermat's theorem.

DEFINITION 9.3.3 If n is a composite number such that $a^{n-1} \equiv 1 \pmod{n}$ for all $a \in (\mathbb{Z}/(n))^*$, then n is said to be a *Carmichael number*.

There are only 8241 Carmichael numbers below 10^{12}; the smallest is 561.

We now characterize Carmichael numbers n in terms of the factorization of n. Recall from Chapter 2 that the *exponent* of a finite group G, $\exp(G)$, is defined to be the smallest positive integer e such that $a^e = 1$ for all $a \in G$. Define $\lambda(n) = \exp((\mathbb{Z}/(n))^*)$, the *Carmichael lambda* function. Using Exercise 5.1, we have the following

THEOREM 9.3.4 A composite number n is a Carmichael number if and only if $\lambda(n) \mid n - 1$.

We can easily compute $\lambda(n)$, given the factorization of n:

THEOREM 9.3.5 The function $\lambda(n)$ has the following properties:

(i) $\lambda(1) = 1$;

(ii) $\lambda(p^e) = p^{e-1}(p-1)$, if p is an odd prime;

(iii) $\lambda(2) = 1$, $\lambda(4) = 2$, and $\lambda(2^e) = 2^{e-2}$ for $e \geq 3$; and

(iv) if $n = \prod_{1 \leq i \leq k} p_i^{e_i}$, then $\lambda(n) = \operatorname{lcm}_{1 \leq i \leq k} \lambda(p_i^{e_i})$.

Proof Part (i) is clear. For part (ii), we know from Theorem 5.6.2 that $(\mathbb{Z}/(p^e))^*$ is cyclic, and has an element of order $\varphi(p^e) = p^{e-1}(p-1) = \lambda(p^e)$. For part (iii), the same theorem shows that $(\mathbb{Z}/(2^e))^*$ has an element of order $2^{e-2} = \lambda(2^e)$ if $e \geq 3$, of order $1 = \lambda(2)$ if $e = 1$, of order $2 = \lambda(4)$ if $e = 2$, and these orders are the maximum possible.

For part (iv), set $t = \operatorname{lcm}_{1 \leq i \leq k} \lambda(p_i^{e_i})$. Then $x^t \equiv 1 \pmod{p_i^{e_i}}$ for all i; hence by the Chinese remainder theorem, $x^t \equiv 1 \pmod{n}$. Thus $\lambda(n) \mid \operatorname{lcm}_{1 \leq i \leq k} \lambda(p_i^{e_i})$.

We now construct an element of order exactly $\lambda(n)$. Let a_i be of order $\lambda(p_i^{e_i})$ in $(\mathbb{Z}/(p_i^{e_i}))^*$. Let $x \equiv a_i \pmod{p_i^{e_i}}$ for $1 \leq i \leq k$. Then by Exercise 5.26, we know that x is of order $\operatorname{lcm}_{1 \leq i \leq k} \lambda(p_i^{e_i})$. ■

As an application of this theorem, let us show that 561 is a Carmichael number. We have $561 = 3 \cdot 11 \cdot 17$, so $\lambda(561) = \operatorname{lcm}(2, 10, 16) = 80$, and $80 \mid 560$.

We now show that Carmichael numbers must obey certain properties:

THEOREM 9.3.6 If n is Carmichael, then it is odd, squarefree, and divisible by at least 3 distinct primes.

Proof Let n be a Carmichael number.

If $n > 2$, then it is easy to see that $2 \mid \lambda(n)$. Hence if $\lambda(n) \mid n - 1$, we must have $n - 1$ even and so n is odd.

Now suppose $p^2 \mid n$; we may assume p is odd. Then from the theorem, we have $p \mid \lambda(n)$; hence $p \mid n - 1$. Thus $p \mid n$ and $p \mid n - 1$, so $p \mid 1$, a contradiction.

Finally, suppose $n = pq$, the product of distinct odd primes. Then $p - 1 \mid \lambda(n)$, so $pq - 1 = n - 1 \equiv 0 \pmod{p-1}$. But $p \equiv 1 \pmod{p-1}$, so $q \equiv 1 \pmod{p-1}$ and thus $q \geq p$. Similarly, $p \geq q$. Hence $p = q$, a contradiction. ■

It has long been an open problem to prove that there are infinitely many Carmichael numbers. This was recently proved by Alford, Granville, and Pomerance. More precisely, they proved the following theorem, where $C(x)$ denotes the number of Carmichael numbers $\leq x$.

THEOREM 9.3.7 For all sufficiently large x we have $C(x) > x^{2/7}$.

There are two other kinds of pseudoprimes that have been studied. Both types occur even less frequently than ordinary pseudoprimes.

DEFINITION 9.3.8 An odd composite number n such that $\gcd(a, n) = 1$ and

$$a^{(n-1)/2} \equiv \left(\frac{a}{n}\right) \pmod{n}$$

is called an *Euler pseudoprime* to the base a. The set of all such odd composite n is denoted by epsp(a).

The name *Euler pseudoprime* comes from Euler's criterion (Theorem 5.7.3). The intuition behind this definition is that if n is prime, then certainly $a^{(n-1)/2} \equiv \left(\frac{a}{n}\right) \pmod{n}$, while if n is composite, $a^{(n-1)/2}$ behaves more or less "randomly".

DEFINITION 9.3.9 Let n be an odd composite number, and write $n-1 = 2^s \cdot d$, with d odd. Then n is called a *strong pseudoprime* to the base a if $a^d \equiv 1 \pmod{n}$ or $a^{2^r \cdot d} \equiv -1 \pmod{n}$ for some r, $0 \le r < s$. The set of all such odd composite n is denoted by spsp(a).

The intuition behind this definition is that, modulo an odd prime power p^k, the element 1 has exactly two square roots: itself and -1. Modulo a non-prime-power n, however, there will be $2^{\omega(n)}$ distinct square roots of 1, where $\omega(n)$ denotes the number of distinct prime divisors of n. Hence a strong pseudoprime is a restricted kind of Fermat pseudoprime; not only does $a^{n-1} \equiv 1 \pmod{n}$, but square roots of 1 behave "correctly".

For example, $2^{1023} \equiv 1 \pmod{2047}$, and therefore $2047 \in \text{spsp}(2)$. Since $\left(\frac{2}{2047}\right) = 1$, we also see that $2047 \in \text{epsp}(2)$. This is no surprise, as the following theorem shows:

THEOREM 9.3.10 We have

$$\text{spsp}(a) \subseteq \text{epsp}(a) \subseteq \text{psp}(a).$$

Proof The inclusion $\text{epsp}(a) \subseteq \text{psp}(a)$ is clear.

We now prove the inclusion $\text{spsp}(a) \subseteq \text{epsp}(a)$. Let $n \in \text{spsp}(a)$, and suppose $n - 1 = 2^s \cdot d$, where d is odd. Also suppose that the prime factorization of n is given by

$$n = p_1^{e_1} p_2^{e_2} \cdots p_t^{e_t}.$$

By the definition of strong pseudoprime, we know either (i) $a^d \equiv 1 \pmod{n}$, or (ii) $a^{2^{k-1} \cdot d} \equiv -1 \pmod{n}$ for some k, $1 \le k \le s$. Thus $2^k \,||\, \text{ord}_{p^e}\, a$ for all primes $p^e \,||\, n$ (in case (i), we have $k = 0$).

Now by Exercise 5.29, we know that $(\text{ord}_{p^e}\, a)/(\text{ord}_p\, a)$ is either 1 or a power of p, hence odd, so $2^k \,||\, \text{ord}_p a$ for all $p \mid n$. Defining $k_i = \nu_2(p_i - 1)$ for $1 \le i \le t$, we see $k \le k_i$ for all i. Let m be the number of indices i with $k_i = k$. Then $n \equiv (2^k + 1)^m \equiv m2^k + 1 \pmod{2^{k+1}}$. If m is odd, then $2^k \,||\, n - 1$, and if m is even, then $2^{k+1} \mid n - 1$. Thus, if $p^e \,||\, n$, then $a^{(n-1)/2} \equiv -1 \pmod{p^e}$ if m is odd, and $a^{(n-1)/2} \equiv 1 \pmod{p^e}$ if m is even.

On the other hand, $\left(\frac{a}{p_i}\right) = -1$ if and only if $\nu_2(\text{ord}_p\, a) = k_i$. Hence $\left(\frac{a}{n}\right) = (-1)^m$. Thus we see that $a^{(n-1)/2} \equiv \left(\frac{a}{n}\right) \pmod{n}$, and so $n \in \text{epsp}(a)$. ■

We conclude this section with a brief discussion of the distribution of pseudoprimes and Carmichael numbers.

Pomerance has shown that there exists a constant $c(a)$ such that

$$\exp((\log x)^{15/38}) < |\mathrm{psp}(a) \cap [1, x]| < xL(x)^{-1/2},$$

for all $x > c(a)$. Here $L(x) = x^{\log\log\log x/\log\log x}$.

Letting $C(x)$ denote the number of Carmichael numbers $\leq x$, Pomerance has also shown that

$$C(x) < xL(x)^{-a}$$

for all sufficiently large x, where a is any real number less than 1. He conjectures that

$$C(x) \geq x \cdot L(x)^{-1+o(1)}.$$

As mentioned previously, it is now known that $C(x) > x^{2/7}$ for all x sufficiently large.

9.4 Probabilistic Primality Tests

In this section we assume that n is an odd integer.

Recall from Chapter 3 the definitions of the three types of randomized algorithms to recognize elements of a language L:

Given $x \notin L$, a *Monte Carlo* algorithm for L will always correctly reply "$x \notin L$". Given $x \in L$, such an algorithm will correctly reply "$x \in L$" at least 50% of the time. We emphasize that this 50% figure is for each x, and is tied only to the sequence of internal coin tosses, not any assumed distribution of x. We sometimes say that L is accepted with *one-sided* error. The class of languages with polynomial-time Monte Carlo recognition algorithms is denoted by $\mathcal{RP}$.

Given x, an *Atlantic City* algorithm for L will reply correctly (either "$x \in L$" or "$x \notin L$") at least 75% of the time. Again, the 75% figure refers to each x. We sometimes say x is accepted with *two-sided* error. The class of languages with polynomial-time Atlantic City recognition algorithms is denoted by $\mathcal{BPP}$.

Finally, a *Las Vegas* algorithm for L can be viewed as a combination of two algorithms: a Monte Carlo algorithm for L, and another Monte Carlo algorithm for $\overline{L}$, the complement of L. Such an algorithm always answers correctly, but its running time is probabilistic. The class of languages with polynomial-time Las Vegas algorithms is denoted by $\mathcal{ZPP}$.

A *test for L* is defined to be any one of the three types of randomized algorithms for recognizing L. Thus, like a driver's test or college entrance examination, a test for L provides good, but not necessarily conclusive, evidence of membership or nonmembership. Each kind of test can be iterated to decrease the error probability.

We can also classify the three types of randomized algorithms based on whether they produce a *proof* of membership in L or $\overline{L}$. Given $x \in L$, a Monte Carlo algorithm for L

will, with high probability, produce a proof that indeed $x \in L$. The set of computations proving $x \in L$ is sometimes referred to as a *certificate* of membership in L. However, if $x \notin L$, in general, no such proof is produced, and the result should be viewed only as *good evidence* that $x \notin L$.

A Las Vegas algorithm will produce a proof of membership or nonmembership for *all* inputs x, while an Atlantic City algorithm may *never* produce such a proof.

What does this have to do with primality testing? We will see in this section that there exists a fast and simple Monte Carlo algorithm for the set of composite numbers, written in binary. This primality test (using our definition of "test" above), given a composite input n, will with high probability produce a proof that n is composite; furthermore, by iterating the test, we can make this probability as close to 1 as desired. If, however, the input is prime, the test will never produce a proof that this is the case, and therefore a response of "prime" must be viewed only as good evidence that the input actually is prime.

In practice, given a number n of unknown type, one runs a few iterations of the Monte Carlo primality test. If all iterations fail to produce a proof that n is composite, then one has the "moral certainty" that n is prime. In many applications (e.g. cryptography; see Chapter 15), this is sufficient. If, however, a *proof* that n really *is* prime is necessary, one then applies the methods of sections 9.1, 9.6, or Chapter 13.

In this section we discuss two "probabilistic" primality tests. Both of the tests use Monte Carlo algorithms for compositeness, so these tests provide proofs of compositeness, but only good evidence of primality. Both use $O((\lg n)^3)$ bit operations.

The first test is due to R. Solovay and V. Strassen. It is based on $E(n)$, the *group of Euler liars*, defined as follows:

$$E(n) = \{a \in (\mathbb{Z}/(n))^* \ : \ \left(\frac{a}{n}\right) \equiv a^{(n-1)/2} \pmod{n}\}.$$

For example, $E(65) = \{1, 8, 14, 18, 47, 51, 57, 64\}$.

The Solovay-Strassen test is based on the following lemma, which was also discovered (independently) by D. H. Lehmer:

LEMMA 9.4.1 Let n be an odd integer ≥ 3. Then n is a prime iff $E(n) = (\mathbb{Z}/(n))^*$.

Proof If n is an odd prime, then $\left(\frac{a}{n}\right) \equiv a^{(n-1)/2} \pmod{n}$ by Theorem 5.8.1.

To prove the converse, assume that $E(n) = (\mathbb{Z}/(n))^*$, but n is composite. Then

$$a^{n-1} \equiv \left(\frac{a}{n}\right)^2 \equiv 1 \pmod{n}$$

for all $a \in (\mathbb{Z}/(n))^*$. Thus n is Carmichael, and by Theorem 9.3.6, n must be squarefree.

Thus we can write $n = pr$, with p prime, $r > 1$, and $\gcd(p, r) = 1$. Let g be a quadratic nonresidue modulo p and let $a \equiv g \pmod{p}$ and $a \equiv 1 \pmod{r}$.

By Theorem 5.9.2 we have

$$\left(\frac{a}{n}\right) = \left(\frac{a}{pr}\right) = \left(\frac{a}{p}\right)\left(\frac{a}{r}\right) = \left(\frac{g}{p}\right)\left(\frac{1}{r}\right) = (-1)(+1) = -1.$$

Since by assumption $\left(\frac{a}{n}\right) \equiv a^{(n-1)/2} \pmod{n}$ for all $a \in (\mathbb{Z}/(n))^*$, we must have $a^{(n-1)/2} \equiv -1 \pmod{r}$. But this contradicts $a \equiv 1 \pmod{r}$. ■

A nonzero element in $\mathbb{Z}/(n) - E(n)$ is sometimes called an *Euler witness* to the compositeness of n.

We now construct a primality test for odd numbers ≥ 3, based on this theorem:

SOLOVAY-STRASSEN(n)
choose a at random from $\{1, 2, \ldots, n-1\}$
(1) if $\gcd(a, n) \neq 1$
    then return('composite') and stop
else
    (2) if $\left(\frac{a}{n}\right) \not\equiv a^{(n-1)/2} \pmod{n}$
        then return('composite') and stop
    else
        return('prime')

THEOREM 9.4.2 If n is an odd prime, then SOLOVAY-STRASSEN(n) returns 'prime'. If n is an odd composite, then for at least $1/2$ of all a in $\{1, 2, \ldots, n-1\}$, it returns 'composite'. The algorithm uses $O((\lg n)^3)$ bit operations.

Proof If n is an odd prime, then step (1) will never find a nontrivial gcd, and so the algorithm will proceed to step (2). From the definition of the Legendre symbol, this step will always return 'prime'.

If n is an odd composite, then if $a \notin (\mathbb{Z}/(n))^*$, step (1) will return 'composite'. Otherwise, $a \in (\mathbb{Z}/(n))^*$. Using the lemma, $E(n) \neq (\mathbb{Z}/(n))^*$. But it is easy to see that $E(n)$ is a subgroup of $(\mathbb{Z}/(n))^*$, so it must in fact be a *proper* subgroup. Hence

$$|E(n)| \leq |(\mathbb{Z}/(n))^*|/2 = \varphi(n)/2 \leq (n-1)/2.$$

Thus with probability $\geq 1/2$, a is an Euler witness to the compositeness of n, and the algorithm returns 'composite'.

By Theorems 4.2.4 and 5.9.3, we can compute $\gcd(a, n)$ and $\left(\frac{a}{n}\right)$ using $O((\lg n)^2)$ bit operations, and by Theorem 5.4.1, we can compute $a^{(n-1)/2} \pmod{n}$ using $O((\lg n)^3)$ bit operations. ■

Note that it is possible to avoid the extra computation of the gcd in step (1) of SOLOVAY-STRASSEN; see Exercise 11.

With a minor addition to the algorithm to handle the case of the unit 1 and even numbers, we have shown

THEOREM 9.4.3 COMPOSITES $\in \mathcal{RP}$.

We now turn to a second probabilistic prime test, the Miller-Rabin test. It is based on the concept of "strong pseudoprime", introduced in the previous section.

Let $n - 1 = 2^s \cdot d$, where d is odd, and define $S(n)$, the *set of strong liars*, as follows:

$$S(n) = \{a \in (\mathbb{Z}/(n))^* \ : \ a^d \equiv 1 \pmod n \text{ or } a^{2^r \cdot d} \equiv -1 \pmod n \text{ for some } r,\ 0 \le r < s\}$$

For example, $S(65) = \{1, 8, 18, 47, 57, 64\}$.

LEMMA 9.4.4 Let n be an odd integer ≥ 3. Then n is a prime iff $S(n) = (\mathbb{Z}/(n))^*$. If n is composite, then $|S(n)| \le (n-1)/4$.

A nonzero element of $\mathbb{Z}/(n) - S(n)$ is said to be a *strong witness* to the compositeness of n.

Proof Assume n is prime. Then if $a \in (\mathbb{Z}/(n))^*$, we have $a^{n-1} \equiv 1 \pmod n$. Hence either $a^{(n-1)/2} \equiv -1 \pmod n$, or $a^{(n-1)/2} \equiv 1 \pmod n$. In the first case, we are done. In the second case, we again have $a^{(n-1)/4} \equiv \pm 1 \pmod n$. Continuing in this fashion, eventually either $a^d \equiv 1 \pmod n$, or $a^{2^r \cdot d} \equiv -1 \pmod n$ for some r, $0 \le r < s$.

Now assume n is composite. Recall that $n - 1 = 2^s \cdot d$, with d odd. Write $n = p_1^{e_1} \cdots p_j^{e_j}$.

Let k be the largest integer such that there exists at least one $b \in (\mathbb{Z}/(n))^*$ with $b^{2^k} \equiv -1 \pmod n$. To see that k is well-defined, note that $(-1)^{2^0} \equiv -1 \pmod n$, and so $k \ge 0$. Also, $k \le \nu_2(\lambda(n)) - 1$.

Note that $n \equiv 1 \pmod{2^{k+1}}$ by Theorem 9.1.5, since $p_i \equiv 1 \pmod{2^{k+1}}$ for all i, $1 \le i \le j$.

Let $m = 2^k \cdot d$. Then $2m \mid n - 1$. Now consider the following chain of subgroups:

$$\begin{array}{c} (\mathbb{Z}/(n))^* \\ | \\ J = \{a \in (\mathbb{Z}/(n))^* \ : \ a^{n-1} \equiv 1 \pmod n\} \\ | \\ K = \{a \in (\mathbb{Z}/(n))^* \ : \ a^m \equiv \pm 1 \pmod{p_i^{e_i}}\ \forall i\} \\ | \\ L = \{a \in (\mathbb{Z}/(n))^* \ : \ a^m \equiv \pm 1 \pmod n\} \\ | \\ M = \{a \in (\mathbb{Z}/(n))^* \ : \ a^m \equiv 1 \pmod n\} \end{array}$$

It is easy to see that each of these sets is actually a group, and the containments are as indicated. Note that every strong liar in $(\mathbb{Z}/(n))^*$ is a member of L, since if $a^d \equiv 1 \pmod{n}$, then clearly $a^m \equiv 1 \pmod{n}$, while if $a^{2^t \cdot d} \equiv -1 \pmod{n}$ for some t, then $t \leq k$ by the definition of k. We will show that, provided $n \neq 9$, L is a subgroup of index at least 4 in $(\mathbb{Z}/(n))^*$. From this it will follow that $|S(n)| \leq (n-1)/4$.

Now every element of

$$G = \{a \in (\mathbb{Z}/(n))^* \ : \ a \equiv \pm 1 \pmod{p_i^{e_i}}\ \forall i\}$$

is a 2^k-th power; hence an m-th power. (To see this, put $x \equiv b$ or $x \equiv b^2 \pmod{p_i^{e_i}}$; then $x^{2^k} \equiv \pm 1 \pmod{p_i^{e_i}}$ for each i.) Thus M has index 2^j in

$$K = \{a \in (\mathbb{Z}/(n))^* \ : \ a^m \in G\}.$$

Similarly, M has index 2 in L. Thus $(K : L) = 2^{j-1}$.

From the diagram, we see that

$$((\mathbb{Z}/(n))^* : L) \geq ((\mathbb{Z}/(n))^* : J)(K : L) = 2^{j-1}((\mathbb{Z}/(n))^* : J).$$

If $j \geq 3$, then it follows that that $((\mathbb{Z}/(n))^* : L) \geq 4$.

If $j \geq 2$, then by Theorem 9.3.6, n cannot be Carmichael. Thus there exists $a \in (\mathbb{Z}/(n))^*$ with $a^{n-1} \not\equiv 1 \pmod{n}$; thus J is a proper subgroup of $(\mathbb{Z}/(n))^*$ and so $((\mathbb{Z}/(n))^* : J) \geq 2$. Thus $((\mathbb{Z}/(n))^* : L) \geq 4$.

Finally, if $j = 1$, then $n = p^e, e \geq 2$. But then $|J| = p-1$, so $((\mathbb{Z}/(n))^* : J) = p^{e-1} \geq 4$, except when $p^e = 9$. When $n = 9$, however, there are exactly 2 strong liars: 1 and -1, and so the number of strong liars is nevertheless $\leq (n-1)/4$. ■

We now show how to use the result of Lemma 9.4.4 in a Monte Carlo primality test:

MILLER-RABIN(n)

choose a at random from $\{1, 2, \ldots, n-1\}$
express $n - 1 = 2^s \cdot d$, d odd
compute successively $a_0 = a^d \pmod{n}$, $a_1 = a_0^2 \pmod{n}, \ldots, a_k = a_{k-1}^2 \pmod{n}$
  until $k = s$ or $a_k \equiv 1 \pmod{n}$
if ($k = s$) and $a_k \not\equiv 1$ then return('composite')
  else if ($k = 0$) then return ('prime')
    else if $a_{k-1} \not\equiv -1$ then return('composite')
      else return('prime')

THEOREM 9.4.5 If n is an odd prime, then MILLER-RABIN(n) always returns 'prime'. If n is an odd composite, then MILLER-RABIN(n) returns 'composite' for at least $3/4$ of all a, $1 \leq a \leq n-1$. The algorithm uses $O((\lg n)^3)$ bit operations.

Proof If n is an odd prime, then step (4) will never return 'composite'. Otherwise, let k be the least index such that $a_k \equiv 1 \pmod{n}$. Then if $k = 0$, the algorithm will return 'prime' in line (5). Otherwise we must have $a_{k-1} \equiv -1 \pmod{n}$, and the algorithm will return 'prime' in line (6).

Now assume n is an odd composite number.

If $a \notin (\mathbb{Z}/(n))^*$, then we can never have $a^{2^s \cdot d} \equiv 1 \pmod{n}$, so we will find $k = s$ and $a_k \not\equiv 1$. Hence step (4) declares n to be composite.

Otherwise, $a \in (\mathbb{Z}/(n))^*$. Using Lemma 9.4.4, we know that at least $3/4$ of all $a \in (\mathbb{Z}/(n))^*$ are strong witnesses, and for these values of a it is easy to verify that the algorithm returns 'composite'.

To prove the running time bound of $O((\lg n)^3)$ bit operations, we observe that the most time-consuming part of the algorithm is the computation of $a^d, a^{2d}, \ldots, a^{2^k \cdot d} \pmod{n}$. We can compute $a^d \pmod{n}$ using $O((\lg n)^3)$ bit operations via the power algorithm, and we then compute the remaining powers by squaring modulo n, again using $O((\lg n)^3)$ bit operations. ■

Which of the two algorithms presented in this section should be used in practice? The answer is probably the Miller-Rabin algorithm, for at least three reasons. First, the Solovay-Strassen algorithm is harder to program, as it involves the computation of the Jacobi symbol. Second, the error probability for Miller-Rabin is bounded above by $1/4$, while that for Solovay-Strassen is bounded by $1/2$. Third, by Theorem 9.3.10, any Euler witness is also a strong witness to the compositeness of n.

9.5 ERH-Based Methods

In this section, we show that by assuming the validity of the Extended Riemann Hypothesis (ERH), the Monte Carlo primality tests of the previous section can be made deterministic. The reader should review Sections 8.4 and 8.5 for a discussion of the ERH.

The following theorem gives a deterministic version of the Solovay-Strassen test:

THEOREM 9.5.1 Assume ERH. If n is an odd composite integer, then there is a positive integer $a < n$, with $a \leq 2(\log n)^2$, for which either (i) $\gcd(a, n) \neq 1$, or (ii) $\left(\frac{a}{n}\right) \not\equiv a^{(n-1)/2} \pmod{n}$.

Proof By Lemma 9.4.1,

$$E(n) = \{a \in (\mathbb{Z}/(n))^* : \left(\frac{a}{n}\right) \equiv a^{(n-1)/2} \pmod{n}\}$$

is a nontrivial subgroup of $(\mathbb{Z}/(n))^*$ when n is an odd composite. Ankeny's theorem (Theorem 8.5.3) implies that there must be some $a \in \mathbb{Z}/(n) - E(n)$ that is $O((\log n)^2)$. This element must either lie outside of $(\mathbb{Z}/(n))^*$ (in which case $\gcd(a, n) \neq 1$), or it must satisfy $\left(\frac{a}{n}\right) \not\equiv a^{(n-1)/2} \pmod{n}$. The explicit bound of $2(\log n)^2$ is obtained from Theorem 8.8.17. ■

This leads to a simple deterministic prime test: We replace the test of a random a in the algorithm SOLOVAY-STRASSEN with a series of tests, one for each a up to $2(\log n)^2$. We give the algorithm below.

```
DETERMINISTIC SOLOVAY-STRASSEN(n)
for a ← 2 to min(n − 1, 2(log n)^2) do
   if gcd(a, n) ≠ 1
         then return('composite')
   else
         if (a/n) ≢ a^((n−1)/2) (mod n)
               then return('composite')
return('prime')
```

THEOREM 9.5.2 Assume ERH. For $n \geq 3$, DETERMINISTIC SOLOVAY-STRASSEN returns 'prime' iff n is prime. The algorithm uses $O((\lg n)^5)$ bit operations.

Proof The correctness is a consequence of Theorem 9.5.1 and Euler's criterion. Since we always have $a < n$, the conditions on a can be tested using $O((\lg n)^3)$ bit operations, by Corollary 4.2.4, Theorems 5.9.3, and Theorem 5.4.1. (The time to compute the upper loop limit is negligible compared to this.) We test at most $O((\lg n)^2)$ values of a, so the running time follows. ■

One can save a factor of $\lg \lg n$ in the running time by only testing prime values of a. We also note that the gcd computation can also be eliminated, as in the randomized version of this test. (See Exercise 11.)

Theorem 9.5.2 has the following important consequence, which was proved by G. Miller in 1976.

THEOREM 9.5.3 Assume ERH. Then PRIMES $\in \mathcal{P}$.

This result can also be proved by showing that the analog of Theorem 9.5.2 is true for other randomized prime tests, including the Miller-Rabin test. For this, we refer the reader to Exercises 50 and 51.

9.6 Primality Testing Using Algebraic Number Theory

In this section, we discuss primality tests based on algebraic number theory. These tests, developed over the past ten years, have greatly increased our ability to feasibly prove the primality of numbers with hundreds of digits. While a knowledge of elementary algebraic number theory may be helpful in understanding this section, it is not essential.

We will present a primality test due to H. W. Lenstra, Jr., based on ideas of Adleman, Pomerance and Rumely, which we call the "cyclotomic rings" test. There are actually many versions of this test; one form provides a deterministic algorithm for determining if n is prime that uses

$$O((\lg n)^{c \lg \lg \lg n}) \tag{9.1}$$

bit operations, for some constant c. We stress that this test, unlike the test of section 9.4, actually provides a *proof* of primality, not just good evidence.

However, the simplified version that we present here will, given a prime n, provide a proof that n is prime in time bounded by (9.1) *if* a certain unproved hypothesis ("Hypothesis 9.6.7") about the distribution of primes is true. This hypothesis is a consequence of the Generalized Riemann Hypothesis (GRH), as discussed in Section 8.7. We emphasize that this algorithm is deterministic, and unlike the ERH-dependent tests of section 9.5, the unproved hypothesis affects only the running time bound we can prove, and *not* the correctness of the proof provided by the test.

The test is based on checking identities in certain cyclotomic rings. If n is prime, these identities will be true. If n is composite, however, these identities are likely to fail. If, however, all these identities are verified, then we get some "extra information" about the possible divisors of n. More precisely, we will show that the possible divisors of n must fall into a relatively small number of congruence classes, and these are easily checked.

The following simple example roughly illustrates the flavor of the "extra information" obtained. Suppose for some integer a we have $a^{(n-1)/2} \equiv -1 \pmod{n}$. Then it can be shown (see Exercise 46) that $\nu_2(r-1) \geq \nu_2(n-1)$ for every $r \mid n$.

The version of the cyclotomic rings test we present here is not really practical, since it involves arithmetic operations on polynomials of rather large degree. A more practical test, using Jacobi sums instead of Gauss sums, was formulated by H. Cohen and H. W. Lenstra, Jr. This test, implemented by A. K. Lenstra, can test a number of several hundred digits for primality in just a few minutes on a medium-size computer.

Before we describe the cyclotomic rings test, we state some notation and prove some lemmas. In what follows, we assume n is the odd number to be tested, p and q are primes not dividing n, and $p \mid q-1$. Let ζ_p and ζ_q be primitive p-th and q-th roots of unity in $\mathbb{C}$. Let g be a generator of $(\mathbb{Z}/(q))^*$.

Define the character $\chi = \chi_{p,q}$, which maps $(\mathbb{Z}/(q))^*$ to $\langle \zeta_p \rangle$, the multiplicative group of the p-th roots of unity, as follows: $\chi(g^j) = \zeta_p^j$. The reader should check that χ is indeed a homomorphism, and does not depend on the particular representative chosen mod q. By $\overline{\chi}(a)$ we mean $(\chi(a))^{-1}$.

Finally, we define the *Gauss sum* $\tau(\chi)$ as follows:

$$\tau(\chi) = \sum_{1 \le a \le q-1} \chi(a)\zeta_q^a.$$

Note that the Gauss sum takes its value in the ring $R = R_{p,q} = \mathbb{Z}[\zeta_p, \zeta_q]$.

LEMMA 9.6.1

$$\tau(\chi)\tau(\overline{\chi}) = \chi(-1)q.$$

Proof We have

$$\begin{aligned}\tau(\chi)\tau(\overline{\chi}) &= \left(\sum_{1 \le a \le q-1} \chi(a)\zeta_q^a\right)\left(\sum_{1 \le b \le q-1} \overline{\chi}(b)\zeta_q^b\right) \\ &= \sum_{1 \le a \le q-1}\ \sum_{1 \le b \le q-1} \chi(a)\overline{\chi}(b)\zeta_q^{a+b} \\ &= \sum_{1 \le a \le q-1}\ \sum_{1 \le b \le q-1} \chi(ab)\overline{\chi}(b)\zeta_q^{ab+b}.\end{aligned}$$

This last equality comes from observing that, for fixed b, both a and ab run over all the residue classes in $(\mathbb{Z}/(q))^*$. Hence

$$\begin{aligned}\tau(\chi)\tau(\overline{\chi}) &= \sum_{1 \le a \le q-1}\ \sum_{1 \le b \le q-1} \chi(a)\zeta_q^{b(a+1)} \\ &= \sum_{1 \le a \le q-1} \chi(a) \sum_{1 \le b \le q-1} \zeta_q^{b(a+1)}.\end{aligned}$$

Now

$$\begin{aligned}\sum_{1 \le b \le q-1} \zeta_q^{b(a+1)} &= \zeta_q^{a+1} + \zeta_q^{2(a+1)} + \cdots + \zeta_q^{(q-1)(a+1)} \\ &= \begin{cases} -1, & \text{if } a \not\equiv -1 \pmod{q}, \\ q-1, & \text{if } a \equiv -1 \pmod{q}. \end{cases}\end{aligned}$$

Hence

$$\tau(\chi)\tau(\overline{\chi}) = (q-1)\chi(-1) - \sum_{1 \le a \le q-2} \chi(a). \tag{9.2}$$

Now

$$\sum_{1\le a\le q-1} \chi(a) = \sum_{1\le a\le q-1} \chi(ga) = \chi(g) \sum_{1\le a\le q-1} \chi(a);$$

since $\chi(g) \neq 0$, we must have

$$\sum_{1\le a\le q-1} \chi(a) = 0. \tag{9.3}$$

Thus, combining equations (9.2) and (9.3), we find

$$\tau(\chi)\tau(\overline{\chi}) = (q-1)\chi(-1) - (-\chi(-1)) = q\chi(-1),$$

which is the desired result. ∎

LEMMA 9.6.2 Let $R = \mathbb{Z}[\zeta_p, \zeta_q]$. If n is a prime, then

$$\tau(\chi)^{n^{p-1}-1} \equiv \chi(n) \pmod{nR}.$$

Proof Using Exercise 5.37, we have

$$\begin{aligned}
\tau(\chi)^{n^{p-1}} &= \left(\sum_{1\le a\le q-1} \chi(a)\zeta_q^a\right)^{n^{p-1}} \\
&\equiv \sum_{1\le a\le q-1} \chi(a)^{n^{p-1}} \zeta_q^{an^{p-1}} \pmod{nR} \\
&\equiv \sum_{1\le a\le q-1} \chi(a)\zeta_q^{an^{p-1}} \pmod{nR}.
\end{aligned}$$

Now define $b \equiv (n^{p-1})^{-1} \pmod q$. Then as a runs over the residue classes in $(\mathbb{Z}/(q))^*$, so does ab. Hence

$$\begin{aligned}
\tau(\chi)^{n^{p-1}} &\equiv \sum_{1\le a\le q-1} \chi(ab)\zeta_q^{abn^{p-1}} \pmod{nR} \\
&\equiv \sum_{1\le a\le q-1} \chi(ab)\zeta_q^a \pmod{nR} \\
&\equiv \chi(b) \sum_{1\le a\le q-1} \chi(a)\zeta_q^a \pmod{nR} \\
&\equiv \chi(b)\tau(\chi) \pmod{nR} \\
&\equiv \chi(n)^{1-p}\tau(\chi) \pmod{nR} \\
&\equiv \chi(n)\tau(\chi) \pmod{nR},
\end{aligned}$$

since $\chi(n)$ is a p-th root of unity.

Now Lemma 9.6.1 shows that $\tau(\chi)$ is a unit in R/nR, so we may "cancel" a $\tau(\chi)$ from both sides, and the result follows. ■

LEMMA 9.6.3 If for some prime $r \neq p$ we have $\zeta_p^i \equiv \zeta_p^j \pmod{r\mathbb{Z}[\zeta_p]}$, then $\zeta_p^i = \zeta_p^j$.

Proof Let $k = i - j$. Then $\zeta_p^k \equiv 1 \pmod{r\mathbb{Z}[\zeta_p]}$, and hence

$$\Phi_p(\zeta_p^k) \equiv \Phi_p(1) \equiv p \not\equiv 0 \pmod{r\mathbb{Z}[\zeta_p]},$$

where $\Phi_p(X)$ is the p-th cyclotomic polynomial, as discussed in Exercise 6.34. Thus ζ_p^k cannot be a primitive p-th root of 1. Hence $\zeta_p^k = 1$, and hence $\zeta_p^i = \zeta_p^j$. ■

Now let P and Q be finite sets of primes not dividing n with the following property: if $q \in Q$, then $q - 1$ is the product of *distinct* primes from P. (An example is $P = \{2, 3, 5, 7\}$, $Q = \{2, 3, 7, 11, 31, 43, 71, 211\}$.)

Define $z = \prod_{p\in P} p$ and $w = \prod_{q\in Q} q$. Then $q - 1 \mid z$ for all $q \in Q$, and so $(\operatorname{ord}_w a) \mid z$ for all $a \in (\mathbb{Z}/(w))^*$. Thus $a^z \equiv 1 \pmod{w}$ for all $a \in (\mathbb{Z}/(w))^*$.

The following theorem forms the basis of our cyclotomic rings test:

THEOREM 9.6.4 Suppose n is an odd integer satisfying the following two hypotheses:

(i) For all primes p and q with $q \in Q$ and $p \mid q - 1$, we have

$$\tau(\chi_{p,q})^{n^{p-1}-1} \equiv \chi_{p,q}(n) \pmod{nR_{p,q}};$$

(ii) For each prime $p \in P$ and each prime $r \mid n$ we have $\nu_p(r^{p-1} - 1) \geq \nu_p(n^{p-1} - 1)$.

Then each prime $r \mid n$ is congruent to $n^i \pmod{w}$ for some i, $0 \leq i < z$.

Proof By hypothesis (ii) we know that for all $p \in P$ and primes $r \mid n$ there exist integers a_p, $b_p(r)$, and e_p such that

$$n^{p-1} = 1 + a_p p^{e_p};$$
$$r^{p-1} = 1 + b_p(r) p^{e_p},$$

where $p \nmid a_p$.

Now for each pair of primes p, q with $q \in Q$ and $p \mid q - 1$, we have (by Lemma 9.6.2)

$$\chi(r) \equiv \tau(\chi)^{r^{p-1}-1} \pmod{rR}$$

for each prime $r \mid n$. Hence, raising each side to the a_p power, we get

$$\chi(r)^{a_p} \equiv \tau(\chi)^{(r^{p-1}-1)a_p} \pmod{rR}. \tag{9.4}$$

Since

$$(r^{p-1} - 1)a_p = (n^{p-1} - 1)b_p(r),$$

we have

$$\tau(\chi)^{(r^{p-1}-1)a_p} = \tau(\chi)^{(n^{p-1}-1)b_p(r)}. \tag{9.5}$$

By condition (i) we see

$$\begin{aligned}\tau(\chi)^{n^{p-1}-1} &\equiv \chi(n) \pmod{nR}\\ &\equiv \chi(n) \pmod{rR}.\end{aligned}$$

Hence, raising both sides to the $b_p(r)$ power, we get

$$\tau(\chi)^{(n^{p-1}-1)b_p(r)} \equiv \chi(n)^{b_p(r)} \pmod{rR}. \tag{9.6}$$

Now, combining (9.4–9.6), we see

$$\chi(r)^{a_p} \equiv \chi(n)^{b_p(r)} \pmod{rR}.$$

Using Lemma 9.6.3, we find that $\chi(r)^{a_p} = \chi(n)^{b_p(r)}$. Since $p \nmid a_p$, we see that $\chi(r) = \chi(n)^{\ell_p(r)}$, where $\ell_p(r)$ is defined by

$$a_p\ell_p(r) \equiv b_p(r) \pmod{p}.$$

Using the Chinese remainder theorem, we now determine an integer $\ell(r)$ with $0 \le \ell(r) < z$ such that

$$\ell(r) \equiv -\ell_p(r) \pmod{p}$$

for all $p \in P$. Then

$$\begin{aligned}\chi(rn^{\ell(r)}) &= \chi(r)\chi(n)^{\ell(r)}\\ &= \chi(n)^{\ell_p(r)}\chi(n)^{\ell(r)}\\ &= 1.\end{aligned}$$

Thus we have shown that $rn^{\ell(r)}$ lies in the kernel of $\chi_{p,q}$ for all $q \in Q$ and all $p \mid q-1$. Consider this for a fixed q. Since $rn^{\ell(r)}$ lies in the kernel of $\chi_{p,q}$ for all $p \mid q-1$, and $q-1$ is squarefree, we see that $rn^{\ell(r)} \equiv 1 \pmod{q}$. (For if t is in the kernel of χ, then $t = g^a$ for a generator g, and $a \equiv 0 \pmod{p}$. Since this is true for all $p \mid q-1$, we see that $a \equiv 0 \pmod{q-1}$.) Hence $rn^{\ell(r)} \equiv 1 \pmod{w}$; hence $r \equiv n^i \pmod{w}$ with $i = z - \ell(r)$. ■

We cannot immediately design a primality test based on this theorem, since it seems difficult to check condition 9.6.4 (ii) without knowing the divisors of n. However, we can use the following:

THEOREM 9.6.5 Suppose for primes p and q (not necessarily in Q) the following hold: $p \mid q-1$, $\chi_{p,q}(n) \neq 1$, and $\tau(\chi_{p,q})^{n^{p-1}-1} \equiv \chi_{p,q}(n) \pmod{nR}$. Then $\nu_p(r^{p-1}-1) \geq \nu_p(n^{p-1}-1)$ for all primes $r \mid n$.

Proof Let h be the order of the element $\tau(\chi_{p,q})$ in $(R_{p,q}/rR_{p,q})^*$. Then

$$\tau(\chi)^{p(r^{p-1}-1)} \equiv \chi(r)^p \equiv 1 \pmod{rR}$$

by Lemma 9.6.2. Thus $h \mid p(r^{p-1}-1)$.

On the other hand,

$$\begin{aligned}\tau(\chi)^{n^{p-1}-1} &\equiv \chi(n) \pmod{nR}\\ &\not\equiv 1 \pmod{rR},\end{aligned}$$

by Lemma 9.6.3. Also $(\tau(\chi)^{n^{p-1}-1})^p \equiv 1 \pmod{rR}$, so $\nu_p(h) = e_p+1$. Hence $\nu_p(r^{p-1}-1) \geq e_p$, and the result follows. ■

We can now state our version of the cyclotomic test:

CYCLOTOMIC RINGS TEST(n)

(1) if n is a perfect power,
 then return ("composite")

$z \leftarrow 1$

(2) repeat
 $z \leftarrow$ smallest squarefree integer $> z$
 factor $z = \prod_i p_i$
 $P \leftarrow \{p : p = p_i \text{ for some } i\}$
 $Q \leftarrow \{q \leq z+1 : q \text{ is prime and } q-1 \mid z\}$
 $w \leftarrow \prod_{q \in Q} q$

until
 $w > \sqrt{n}$

(3) if $n \in P$ or $n \in Q$
 then return("prime")

(4) if n is divisible by a prime in P or Q
 then return("composite")

(5) for all $(p, q) \in P \times Q$ with $p \mid q-1$ do
 select $\chi_{p,q}$
 if $\tau(\chi_{p,q})^{n^{p-1}-1} \not\equiv \chi_{p,q}(n) \pmod{n}$
 then return ("composite")

```
(6)  for all p ∈ P do
        (7)  q ← the least prime congruent to 1 mod p
               such that n^((q−1)/p) ≢ 1 (mod q).
        (8)  if n = q
               then return("prime")
        (9)  if q | n
               then return("composite")
        select χ_{p,q}
        (10)  if τ(χ_{p,q})^(n^(p−1) − 1) ≢ χ_{p,q}(n) (mod n)
               then return ("composite")
(11)  for i ← 0 to z − 1 do
        r_i ← n^i mod w
        if r_i is a proper divisor of n
               then return ("composite")
return("prime");
```

Before we give our analysis of this algorithm, let us sketch a heuristic argument why loop (1) does not take too long, i.e. why z is not too large. Suppose P is the set of the first t prime numbers, $p_1, p_2, \ldots, p_t$. From these t primes, we can form S, the set of 2^t numbers m such that $m - 1$ is a product of primes chosen from P. Using Theorems 8.2.1 and 8.2.4, which say that $\sum_{p \le x} \log p \sim x$, we see that the product of all elements of P is about $e^{t \log t}$, so the "average" member of S has size approximately $e^{\frac{1}{2} t \log t}$. If one makes the assumption that elements of S factor like randomly chosen numbers, we would expect approximately $2^t / (\frac{1}{2} t \log t)$ of them to be prime. Thus we expect

$$\begin{aligned} w &\approx (e^{\frac{1}{2} t \log t})^{2^t / (\frac{1}{2} t \log t)} \\ &= e^{2^t}. \end{aligned}$$

Now $w \approx n^{1/2}$, so solving for t we get $t \approx (\log \log n)/(\log 2)$. Since z is the product of the first t primes, we see that

$$\begin{aligned} z &\approx e^{t \log t} \\ &\approx (\log n)^{(\log \log \log n)/\log 2}. \end{aligned} \tag{9.7}$$

At present, no estimate for the size of z as good as (9.7) has been rigorously proved. However, we do have the following theorem:

THEOREM 9.6.6 (ODLYZKO AND POMERANCE) Let $f(n)$ denote the least positive squarefree integer such that the product of the primes q with $q - 1 \mid f(n)$ exceeds $\sqrt{n}$.

Then there exist positive, absolute, effective constants c_1 and c_2 such that for all integers $n \geq 2$,

$$(\lg n)^{c_1 \lg\lg\lg n} < f(n) < (\lg n)^{c_2 \lg\lg\lg n}.$$

Previously we mentioned that analysis of our algorithm depends on a hypothesis about the distribution of primes. The sticky point is step (7), where we search for a prime q with certain properties. More formally, we make the following

HYPOTHESIS 9.6.7 The least prime $q \equiv 1 \pmod{p}$ for which n is not a p-th power mod q is $O(p^2(\lg np)^2)$.

It was proved in Lemma 8.7.13 that this hypothesis is a consequence of the GRH. A heuristic argument (see Exercise 47) suggests that the correct bound may in fact be $O(p \lg p)$.

We now prove

THEOREM 9.6.8 Algorithm CYCLOTOMIC RINGS TEST correctly determines whether its input n is prime or composite. If Hypothesis 9.6.7 is correct, then the algorithm uses $O((\lg n)^{c_3 \lg\lg\lg n})$ bit operations for some constant c_3.

Proof First, let us see why the algorithm terminates. Loop (2) terminates by the estimate in Theorem 9.6.6. Loop (6) terminates because n is not a perfect power (by step (1)) and hence, by the Chebotarev density theorem, there exist infinitely many primes q for which $q \equiv 1 \pmod{p}$ and for which n is not a p-th power modulo q. (See Theorem 8.7.12.)

Now let us prove that the algorithm gives the correct output. Suppose n is prime. Then either $n \in P \cup Q$, or by Lemma 9.6.2, n must pass the tests in step (5). Similarly, in loop (6), we either discover that $n = q$ for some prime q, or that n passes the test in step (10). Finally, if n is prime, then part (4) cannot find a proper divisor. Hence the algorithm declares n prime.

Now suppose n is composite. If it is a perfect power, this is detected in step (1). If n is properly divisible by one of the "auxiliary" primes in the algorithm, this is detected in step (4) or (9). Hence assume n passes all the tests in steps (5) and (10). Since it passes the tests in step (5), hypothesis (i) of Theorem 9.6.4 is satisfied. Since it passes the tests in step (10), by Theorem 9.6.5, hypothesis (ii) of Theorem 9.6.4 is also satisfied. We then conclude that each prime $r \mid n$ is congruent to $n^i \pmod{w}$. Since n is composite, it is divisible by a prime $\leq \sqrt{n}$. Since $w > n^{1/2}$, step (11) suffices to find a proper divisor of n.

Finally, let us examine the running time of the algorithm. By exercise 3.4, step (1) can be done using $O((\lg n)^3)$ bit operations. By Theorem 9.6.6, loop (2) performs at most $O((\lg n)^{c_2 \lg\lg\lg n})$ iterations. Each iteration of loop (2) involves factoring z, which can be done by brute force in $O(z^{1/2}(\lg z)^2)$ bit operations. To construct the set Q, we can find all the primes $q \leq z + 1$ and test for which $q - 1 \mid z$ (for example, using the sieve of Eratosthenes—see section 9.8) in $O(z \lg z)$ bit operations. Thus, the total cost of loop (2) is $O((\lg n)^{O(\lg \lg \lg n)})$ bit operations.

Now each element of P is a prime divisor of z, which is bounded by $(\lg n)^{c_2 \lg \lg \lg n}$, so it follows from Exercise 9.4 that $|P| = O(\lg \lg n)$. Every element q of Q satisfies $q - 1 \mid z$, so $|Q|$ is bounded by $d(z)$, where d is the number of divisors function. By the estimate in Theorem 8.8.9, it follows that $|Q| = O((\lg n)^{c_4})$ for some constant c_4. Hence steps (3) and (4) can be done using $O((\lg n)^{c_4+2})$ bit operations.

Step (5) involves computations in the ring $R = \mathbb{Z}[\zeta_p, \zeta_q]$. Elements of R can be represented by polynomials in $\mathbb{Z}[X]$ of degree less than $(p - 1)(q - 1)$. Arithmetic operations on these polynomials is done modulo n and $\Phi_{pq}(X)$. Checking the identities involves the computation of $\tau(\chi)^{n^{p-1}-1}$, which can be done using the power algorithm of section 5.4. We leave it to the reader to verify that these computations can be done using $O((\lg n)^{O(\lg \lg \lg n)})$ bit operations. A similar analysis applies to step (10).

By Hypothesis 9.6.7, step (7) can be done using $O(p^3(\lg np)^4)$ bit operations for each p.

Finally, step (11) can be done (again using the power algorithm) using $O((\lg n)^{O(\lg \lg \lg n)})$ bit operations. ■

9.7 Generation of "Random" Primes

In this section we consider the generation of "random" primes smaller than a given bound x. The generation of large random primes is an integral part of many cryptographic schemes, such as RSA (see Chapter 15).

If our primality tests are deterministic or Las Vegas, then the following algorithm correctly generates a random odd prime. The expected number of primality tests used is $O(\lg x)$.

```
RANDOM PRIME(x)
repeat
        choose an integer j at random from the interval [1, (x − 1)/2]
        n ← 2j + 1
        If PRIMETEST(n) = 'prime' then test ← true
until
        test
return(n)
```

This method clearly generates a random odd prime in the given interval, and each odd prime is equally likely to be generated. If PRIMETEST is taken to be the deterministic cyclotomic rings test, then the algorithm runs in expected time $(\lg x)^{O(\lg\lg\lg x)}$. If PRIMETEST is taken to be the Las Vegas test of Adleman-Huang, then RANDOM PRIME runs in expected polynomial time. However, this method does not appear to be practical.

Because of their efficiency and ease of implementation, we would like to use one of the probabilistic tests of section 9.4 as our procedure PRIMETEST. In this case, of course, the result cannot be guaranteed to be a prime, only a "probable prime". Using a probabilistic test, however, involves a subtlety that may not be immediately apparent.

Suppose we define PRIMETEST as follows:

```
PRIMETEST(n, k)
i ← 0
repeat
        i ← i + 1
        r ← MILLER-RABIN(n)
until
        (i = k) or r = 'composite'
if (r = 'composite')
        then return('composite')
else
        return('prime')
```

Here n is the number to be tested, and k indicates the number of rounds of the Miller-Rabin test to perform.

By Theorem 9.4.5 we know that if n is composite, then PRIMETEST(n, k) will mistakenly return 'prime' with probability $\leq 4^{-k}$. It is now seductive (but incorrect) to deduce that RANDOM PRIME will return a prime with probability $\geq 1 - 4^{-k}$.

A little thought will show why this is so. Let S be a set of odd numbers ≥ 3, containing at least one prime, and choose a random element n of S. Let X denote the event that n is composite, let X' denote the event that n is prime, and let Y_k represent the event that PRIMETEST(n, k) declares that n is prime. Then Theorem 9.4.5 states that $\Pr[Y_k \mid X] \leq 4^{-k}$, but what we *really* want is $\Pr[X \mid Y_k]$, which depends strongly on the number of primes in S.

Then we have

LEMMA 9.7.1

$$\Pr[X \mid Y_k] \leq \frac{\Pr[Y_k \mid X]}{\Pr[X']}. \tag{9.8}$$

Proof Note that $\Pr[Y_k] \geq \Pr[X']$. Hence

$$\Pr[X \mid Y_k] = \frac{\Pr[Y_k \mid X]\Pr[X]}{\Pr[Y_k]} \leq \frac{\Pr[Y_k \mid X]}{\Pr[Y_k]} \leq \frac{\Pr[Y_k \mid X]}{\Pr[X']}.$$ ■

Now let S be the set of odd numbers in the interval $[3, x]$.

COROLLARY 9.7.2 For $x \geq 3$,

$$\Pr[X \mid Y_k] \leq 4^{-k} \log x.$$

Proof Using Theorem 8.8.1, we see

$$\Pr[X'] \geq \frac{\pi(x) - 1}{\lfloor (x-1)/2 \rfloor} \geq \frac{\frac{x}{\log x} - 1}{(x-1)/2} \geq \frac{1}{\log x},$$

for $x \geq 17$. As can easily be checked, the inequality $\Pr[X'] \geq 1/(\log x)$ is also valid for $3 \leq x \leq 17$, so it is valid for all $x \geq 3$. Thus

$$\Pr[X \mid Y_k] \leq 4^{-k} \log x$$

for $x \geq 3$. ■

To achieve an error probability smaller than ϵ, choose

$$k \geq \frac{\log \log x - \log \epsilon}{\log 4}.$$

We have proven

THEOREM 9.7.3 Algorithm RANDOM PRIME, using

$$\text{PRIMETEST}(x, (\log \log x - \log \epsilon)/(\log \log 4))$$

as a subroutine, correctly produces a prime output with probability $\geq 1 - \epsilon$, for $x \geq 3$. Furthermore, if the result is prime, then it is equally likely to be any odd prime in the interval $[1, x]$. The algorithm uses an expected number of $O(\lg x)$ calls to PRIMETEST, and the expected number of bit operations used is $O((\lg x)^4 \lg \lg x)$, for any fixed error bound ϵ.

By first trial dividing by the primes $\leq (\log x)^2/(\log \log x)$, we can improve the expected number of bit operations to $O((\lg x)^4)$. See Exercise 12.

The result in Theorem 9.7.3 can be improved, since for most composite numbers, the fraction of numbers that are strong witnesses is *much* larger than $3/4$. Recently, Burthe has shown that the error probability in (9.8) is actually less than 4^{-k} for all $x \geq 3$.

9.8 Prime Number Sieves

Frequently one needs to generate the prime numbers, from 1 to some large bound n. Perhaps the primes themselves are needed, or perhaps one merely wants to assemble statistics about the primes—the largest difference between successive primes in some interval, for example.

One strategy would be to test $1, 2, 3, \ldots$ successively for primality, but this leads to an algorithm whose running time might be as much as $\Theta(n^{3/2}(\lg n)^2)$ (if trial division is used), or $\Theta(n(\lg n)^3)$ (if probabilistic prime tests are used). Sieving methods give better results, both in theory and in practice.

Let us begin with the most ancient of methods consecutively, the *sieve of Eratosthenes*. In this method, we imagine the positive integers written down on a piece of paper. We circle 2 as prime, and cross off every second number beginning with 4. The next uncircled number is 3, so it must be prime. We circle it, and cross off every third number beginning with 9. We continue in this fashion; every composite number will eventually be crossed off, and the circled numbers are the primes.

In a real computer implementation of this method, we must specify an upper bound for the sieve, say n. We can avoid storing the integers themselves (which would require $\Omega(n \lg n)$ storage) by storing a 1 in location i if i is prime, and 0 otherwise. This leads to the following algorithm:

ERATOSTHENES(n)
{ initialize }
(1) $a[1] \leftarrow 0$
(2) for $i \leftarrow 2$ to n do $a[i] \leftarrow 1$
(3) $p \leftarrow 2$
(4) while $p^2 \leq n$ do
{ sift out multiples of p }
(5) for $j \leftarrow p$ to $\lfloor n/p \rfloor$ do $a[jp] \leftarrow 0$
{ now find the next prime }
(6) repeat $p \leftarrow p + 1$ until $a[p] = 1$
(7) return(a)

THEOREM 9.8.1 On input n, ERATOSTHENES returns an array a such that for all i with $1 \leq i \leq n$, we have $a[i] = 1$ if i is prime, and $a[i] = 0$ otherwise. It uses $O(n(\lg n)^2 \lg \lg n)$ bit operations and $O(n)$ space.

Proof Let us first sketch a proof that the algorithm is correct. It suffices to prove by induction on k that after k passes through the while loop starting at line (4), the numbers that remain unsifted (i.e. those i with $a[i] = 1$) are precisely $p_1, p_2, \ldots, p_k$ and all numbers $1 \leq r \leq n$ with $\gcd(r, p_1 p_2 \cdots p_k) = 1$. For if r is not of this form, then we can write $r = p_i s$ with $1 \leq i \leq k$ and $s \geq p_i$. But then r is sifted out on the i-th pass through the loop.

There is one subtle point that needs to be addressed: what guarantees that the repeat loop in line (6) will actually find a prime, and not fall off the right end of the array? Consider, for example, performing the sieve on the integers from 1 to p_k^2. After p_k is found, we move to the right, looking for the next prime. To ensure that we find it, we must have $p_{k+1} \leq p_k^2$. This follows easily from results in Chapter 8; see Exercise 31. We could, of course, modify the repeat loop slightly to avoid having the algorithm's correctness depend on this theorem.

Now consider the time and space requirements. Space is easiest: we need an array of length n to hold 0's and 1's, and a constant number of extra registers to hold numbers no bigger than $2n$. This gives a space bound of $O(n)$. For the time bound, observe that for each prime p we perform $O(n/p)$ multiplications; this gives a bound of

$$\sum_{p \leq \sqrt{n}} O(n/p)(\lg n)^2 = O(n(\lg n)^2 \lg \lg n),$$

where we have used Theorem 8.8.5. ■

Note that the multiplications used can easily be replaced by additions; this is an example of the program transformation known as *strength reduction*. This simple change (see Exercise 32) gives an algorithm that runs in the same space bound and uses only $O(n(\lg n) \lg \lg n)$ bit operations.

In practice, the real bottleneck in the sieve of Eratosthenes is not time, but rather space. A significant improvement to the $O(n)$ space bound was found by Bays and Hudson; it is called the *segmented* sieve.

The idea is to break the interval from 1 to n into $\lceil n/\Delta \rceil$ intervals of size Δ and sieve each of these intervals separately. The algorithm is as follows:

SEGMENT(n, Δ)
Find all the primes $\leq \sqrt{n}$
$r \leftarrow \lfloor \sqrt{n} \rfloor$
while $r < n$ do
    Sieve the interval $[r + 1, r + \Delta]$
    Output the primes $\leq n$ found in this interval
    $r \leftarrow r + \Delta$

By "sieve the interval" we mean, for each prime $p \leq \sqrt{n}$, cross off the integers in that interval that are multiples of p.

Again, it is not difficult to arrange the computation such that the main loop requires only additions, not multiplications (see Exercise 33). We then have the following theorem:

THEOREM 9.8.2 The algorithm SEGMENT(n, Δ) prints out the primes between 1 and n, and uses $O(\sqrt{n} + \Delta)$ space and $O(n(\lg n)(\lg\lg n) + (n^{3/2}\lg n)/\Delta)$ bit operations.

Proof We leave the proof of correctness to the reader.

For the resource bounds, notice that the first step (computing the primes up to $\sqrt{n}$) can be done in $o(n)$ time and $O(\sqrt{n})$ space, by the previous theorem.

The sieving loop uses $O(\Delta)$ space (once again, we store a list of 1's and 0's, not the integers themselves) for a total space consumption of $O(\sqrt{n} + \Delta)$.

We can sieve each interval by p using $O(\lceil \Delta/p \rceil)$ additions, so the total cost of the loop is

$$\left\lceil \frac{n}{\Delta} \right\rceil \sum_{p \le \sqrt{n}} O(\lceil \Delta/p \rceil) O(\log n) = O(n(\lg n)(\lg\lg n) + n^{3/2}(\lg n)/\Delta)$$

bit operations. ■

Now if we take $\Delta = \lceil n^{1/2} \rceil$, we get an algorithm that uses $O(n(\lg n)(\lg\lg n))$ time and $O(\sqrt{n})$ space. With this algorithm, one can easily create a table of primes up to 10^9 on even a very small computer.

Further improvements to these running times are possible. For example, Pritchard's "linear segmented wheel sieve" uses $O(n\lg n)$ bit operations and $O(\sqrt{n}/(\lg\lg n))$ space.

Lest we give the reader the impression that sieve methods are good only for making tables of primes, we conclude this section with another example. Suppose we want to tabulate the values of the function $d(k)$ for $1 \le k \le n$. (Recall that $d(k)$ denotes the number of divisors of k.) We could factor each number and use the formula for $d(n)$ given in Exercise 2.8, but that is far too slow in practice. Instead, we sieve the integers from 1 to n by *every* integer $\le \sqrt{n}$.

```
NUMBER OF DIVISORS TABLE(n)
for i ← 1 to n do
      x[i] ← 0
for j ← 1 to ⌊√n⌋ do
      x[j²] ← x[j²] + 1
      for k ← j + 1 to ⌊n/j⌋ do
            x[kj] ← x[kj] + 2
return(x)
```

THEOREM 9.8.3 Algorithm NUMBER OF DIVISORS TABLE(n) correctly computes $d(k)$ for $1 \le k \le n$. It uses $O(n(\lg n)^3)$ bit operations and $O(n\lg\lg n)$ space.

Proof Left to the reader as Exercise 39. ∎

It is also easy to modify algorithm NUMBER OF DIVISORS TABLE so that all of the multiplications are replaced by additions, improving the running time to $O(n(\lg n)^2)$ bit operations. Even more improvement is possible; see Exercise 40.

9.9 Computing $\pi(x)$ and p_n

In this section, we discuss algorithms for computing $\pi(x)$, the number of primes $\leq x$, and p_n, the n-th prime. Although it is not obvious at first glance, these functions can be computed in significantly less time than it costs to actually list the primes up to x.

We emphasize that the task is to compute the *exact value* of $\pi(x)$, not an approximation; if an approximation is desired, one can use the prime number theorem with error term given in Chapter 8.

Our first observation is fairly simple: algorithms for computing these functions are polynomial-time Turing interreducible. We prove that computing p_n can be done quickly, given an oracle for $\pi(x)$; the other direction follows in the same manner.

To see this, we observe that Theorem 8.8.4 says that for $n \geq 6$,

$$p_n \in (n \log n,\ n(\log n + \log\log n)),$$

an interval of length $n \log\log n$. Since $\pi(x)$ is a monotone increasing function, to find p_n we can perform a binary search in this interval, looking for r such that $\pi(r) = n$ and $\pi(r-1) = n-1$. This takes $O(\log_2(n \log\log n)) = O(\lg n)$ queries to our oracle for $\pi(x)$, and when we find such an r, we have $p_n = r$. Thus, for the remainder of this section, we will restrict our attention to the efficient computation of $\pi(x)$.

In many texts, one sees statements like, "There is no known formula for $\pi(x)$." Since we can write

$$\pi(x) = \sum_{p \leq x} 1,$$

it is clear that the truth of this statement depends on one's interpretation of the word "formula".

The conventional interpretation of "formula" involves familiar functions, such as addition and multiplication, possibly combined with less "legitimate" functions such as factorial, greatest integer, summation, and so forth. But the statement is still false under such an interpretation, as the "formula" for p_n in Exercise 28 shows.

Another interpretation of "formula" was given by H. Wilf. In this interpretation, a "formula" for counting the elements of a combinatorial class means an algorithm whose running time is asymptotically negligible compared to the effort of actually *constructing* the elements of that class. A "good formula" is one that can be evaluated in polynomial time. Recently, several researchers have developed "formulas" in Wilf's sense for $\pi(x)$, and hence p_n. However, we still do not have "good formulas" for these functions.

The computation of $\pi(x)$ for ever-larger values of x has received much attention in the past 150 years. The German astronomer E. D. F. Meissel began his computations in 1870 and in 1885 announced $\pi(10^9) = 50{,}847{,}478$. However, this value was too small by 56. Meissel's method was improved by D. H. Lehmer, who computed $\pi(10^{10}) = 455{,}052{,}512$, but his value was too large by 1. J. Bohman, using the same method, computed $\pi(10^{13}) = 346{,}065{,}535{,}898$, but his value was too small by 941.

Currently, the best "formulas" for $\pi(x)$ are a class of algorithms developed by J. C. Lagarias and A. M. Odlyzko. These use $O(x^{3/5-2b/5+\epsilon})$ bit operations and $O(x^{b+\epsilon})$ space, for any b with $0 \le b \le 1/4$, and all $\epsilon > 0$. By setting $b = 0$, this gives an algorithm that uses $O(x^{3/5+\epsilon})$ bit operations and $O(x^{\epsilon})$ space. By setting $b = 1/4$, we get the fastest-known algorithm, which uses $O(x^{1/2+\epsilon})$ bit operations and $O(x^{1/4+\epsilon})$ space. The algorithms are based on numerical integration of transforms of the Riemann zeta function, and we will not discuss them in this book.

Lagarias, Odlyzko, and V. Miller discovered another "formula" for $\pi(x)$ that uses $O(x^{2/3+\epsilon})$ bit operations and $O(x^{1/3+\epsilon})$ space. Using their method, they computed $\pi(10^{16}) = 27923\ 83410\ 33925$ in approximately 12 hours on an IBM 3081 Model K. Recently, M. Deleglise and J. Rivat [1995] computed $\pi(10^{17}) = 2\ 62355\ 71576\ 54233$ and $\pi(10^{18}) = 24\ 73995\ 42877\ 40860$.

In the remainder of this section, we describe a simplified version of the Lagarias-Miller-Odlyzko algorithm that uses $O(x^{2/3+\epsilon})$ space.

We introduce two functions: $\phi(x, a)$ denotes the number of positive integers $\le x$ with all prime factors $> p_a$, the a-th prime. (To make $\phi(x, 0) = \lfloor x \rfloor$, we set $p_0 = 1$, but emphasize again that 1 is not a prime.) By $P_k(x, a)$ we will mean the number of positive integers $\le x$ that are divisible by exactly k (not necessarily distinct) primes, each of which is $> p_a$. (We define $P_0(x, a) = 1$.) Clearly

$$\phi(x, a) = P_0(x, a) + P_1(x, a) + P_2(x, a) + \cdots. \tag{9.9}$$

We now prove some lemmas:

LEMMA 9.9.1 If $a = \pi(x^{1/3})$ then

$$\phi(x, a) = 1 + \pi(x) - a + P_2(x, a).$$

Proof If $a = \pi(x^{1/3})$, then $P_k(x, a) = 0$ for all $k \ge 3$, for otherwise there would be an integer $n \le x$ that is the product of ≥ 3 primes, each greater than $x^{1/3}$. Also $P_1(x, a)$ counts the number of positive integers $\le x$ which are divisible by exactly 1 prime; that prime must be $> p_a$. Hence $P_1(x, a) = \pi(x) - a$ for $x \geq p_a$. Substituting these two results in equation (9.9) gives the result. ∎

LEMMA 9.9.2 For all $x, a \geq 1$ we have

$$P_2(x, a) = \sum_{p_a < p \leq \sqrt{x}} (\pi(x/p) - \pi(p) + 1).$$

Proof The function $P_2(x, a)$ counts the positive integers $\leq x$ that are the product of exactly two primes, each greater than p_a. Each term in the sum clearly corresponds to an integer $n = pq$, where $p_a < p \leq q$. Also, since $n \leq x$, we may assume that $p \leq \sqrt{x}$. For each such prime p, there are exactly $\pi(x/p) - \pi(p) + 1$ suitable q's that give a product $pq \leq x$; the addition of 1 is necessary since we don't want to omit the case $q = p$. ∎

LEMMA 9.9.3 For $x, a \geq 1$ we have

$$\phi(x, a) = \phi(x, a - 1) - \phi(x/p_a, a - 1). \tag{9.10}$$

Proof The numbers counted by $\phi(x, a - 1)$ are of two kinds: those not divisible by p_a, of which there are $\phi(x, a)$, and those divisible by p_a, of which there are $\phi(x/p_a, a - 1)$. ∎

We can use the recurrence (9.10) to compute $\phi(x, a)$. For example, let us compute $\phi(300, 3)$:

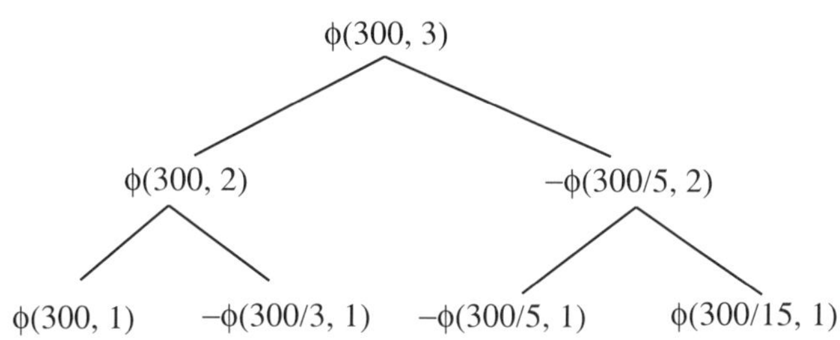

Figure 9.2
The computation of $\phi(300, 3)$

At this point we can stop, since $\phi(x, 1) = \lfloor (x + 1)/2 \rfloor$. Thus we find

$$\phi(300, 3) = \phi(300, 1) - \phi(100, 1) - \phi(60, 1) + \phi(20, 1) = 80.$$

The computation of $\phi(x, a)$ by repeated use of (9.10) can be represented by a binary tree. The root is labeled $\phi(x, a)$, and each node with label $\pm\phi(y, b)$ has two children, the left labeled $\pm\phi(y, b - 1)$, and the right labeled $\mp\phi(y/p_b, b - 1)$. The tree terminates at leaves which are evaluated. By "expanding" a node, we mean the process of adding two children using (9.10).

In order to compute $\phi(x, a)$ in a reasonable length of time, we cannot expand the root down to nodes of the form $\pm\phi(y, 1)$; such a tree would have 2^{a-1} leaves. Instead, we use the following "truncation" rule, which tells us when to stop expanding a node:

Do not split a node labeled $\pm\phi(x/n, b)$ if

(i) $b = 1$ and $n \le x^{1/3}$; or

(ii) $n > x^{1/3}$.

Nodes of the first type are called *ordinary leaves*, and the second type are called *special leaves*.

LEMMA 9.9.4 There are at most $x^{1/3}$ ordinary leaves. There are at most $x^{2/3}$ special leaves.

Proof The first statement follows because $n \le x^{1/3}$ by the definition of ordinary leaf, and no two leaves can have the same value of n.

To prove the second statement, we see that a special leaf $\pm\phi(x/n, b)$ has $n > x^{1/3}$. Then we must have $n = mp_{b+1}$, where $m \le x^{1/3}$, since its parent $\mp\phi(x/m, b+1)$ is *not* a special leaf. But then there are at most $x^{2/3}$ such parent nodes, since $m \le x^{1/3}$ and $b + 1 < x^{1/3}$, and the result follows. ■

We can evaluate the contribution from the ordinary leaves in $O(x^{1/3} \lg x)$ bit operations, since $\phi(x, 1) = \lfloor (x + 1)/2 \rfloor$. To evaluate the contribution from the special leaves requires more work, as there are $O(x^{2/3})$ of them.

The necessary trick is a clever data structure, which allows us to perform both of the following operations on an array of n elements, using only $O(\lg n)$ additions: increment a position (i.e. set $y[i] \leftarrow y[i] + c$); and find a subrange sum of the array (i.e. compute $\sum_{r \le k \le s} y[k]$). The data structure uses $O(n)$ storage locations. The details of this data structure are left to the reader as Exercise 42.

The array we will use will contain a 1 in position $y[i]$ if i is not divisible by $2, 3, \ldots, p$, and a 0 otherwise. Before the first pass, we set all array elements to 1. Then we sieve by 2, changing all locations i that are multiples of 2 to 0. At this point, we can evaluate $\pm\phi(x/n, 1)$ for all the relevant special leaves. Then we sieve by 3 and evaluate $\pm\phi(x/n, 2)$, etc. (To avoid repeatedly scanning the list of special leaves, we sort it before we begin evaluation. This can be done in $O(x^{2/3}(\lg x)^2)$ steps.) We continue sieving over all primes $p < x^{1/3}$.

We now summarize the various steps in the computation of $\pi(x)$ with the following algorithm sketch:

LMO(x)
(1) Use the sieve of Eratosthenes, as discussed in the previous
section, to compute the primes $\le x^{2/3}$. By one pass through

a suitable table produced by ERATOSTHENES, also compute $\pi(n)$ for $1 \le n \le x^{2/3}$.

(2) From this table, compute $a = \pi(x^{1/3})$.

(3) Evaluate $P_2(x, a)$ by computing the $O(\sqrt{x})$ terms of this sum.

(4) Use the recurrence (9.10) for $\phi(x, a)$ to construct a binary tree with $\phi(x, a)$ as the root, and nodes labeled $\pm\phi(x/n, b)$. Continue expanding until $b = 1$ and $n \le x^{1/3}$ (an "ordinary" leaf) or $n > x^{1/3}$ (a "special" leaf).

(5) Evaluate the ordinary leaves using the fact that $\phi(x, 1) = \left\lfloor \frac{x+1}{2} \right\rfloor$.

(6) Sort the special leaves by order of increasing b, and evaluate them using sieving and the special data structure described above.

(7) Return $\pi(x) = \phi(x, a) - P_2(x, a) + a - 1$.

THEOREM 9.9.5 The algorithm LMO correctly computes $\pi(x)$. It uses

$$O(x^{2/3}(\lg x)^2(\lg\lg x))$$

bit operations and $O(x^{2/3} \lg x)$ space.

Proof The proof of correctness follows from Lemmas 9.9.1–9.9.3.

To see the time bound, we observe that the most expensive step is step (6). Here we must sieve an array of size $x^{2/3}$ by the primes $p \le x^{1/3}$. When we sieve by p, we decrement the locations in the array that correspond to multiples of p, and each decrement requires $\lg x$ additions on numbers with $O(\lg x)$ bits. Thus the total cost is given by

$$\sum_{p<x^{1/3}} \left\lceil \frac{x^{2/3}}{p} \right\rceil O((\lg x)^2),$$

which by Theorem 8.8.5 is $O(x^{2/3}(\lg x)^2(\lg\lg x))$.

The correctness of the space bound is left to the reader. ■

9.10 Exercises

1. (Euclid; Euler.) A perfect number n is such that $\sigma(n) = 2n$. Show that if $2^p - 1$ is prime (a Mersenne prime), then $2^{p-1}(2^p - 1)$ is perfect. Further show that every even perfect number is of this form.

2. Prove *Wilson's Theorem*: if p is prime then $(p - 1)! \equiv -1 \pmod p$.

3. Prove the converse of Wilson's theorem: if n is composite and $n > 4$, then $(n-1)! \equiv 0 \pmod{n}$.

4. Show that $\liminf_{p\to\infty} T(p)/(\log_2 p) < 2$.

5. Show how to improve Corollary 9.1.6 so that it suffices that $f > \sqrt[3]{n}$. (Hint: if n has a factorization, then it must be of the form $(xf+1)(yf+1)$. Show how to compute $x+y$; then recover x and y via the quadratic formula.)

6. (Chernick.) Show that if $6n+1$, $12n+1$, and $18n+1$ are all primes, then $(6n+1)(12n+1)(18n+1)$ is a Carmichael number.

7. Show that the following set can be recognized in deterministic polynomial time: $\{n : \sigma(n) \text{ is a power of } 2\}$.

8. Show that the following set can be recognized in deterministic polynomial time: $\{n : \varphi(n) \text{ is a power of } 2\}$.

9. Show, by giving a counterexample, that the set $S(n)$ does *not* necessarily form a group under multiplication modulo n.

10. (Gerlach) Prove that $S(n)$ is a group under multiplication modulo n if and only if n is a prime power or n has a prime divisor congruent to 3 mod 4.

11. Show how to modify the algorithm SOLOVAY-STRASSEN to avoid the need for the gcd computation in step (1).

12. Show how to reduce the expected number of bit operations in Theorem 9.7.3 to $O((\lg x)^4)$, by first trial dividing by the primes $\le (\log x)^2/(\log\log x)$.

13. [Open] Show that the error probability of $1/2$ in the Solovay-Strassen test cannot be improved, in the sense that there is an infinite sequence of distinct integers n_i such that $\lim_{i\to\infty} |E(n_i)|/n_i = 1/2$.

14. [Open] Show that the error probability of $1/4$ in the Miller-Rabin test cannot be improved, in the sense that there is an infinite sequence of distinct integers n_i such that $\lim_{i\to\infty} |S(n_i)|/n_i = 1/4$.

15. Using the methods of this chapter and a computer, prove or disprove the primality of the following numbers: $30^{32}+1$; $2^{127}-1$; $2^{101}-1$.

16. (Erdős.) Let p be a prime number > 3. Show that $(2^{2p}-1)/3$ is a pseudoprime to the base 2.

17. (Rotkiewicz.) Let p be a prime number > 5. Show that $(2^{2p} + 1)/5$ is a pseudoprime to the base 2.

18. (Malo; Sierpiński.) Let n be a pseudoprime to the base 2. Show that $2^n - 1$ is also a pseudoprime to the base 2.

19. Show that there are infinitely many pseudoprimes to the base a. (Hint: consider the numbers

$$\frac{a^{a^{n^2}} - 1}{a^{a^n} - 1}.)$$

20. Show that there are infinitely many Euler pseudoprimes to the base a. (Hint: consider the numbers

$$\frac{a^{a^{2(2a)^n}} - 1}{a^{a^{(2a)^n}} - 1}.)$$

21. (Malm.) Suppose $n \equiv 3 \pmod 4$. Show that $n \in \mathrm{epsp}(a)$ iff $n \in \mathrm{spsp}(a)$.

22. (Shanks.) Let $n = (12k + 1)(24k + 1)$. Show that if both factors are prime, then $n \in \mathrm{psp}(2)$ and $n \in \mathrm{psp}(3)$.

23. Let L_n be the *Lucas sequence*, defined by $L_0 = 2; L_1 = 1$; and $L_n = L_{n-1} + L_{n-2}$ for $n \geq 2$. Show that if n is prime, then $L_n \equiv 1 \pmod n$. Is the converse true?

24. (Perrin.) Let $u_0 = 3$, $u_1 = 0$, $u_2 = 2$, and for $n \geq 3$ define $u_n = u_{n-2} + u_{n-3}$. Show that if n is a prime, then $u_n \equiv 0 \pmod n$. Is the converse true?

25. (Monier; Baillie and Wagstaff.) Let n be odd, and recall the definition of $F(n)$, the group of Fermat liars:

$$F(n) = \{a \in (\mathbb{Z}/(n))^* \ : \ a^{n-1} \equiv 1 \pmod n\}.$$

Show that

$$|F(n)| = \prod_{p \mid n} \gcd(n - 1, p - 1).$$

26. Show that if p is a prime, and n is a positive integer such that $p \mid n$ and $n + 1 \mid p^2 - 1$, then in fact $p = n$.

27. (D. J. Lehmann.) Prove that the following gives a polynomial-time Atlantic City algorithm for primality with error probability $1 - 2^{-k}$: draw k random elements a_1, $a_2, \ldots, a_k$, from $(\mathbb{Z}/(n))^*$, and for each one compute $b_i = a_i^{(n-1)/2} \pmod n$. Say 'prime' if all the b_i are ± 1 and at least one $b_i = -1$, and 'composite' otherwise.

28. (J. M. Gandhi.) Let $P_n = p_1 p_2 \cdots p_n$ be the product of the first n primes. Prove that the next prime, p_{n+1}, is the unique integer m such that

$$1 < 2^m \left(\sum_{d \mid P_n} \frac{\mu(d)}{2^d - 1} - \frac{1}{2} \right) < 2.$$

(Hint: perform the sieve of Eratosthenes on the binary number $0.111111111\cdots$).

29. (Mills.) Show that there is a real number x such that $\lfloor x^{3^k} \rfloor$ is a prime for all $k \geq 0$. Hint: use Theorem 8.6.1.

30. Prove that if $f(x)$ is a nonconstant polynomial with integer coefficients, then there exist infinitely many integers n such that $|f(n)|$ is not a prime.

31. Using Lemma 8.8.4, show that $p_{n+1} \leq p_n^2$ for $n \geq 1$.

32. Show how to modify the algorithm ERATOSTHENES so that no multiplications are necessary.

33. Show how to replace most of the multiplications in the segmented sieve of Eratosthenes with additions.

34. (Legendre.) Prove the following formula:

$$\pi(x) - \pi(\sqrt{x}) + 1 = \lfloor x \rfloor - \sum_{p \leq \sqrt{x}} \left\lfloor \frac{x}{p} \right\rfloor + \sum_{p < q \leq \sqrt{x}} \left\lfloor \frac{x}{pq} \right\rfloor - \cdots. \tag{9.11}$$

Hint: use the principle of inclusion-exclusion.

35. Prove that the number of positive integers $\leq x$ with a prime factor $> \sqrt{x}$ is asymptotically $(\log 2)x$. Hint: a number $\leq x$ has at most one prime factor $> \sqrt{x}$.

36. Let $S_2(x)$ denote the number of squarefree positive integers $\leq x$. Prove that $S_2(x) = (6/\pi^2)x + O(\sqrt{x})$. (Hint: use Exercise 2.19.)

37. Using the previous two exercises, prove that the number of nonzero terms in Legendre's formula (9.11) is asymptotically equal to $(1 - \log 2)(6/\pi^2)x$, or about $.187x$.

38. Use the principle of inclusion-exclusion and sieving techniques to show how to calculate the exact value of $S_2(x)$ using $O(x^{1/2+\epsilon})$ bit operations and $O(x^{1/4+\epsilon})$ space, for all $\epsilon > 0$.

39. Prove the correctness of algorithm NUMBER OF DIVISORS TABLE. Show that it uses $O(n(\lg n)^3)$ bit operations and $O(n \lg \lg n)$ space. (Hint: for the space bound, use Theorem 2.5.4 and the inequality between the arithmetic and geometric mean.) Show how

to replace the multiplications with additions and therefore improve the running time to $O(n(\lg n)^2)$ bit operations.

40. (Continuation.) Show how to improve the time bound for computing a table of $d(k)$ for $1 \leq k \leq n$ to $O(n(\lg n)(\lg \lg n))$ bit operations by sieving by the prime powers $\leq n$.

41. Using sieving, show how to tabulate the values of $\sigma(k)$, $\varphi(k)$, and $\mu(k)$ for $1 \leq k \leq n$, using $O(n(\lg n)^3)$ bit operations.

42. Show how to implement a data structure that allows the following two operations to be performed on an array of n elements, using only $O(\lg n)$ additions: (i) increment a position (i.e. set $y[i] \leftarrow y[i] + c$); (ii) find a subrange sum of the array (i.e. compute $\sum_{r \leq k \leq s} y[k]$). Hint: expand the array by a factor of $\log_2 n$, adding extra rows for the sum of pairs of elements, sum of quadruples of elements, etc.

43. (Selfridge.) Prove the following improvement to Theorem 9.1.1: If for every $q \mid n-1$ there exists a_q such that $a_q^{n-1} \equiv 1 \pmod{n}$ and $a_q^{(n-1)/q} \not\equiv 1 \pmod{n}$, then n is prime. Show that if n is prime, and the factorization of $n-1$ is given, then an algorithm to prove n prime based on this idea uses an expected number of $O((\lg n)^4/(\lg \lg n))$ bit operations.

44. Let n be an odd number ≥ 3, and define

$$D(n) = \{x \in (\mathbb{Z}/(n))^* \ : \ x^{\lambda(n)/2} \equiv \pm 1 \pmod{n}\}.$$

Show that $D(n) = (\mathbb{Z}/(n))^*$ if and only if n is a prime power.

45. [Open] Prove or disprove the existence of an algorithm to determine all primes $\leq n$ using $O(n)$ bit operations.

46. Let a be an integer, and suppose that $a^{(n-1)/2} \equiv -1 \pmod{n}$. Show that $\nu_2(r-1) \geq \nu_2(n-1)$ for every $r \mid n$, with equality holding if and only if $\left(\frac{a}{r}\right) = -1$.

47. Give a heuristic argument that the least prime $q \equiv 1 \pmod{p}$ such that n is not a p-th power mod q is $O(p \lg p)$.

48. Show that the language

$$\texttt{FACTOR} = \{(n, k) \ : \ n \text{ has a nontrivial positive divisor} \leq k\}$$

is in $\mathcal{NP} \cap \text{co-}\mathcal{NP}$.

49. Assume ERH. Show that to prove n prime, it suffices to test the Solovay-Strassen predicate for all *prime* $a \leq 2(\log n)^2$. Use the prime number theorem to show that the running time for testing primality can be lowered to $O((\lg n)^5/(\lg \lg n))$.

50. Let $S(n)$ denote the set of strong liars, defined in Section 9.4. Assuming ERH, prove the following.

    (a) If n is an odd composite integer, greater than 1, then n is prime iff for every a with $1 < a < n$ and $a \leq 2(\log n)^2$, $a \in S(n)$.

    (b) Conclude that PRIMES belongs to $\mathcal{P}$.

51. Let n denote an odd integer. Prove the following result, and from this conclude that (assuming ERH) PRIMES belongs to $\mathcal{P}$:

$$n \text{ is prime} \iff \{x^{(n-1)/2} : x \in (\mathbb{Z}/(n))^*, \quad x \leq 2(\log n)^2\} = \{\pm 1\}.$$

52. (H. W. Lenstra) Prove the following results.

    (a) Show that if n is a prime power but not prime, the least a for which $a^{n-1} \not\equiv 1 \pmod{n}$ is $O((\log n)^2)$.

    (b) Show that if n an odd composite integer with at least two prime factors, then there is a nontrivial subgroup of index 2 in $(\mathbb{Z}/(n))^*$ containing $S(n)$, the set of strong liars.

    (c) Conclude that to prove PRIMES $\in \mathcal{P}$, it suffices to assume the ERH for L-functions of *quadratic* characters only.

53. [Open] What is the average running time of DETERMINISTIC SOLOVAY-STRASSEN?

54. (Jaeschke) Find a number that is a strong pseudoprime to all prime bases p with $2 \leq p \leq 31$.

9.11 Notes on Chapter 9

Currently, the largest known prime has 258716 decimal digits. The record-setting primes throughout history are given below in Table 9.1.

Some of the entries in the table require explanation. For example, Picutti [1989] has observed that the anonymous author of folio 28r of the Codex Palatino 573 in the Biblioteca Nazionale in Florence claimed that $2^{12}(2^{13} - 1)$ is the fifth perfect number (and by implication that $2^{13} - 1 = 8191$ is prime). Johannes Müller (Regiomontanus) correctly gave the fifth perfect number in the Codex Vindobonensis 5203, written in 1461 or before; see Curtze [1899]. The anonymous author of Codex Latinus Monacensis 14908 also gave the fifth perfect number in 1461. See Curtze [1895]. Similarly, Picutti observed that the anonymous author of folio 24r of the Codex Ottobonianus Latinus 3307 of the Biblioteca Apostolica Vaticana observed in 1460 that $2^{16}(2^{17} - 1)$ is the sixth perfect number.

Cataldi's work was composed in 1588, but not published until 1603.

Looff's prime was marked with a question mark in his table, but Reuschle claimed that Looff had actually proved it prime.

Table 9.1
Largest Known Primes Throughout History

Date	Prime	No. of Decimal Digits	Discoverer	References
< 1458	$2^{13} - 1$	4	Codice Palatino 573	Picutti [1989]
1460	$2^{17} - 1$	6	Codex Ottob. Lat. 3307	Picutti [1989]
1588	$2^{19} - 1$	6	Cataldi	Dickson [1919, p. 10] Wertheim [1902]
1772	$2^{31} - 1$	10	Euler	Dickson [1919, p. 19] Euler [1849, p. 584]
1851	999999000001	12	Looff	Looff [1851] Reuschle [1856, pp. 3,18]
1 Jan 1855	67280421310721	14	Clausen	K. Biermann [1964]
1876	$2^{127} - 1$	39	Lucas	Lucas [1876a, 1877a]
7 Jun 1951	$934(2^{127} - 1) + 1$	42	Miller & Wheeler	J. Miller and Wheeler [1951] J. Miller [1951]
14 Jul 1951	$(2^{148} + 1)/17$	44	Ferrier	D. H. Lehmer [1952a] Ferrier [1952]
July 1951	$180(2^{127} - 1)^2 + 1$	79	Miller & Wheeler	J. Miller and Wheeler [1951] J. Miller [1951]
30 Jan 1952	$2^{521} - 1$	157	Robinson	D. H. Lehmer [1952a]
	$2^{607} - 1$	183		Robinson [1954]
25 Jun 1952	$2^{1279} - 1$	386	Robinson	D. H. Lehmer [1952b] Robinson [1954]
7 Oct 1952	$2^{2203} - 1$	664	Robinson	D. H. Lehmer [1953]
9 Oct 1952	$2^{2281} - 1$	687	Robinson	Robinson [1954]
8 Sep 1957	$2^{3217} - 1$	969	Riesel	Riesel [1958a, 1958b]
3 Nov 1961	$2^{4253} - 1$	1281	Hurwitz & Selfridge	Hurwitz and Selfridge [1961]
	$2^{4423} - 1$	1332		Hurwitz [1962]
11 May 1963	$2^{9689} - 1$	2917	Gillies	Gillies [1964]
16 May 1963	$2^{9941} - 1$	2993	Gillies	
2 Jun 1963	$2^{11213} - 1$	3376	Gillies	
4 Mar 1971	$2^{19937} - 1$	6002	Tuckerman	Tuckerman [1971]
30 Oct 1978	$2^{21701} - 1$	6533	Noll & Nickel	Noll and Nickel [1980]
9 Feb 1979	$2^{23209} - 1$	6987	Noll	Noll and Nickel [1980]
8 Apr 1979	$2^{44497} - 1$	13395	Nelson & Slowinski	Slowinski [1979] Nelson [1980]
25 Sep 1982	$2^{86243} - 1$	25962	Slowinski	Ewing [1983]
20 Sep 1983	$2^{132049} - 1$	39751	Slowinski	Haworth [1989, p. 8]
6 Sep 1985	$2^{216091} - 1$	65050	Slowinski	Peterson [1985]
6 Aug 1989	$391581 \cdot 2^{216193} - 1$	65087	Brown et al.	Brown et al. [1990]
19 Feb 1992	$2^{756839} - 1$	227832	Slowinski & Gage	Ewing [1992]
10 Jan 1994	$2^{859433} - 1$	258716	Slowinski	Ewing [1994]

According to K. Biermann [1964], the German mathematician and astronomer Thomas Clausen provided the factorization $2^{64} + 1 = 274177 \cdot 6728\,04213\,10721$ in a letter to Gauss dated 1 January 1855, and he knew both factors were prime.

According to Dickson [1919, Chapter XIII], in the 13th century, Fibonacci gave a list of the primes from 11 to 97. Cataldi in 1603 listed the primes up to 750. In 1657, Frans van Schooten gave a table of primes to 9979. In 1746, J. G. Krüger gave a table of primes to 100999; it was computed by P. Jäger. In 1772, A. F. Marci gave a list of primes to 400000.

For tables of primes constructed in the 20th century, see D. N. Lehmer [1914]; C. Baker and Gruenberger [1959]; Lifchitz [1971]. Bays and Hudson computed a table of primes up to 1.2×10^{12}; see Ribenboim [1988b, p. 177].

The lower bound on odd perfect numbers is due to Brent and Cohen [1989] and Brent, Cohen, and te Riele [1989].

For the five problems on primality testing, see Lucas [1891, p. 354]. Most of these problems also appear in Carmichael [1914, p. 29] and G. Hardy and Wright [1985, pp. 5–6]. In a 1956 letter to von Neumann, Gödel asked about the computational complexity of testing primality; see Sipser [1992] for a reproduction of the letter (written in German) and an English translation.

The following are surveys or popular articles on primality testing: Selfridge and Guy [1971]; H. Williams [1978a]; Pomerance [1981a]; Richards [1982]; Couvreur and Quisquater [1982]; H. W. Lenstra [1983, 1986a]; Pomerance [1983]; Rumely [1983]; Dixon [1984]; Pomerance [1984a]; Nicolas [1984, 1985b]; Wagon [1986]; Balasubramanian [1986]; Prasad [1986]; Condie [1988]; Keller [1991]. Pin [1981] discussed primality testing on hand-held calculators.

In this chapter, we are most interested in whether $\texttt{PRIMES} \in \mathcal{P}$, but several investigators have examined the set PRIMES relative to other complexity classes. For example, Minsky and Papert [1966] proved that PRIMES is not a regular set (i.e., not accepted by a finite automaton). Hartmanis and Shank [1968], and independently, Schützenberger [1968], proved that PRIMES is not a context-free language (i.e., not accepted by a pushdown automaton). Hsia and Yeh [1973] showed that the language $\{0^p : p \text{ is prime}\}$ is accepted by a deterministic, halting 3-marker automaton.

9.1

The phrase "moral certainty" was used by Kraitchik [1929, p. 160] to describe his belief that $N = 16143694150072550161$ was a prime number. After twenty pages of calculations, he says [1929, p. 179], "It remains to appreciate the degree of 'moral certainty' we have in affirming that N is prime... We consider the proof that N is prime as sufficient." Nevertheless, a simple calculation shows that $2^{N-1} \not\equiv 1 \pmod{N}$, and so N is composite. The reader may enjoy trying to find the factors of N using the methods of Chapter 11. The moral of the story is that "moral certainty" is no substitute for proof.

For the primality test of Lehmer-Kraitchik, see Kraitchik [1926, p. 135] and D. H. Lehmer [1927]. Earlier Lucas had published weaker primality tests, also based on Fermat's theorem (Lucas [1878a]).

The improvement to the Lehmer-Kraitchik test is due to J. Selfridge; see Brillhart and Selfridge [1967]. Theorem 9.1.5 is due to Pocklington [1914].

For other papers on primality testing using the converse of Fermat's theorem, see D. H. Lehmer [1928a, 1936a, 1939, 1949]; Brillhart, Lehmer, and Selfridge [1975].

Similar tests, involving Lucas sequences, depend on the factorization of $n + 1$, $n^2 + n + 1$, etc. For these, see M. Morrison [1975]; H. Williams and Judd [1976a, 1976b]; H. Williams and Holte [1978]; Wunderlich [1983a].

For Theorem 9.1.3, see Fellows and Koblitz [1992]; Konyagin and Pomerance [1994].

Theorem 9.1.4 is from Pratt [1975]. A popular treatment of Pratt's proof can be found in Higgins and Campbell [1994]. However, the desire for a succinct certificate of primality goes back quite far. For example, F. Landry [1867] proved that 1133836730401, a factor of $2^{75} + 1$, is prime. After stating this result, he went on to say

> At this point we are, if not uneasy, then at least somewhat embarrassed.
>
> Indeed, when one has succeeded in factoring a number, and has given its factors, this can be verified immediately. But it is a different matter when the methods used fail to discover any factor, and one then asserts that the number is *prime*. How could one then transmit to another such a totally personal conviction? Who would be convinced, without having redone all the calculations, and without having understood the principles on which those calculations were based?
>
> We understand well that our claim is valid only as an assertion, worthwhile until someone proves the contrary, or until we make known our methods and enable others to apply them.

Pomerance [1987] has shown that there exist certificates of primality that can be verified in only $O(\log p)$ multiplications modulo p. For other types of short proofs of primality, see Guy, Lacampagne, and Selfridge [1987]. J. Jones, Sato, Wada, and Wiens [1976] prove the amusing result that if p is prime, there exists a proof of that fact that requires only 87 integer additions and multiplications (on numbers of unbounded size).

Pollard [1971b] gave a deterministic algorithm for testing primality that uses $O(n^{1/4}(\lg n)^2)$ bit operations. Later he gave a similar algorithm that achieves a running time of $O(n^{1/8+\epsilon})$ bit operations, via "fast Fourier" multiplication. See Pollard [1974]. Adleman and Leighton [1981] gave an algorithm that runs in $O(n^{1/10.89})$ bit operations.

Other papers giving deterministic algorithms for primality testing include Dijkstra [1957]; Strassen [1976].

Theorem 9.1.7 is a particular case of a more general method due to H. W. Lenstra, Jr. [1982].

Pintz, Steiger, and Szemerédi [1989] have shown the existence of an *infinite* set of primes P whose membership can be tested in deterministic polynomial time. In particular, they showed that there are an infinite number of primes p such that $3^k \,||\, p - 1$, with $3^k > p^{1/3}$.

9.2

Heuristic arguments suggest that there are only a finite number of prime Fermat numbers, but infinitely many Mersenne primes. See Exercise 8.44.

We use F_n to denote the n-th Fermat number here, which is a time-honored notation. There should be no confusion with the use of F_n to denote the n-th Fibonacci number. For these tests, see Pepin

[1877]; Proth [1878]; Lucas [1879]; Nazarevsky [1904]; Sierpiński [1964b].Note that Pepin's test is essentially the Solovay-Strassen test applied to F_n; it follows that any nonresidue of F_n, such as 3 or 10, will work in place of 5 in the test.

The following generalization of Pepin's test was given by Hurwitz [1896] and Carmichael [1913]: let $\Phi_n(x)$ denote the n-th cyclotomic polynomial. Then n is prime if and only if there exists an integer q such that $\Phi_{n-1}(q) \equiv 0 \pmod{n}$. Pepin's test corresponds to the case $n = 2^{2^k} + 1$, $\Phi_n(x) = x^{2^{k-1}} + 1$.

Lucas [1878a, pp. 238, 292; 1896, p. 234] used a test similar to Pepin's to prove that F_6 is composite. Morehead [1905] and Western [1905] independently used Pepin's test to prove that F_7 is composite. Four years later, they joined forces to prove that F_8 is composite, "each doing a half of the whole work." (This may be the first example of a distributed computation in number theory!) According to Morehead and Western [1909], the computations "were wholly performed on a 10-figure arithmometer." See Archibald [1914] and Carmichael [1919] for a discussion of early work on Fermat numbers.

Moving to the age of the digital computer, Robinson [1954] examined F_{10}, Paxson [1961] examined F_{13}, and Hurwitz and Selfridge [1961] proved that F_{14} is composite; also see Selfridge and Hurwitz [1964]. Young and Buell [1988] proved that F_{20} is composite. Of this last computation, which required 10 CPU days on a Cray-2, the authors say, "Never have so many circuits labored for so many cycles to produce so few output bits." Recently, Crandall, Doenias, Norrie, and Young [1995] proved that F_{22} is composite. This was later verified by Trevisan and Carvaho.

The primality test for Mersenne numbers was originally due to Lucas [1878a, p. 316]. In this paper Lucas proved that the condition $S_{p-1} \equiv 0 \pmod{n}$ is sufficient for the primality of $n = 2^p - 1$, when $p \equiv 1 \pmod{4}$. D. H. Lehmer proved that the condition is necessary and sufficient for all exponents; see D. H. Lehmer [1930a, 1935]. Also see D. H. Lehmer [1932a]. Errors in Lucas' work were pointed out by Carmichael [1913]. Our proof is based on Brewer [1951]. For other proofs, see Western [1932]; Kaplansky [1945]; Knuth [1981, pp. 389–394]; M. Rosen [1988]. Since the numbers involved in the Lucas-Lehmer test for large Mersenne primes are huge (e.g. 227,000 decimal digits) the use of "fast" multiplication methods is essential.

For similar tests, see Pepin [1878]; Ward [1959]; Inkeri [1960]; H. Williams and Zarnke [1968]; Riesel [1969]; Stechkin [1971]; Zarnke and Williams [1971]; H. Williams [1971, 1972]; Jönsson [1972]; Cormack and Williams [1980]; H. Williams [1981, 1982b]; Borho, Buhl, Hoffmann, Mertens, Nebgen, and Reckow [1983]; Akushskiĭ and Burtsev [1986]; H. Williams [1987, 1988]; Guthmann [1992]; and Bosma [1993].

Similar tests have also been constructed for the numbers $k! \pm 1$ and $2 \cdot 3 \cdot 5 \cdots p \pm 1$; see Borning [1972]; Templer [1980]; Buhler, Crandall, and Penk [1982]; and Caldwell [1995].

By far the most detailed source of information on Mersenne numbers is the work of Haworth [1989].

Another class of numbers that has attracted attention, particularly among recreational mathematicians, is the so-called *repunits*, i.e. numbers $R_n = (10^n - 1)/9$. (The term was introduced by Beiler [1964].) Presently, it is known that R_2, R_{19}, R_{23}, R_{317}, and R_{1031} are the only primes of this form for $n \leq 2000$. See, for example, D. H. Lehmer [1927, 1929]; H. Williams [1978b]; H. Williams and Seah [1979]; H. Williams and Dubner [1986]; Dubner [1993].

Recently, Parady, Smith, and Zarantonello [1990] found the largest known twin primes using Proth's theorem for numbers of the form $3a2^n + 1$ and a theorem of Riesel for numbers of the form $3a2^n - 1$. They are $1706595 \cdot 2^{11235} \pm 1$, numbers of 3389 decimal digits.

9.3

The function we have called $\lambda(n)$ was first discussed by A. Cauchy [1841], who called it the "maximum indicator". However, the general idea of the λ function appears in Gauss [1801, §92]. The average order of $\lambda(n)$ is discussed in Erdős, Pomerance, and Schmutz [1991].

Carmichael numbers are named after R. D. Carmichael [1910, 1912], the first to discuss them in detail. He proved Theorem 9.3.5. (Note, however, that essentially the same result was previously stated by Korselt [1899].) The largest table of Carmichael numbers currently in existence is that of Pinch, who has calculated the Carmichael numbers up to 10^{16}. Pinch and Keller [1992] calculated the Carmichael numbers up to 10^{15}; they determined that $C(10^{15}) = 105212$. Also see Pinch [1993a]. The table of Jaeschke [1990] contained errors, most notably the assertion that $C(10^{12}) = 8238$, when in fact $C(10^{12}) = 8241$. For some other papers on Carmichael numbers, see Chernick [1939]; Beeger [1950]; Duparc [1951]; Knödel [1953a, 1953b]; Erdős [1956]; Swift [1975]; Yorinaga [1978]; Hill [1979]; Yorinaga [1979a, 1979b]; Wagstaff [1980]; Yorinaga [1980]; Rougon and Dussaud [1981]; Yorinaga [1981]; Woods and Huenemann [1982]; Clarke and Shannon [1983]; Dubner [1989]; Keller [1988]; Löh [1988]; Löh and Niebuhr [1989]; Shanks [1990a, 1990b]; Zhang [1991]; Guillaume and Morain [1992a, 1992b]; Löh [1993]; S. Graham [1994].

The proof that there are infinitely many Carmichael numbers is given in Alford, Granville, and Pomerance [1994a]. Also see Granville [1992]; Pomerance [1993].

The notion of Carmichael numbers was generalized by H. Williams [1977] to the case of Lucas sequences.

It was reported by Jeans [1897] that an ancient Chinese manuscript contained the assertion that n is a prime if and only if $2^n \equiv 2 \pmod{n}$. However, this seems to have been a misapprehension; see Needham [1959, p. 54, footnote d]. It has nevertheless been repeated by many authors (cf. Dickson [1919, p. 91]; Pomerance [1982b], etc.)

The first person to investigate pseudoprimes seems to have been F. Sarrus [1819]. He showed that $341 \in \text{psp}(2)$.

The term "pseudoprime" is apparently due to D. H. Lehmer; see Erdős [1949]. The term has not been used consistently in the literature. Sometimes a "pseudoprime" can be prime (as in Shanks [1978]), leading to the strange term "composite pseudoprime".

The term "strong pseudoprime", as we have used it, is apparently due to Selfridge in the mid-1970's (unpublished). The term apparently first appeared in print in in H. Williams [1978a] and Shanks [1978]. However, the idea behind the strong pseudoprime test appeared earlier in the little-known booklet of Dubois [1971, p. 28].

Earlier, Lehmer [1976b] used the term "strong pseudoprime" for what we call *Euler pseudoprime*. The term "Euler pseudoprime" is due to Shanks, and first appeared in H. Williams [1978a] and Shanks [1978].

The term "even pseudoprime" has been used in the literature to mean an even composite number n such that $2^n \equiv 2 \pmod{n}$. D. H. Lehmer showed that $n = 161038$ is an even pseudoprime; see Erdős [1950]. Beeger [1951] proved that there are infinitely many.

For the use of pseudoprime tests in primality testing, see Norman [1979]; Pomerance, Selfridge, and Wagstaff [1980].

For other papers on pseudoprimes, see Malo [1903]; Cipolla [1904b]; Escott [1906]; Poulet[1938]; Sispánov [1941]; Sierpiński [1947]; Rotkiewicz [1962, 1963ab, 1964abcdefg]; Szymiczek [1965]; Szymiczek [1967]; Rotkiewicz [1967]; Mąkowski and Rotkiewicz [1969]; Rotkiewicz [1969, 1972a, 1972b]; Rotkiewicz and Wasén [1979]; Rotkiewicz [1979, 1980]; van der Poorten and Rotkiewicz [1980]; Wagstaff [1982]; Somer [1989]; McDaniel [1989]; Matthews [1994]; and Bleichenbacher and Maurer [1994].

The notion of "pseudoprime" has been generalized to Lucas sequences, higher-order linear recurrences, and elliptic curves. The first paper in this direction seems to be that of E. Lehmer [1964]. Later papers include Lind [1967]; E. Parberry [1970]; Rotkiewicz [1973]; Hoggatt and Bicknell [1974]; White, Hunt and Dresel [1977]; Baillie and Wagstaff [1980]; Pollin and Schoenberg [1980]; Owings [1982]; W. Adams and Shanks [1982]; Rotkiewicz [1982]; Singmaster [1983]; Kurtz, Shanks and Williams [1986]; Kiss, Phong, and Lieuwens [1986]; W. Adams [1987]; Phong [1988]; Gurak [1989]; Ito [1989]; Di Porto and Filipponi [1989]; D. Gordon [1989]; Miyamoto and Ram Murty [1989]; Filipponi [1990]; Gurak [1990]; Lidl and Müller [1990, 1993]; Lidl, Müller, and Oswald [1990]; D. Gordon and Pomerance [1991]; Arno [1991]; Somer [1991]; Müller and Oswald [1991]; Kowol [1992]; Di Porto [1993]; and Bruckman [1994a, 1994b, 1994c]. Recently, Mo and Jones [1995] have developed a new randomized primality test based on Lucas sequences that has an error probability of $\leq 1/8$.

Estimates on the number of pseudoprimes and Carmichael numbers less than a given bound can be found in Erdős [1956]; Pomerance, Selfridge, and Wagstaff [1980]; Pomerance [1981b]; Pomerance [1982c]; Erdős and Pomerance [1986]; Pomerance [1989]; and Kim and Pomerance [1989]. Also see Jaeschke [1990], where a more complicated heuristic formula from Pomerance [1981b] is compared with the actual distribution of Carmichael numbers.

The term "false witness" is sometimes used in the literature for "liar."

The book of Rotkiewicz [1972c] contains hundreds of problems and references about pseudoprimes.

9.4

The first probabilistic primality test was Solovay and Strassen [1977]. (Their paper was actually submitted in 1974, however. It contains a small flaw, which they later repaired.) D. H. Lehmer [1976b] independently proved Lemma 9.4.1.

For the Miller-Rabin test, see G. Miller [1976]; Rabin [1976]; Rabin [1980b]. Miller's original paper did not refer to probabilistic tests, but rather to an ERH-based test.

The improvement in the Miller-Rabin error probability to $1/4$ was discovered by Rabin [1980b]; Monier [1980]; and Atkin and Larson [1982], independently. Monier also found closed-form expressions for $|E(n)|$ and $|S(n)|$. Our proof of Lemma 9.4.4 is due to H. W. Lenstra, Jr.; it appears in Bosma and van der Hulst [1990b]. Also see J. Davenport and Smith [1987].

Pinch [1993b] discusses the implementation of probabilistic primality tests in several computer algebra systems. Also see Arnault [1991, 1993, 1995]. For other papers on probabilistic primality tests, see Malm [1977]; Baratz [1978]; Chaitin and Schwartz [1978]; Herlestam [1980]; Mignotte [1980b]; Lehmann [1982]; Finn [1982b, 1982c]; Finn and Lieberherr [1983]; Kranakis [1984a, 1984b]; Calude and Zimand [1984]; Fürer [1985]; Boyd [1988]. The papers of Haghighi [1988] and Jammalamadaka and Uppuluri [1989] make no new contribution.

9.5

Theorem 9.5.3 was first proved by G. Miller [1976], by analyzing a deterministic version of the Miller-Rabin test. Our proof follows Vélu [1978].

The question of the least witness for the compositeness of n is of great interest, and one can ask how the bound of $2(\log n)^2$ compares with actual data about the problem. This has been investigated for the Miller-Rabin test. Alford, Granville, and Pomerance [1994a] have shown that no fixed collection of bases are sufficient to test primality for all n, using the strong pseudoprime test. By Theorem 9.3.10, this also holds for the deterministic Solovay-Strassen algorithm. Alford, Granville, and Pomerance [1994b] show that there are infinitely many Carmichael numbers n with least strong witness larger than $(\log n)^{1/(3 \log\log\log n)}$.

Numerical data suggests that $O((\log n)^2)$ is not a tight bound for the number of bases needed by the deterministic Miller-Rabin test. Pomerance, Selfridge, and Wagstaff [1980] found that the bases $a \le 11$, used in the Miller-Rabin test, are sufficient to test the primality of any $n \le 2^{32}$. Using a table of 2-pseudoprimes computed by Pinch (personal communication), we found that bases $a \le 37$ are sufficient for $n \le 2^{32}$, using the Solovay-Strassen test.

Table 9.2 lists, for prime a, the minimum odd n such that the bases $\le a$ are needed to correctly decide primality using the Miller-Rabin test. The data are from Pomerance, Selfridge, and Wagstaff [1980]; Wagon [1986]; and Jaeschke [1993]. (The value $a = 19$ has been skipped, because the least strong pseudoprime for $a \le 17$ is also a strong pseudoprime for $a \le 19$.)

The following heuristic argument suggests that the number of bases truly required by the Solovay-Strassen algorithm is $\Theta((\lg n)(\lg\lg n))$. There are $O(n^\alpha)$ 2-pseudoprimes $\le n$, for some α between 0 and 1. Starting with this set, we successively remove the m for which $\left(\frac{p}{m}\right) \not\equiv p^{(m-1)/2}$, for $p = 2, 3, 5, \ldots$. We expect each prime to eliminate about half of the m. (Examination of Pinch's data

Table 9.2
Composite n Requiring Many Bases to Test Primality.

a	n	$\log n$	$2(\log n)^2$
2	9	2.197	9.656
3	2047	7.624	116.255
5	1373653	14.133	399.482
7	25326001	17.047	581.224
11	3215031751	21.891	958.441
13	2152302898747	28.398	1612.843
17	3474749660383	28.877	1677.710
23	341550071728321	33.465	2239.748

supports this hypothesis.) Hence the set should be empty once we have examined approximately $\log_2 n^\alpha$ bases. Results of Alford, Granville, and Pomerance [1994a] imply that $\alpha = \Theta(1)$, so we should have $p = \Theta((\lg n)(\lg\lg n))$.

An analogous argument can be applied to the Miller-Rabin test.

9.6

The original cyclotomic rings test, based on higher reciprocity laws, was first sketched in the article of Adleman [1980]. The article of Adleman, Pomerance, and Rumely [1983] filled in the missing details; it is there that the reader will find the proof of Theorem 9.6.6. The cyclotomic rings test is frequently called the "APR" test, for the initials of its inventors.

The heuristic argument for the size of z is given in Adleman [1980] and Rumely [1983].

The algorithm we presented is due to H. W. Lenstra [1981, 1986a]. We have closely followed the expository articles of Rumely [1983] and Dixon [1984]. Also see Nicolas [1981]. We note that the discussion in Riesel [1985a, pp. 138–144] does not adequately discuss how hypothesis (ii) of Theorem 9.6.4 can be verified without prior knowledge of the divisors of n.

Adleman has conjectured that we can take the set P to be the first t primes. In practice, to test all primes of reasonable size, it suffices to precompute the sets P and Q once. For example, as shown in the article of Adleman, Pomerance, and Rumely [1983], taking P to be the first 9 primes suffices to test all primes with fewer than 621 decimal digits.

Another practical improvement to the test we have presented is as follows: if $n^{p-1} \not\equiv 1 \pmod{p^2}$, then hypothesis (ii) of Theorem 9.6.4 is satisfied for p. Since for a given p this occurs for $1 - 1/p$ of all possible n, we can frequently bypass the search for q in step (7) of the algorithm.

For a much more practical test, based on Jacobi sums instead of Gauss sums, see H. W. Lenstra [1981]; H. Cohen and H. W. Lenstra [1984c]. The implementation of the Jacobi sum test is discussed in a paper of H. Cohen and A. K. Lenstra [1987]. The largest number proved prime with this test is $(2^{3539} + 1)/3$, a number with 1065 decimal digits. See Bosma and van der Hulst [1990b].

For other "algebraic" primality tests, see the papers of H. W. Lenstra [1982, 1985]; Huang [1984b]; Mihailescu [1988]; Bosma and van der Hulst [1990a, 1990b]. The paper of Harari [1983] makes no new contribution.

9.7

The material in this section is based on Beauchemin, Brassard, Crépeau, Goutier, and Pomerance [1988]. Also see Beauchemin, Brassard, Crépeau and Goutier [1987]; D. Bond [1984]. Gerlach [1992] estimated the number of witnesses providing a proper divisor of n.

Kim and Pomerance [1989] discussed the probability $P(x)$ that n is composite if $a^{n-1} \equiv 1 \pmod n$, where n is an odd number $\le x$, and a is randomly chosen in the range $1 < a < n-1$. They proved, among other things, that $P(x) \le (\log x)^{-197}$ for $x \ge 10^{10^5}$. Also see Damgård and Landrock [1993];

Damgård, Landrock, and Pomerance [1993]; Burthe [1995]. See Rivest [1991] for a discussion of generating many large random primes.

Plaisted [1979] discussed how to generate large primes, together with a short, easy-to-verify certificate of primality. Also see Maurer [1990, 1992].

The generation of random "strong primes" has been studied in the literature. (A "strong prime" is essentially a prime p such that $p \pm 1$ contains large factors, so that $N = pq$ cannot be easily factored using the "$p \pm 1$" methods discussed in Chapter 13.) For these papers, see J. Gordon [1984a, 1984b]; Shawe-Taylor [1986, 1992]; Ogiwara [1989]; Demytko [1989]; Condie [1993].

Pintz, Steiger, and Szemerédi [1989] showed how to generate a k-digit prime, together with a short, easily-verified certificate of primality, in $O(k^4)$ bit operations.

9.8

There is no surviving work of Eratosthenes (c. 276–195 B.C.). His sieve was described by Nicomachus (c. 100 A.D.). See D'ooge, Robbins, and Karpinski [1926].

Writing a program to find all the primes $\leq n$ is frequently assigned in introductory programming courses. Nevertheless, some authors have managed to publish incorrect programs. For example, the program by Wirth [1973, p. 141] contained errors that were found by Pritchard [1980, 1984].

Early computer implementations of the sieve of Eratosthenes were presented by Wood [1961]; Chartres [1967a, 1967b]; and Singleton [1969a, 1969b].

For an inefficient way to compute p_n using an algorithm for $\pi(x)$, see Brun [1931] and D. H. Lehmer [1932c]. For related material, see Lambek and Moser [1954]; Golomb [1976b].

The segmented sieve was first analyzed by Bays and Hudson [1977], although it certainly was used much earlier than that; for example, see Brent [1973]. Pritchard [1983a] gave an algorithm that uses $O(n \lg n)$ bit operations and runs in $O(\sqrt{n}/(\lg \lg n))$ space.

In the literature, it has been traditional to analyze sieve algorithms using a model where an arithmetic operations takes 1 unit of time. This model is perhaps more natural than bit complexity, for one is rarely interested in constructing a table of primes whose number of entries exceeds 2^w, where w is the computer's word size. In this model, the traditional sieve of Eratosthenes uses $O(n \lg \lg n)$ arithmetic operations, and it was an open question to lower this. Mairson [1977] gave the first algorithm which ran in arithmetic linear time. Another "linear" sieve algorithm was given by Gries and Misra [1978].

Pritchard [1981] gave the first "sublinear" sieve; it uses $O(n/(\lg \lg n))$ arithmetic operations (but $O(n(\lg n)/(\lg \lg n))$ bit operations). Also see Pritchard [1982]. Recently, Dunten, Jones, and Sorenson [1994] found a new sublinear sieve that uses $O(n/(\lg \lg n))$ arithmetic operations, but less space than Pritchard's.

Bengelloun's sieve [1986] has the unusual feature of being *incremental*; that is, after the primes from 1 to n have been determined, the primality of $n + 1$ can be decided using only a *constant* number of arithmetic operations. In practice, this may be useful in searching the primes for some property when one does not know a priori how long the search may continue, but its linear space requirement could be prohibitive. Also see Pritchard [1994].

For a comparison of recent sieve algorithms, see Pritchard [1987]. Other papers on the generation of the primes $\leq n$ include Fischer [1965c]; Misra [1981]. The paper of Luo [1989] makes no new contribution; comments about this article may be found in Quesada, Pritchard, and James [1992].

Sorenson [1990b] compared several sieve algorithms and found that Pritchard's sublinear sieve [1981] ran approximately twice as fast as the ordinary Eratosthenes sieve for $n = 10^6$. For segmented sieves, Pritchard's linear segmented wheel sieve [1983] was approximately 40% faster than the Bays-Hudson method for $n = 10^9$.

On the other hand, Sorenson [1991a] has shown that the ordinary sieve of Eratosthenes is better than Pritchard's linear sieve [1987, Algorithm 3.3, p. 23] for all practical values of n, even if a wheel is used.

For parallel versions of Eratosthenes' sieve, see Hikita and Kawai [1980]; I. Parberry [1981]; Bokhari [1987]; Cosnard and Philippe [1989]; Sorenson and Parberry [1994].

Sieving methods are also extremely useful for computing other tables of number theoretic functions. We will use sieve methods again in Chapter 11. For a survey of sieving techniques, see Wunderlich [1967].

9.9

For a discussion of what it means to be a "good formula", see Wilf [1982].

For some computationally useless (but amusing) formulas that give only prime numbers, or express each prime in terms of preceding primes, see Isenkrahe [1900]; Buck [1946]; Mills [1947]; Kuipers [1950]; Moser [1950]; Ansari [1951]; Niven [1951]; Wright [1951]; Ore [1952]; Bang [1952]; Sierpiński [1952]; Wright [1954]; Srinivasan [1961]; Willans [1964]; Goodstein and Wormell [1967]; V. Harris [1969]; Dudley [1969]; Gandhi [1971]; Vanden Eynden [1972]; Golomb [1974]; Papadimitriou [1975]; J. Jones [1975]; Regimbal [1975]; Golomb [1976a]; Tsangaris and Jones [1992].

Early work on $\pi(x)$ is contained in the papers of Meissel [1870, 1871, 1883, 1885]. For a discussion of Meissel's method in English, see Mathews [1892, pp. 272–278]. The first computer implementation of Meissel's method was by D. H. Lehmer [1959]. Also see Piarron de Mondesir [1878]; Mapes [1963]; Lindgren [1963]; Bohman [1972]; Shiu [1987].

For the $O(x^{2/3+\epsilon})$ Meissel-Lehmer method, see Lagarias and Odlyzko [1984]; Lagarias, Miller, and Odlyzko [1985]. For the analytic method, see Lagarias and Odlyzko [1987]. For the recent computation of $\pi(10^{17})$ and $\pi(10^{18})$, see Deleglise and Rivat [1995].

For counting primes in arithmetic progressions, see Hudson [1977]; Hudson and Brauer [1977]. For counting numbers that are the sum of two squares, see Shiu [1986].

A Solutions to Exercises

This chapter presents solutions and comments to the exercises. (No solution is available to those exercises marked "Open.")

Chapter 2

1. Let $d = \gcd(m, n)$. The only primes that can divide d are the p_i. If $p_i{}^{g_i} \,\|\, d$, and $g_i > \min(e_i, f_i)$, then $g_i > e_i$ or $g_i > f_i$. Hence $d \nmid m$ or $d \nmid n$, a contradiction. If $g_i < \min(e_i, f_i)$, then $p_i g_i \mid m$ and $p_i g_i \mid n$, a contradiction. Hence $g_i = \min(e_i, f_i)$. A similar argument works for $\operatorname{lcm}(m, n)$.

2. Use the previous exercise.

3. One possibility is to extend the definition of Exercise 1, allowing the exponents in the "prime factorization" to be any integers. Another is to use Exercise 2, reasoning as follows: $\gcd(a/b, c/d) = \gcd(ad, bc)/bd$ and $\operatorname{lcm}(a/b, c/d) = \operatorname{lcm}(ad, bc)/bd$. It is easy to see that these definitions coincide.

4. Use Exercise 1 and the observation that $\min(e_i, f_i) + \max(e_i, f_i) = e_i + f_i$.

5. One possibility is the following:

$$\gcd(1/a_1, 1/a_2, \ldots, 1/a_n) = 1/\operatorname{lcm}(a_1, a_2, \ldots a_n).$$

For other solutions, see Lebesgue [1837, p. 258]; Serret [1852]; Kesava Menon [1969]; Marsh and Edgar [1971]; Callan [1991].

6. Following the hint, let n be the least positive integer with two distinct factorizations. Then if p is a prime which appears in two distinct factorizations of n, then n/p would be a smaller number with two distinct factorizations. Thus $n = p_1 p_2 \cdots p_k = q_1 q_2 \cdots q_\ell$, where $p_i \neq q_j$ for all i, j. Without loss of generality, take the p_i and q_j to be listed in increasing order. Then $p_1^2 \leq n$ and $q_1^2 \leq n$, since n is composite. Let $m = n - p_1 q_1$. Since $p_1 \neq q_1$, we have $0 < m < n$, so m has a single factorization into primes. But $p_1 \mid m$ and $q_1 \mid m$. Hence the factorization of m contains p_1 and q_1, and $p_1 q_1 \mid m$. But then $p_1 q_1 \mid n$. Hence $q_1 \mid n/p_1$. But $n/p_1 < n$, and so it has a unique factorization $p_2 \cdots p_k$. This is a contradiction.

This proof was first given by Lindemann [1933]; also see H. Davenport [1983, pp. 19–20].

7. Use the inclusion-exclusion principle to deduce that the number of integers $\leq x$ which are not divisible by any of a set of pairwise relatively prime integers $a_1, a_2, \ldots, a_k$ is

$$\lfloor x \rfloor - \sum_{1 \leq i \leq k} \left\lfloor \frac{x}{a_i} \right\rfloor + \sum_{1 \leq i, j \leq k} \left\lfloor \frac{x}{a_i a_j} \right\rfloor - \cdots.$$

Now let $x = n$, and $a_i = p_i$. We deduce that $\varphi(n) = n \prod_{1 \leq i \leq k} (1 - 1/p_i)$.

8. Any divisor of n must be of the form $\prod_{1 \leq i \leq k} p_i{}^{d_i}$, where $0 \leq d_i \leq e_i$. We can choose the d_i in $\prod_{1 \leq i \leq k} (e_i + 1)$ ways, and all of these choices result in distinct divisors.

9. First show that σ_r is multiplicative, and then compute $\sigma_r(p^e)$.

10. Since $n \mid p^e r$, there exists a positive integer a such that $p^e r = an$. Then $p^{e-1} r = (a/p)n$. Since $n \nmid p^{e-1} r$, a/p cannot be an integer. Hence $p \nmid a$. But since $p \nmid r$, it follows that $p^e \,\|\, n$.

11. Clearly if $a = r^s$, $b = r^t$, then $a^t = r^{st} = b^s$.

To prove the other direction, let $p_1, p_2, \ldots, p_m$ represent all the distinct primes dividing either a or b. Then we can write $a = p_1^{e_1} \cdots p_m^{e_m}$, $b = p_1^{f_1} \cdots p_m^{f_m}$. Then since $a^i = b^j$, we have $ie_k = jf_k$ for $1 \le k \le m$. Let $g = \text{lcm}(i, j)$. Define $r = a^{i/g}$; and observe that r is an integer, since $g \mid ie_k$ for $1 \le k \le m$. Then $a = r^{g/i}$ and $b = r^{g/j}$.

12. If $k \ne 0, p$, then $\binom{p}{k} = p!/(k!(p-k)!)$ has a numerator which is divisible by p and a denominator that isn't.

13. We show that $a^p \equiv a \pmod p$ by induction on a. Clearly this is true for $a = 1$. Now assume true for a; we then have $(a+1)^p \equiv a^p + 1^p \pmod p$ by the previous exercise. By induction, this is $a + 1 \pmod p$. Now multiply by a^{-1} to obtain the desired result.

14. Using the hint, suppose $x_1, x_2, \ldots, x_k$ are the positive integers $\le n$ that are relatively prime to n. If $\gcd(a, n) = 1$, then $ax_i \bmod n$ is relatively prime to n. All these $ax_i \bmod n$ are in fact distinct, for if $ax_i \equiv ax_j \pmod n$, then $n \mid a(x_i - x_j)$; since $\gcd(a, n) = 1$, we must have $n \mid x_i - x_j$. Thus we see that, modulo n,

$$\prod_{1 \le i \le k} x_i \equiv \prod_{1 \le i \le k} ax_i \equiv a^{\varphi(n)} \prod_{1 \le i \le k} x_i,$$

so $a^{\varphi(n)} \equiv 1 \pmod n$.

15. Count the number of multiples of $p, p^2, p^3, \ldots$ in the list of integers $1, 2, \ldots, n$. (This exercise is from Legendre [1830, Vol. I, p. 10].)

16. This result has been found independently by several writers. For example, see I. Williams [1976]; Breusch [1979]; and Shallit [1991].

17. Suppose $n = \prod_{1 \le i \le k} p_i^{e_i}$. Then, as in the solution to Exercise 8 above, any divisor d of n must be of the form $\prod_{1 \le i \le k} p_i^{d_i}$, where $0 \le d_i \le e_i$. Since φ is multiplicative, we have

$$\begin{aligned}\sum_{d \mid n} \varphi(d) &= \prod_{1 \le i \le k} \sum_{0 \le j \le e_i} \varphi(p_i^j) \\ &= \prod_{1 \le i \le k} \left(1 + \sum_{1 \le j \le e_i} p_i^{j-1}(p_i - 1)\right) \\ &= \prod_{1 \le i \le k} p_i^{e_i} \\ &= n.\end{aligned}$$

Now, applying Möbius inversion, we get $\varphi(n) = \sum_{d \mid n} \mu(d) n/d$.

18. Define $f(x) = \sum_{1 \le n \le x} \mu(n) \lfloor x/n \rfloor$. Then

$$\begin{aligned}f(x) - f(x-1) &= \sum_{1 \le n \le x} \mu(n)(\lfloor x/n \rfloor - \lfloor (x-1)/n \rfloor) \\ &= \sum_{d \mid \lfloor x \rfloor} \mu(d),\end{aligned}$$

which, by Theorem 2.3.3, is 1 if $\lfloor x \rfloor = 1$ and 0 otherwise. It follows that $f(x) = 1$ for $x \ge 1$.

19. We have

$$\sum_{1\le k\le x} \mu(k)g(x/k) = \sum_{1\le k\le x} \mu(k) \sum_{1\le j\le x/k} f(x/jk)$$
$$= \sum_{1\le k\le x} f(x/k) \sum_{d\,|\,k} \mu(d)$$
$$= f(x),$$

where in the last step we have used Lemma 2.3.3.

Similarly,

$$\sum_{1\le k\le x} f(x/k) = \sum_{1\le k\le x} \sum_{1\le j\le x/k} \mu(j)g(x/jk)$$
$$= \sum_{1\le k\le x} g(x/k) \sum_{d\,|\,k} \mu(d)$$
$$= g(x).$$

20. It is easy to see that

$$\psi(x) = \sum_{1\le k\le \log_2 x} \vartheta(x^{1/k}). \tag{A.1}$$

Then, by setting $x = 2^y$ in (A.1), we get $\psi(2^y) = \sum_{1\le k\le y} \vartheta(2^{y/k})$. Now let $f(y) = \vartheta(2^y)$ and $g(y) = \psi(2^y)$; then $g(y) = \sum_{1\le k\le y} f(y/k)$. Now use the result of the previous exercise to get $f(y) = \sum_{1\le k\le y} \mu(k)g(y/k)$; in other words, $\vartheta(2^y) = \sum_{1\le k\le y} \mu(k)\psi(2^{y/k})$. Hence $\vartheta(x) = \sum_{1\le k\le \log_2 x} \mu(k)\psi(x^{1/k})$.

21. Set $f(x) = x^2$, and use Exercise 2.19. We get

$$g(x) = \sum_{1\le k\le x} (x/k)^2$$
$$= x^2\left(\sum_{1\le k<\infty} 1/k^2 - \sum_{k>x} 1/k^2\right)$$
$$= x^2\left(\frac{\pi^2}{6} + O(1/x)\right).$$

From this, we get

$$x^2 = \sum_{1\le k\le x} \mu(k)\left(\frac{\pi^2}{6}\left(\frac{x^2}{k^2} + O\left(\frac{x}{k}\right)\right)\right)$$

and desired result follows by rearranging, with the help of Exercise 2.18.

22. Using Exercise 2.17, we get

$$\sum_{1\le n\le x} \varphi(x) = \sum_{1\le n\le x} \sum_{d\,|\,n} \mu(d)(n/d)$$

$$
\begin{aligned}
&= \sum_{1\le d\le x} \mu(d) \sum_{1\le r\le x/d} r \\
&= \sum_{1\le d\le x} \mu(d) \left(\frac{x^2}{2d^2} + O\left(\frac{x}{d}\right)\right) \\
&= \frac{x^2}{2} \sum_{1\le d\le x} \frac{\mu(d)}{d^2} + O(x \log x) \\
&= \frac{3}{\pi^2} x^2 + O(x \log x),
\end{aligned}
$$

where we have used the result of the previous exercise. This result is due to Dirichlet [1849].

23. Use Euler's summation formula with $f(x) = \log x$.

A stronger version of this result, known as *Stirling's approximation*, says that

$$n! = n^n e^{-n} \sqrt{2\pi n} \left(1 + \frac{1}{12n} + O\left(\frac{1}{n^2}\right)\right).$$

See, for example, Knuth [1975, §1.2.11.2].

24. We have

$$
\begin{aligned}
\sum_{n\le x} \sigma(n) &= \sum_{n\le x} \sum_{d\,|\,n} d = \sum_{cd\le x} d \\
&= \sum_{1\le c\le x} \sum_{1\le d\le \lfloor \frac{x}{c} \rfloor} d \\
&= \sum_{1\le c\le x} \frac{1}{2} \left\lfloor \frac{x}{c} \right\rfloor \left(\left\lfloor \frac{x}{c} \right\rfloor + 1 \right) \\
&= \frac{1}{2} \sum_{1\le c\le x} \left(\frac{x}{c} + O(1)\right) \left(\frac{x}{c} + O(1)\right) \\
&= \left(\frac{x^2}{2} \sum_{1\le c\le x} \frac{1}{c^2} \right) + O(x \log x).
\end{aligned}
$$

Now estimate $\sum_{1\le c\le x} \frac{1}{c^2}$ by approximating the sum with an integral; we find

$$
\begin{aligned}
\sum_{1\le c\le x} \frac{1}{c^2} &= \left(\sum_{c\ge 1} \frac{1}{c^2} \right) + O\left(\frac{1}{c}\right) \\
&= \zeta(2) + O\left(\frac{1}{c}\right).
\end{aligned}
$$

Combining this with the previous result gives the desired conclusion.

25. We have

$$\sum_{1\le k\le n} \frac{\log k}{k} = \int_1^n \frac{\log t}{t}\,dt + \int_1^n (t - \lfloor t \rfloor) f'(t)\,dt,$$

where $f'(t) = \frac{1-\log t}{t^2}$. The first integral is easily seen to be $(\log n)^2/2$. The second integral can be written as

$$\int_1^\infty (t - \lfloor t \rfloor) f'(t)\,dt - \int_n^\infty (t - \lfloor t \rfloor) f'(t)\,dt = A - B.$$

Now term A represents a constant, since its absolute value is bounded by

$$\int_1^\infty \frac{|1-\log t|}{t^2}\,dt = \int_1^e \frac{1-\log t}{t^2}\,dt - \int_e^\infty \frac{1-\log t}{t^2}\,dt = \frac{2}{e}.$$

On the other hand, the term B is easily seen to be $O(\frac{\log n}{n})$. Thus, we obtain the desired result.

How do we evaluate A? Letting $n \longrightarrow \infty$, we see

$$A = \lim_{n\to\infty}\left(\left(\sum_{1\le k\le n} \frac{\log k}{k}\right) - \frac{1}{2}(\log n)^2\right);$$

setting $n = 10000$ gives $A \approx -0.073$.

26. Let $f(t) = \frac{1}{t\log t}$. By Euler's summation formula,

$$\sum_{2\le k\le n} \frac{1}{k\log k} = \int_2^n \frac{dt}{t\log t} + \int_2^n (t - \lfloor t \rfloor) f'(t)\,dt + f(2)$$

where $f'(t) = -\frac{1+\log t}{t^2(\log t)^2}$.

Now the first integral is easily seen to be $\log\log n - \log\log 2$. The second integral can be written

$$\int_2^n (t - \lfloor t \rfloor) f'(t)\,dt = \int_2^\infty (t - \lfloor t \rfloor) f'(t)\,dt - \int_n^\infty (t - \lfloor t \rfloor) f'(t)\,dt = B - C.$$

Now the term B represents a constant, since it is bounded in absolute value by

$$-\int_2^\infty f'(t)\,dt = f(2).$$

On the other hand, the term C is easily seen to be $O(\frac{1}{n\log n})$. Thus the result follows.

To estimate B, we use the same trick as in the previous exercise. For $n = 10000$, we find $B \approx .7947$.

This exercise, and the previous one, appeared without solution in Apostol [1976].

27. First, as noted in the proof of Theorem 2.5.4, we have

$$\sum_{1\le k\le n} \left\lfloor \frac{n}{k} \right\rfloor = \sum_{1\le k\le n} d(k).$$

Similarly, it is easy to see that

$$\sum_{1\le k\le (n+1)/2} \left\lfloor \frac{n}{2k-1} \right\rfloor = \sum_{1\le k\le n} d_{2,1}(k).$$

Hence

$$\begin{aligned}\sum_{1\le k\le n} (d_{2,1}(k) - d_{2,0}(k)) &= \sum_{1\le k\le n} (2d_{2,1}(k) - d(k)) \\ &= 2 \sum_{1\le k\le (n+1)/2} \left\lfloor \frac{n}{2k-1} \right\rfloor - \sum_{1\le k\le n} \left\lfloor \frac{n}{k} \right\rfloor \\ &= \sum_{1\le k\le n} (-1)^{k+1} \left\lfloor \frac{n}{k} \right\rfloor .\end{aligned}$$

To estimate this sum, split it into two parts: $1 \le k \le t$, and $t < k \le n$, where $t = \lfloor \sqrt{n} \rfloor$. Then the first part of the sum is $n \log 2 + O(\sqrt{n})$, and the second part of the sum is $O(\sqrt{n})$. This result was first stated by Cesàro [1881a].

28. This exercise is based on E. Landau's first published paper [1899]. Also see Ribenboim [1985].

29. We have

$$\sum_{x_1x_2x_3\le n} 1 = \sum_{x_3\le n} \sum_{x_1x_2\le n/x_3} 1 = \sum_{x_3\le n} \left(\frac{n}{x_3} \log\left(\frac{n}{x_3}\right) + O\left(\frac{n}{x_3}\right)\right),$$

where we have used Theorem 2.5.4. Thus

$$\sum_{x_1x_2x_3\le n} 1 = n\log n \sum_{x_3\le n} \frac{1}{x_3} - n \sum_{x_3\le n} \frac{1}{x_3 \log x_3} + n \sum_{x_3\le n} O\left(\frac{1}{x_3}\right).$$

Now the first sum is

$$\log n + \gamma + O\left(\frac{1}{n}\right)$$

by the result in Section 2.5. The third sum can be evaluated in the same way. The second sum is

$$\frac{1}{2}(\log n)^2 + A + O\left(\frac{\log n}{n}\right)$$

from Exercise 26. Putting this all together, we find

$$\sum_{x_1x_2x_3\le n} 1 = \frac{1}{2} n(\log n)^2 + O(n\log n).$$

30. Based on the result of the previous exercise, it seems reasonable to conjecture that

$$\sum_{x_1x_2\cdots x_k\le n} 1 = \frac{1}{(k-1)!} n(\log n)^{k-1} + O(n(\log n)^{k-2}).$$

It's not hard to prove this formula by induction. In the crucial step, use Euler's summation formula to prove that

$$\sum_{c \le n} \frac{1}{c}(\log(n/c))^{k-1} = \frac{1}{k}(\log n)^k + O((\log n)^{k-1}).$$

31. Pair each divisor $d \le \sqrt{n}$ with n/d. This pairing accounts for each divisor exactly once, unless n is a square.

32. The solution can be found in Moser and Kugelmass [1964]; see also Niven [1984].

33. We have

$$\begin{aligned}\sum_{d \mid n} \mu(d) \sum_{1 \le k \le n/d} f(kd) &= \sum_{1 \le k \le n} f(k) \sum_{d \mid \gcd(k,n)} \mu(d) \\ &= \sum_{\substack{1 \le k \le n \\ \gcd(k,n)=1}} f(k),\end{aligned}$$

where we have used Lemma 2.3.3.

34. Let $f(n) = 1$ if n is odd, and n if n is even. Let $g(n) = n$ if n is odd, and 1 if n is even. Then it is easy to see that $f(n) \ne O(g(n))$ and $g(n) \ne O(f(n))$.

To construct an example where the functions are nondecreasing, let $r(2n - 1) = r(2n) = (2n - 1)!$, and $s(2n) = s(2n + 1) = (2n)!$. Then these functions are clearly nondecreasing, and $r(n)/s(n) = n$ if n is odd, and $1/n$ if n is even.

35. Using the hint, we see that

$$\begin{aligned}\sum_{1 \le n \le x} d(n) &= \left(2 \sum_{1 \le d \le \sqrt{x}} (\lfloor x/d \rfloor - d)\right) + \lfloor \sqrt{x} \rfloor \\ &= \left(2x \sum_{1 \le d \le \sqrt{x}} \frac{1}{d}\right) - 2 \sum_{1 \le d \le \sqrt{x}} d + O(\sqrt{x}) \\ &= x \log x + (2\gamma - 1)x + O(\sqrt{x}).\end{aligned}$$

36. Use Theorems 2.6.1 and 2.7.1 to obtain the estimate $\sum_{p \le x} p^{-1/2} = 2\sqrt{x}/\log x + o(\sqrt{x}/\log x)$.

Chapter 3

1. For (a), use the definition, Eq. (2.1). For (b) and (c), see Bach and Sorenson [1993]. Fitch [1974] gave an algorithm for k-th roots but did not analyze it.

2. The base-b digits of n can be computed (starting with the least significant digit) by successively determining $n \bmod b$ and then setting $n \leftarrow \lfloor n/b \rfloor$, until $n = 0$. The bit complexity is bounded by $\sum_{0\le k\le(\log_b n)+1} O((\lg b)(\lg n/b^k))$, which is $O((2 + \log_b n)(1 + \log b)(1 + \log n))$. This is $O((\lg n)^2)$.

3. The cost of the multiplication is given by $(\lg m_1)(\lg m_2) + (\lg m_1 m_2)(\lg m_3) + \cdots + (\lg m_1 m_2 \cdots m_{k-1})(\lg m_k)$. This is bounded by $(\lg m)(\lg m_2 + \lg m_3 + \cdots + \lg m_k)$, which (since $k = O(\lg m)$) is $O((\lg m)^2)$.

4. See e.g., Bach and Sorenson [1993]. Recently, H. W. Lenstra, Jr. and J. Pila (personal communication) have shown that this problem can be solved using $O((\lg n)^2)$ bit operations. Their solution uses a gcd-free basis and the fast arithmetic of Karatsuba and Ofman [1962]. D. Bernstein [1995] showed that a running time of $(\lg n)^{1+o(1)}$ can be achieved asymptotically.

5. For the first part, using Exercise 2.11, determine r such that $a = r^s$, $b = r^t$, if such r exists. For the second part, do the same thing, then check if $is = jt$.

6. See Borwein [1985].

7. Compute F_k iteratively, using the formula $F_k = F_{k-1} + F_{k-2}$. It is easy to prove by induction that $F_k \le 2^k$; hence the total cost is $\sum_{1\le j\le k} O(j) = O(k^2)$.

8. See, e.g., Pollard [1971b, p. 339].

9. Use the greedy algorithm: let $x_1 = \lfloor \sqrt{n} \rfloor$, and then express $n - x_1^2$ as a sum of squares. In general, define $n_0 = n$, and for $i \ge 1$, define $x_i = \lfloor \sqrt{n_{i-1}} \rfloor$ and $n_i = n_{i-1} - x_i^2$. It is easy to see that $n_1 \le 2\sqrt{n}$. Using this, one can easily prove by induction that $n_i \le 4n^{1/2^i}$. After $\le \log_2 \log_2 n$ steps, the original number n has been reduced to $n_j \le 8$. But each positive integer ≤ 8 can be represented as a sum of at most four squares. From this, the result follows.

What is the running time of this algorithm? By Exercise 3.1 (b), we can compute x_i in $O((\lg n_i)^2)$ bit operations. Summing this from $i = 0$ to $\log_2 \log_2 n$, we find that the number of bit operations is $O((\lg n)^2)$.

For the smallest number requiring n squares in its greedy representation, see Lemoine [1882].

10. For multiplication, use divide-and-conquer, and the formula $(a \cdot 2^k + b)(c \cdot 2^k + d) = (2^{2k} + 2^k)ac + 2^k(a - b)(d - c) + (2^k + 1)bd$. See Karatsuba and Ofman [1962]. Division can be done within the required time bound by using Newton's method to approximate the reciprocal of the dividend. See Aho, Hopcroft, and Ullman [1974].

11. We can check whether n is even by computing $n \bmod 2 = n - 2\lfloor n/2 \rfloor$. By naive bit complexity, this costs $O(\log n)$. The same cost holds in our formalized model.

12. One way to do this is to encode 0 as 00, 1 as 01, and a separator between strings as 11. Thus the pair of strings (011, 10) would be encoded as 000101110110. Clearly, this encoding is $1-1$, and furthermore, it and its inverse can be computed in polynomial time and constant space.

13. Counting to an arbitrary limit will overflow any fixed register. Therefore, one must represent large numbers, e.g., by a linked list of registers. But a sufficiently large number will use so many registers that the pointers themselves will need multiple registers to be represented, etc.

14. For one direction, use the CHOICE command to guess the bits of the "proof of membership" y corresponding to string x; then verify these bits in polynomial time using the membership algorithm for (x, y). For the other direction, assume that a set S is accepted by such a machine, and create the "proof of membership" y by concatenating together the indices of the choices made (either m_0 or m_1).

15. For Turing machines, this was done by Gill [1977], and exactly the same techniques can be used for our polynomial-cost RAM model.

16. One way to do this is to suppose that the random access machine may use the following type of instruction:

$$R_i \leftarrow \text{ORACLE } R_j.$$

Let A be a fixed, but unspecified, set of integers. Upon encountering the above instruction, R_i is set to 1 if $R_j \in A$, and 0 otherwise; the process takes $\lg R_j$ steps. See, for example, Rogers [1967].

17. We provide an algorithm for the case $n > 0$, leaving the case $n \leq 0$ to the reader.

```
PARITY(n)
{ parity of a positive integer, using only additions, subtractions, and comparisons }
while (n > 0) do
        a ← 1
        while (a + a ≤ n) do
                a ← a + a
        n ← n − a
if (a = 1) then return("odd") else return("even")
```

The naive bit complexity of this method is $O((\lg n)^3)$.

18. All four operations can be performed in the stated time bounds by the traditional pencil-and-paper methods. Addition and subtraction, suitably coded, can actually be done by finite automata; see e.g., Minsky [1967, pp. 22–23].

19. If we restrict ourselves to instructions 1,2,3,6,7,11,13,14 in Table 3.2, we obtain the model of Cook and Reckhow [1973]. Using their Theorem 2, we can simulate a multi-tape Turing machine with polynomial overhead. For the other direction, we must show how a Cook-Reckhow machine can model ours. For instructions 4 and 5 (multiplication and division), we use the traditional pencil-and-paper methods. Instruction 12 can be simulated by instruction 11. Instruction 15 just adds a cost of 1 to every program. To simulate instructions 9 and 10 (LENGTH and ENCODE), we use a variant of the method given in Exercise 17. To simulate instruction 8 (DECODE), we treat the contents of the registers as coefficients of a polynomial in b, and use Horner's rule. It is readily verified that all these can be done with at most polynomial blow-up.

20. Any computation performed on a polynomial-cost RAM can be simulated by a multi-tape Turing machine that uses one tape to hold the contents of each register used by the RAM, separated by delimiters. It is readily verified that each of the instructions in the polynomial-cost RAM's instruction set can be performed by the Turing machine (on auxiliary work tapes) with at most a linear blow-up in space. Thus, if the polynomial-cost RAM uses $S(n)$ space, then the simulating TM uses at most $O(S(n))$ space.

On the other hand, we may simulate the computations of a Turing machine with a polynomial-cost RAM by following the ideas in Minsky [1967]. Again, the simulation causes at most a linear blow-up in space.

21. Consider the following program:

$$\begin{array}{l} R_1 \leftarrow 1 \\ \text{L1: } R_1 \leftarrow R_1 + R_1 \\ R_{R_1} \leftarrow 1 \\ R_0 \leftarrow R_0 - 1 \\ \text{IF } R_0 > 0 \text{ GOTO L1} \\ R_1 \leftarrow 1 \\ \text{END} \end{array}$$

Its time consumption is $\Theta(n^2)$, but its space consumption is $\Theta(2^n)$.

22. For $x \neq 0$, we have $\lg_b x = 1 + \lfloor \log_b |x| \rfloor$, and $\log_b |x| \leq 1 + \lfloor \log_b |x| \rfloor \leq 1 + \log_b |x|$ for all bases $b \geq 2$ and $x \neq 0$. It now follows that

$$\frac{(\lg_2 x) - 1}{\log_2 b} \leq \lg_b x \leq \frac{\lg_2 x}{\log_2 b} + 1$$

for $x \neq 0$. Hence $\lg_b x = \Theta(\lg_2 x)$.

If $b = 1$, then "base-b representation" is unary. In this case, $\lg x = o(\lg_b x)$.

23. See, for example, Cormen, Leiserson, and Rivest [1990, §36.4].

24. **(a)** The probability that $\beta_i \neq \alpha_i$ is $p = 1/2$, so the expected number of flips until this occurs is $1/p$, or 2.

(b) Choose the minimal n such that $b < 2^n$. Flip n bits, which defines an interval I of measure 2^{-n} in $[0, 1]$. If I lies completely within some interval $[i/b, (i+1)/b]$, output i. Otherwise I is split by some multiple of $1/b$; now apply the solution to part (a).

This exercise seems to be folklore; an analysis taking into account the time for approximating α appears in Bach [1988]. See Knuth and Yao [1976] for analyses of the randomness consumption of various strategies for random variate generation.

25. Suppose $x \in L$. (The case $x \notin L$ is symmetric and can be handled in the same way.) Then the algorithm accepts x with probability $p = 1/2 + \epsilon + t$ and rejects x with probability $q = 1 - p$, where $0 \leq t \leq 1/2$. Note that $pq = (1/4 - \epsilon^2) - (2\epsilon t + t^2) \leq 1/4 - \epsilon^2$. If n trials are done, then the probability of getting r accepts and $n - r$ rejects is exactly $\binom{n}{r} p^r q^{n-r}$. The probability that a wrong decision is made is then the probability that fewer than $n/2$ accepts occur, which is

$$\begin{aligned} \sum_{0 \leq r < n/2} \binom{n}{r} p^r q^{n-r} &\leq p^{n/2} q^{n/2} \sum_{0 \leq r < n/2} \binom{n}{r} \\ &\leq (pq)^{n/2} 2^n \\ &= (1/4 - \epsilon^2)^{n/2} 2^n \end{aligned}$$

$$= (1 - 4\epsilon^2)^{n/2}.$$

This exercise is a special case of an inequality of Chernoff [1952].

26. See Minsky [1967, Chap. 14].

27. See, for example, Cook [1985]; Karp and Ramachandran [1990]; Leighton [1992].

28. **(a)** See Cook [1985]; Karp and Ramachandran [1990].

(b) The $\mathcal{P}$-completeness of the circuit value problem is from Ladner [1975]. For a $\mathcal{P}$-complete number-theoretic problem, see Karloff and Ruzzo [1989].

29. Clearly `SUBSET SUM` is in $\mathcal{NP}$, as we can quickly compute $\sum_{i\in I} a_i$ and verify that it equals t. Now reduce from `PARTITION`: given a set $A = \{a_1, \ldots, a_n\}$, transform it to a subset-sum problem by computing $t = \frac{1}{2}\sum_{1\le i\le n} a_i$.

To show that `SUBSET SUM` can be solved in $O(n\, t \lg t)$ bit operations (pseudo-polynomial time), we use a simple form of dynamic programming. Make an array b indexed from 0 to t. The meaning of b is that $b[i] = 1$ iff there is a subset of A that sums to i. To start, set $b[0] \leftarrow 1$. Now, considering each a_i in turn, set $b[a_i + k] \leftarrow 1$ for each k, $0 \le k \le t$ such that $b[k] = 1$. After all the a_i's have been processed, which uses $O(n\, t \lg t)$ bit operations, we accept if there is a 1 in $b[t]$.

30. Let A be a multiset of positive integers, $\{a_1, a_2, \ldots, a_n\}$, where the a_i need not be distinct. Then we can transform an instance (A, t) of `KNAPSACK` to an instance of `SUBSET SUM` in polynomial time as follows: for $1 \le i \le n$, let $b_i = 3^n a_i + 3^{i-1}$. Define $B = \{b_i : 1 \le i \le n\}$, $C = \{3^{i-1} : 1 \le i \le n\}$, $t' = 3^n t + (3^n - 1)/2$. and consider $(B \cup C, t')$.

31. (Solution by Anna Lubiw.) By Exercise 3, we can quickly compute $\prod_{i\in I} a_i$ and verify that it equals t. Hence `SUBSET PRODUCT` is in $\mathcal{NP}$. Now reduce from `3-SAT`: given a formula with n clauses over variables $x_1, x_2, \ldots, x_m$, let $e_{ij} = 1$ if variable x_i appears unnegated in clause j, and 0 otherwise. Let $f_{ij} = 1$ if variable x_i appears negated in clause j, and 0 otherwise. Now, for $1 \le i \le m$, define

$$y_i = 2^{e_{i1}} 3^{e_{i2}} \cdots p_k^{e_{ik}} \cdots p_n^{e_{in}} p_{n+i},$$

where, as usual, p_k denotes the k-th prime, with $p_1 = 2$. Also define

$$z_i = 2^{f_{i1}} 3^{f_{i2}} \cdots p_k^{f_{ik}} \cdots p_n^{f_{in}} p_{n+i}.$$

Let

$$A = \{y_i : 1 \le i \le m\} \cup \{z_i : 1 \le i \le m\} \cup \{p_i, p_i^2 : 1 \le i \le n\}$$

and let the target be $t = 2^4 3^4 \cdots p_n^4 p_{n+1} p_{n+2} \cdots p_{n+m}$. By the prime number theorem, all numbers involved have $O((4n + m)\lg(n + m))$ bits. Correctness is left to the reader.

Chapter 4

1. The following iterative algorithm computes gcd(u, v):

```
EUCLID2(u, v)
while (v ≠ 0) do
        t ← u
        u ← v
        v ← t mod v
return(u)
```

2. Use Exercise 2.4.

3. Use induction on n. This exercise is due to de Moivre [1730]. Also see Binet [1843]. The formula is sometimes called the "Binet form" for Fibonacci numbers.

4. We have $F_{k+2}/F_{k+1} = 1 + F_k/F_{k+1}$. In order that $\lfloor F_{k+2}/F_{k+1} \rfloor = 1$, we must have $0 < F_k/F_{k+1} < 1$; i.e. $F_k < F_{k+1}$. This is true for all $k \geq 2$. It immediately follows that $F_{k-1} = F_{k+1} \bmod F_k$ for all $k \geq 2$.

5. From Eq. (4.3) we see $a_0 u_1 \leq u_0, a_1 u_2 \leq u_1, \ldots, a_{n-2} u_{n-1} \leq u_{n-2}, a_{n-1} u_n \leq u_{n-1}$. Now use induction on k to verify that $a_0 a_1 \cdots a_{k-1} u_k \leq u_0 = u$. The same method shows $a_1 a_2 \cdots a_{n-1} \leq v$.

6. Use induction on n.

7. First show that $p_n = X_n p_{n-1} + Y_n p_{n-2}$ and $q_n = X_n q_{n-1} + Y_n q_{n-2}$ for all $n \geq 2$.

8. Use induction on n.

9. Think of feeding the nearest-integer continued fraction algorithm with the interval $(-\infty, \infty)$ and observing what happens at each step. The set of possible values for $a_0 = a_0(x)$ is simply $\mathbb{Z}$, and the set of possible values for $x_0 - a_0$ is $(-1/2, 1/2]$. Hence either $x_0 = a_0$, and the continued fraction terminates, or $x_1 = 1/(x_0 - a_0) \in (-\infty, -2) \cup [2, \infty)$. If $a_1 = \text{round}(x_1) \leq -3$, then set of possible values for $x_1 - a_1$ is again $(-1/2, 1/2]$. If $a_1 = -2$, however, then the set of possible values for $x_1 - a_1$ is $(-1/2, 0)$. In this case, the set of possible values for $x_2 = 1/(x_1 - a_1)$ is $(-\infty, -2)$, and so we see that $a_2 \leq -2$. Furthermore, if $a_2 = -2$, then the expansion cannot terminate with a_2.

Continuing in this fashion, we conclude that the set

$$P(b_0, b_1, \ldots, b_i) = \{x_i - a_i \ : \ a_0(x) = b_0, a_1(x) = b_1, \ldots, a_i(x) = b_i\}$$

is precisely one of $I_1 = (-1/2, 1/2]$, $I_2 = (-1/2, 0)$, or $I_3 = [0, 1/2]$. Furthermore, which one depends only on a "finite-state" function of the b_j. More precisely, we can construct a finite-state machine, whose states are the three intervals I_1, I_2, I_3, and a "start state" S_0, and whose transitions are sets of integers. The accepting states correspond to those intervals that contain 0. This machine accepts precisely those finite strings of integers that represent the nearest-integer continued fraction expansion of all rational numbers. The transition function, mapping $\{S_0, I_1, I_2, I_3\} \times \mathbb{Z}$, is defined as follows: $\delta(S_0, \mathbb{Z}) = I_1$; $\delta(I_1, A_1 \cup A_2) = I_1$; $\delta(I_1, -2) = I_2$; $\delta(I_1, 2) = I_3$; $\delta(I_2, -2) = I_2$; $\delta(I_2, A_1) = I_1$; $\delta(I_3, 2) = I_3$; $\delta(I_3, A_2) = I_1$. Here $A_1 = \{j \in \mathbb{Z} : j \leq -3\}$, and $A_2 = \{j \in \mathbb{Z} : j \geq 3\}$. The states I_1 and I_3 represent accepting states. The "rules" given in Theorem 4.6.1 correspond precisely to this machine.

The output of the nearest-integer continued fraction algorithm is completely characterized by this finite-state machine. Since the transformations made at each step are invertible, two different expansions cannot correspond to the same rational number.

The same construction can be used to generate "rules" for the continued fraction expansion using *any* integer function f whose kernel $f^{-1}[0]$ is the finite union of intervals whose endpoints are rational or the roots of quadratic equations.

10. Use induction on n.

11. We leave it to the reader to verify that *if the least-remainder algorithm performs n division steps on input (u, v), then it also performs n division steps on input $(u, -v)$.*

Next, we prove the following lemma: *If $u \geq 2v > 0$, and the least-remainder algorithm performs n division steps, then $u \geq A_{n+1}$ and $v \geq A_n$.* The proof is by induction on n. The lemma is true for $n = 1$, since by hypothesis we must have $u \geq 2$, $v \geq 1$. Now assume it is true for all $k < n$; we prove it for n. Since $u \geq 2v$, we can write $u = qv + r$, where $q \geq 2$, and $|r| \leq v/2$. Now the algorithm performs $n - 1$ division steps on input (v, r); hence by the previous lemma, it performs $n - 1$ division steps on input $(v, |r|)$. Since $v \geq 2|r|$, the induction hypothesis applies, and we can conclude that $v \geq A_n$ and $|r| \geq A_{n-1}$. Thus if $q \geq 3$, then $u \geq \frac{5}{2}A_n \geq A_{n+1}$. If $q = 2$, we must have $r \geq 0$, since $u \geq 2v$; hence $u \geq 2A_n + A_{n-1} = A_{n+1}$.

Finally, to complete the proof of the exercise, write $u = qv + r$, with $|r| \leq v/2$, and $q \geq 1$. Then by combining the two lemmas above, we see that if the algorithm performs $n - 1$ division steps on input (v, r), then $v \geq A_n$ and $r \geq A_{n-1}$. If $q \geq 2$, then $u \geq \frac{3}{2}A_n \geq A_n + A_{n-1}$, while if $q = 1$, then $r \geq 0$, and hence $u \geq A_n + A_{n-1}$.

12. Let $m = \sum_i \lg m_i$. Using the identities $\gcd(m_1, m_2, m_3) = \gcd(\gcd(m_1, m_2), m_3)$ and $\operatorname{lcm}(m_1, m_2, m_3) = \operatorname{lcm}(\operatorname{lcm}(m_1, m_2), m_3)$, we can iteratively compute the gcd and lcm of the k numbers using $O(m^2)$ bit operations.

13. Using Exercise 2.11, we see that if there exists i, j such that $a^i = b^j$, then there exist positive integers r, s, t such that $a = r^s$ and $b = r^t$. Thus if $a \geq b$, we must have $b \mid a$, and furthermore $a/b = r^{s-t}$. From this, the correctness of the suggested algorithm easily follows.

We now prove the time bound. For simplicity, we assume that we can divide a by b, obtaining a quotient and remainder, using $(\log a/b)(\log b)$ bit operations. This estimate differs from the true cost by only a constant factor, provided $b \neq 1$ and $a \neq b$, and these exceptions occur at most once each during the execution of the algorithm. Let $T(a, b)$ denote the number of bit operations used by the algorithm on input (a, b) with $a \geq b$, not including the time needed to handle the cases $a = b$, $b = 1$. Then we have:

$$T(a, b) = \begin{cases} T(a/b, b) + \log(a/b)(\log b), & \text{if } a \geq b^2 \\ T(b, a/b) + \log(a/b)(\log b), & \text{if } a < b^2. \end{cases}$$

It now follows by induction that $T(a, b) \leq \frac{1}{2}(\log a)^2$.

14. This exercise is implicit in Dirichlet [1849]; also see Mertens [1874b, pp. 289–291]. Cesàro [1881b; 1883, Note XIV; 1885, p. 237] was apparently the first to state the result in these picturesque terms. Also see Sylvester [1883].

15. Suppose $f(x, y)$ is such a polynomial. Then $f(2, y)-2$ is 0 for infinitely many y; hence $f(2, y)-2$ must be the constant polynomial 0. But $f(2, 1) - 2 = -1$, a contradiction.

16. See Bach, Driscoll, and Shallit [1993].

17. See Gardner [1971].

18. Use Exercise 5, and note that the number of division steps in the Euclidean algorithm is actually $O(\lg u/d)$.

19. First, prove by induction that

$$\frac{p_n}{q_n} = a_0 + \frac{1}{q_0 q_1} - \frac{1}{q_1 q_2} + \cdots + \frac{(-1)^{n-1}}{q_{n-1} q_n}.$$

Now observe that $q_i \geq F_{i+1}$, the $i + 1$'st Fibonacci number.

20. See, e.g. G. Hardy and Wright [1985, Theorem 169].

21. Let $f(x) = \lfloor x + a \rfloor$, with $0 \leq a < 1$. Then $f(x+j) = \lfloor x + j + a \rfloor = \lfloor x + a + j \rfloor = f(x+a)+j$, if j is an integer. Similarly, $\lfloor x + a \rfloor$ is either $\lfloor x \rfloor$ or $\lfloor x \rfloor + 1$; the latter is impossible if x is an integer. Hence $\left| \lfloor x + a \rfloor - x \right| < 1$.

22. Let S be a subset of $[0, 1)$. Then we may define $f(x)$ on $[0, 1)$ as follows: $f(x) = 0$ if $x \in S$, and $f(x) = 1$ otherwise. Now extend the domain of f to all of $\mathbb{R}$ using translation. It is easy to see that f is an integer function.

Conversely, let $f(x)$ be an integer function. Let $S = [0, 1) \cap f^{-1}[0]$. Thus we have given a 1–1 correspondence between subsets of $[0, 1)$ and integer functions.

23. We have $\hat{f}(x + j) = -f(-x - j) = -f(-x) + j = \hat{f}(x) - j$, so the translation property holds. Similarly $-1 < f(x) - x < 1$ for all x, so $-1 < f(-x) + x < 1$, and hence $-1 < \hat{f}(x) - x < 1$.

To show $f \neq \hat{f}$, note that $\hat{f}(1/2) = -f(-1/2) = -f(1/2) + 1$. If this were equal to $f(1/2)$, then $f(1/2) = 1/2$, a contradiction, since $\mathbb{Z}$ is the range of f.

24. Consider the case $u = 2^n$, $v = 1$. Then the binary gcd algorithm performs n divisions by 2 on input (u, v), forming the quotients $2^{n-1}, 2^{n-2}, \ldots, 1$. By Table 3.1, it costs $2k$ to divide 2^k by 2. Thus the total cost is $\Omega((\lg u)^2)$, which is not $O((\lg u)(\lg v))$.

25. Let $R(u, v)$ denote the number of reduction steps. We will show that $R(u, v) \leq \log_2(u + v)$. The proof is by induction on $u + v$. It is easy to verify the result is true if $u + v = 1$. To verify the induction step, if u and v are both even, we have $R(u, v) = R(u/2, v/2) \leq \log_2(u/2 + v/2) \leq \log_2(u + v)$. A similar inequality holds if u is even and v is odd, or if u is odd and v is even. Finally, if u and v are both odd, and $u \geq v$, then $R(u, v) = 1 + R((u - v)/2, v) \leq 1 + \log_2((u - v)/2 + v) = \log_2(u + v)$. A similar result holds when $u < v$. Thus we have shown $R(u, v) \leq \log_2(u + v)$; since $R(u, v)$ is an integer, it follows that $R(u, v) \leq \lfloor \log_2(u + v) \rfloor$. For a similar result, see Knuth [1981, Exercise 4.5.2.28, p. 598].

26. Observe that $u \geq 2^{n-1}$, for otherwise by the previous exercise we have $R(u, v) < n$. But u cannot equal 2^{n-1}, for then the first step performed is to divide u by 2, and subsequent processing cannot result in more than $n - 1$ reduction steps. Thus $u \geq 2^{n-1} + 1$. It is easy to see that $R(2^{n-1} + 1, 2^{n-1} - 1) = n$. It remains to see that if $n \geq 3$, then $R(2^{n-1} + 1, v) < n$ for all v with $1 \leq v < 2^{n-1} + 1$. The case where v is even is left to the reader. Otherwise, v is odd, and after one reduction step we are left with $((2^{n-1} + 1 - v)/2, v)$; by the lemma above we have $R((2^{n-1} + 1 - v)/2, v) \leq \log_2(2^{n-1} - 1) \leq n - 2$. Hence $R(2^{n-1} + 1, v) \leq n - 1$.

27. Consider the bound function $f(u, v) = u^2 + v^2$. Then note that f strictly decreases with each step of the algorithm. For $f(u/2, v) = (u/2)^2 + v^2 < f(u, v)$ if $u \neq 0$, and similarly for $f(u, v/2)$. Also,

$$f((u + v)/2, (u - v)/2) = (u^2 + v^2)/2 < f(u, v).$$

Thus eventually $u = 0$ or $v = 0$.

To prove that the algorithm is correct, we only need to see that each transformation preserves the gcd. This is certainly true for the shift steps (a) and (b). To see this for the reduction step (c), note that

$$(d \mid u) \text{ and } (d \mid v) \Longleftrightarrow d \mid ((u + v)/2) \text{ and } d \mid ((u - v)/2).$$

28. The estimate $P_r(u, v) = O(\lg uv)$ follows immediately from considering the bound function $f(u, v) = u^2 + v^2$. Each time step (a) or (b) is performed, f decreases, and each time step (c) is performed, f decreases by a factor of 2. Hence after at most $1 + \lg(u^2 + v^2)$ reduction steps, the algorithm terminates. The result follows upon noting that $2u^2v^2 \geq 2\max(u^2, v^2) \geq u^2 + v^2$.

To see that $P_s(u, v) = O((\lg uv)^2)$, think of the algorithm as performing a series of shift steps followed by a single reduction step, followed by more shift steps. Between any two reduction steps, we can perform at most $\lg u + \lg v$ shift steps. From the result in the paragraph above, there are at most $O(\lg uv)$ reduction steps done, and the result follows.

To see that there exist infinitely many inputs (u, v) with $P_s(u, v) = \Omega((\lg uv)^2)$, count the number of steps when $u = 2^n + 1$, $v = 1$.

Finally, we show that $P_s(u, v) = P_s(v, u)$ and $P_r(u, v) = P_r(v, u)$. We can prove both of these assertions by induction on $u^2 + v^2$. The base cases are left to the reader.

We have assumed that both u and v are odd. Hence the first step performed must be a reduction step. Thus (u, v) is replaced by $((u+v)/2, (u-v)/2)$. By considering u and v (mod 4), we see that exactly one of $(u + v)/2$, $(u - v)/2$ must be even. Without loss of generality, assume $(u + v)/2$ is even. Then a series of shift steps is performed, leading us to $((u + v)/2^k, u - v)$. Now another reduction step is performed, leading us to

$$(((1 + 2^{k-1})u + (1 - 2^{k-1})v)/2^k, ((1 - 2^{k-1})u + (1 + 2^{k-1})v)/2^k).$$

If we perform the same series of steps starting with (v, u), we end with the pair

$$(((1 - 2^{k-1})u + (1 + 2^{k-1})v)/2^k, ((1 + 2^{k-1})u + (1 - 2^{k-1})v)/2^k).$$

Now the induction hypothesis applies, and so the result follows.

This exercise and the previous one are based on the paper of G. Purdy [1983]. It is still open to determine the worst-case of this algorithm (i.e. determine the lexicographically least pair (u, v) such that the algorithm performs n steps on this input).

29. See Abbott et al. [1980].

30. See Knuth [1981, Exercise 4.5.2.35, p. 599] where the method is attributed to M. Penk; also see Norton [1985]; J. Gordon [1989].

31. Since $\gcd(a, b) = 1$, by using the Extended Euclidean algorithm, we can find, in polynomial time, integers r, s such that $ar + bs = 1$. Then $arn + bsn = n$. Let k be any integer. Then $a(rn - bk) + b(sn + ak) = n$. Since $n > ab$, $\frac{n}{ab} > 1$ and so we can choose k such that

$$\frac{-sn}{a} < k < \frac{-sn}{a} + \frac{n}{ab}.$$

Thus $-sn < ak < \frac{n}{b} - sn$. Thus $sn + ak > 0$ (so the coefficient of $b > 0$). Also $\frac{n}{b} - sn - ak > 0$; hence $\frac{n}{b}(1 - bs) - ak > 0$; and since $ar = 1 - bs$ we have $\frac{n}{b}ar - ak > 0$. Now multiply by $\frac{b}{a}$ to get $nr - bk > 0$, so the coefficient of a is also positive. This completes the proof, with $x = rn - bk$ and $y = sn + ak$.

32. Let $d_1, d_2, \ldots, d_k$ be the positive divisors of n. Then $ad_1, ad_2, \ldots, ad_k$ are distinct divisors of an. Thus $\sigma(an) \geq \sum_{1 \leq i \leq k} ad_k \geq a\sigma(n) > 2an$.

33. It is easily verified that 88 and 945 are abundant and $\gcd(88, 945) = 1$. Therefore, by Exercise 31, we know that every integer $> 83160 = 88 \cdot 945$ can be written as $88a + 945b$ for positive integers a, b. Using Exercise 32, we see that $88a$ and $945b$ are abundant. It remains to verify that every integer between 20162 and 83160, inclusive, can be so represented. This is easily done by computer. See the note of Moser [1949].

34. Using the same idea as in the previous exercise, every integer > 36 can be written as $4a + 9b$, which is the sum of two composite integers. It remains to check that every integer between 12 and 36, inclusive, can be written as the sum of two composite integers.

35–41. See Shallit [1986]. The best upper bound on $\alpha(r, s)$ is $O(r^{1/3+\epsilon})$; see Erdős and Shallit [1991].

42. First, some useful identities. It is easy to prove by induction on n that

$$\begin{pmatrix} 1 & 1 \\ 1 & 0 \end{pmatrix}^k = \begin{pmatrix} F_{k+1} & F_k \\ F_k & F_{k-1} \end{pmatrix}.$$

By taking determinants, we find $(-1)^k = F_{k+1}F_{k-1} - F_k{}^2$. Thus

$$\begin{aligned} (-1)^k &= F_{k+1}F_{k-1} - F_k{}^2 = F_{k+1}F_{k-1} - F_k(F_{k+1} - F_{k-1}) \\ &= F_{k-1}(F_{k+1} + F_k) - F_kF_{k+1} \\ &= F_{k-1}F_{k+2} - F_kF_{k+1}. \end{aligned} \tag{A.2}$$

Since

$$\begin{pmatrix} 1 & 1 \\ 1 & 0 \end{pmatrix}^k \begin{pmatrix} 1 & 1 \\ 1 & 0 \end{pmatrix}^k = \begin{pmatrix} 1 & 1 \\ 1 & 0 \end{pmatrix}^{2k},$$

we find

$$F_{2k} = F_k(F_{k-1} + F_{k+1}). \tag{A.3}$$

Now write $k = j + 1$, substitute in (A.2) and (A.3), and sum to get

$$F_{2j+2} = F_j(F_{j+1} + F_{j+3}) + (-1)^j. \tag{A.4}$$

Now prove by induction that

$$A_n = 2 + \frac{F_{2^n-2}}{F_{2^n}} \tag{A.5}$$

for $n \geq 1$, using equation (A.3).

Now from equation (A.5) we have

$$A_n = 2 + \frac{F_{2^n-2}}{F_{2^n}} = 2 + \cfrac{1}{2 + \cfrac{F_{2^n-3}}{F_{2^n-2}}} = \left[2, 2, \frac{F_{2^n-2}}{F_{2^n-3}}\right] = [2, 2, \overbrace{1, 1, \ldots, 1}^{2^n-3}],$$

by equation (A.2) and so finally

$$A_n = [2, 2, \overbrace{1, 1, \ldots, 1}^{2^n-5}, 2].$$

Now we know that $\frac{1+\sqrt{5}}{2} = [1, 1, 1, \ldots]$, so we conclude that

$$\lim_{n\to\infty} A_n = [2, 2, 1, 1, \ldots] = \left[2, 2, \frac{1+\sqrt{5}}{2}\right] = \frac{7-\sqrt{5}}{2}.$$

43–44. See Cesàro [1885]; Diaconis and Erdős [1977]; and Apostol and Ford [1991].

45. See Shallit [1979a, 1982].

46–47. See Floyd and Knuth [1990].

48. See Golomb [1973].

Chapter 5

1. Let $d = \operatorname{ord} a$. Then $a^d = 1$. Write $e = qd + r$, with $0 \leq r < d$. Then $a^r = a^{e-qd} = 1$. If $r \neq 0$, then this contradicts our assumption that $d = \operatorname{ord} a$. Hence $r = 0$ and $d \mid e$.
2. By the extended Euclidean algorithm, we know there exist integers b, c such that $be + cf = d$. Then $a^d = a^{be+cf} = 1$.
3. The two different iterative algorithms arise from deciding to compute the binary expansion from n from left to right, or right to left.

The first computes the binary expansion from left to right:

```
POWER2(a, e, n)
t ← 1
f ← 1
while (e ≥ 2f) do
        f ← 2f
while (e ≠ 0) do
        t ← t² mod n
        if e ≥ f then
                t ← at mod n
                e ← e − f
        f ← f/2
return(t)
```

The second computes the binary expansion from right to left:

```
POWER3(a, e, n)
t ← 1
f ← a
while (e ≠ 0) do
        if (e mod 2 = 1)
          then t ← tf mod n
        e ← ⌊e/2⌋
        f ← f² mod n
return(t)
```

4. We can determine if the system S has a solution by checking if $x_i \equiv x_j \pmod{\gcd(m_i, m_j)}$ for $1 \le i < j \le k$. This can be done in $O((\lg m)^2)$ bit operations.

To find the solution, we can use Gauss's method described in the proof to Theorem 5.5.5. First, we combine the first two congruences, obtaining a single congruence. This can be done in $O((\lg m_1)(\lg m_2))$ bit operations. Then we combine this new congruence with the third congruence, which can be done in $O((\lg m_1 m_2)(\lg m_3))$ bit operations, and so forth. The total cost is $O((\lg m)(\lg m_2 + \lg m_3 + \cdots + \lg m_k))$, and since each $m_i \ge 2$, this is $O((\lg m)^2)$.

5. The obvious method of computing $(x^n - 1)(x - 1)^{-1} \pmod m$ does not work if $\gcd(x - 1, m) > 1$, since then $x - 1$ is not invertible mod m. However, provided $x \ne 1$, we can compute $r = (x^n - 1) \bmod m(x - 1)$. Then $x - 1 \mid r$, so $r/(x - 1)$ gives the desired result. If $x = 1$, then the result is $n \bmod m$.

Another approach is to compute, using the power algorithm, the n-th power of the matrix

$$\begin{bmatrix} 1 & 1 \\ 0 & x \end{bmatrix},$$

modulo m. Then the entry in the upper right gives the desired result.

Both methods use $O((\lg n)(\lg m)^2)$ bit operations if $x = O(m)$.

6. It is easy to prove by induction on n that $X_n = a^n X_0 + (a^{n-1} + a^{n-2} + \cdots + 1)c$. Also, if

$$M = \begin{bmatrix} a & c \\ 1 & 0 \end{bmatrix},$$

then

$$M^n = \begin{bmatrix} a^n & (a^{n-1} + \cdots + 1)c \\ 0 & 1 \end{bmatrix}.$$

Thus $X_n = M^n \begin{bmatrix} X_0 \\ 1 \end{bmatrix}$, and this can be computed, modulo M (using the power algorithm) using $O((\lg n)(\lg M)^2)$ bit operations.

7. With a little thought, it is easy to see that the digit in question is

$$x = \left\lfloor \frac{b^n}{a} \right\rfloor - b \left\lfloor \frac{b^{n-1}}{a} \right\rfloor.$$

Now write $b^{n-1} = qa + r$, $0 \le r < a$, so that $r = b^{n-1} \bmod a$. Then $b^n = qab + rb$, so $\left\lfloor \frac{b^n}{a} \right\rfloor = qb + \left\lfloor \frac{rb}{a} \right\rfloor$ and $\left\lfloor \frac{b^{n-1}}{a} \right\rfloor = q + \left\lfloor \frac{r}{a} \right\rfloor = q$.

Then

$$x = qb + \left\lfloor \frac{rb}{a} \right\rfloor - qb = \left\lfloor \frac{rb}{a} \right\rfloor,$$

and, using the fact that $u \bmod v = u - v \lfloor u/v \rfloor$, we get

$$\begin{aligned} \left\lfloor \frac{rb}{a} \right\rfloor &= \frac{rb - (rb \bmod a)}{a} \\ &= \frac{b \cdot (b^{n-1} \bmod a) - b^n \bmod a}{a}. \end{aligned}$$

8. Let the prime factorization of $p - 1$ be $q_1^{e_1} q_2^{e_2} \cdots q_k^{e_k}$. We use the following observation: if $x^{(p-1)/q_i^{f_i}} \equiv 1 \pmod p$, and $f_i = e_i$ or $x^{(p-1)/q_i^{f_i+1}} \not\equiv 1 \pmod p$, then $q^{e_i - f_i} \,\|\, \mathrm{ord}_p\, x$. (This follows by combining Exercises 5.1 and 2.10.) Hence it suffices to find, for each i, the exponent f_i such that the condition above holds.

This can be done as follows: first compute $q_1^{e_1}, q_2^{e_2}, \ldots, q_k^{e_k}$. This can be done using $O((\lg p)^2)$ bit operations. Next, compute $y_1 = (p-1)/q_1^{e_1}, \ldots, y_k = (p-1)/q_k^{e_k}$. This can also be done in $O((\lg p)^2)$ bit operations. Now, using the binary method, compute $x_1 \equiv a^{y_1} \pmod p, \ldots, x_k \equiv a^{y_k} \pmod p$. This can be done in $O(k(\lg p)^3)$ bit operations, and $k = O((\lg p)/(\lg\lg p))$ by Theorem 8.8.10. Finally, for each i, repeatedly raise x_i to the q_i'th power (mod p) (as many as $e_i - 1$ times), checking to see when 1 is obtained. This can be done in $O((\lg p)^3)$ steps. The total cost is dominated by $O(k(\lg p)^3)$, which is $O((\lg p)^4/(\lg\lg p))$.

9. All arithmetic is done in $\mathbb{Z}/(m)$. Let

$$M = \begin{bmatrix} 0 & 1 & 0 & \cdots & 0 & 0 \\ 0 & 0 & 1 & \cdots & 0 & 0 \\ \vdots & \vdots & \vdots & \ddots & 1 & 0 \\ 0 & 0 & 0 & \cdots & 0 & 1 \\ c_k & c_{k-1} & c_{k-2} & \cdots & c_2 & c_1 \end{bmatrix}.$$

Then

$$\begin{bmatrix} A_{n-k+1} \\ A_{n-k+2} \\ \vdots \\ A_{n-1} \\ A_n \end{bmatrix} = M \begin{bmatrix} A_{n-k} \\ A_{n-k+1} \\ \vdots \\ A_{n-2} \\ A_n \end{bmatrix}.$$

It follows that

$$\begin{bmatrix} A_n \\ \vdots \\ A_{n+k-1} \end{bmatrix} = M^n \begin{bmatrix} A_0 \\ \vdots \\ A_{k-1} \end{bmatrix}.$$

Now, we can compute M^n using $O(k^3(\lg n)(\lg m)^2)$ bit operations using the binary method. This method seems to have been first suggested by J. Miller and Spencer Brown [1966]. The time can be improved to $O(k^2(\lg n)(\lg m)^2)$ by the following observation: the matrix M is a companion matrix corresponding to the characteristic polynomial $f(X) = X^k - a_1X^{k-1} - \cdots - a_{k-1}X - a_k$. We can then compute $r(X) = X^n \pmod{f(X), m}$ in $O(k^2(\lg n)(\lg m)^2)$ bit operations using the binary method. Then $M^n \equiv r(M)$. This method was first suggested by Fiduccia [1985].

10. When we use repeated multiplication, we multiply a times a^j for $1 \le j \le e-1$, at a cost of $\sum_{1\le j\le e-1}(\log a)(\log a^j) = \frac{1}{2}e(e-1)(\log a)^2$ steps (using the stated convention).

11. Let $T(e)(\log a)^2$ denote the number of steps used to calculate a^e using the binary method. Then $T(1) = 0$, and $T(2) = 1$. Also, $T(2e) = T(e) + e^2$, since we first calculate a^e and then square it. Similarly, $T(2e+1) = T(e)+e^2+2e$, since we first calculate a^e, then square it to get a^{2e}, and then multiply this by a. It is now easy to prove by induction on e that $(e^2-1)/3 \le T(e) < (e^2+2e)/3$. Hence the binary method uses $(e^2/3 + O(e))(\log a)^2$ steps.

Note that the sequence $(T(e))$ is a "2–regular" sequence, as studied by Allouche and Shallit [1992].

12. We prove the result by induction on n. Define

$$R(n) = \left(\sum_{1\le k\le n} \text{oddpart}(k)\right) - s_2(n).$$

We will show $R(n) = T(n)$. It is clearly true for $n = 1$. Now assume the result is true for all $k < 2n$; we prove it for $2n$ and $2n+1$:

$$\begin{aligned}
R(2n) &= \left(\sum_{1 \le k \le 2n} \frac{k}{2^{\nu_2(k)}} \right) - s_2(2n) \\
&= \left(\sum_{1 \le k \le n} \frac{2k}{2^{\nu_2(2k)}} \right) + \left(\sum_{1 \le k \le n} \frac{2k-1}{2^{\nu_2(2k-1)}} \right) - s_2(n) \\
&= \left(\sum_{1 \le k \le n} \frac{k}{2^{\nu_2(k)}} \right) + \left(\sum_{1 \le k \le n} 2k - 1 \right) - s_2(n) \\
&= T(n) + n^2 = T(2n).
\end{aligned}$$

Also,

$$\begin{aligned}
R(2n+1) &= \left(\sum_{1 \le k \le 2n+1} \frac{k}{2^{\nu_2(k)}} \right) - s_2(2n+1) \\
&= R(2n) + s_2(2n) + 2n + 1 - s_2(2n+1) \\
&= T(2n) + 2n = T(2n+1).
\end{aligned}$$

This exercise is based on an observation of J.-P. Allouche (personal communication). For an asymptotic evaluation of $R(n)$, see Prodinger and Tichy [1983].

13. See R. Graham, Yao, and Yao [1978].

14. We prove the result by induction on d. It is clear for $d = 1$. Now assume it is true for all polynomials of degree $< d$. If $f(X) = 0$ has no roots in the field k, we are done. Hence assume α is a root. Then we may write $f(X) = (X - \alpha)q(X) + r$, where r is constant. By setting $X = \alpha$, we see that in fact $r = 0$. Hence $\deg q = d - 1$, and we may apply induction to conclude that $q(X) = 0$ has at most $d - 1$ roots. But any root of $f(X) = 0$ that is not equal to α must be a root of $q(X) = 0$, which proves that $f(X) = 0$ has at most d roots.

15. We claim that $X^{p-1} - 1 = (X-1)(X-2)\cdots(X-(p-1))$ in $\mathbb{F}_p[X]$. For define $f(X) = X^{p-1} - 1 - (X-1)(X-2)\cdots(X-(p-1))$; then $\deg f < p - 1$ and by Fermat's theorem, $f(X) = 0$ has the $p - 1$ roots $1, 2, \ldots, p-1$. By the previous exercise, f must be identically 0.

16. Let $g(X) = (X^{p-1} - 1)/(X^d - 1)$; it is easily verified that $g(X)$ is a polynomial of degree $p - 1 - d$ with integer coefficients. Then in $\mathbb{F}_p[X]$ we have $X^{p-1} - 1 = (X^d - 1)g(X)$; if $X^d - 1$ had fewer than d zeroes, then by Exercise 14, $X^{p-1} - 1$ would have fewer than $p - 1$ zeroes. But this would contradict Exercise 15.

17. Clearly $(ab)^{mn} = a^{mn}b^{mn} = 1$ (where 1 denotes the group identity). As in Exercise 1, if $d = \operatorname{ord} ab$, then $d \mid mn$. If $d \ne mn$, then there exists a prime $p \mid mn$ such that $(ab)^{mn/p} = 1$. Since $\gcd(m, n) = 1$, it must be the case that p divides exactly one of m, n. Without loss of generality, assume it divides m and not n. Then $1 = (ab)^{mn/p} = a^{mn/p}(b^n)^{m/p} = a^{mn/p}$. But then by Exercise 2, $a^{\gcd(mn/p, m)} = 1$, and so $a^{m/p} = 1$, a contradiction. Thus $d = mn$.

18. Let $p - 1 = \prod_{p^i}^{e_i}$. Then by Exercise 16, $X^{p_i^{e_i}} = 1$ must have $p_i^{e_i}$ solutions over $\mathbb{F}_p$, while $X^{p_i^{e_i - 1}} = 1$ must have $p_i^{e_i - 1}$ solutions. Also, every root of the second equation is a root of the first. Let g_i be a root of the first equation that is not a root of the second; clearly g_i has order $p_i^{e_i}$. Let $g = \prod_i g_i$; by Exercise 17, it has order $p - 1$.

Our solutions to exercises 14–18 are based on the treatment in Ireland and Rosen [1990].

19. It is easy to prove by induction on e, using the binomial theorem, that $(1 + ap)^{p^{e-2}} \equiv 1 + ap^{e-1} \pmod{p^e}$ for all a. Raising both sides to the p power, we see that $(1 + ap)^{p^{e-1}} \equiv 1 \pmod{p^e}$; hence $1 + ap$ has order dividing p^{e-1}, but it is not p^{e-2}. Hence it must be p^{e-1}.

20. Since $g^{p-1} \equiv 1 \pmod{p}$, we know that $g^{p-1} \equiv 1 + ap \pmod{p^e}$, where a is an integer mod p^{e-1}.

Suppose p doesn't divide a. Then by the previous exercise, $1 + ap$ has order p^{e-1} in $(\mathbb{Z}/(p^e))^*$. Thus g has order $p^{e-1}(p - 1)$ and so is a generator $\pmod{p^e}$.

On the other hand, if p does divide a, we show that $g + p$ is a generator for $(\mathbb{Z}/(p^e))^*$. We see that $(g + p)^{p-1} \equiv g^{p-1} + (p - 1)g^{p-2}p \equiv 1 + (p - 1)g^{p-2}p \equiv 1 + bp \pmod{p^2}$, and p doesn't divide b. Hence the previous case applies.

21. It is easy to see that 1 is a primitive root (mod 2) and 3 is a primitive root (mod 4). We now assume $e \geq 3$. Then it is easy to prove by induction on e that $5^{2^{e-3}} \equiv 1 + 2^{e-1} \pmod{2^e}$. From this, it follows that $5^{2^{e-2}} \equiv 1 \pmod{2^e}$ and hence 5 has order 2^{e-2} in $(\mathbb{Z}/(2^e))^*$. Finally, show that the 2^{e-1} numbers of the form $(-1)^a 5^b$ with $0 \leq a < 2$ and $0 \leq b < 2^{e-2}$ are all distinct $\pmod{2^e}$. Thus for $e > 2$, $(\mathbb{Z}/(2^e))^*$ is the direct product of two cyclic groups, one of order 2 (generated by -1) and one of order 2^{e-2} (generated by 5).

22. The first few examples (g, p) are as follows: (14, 29), (18, 37), (19, 43), (11, 71), (43, 103).

23. By Theorem 5.6.3, we need only consider the cases $n = 2, 4, p^e, 2p^e$, where p is an odd prime and $e \geq 1$. The cases $n = 2, 4$ are left to the reader. We know that $(\mathbb{Z}/(p^e))^*$ is a cyclic group of order $m = p^{e-1}(p - 1)$; hence it has $\varphi(m)$ primitive roots, corresponding to a^f, where a is any primitive root, and f ranges over the residue classes prime to m. A similar argument applies to the case $n = 2p^e$.

24. See S. Collins, Reddy and Sloane [1978].

25. See Ecker [1979].

26. Let $d = \text{lcm}(a, b)$. Then $x^d \equiv 1 \pmod{m}$, and $x^d \equiv 1 \pmod{n}$. Hence $x^d \equiv 1 \pmod{mn}$, which shows that $(\text{ord}_{mn}\, x) \mid d$. Suppose $x^{d/k} \equiv 1 \pmod{mn}$ for some integer $k > 1$. Then at least one of a, b does not divide d/k; without loss of generality it is a. Then, by exercise 1 we have $x^{\gcd(a,d/k)} \equiv 1 \pmod{m}$. But $1 \leq \gcd(a, d/k) < a$, a contradiction.

27. Let $m = n/\gcd(n, e)$. Then $r^m = (g^n)^{e/\gcd(n,e)} = 1$. Hence $(\text{ord}\, r) \mid m$. Suppose $r^{m/k} = 1$ for some positive divisor k of m. Then $g^f = 1$, where $f = \gcd(em/k, n)$. But $f = n/k$, so k must be 1.

28. Since G is cyclic, it has a generator g. Then for each i, there exists b_i such that $a_i = g^{b_i}$. Then any element in $H = \langle a_1, \ldots, a_k \rangle$ can be written as $g^{\sum_i e_i b_i}$ for appropriate integers e_i, and furthermore any element of this form is in H. Hence H is generated by $g^{\gcd(b_1, b_2, \ldots, b_k)}$. Let $b = \gcd(b_1, \ldots, b_k)$; then by the previous exercise H has order $s := n/\gcd(b, n)$. On the other hand, we have $t := \text{ord}a_i = n/\gcd(n, b_i)$. By Exercise 2.5, we know that $s = t$.

A counterexample when G is not cyclic is the additive group $\mathbb{Z}/(2) \times \mathbb{Z}/(2)$. Here $(1, 0)$ and $(0, 1)$ both have order 2, but the order of the group they generate is 4.

29. Using exercise 27, let g be a generator for $(\mathbb{Z}/(p^e))^*$. Then since this group is cyclic, $a \equiv g^t$ for some t, $0 \le t < p^{e-1}(p-1)$. Using the previous exercise, $\text{ord}_{p^e}\, a = p^{e-1}(p-1)/\gcd(t, p^{e-1}(p-1))$, and $\text{ord}_p\, a = (p-1)/\gcd(t, p-1)$. Let $t = p^r d$, where $p \nmid d$. Then $(\text{ord}_{p^e}\, a)/(\text{ord}_p\, a) = p^{e-1-r}$.

30. Let the least residues of $a, 2a, 3a, \ldots, ((p-1)/2)a \pmod p$ be $r_1, r_2, \ldots, r_\lambda$ and $-s_1, -s_2, \ldots, -s_\mu$, where $0 < r_i, s_i < p/2$ for all i. Note that all the r_i are distinct, and so are all the s_i. We claim $r_i \ne s_j$ for all i, j. Then there exist $1 \le b, c \le (p-1)/2$ such that $ba \equiv r_i \pmod p$ and $ca \equiv -s_j \pmod p$. Then $ba + ca \equiv 0 \pmod p$; hence $b + c \equiv 0 \pmod p$. But the inequalities on b, c make this impossible. Hence $r_1, \ldots, r_\lambda, s_1, \ldots, s_\mu$ must be a permutation of the numbers $1, 2, \ldots, (p-1)/2$. Therefore,

$$\prod_{1\le k\le (p-1)/2} ka = (-1)^\mu \prod_{1\le k\le (p-1)/2} k \pmod p,$$

and so $a^{(p-1)/2} \equiv (-1)^\mu \pmod p$.

31. Let $a = 2$, and apply the previous exercise. Then it is easy to see that $\mu = (p+1)/4$ if $p \equiv 3 \pmod 4$, and $\mu = (p-1)/4$ if $p \equiv 1 \pmod 4$. Since $\left(\frac{2}{p}\right) \equiv (-1)^\mu \pmod p$, we have $\left(\frac{2}{p}\right) \equiv 1$ if $p \equiv 1, 7 \pmod 8$, and $\left(\frac{2}{p}\right) \equiv -1$ if $p \equiv 3, 5 \pmod 8$.

32. Let $a = p_1 p_2 \cdots p_k$, and $n = q_1 q_2 \cdots q_\ell$, where the p's are (not necessarily distinct) primes, and the same for the q's. Then

$$\begin{aligned}\left(\frac{a}{n}\right)\left(\frac{n}{a}\right) &= \prod_{\substack{1\le i\le k\\ 1\le j\le \ell}} \left(\frac{p_i}{q_j}\right)\left(\frac{q_j}{p_i}\right)\\ &= (-1)^{(\sum_{1\le i\le k} \frac{p_i-1}{2})(\sum_{1\le j\le \ell} \frac{q_j-1}{2})}\\ &= (-1)^{\frac{a-1}{2}\cdot\frac{n-1}{2}},\end{aligned}$$

where in the last line we have used Eq. (5.5).

33. Using the hint, consider the number $4n^2 + 1$, where $n = 2p_1p_2\cdots p_k$. Now n is odd; let p be an odd prime dividing $4n^2+1$; then we have $(2n)^2 \equiv -1 \pmod p$, and hence $\left(\frac{-1}{p}\right) = 1$; thus $p \equiv 1 \pmod 4$. But p cannot be equal to $p_1, p_2, \ldots, p_k$, since $4n^2$ leaves a remainder of 1 when divided by any of these primes. Thus $4n^2 + 1$ is divisible by a prime of the form $4j + 1$ not on the list, a contradiction.

34. Follow the previous exercise, using the quadratic character of 2 instead of -1.

35. Suppose $x^2 \equiv a \pmod n$. Then $x^2 \equiv a \pmod{2^e}$. Case (a) is clear. Suppose $e \ge 3$. Then by the solution to Exercise 5.21, we can write $x \equiv (-1)^\alpha 5^\beta \pmod{2^e}$. Thus $a \equiv x^2 \equiv 25^\beta \pmod{2^e}$. It follows that $a \equiv 1 \pmod 8$. To prove (c), note that if if $x^2 \equiv a \pmod n$, then $x^2 \equiv a \pmod{p_i}$ for $1 \le i \le k$. It then follows by Euler's criterion that $a^{p_i - 1} \equiv 1 \pmod{p_i}$.

To prove the converse, assume conditions (a)–(c) are satisfied. We must show the congruence $x^2 \equiv a \pmod n$ is solvable. Using the Chinese remainder theorem, it suffices to prove that the following $k+1$ congruences are solvable: $x^2 \equiv a \pmod{2^e}$ and $x^2 \equiv a \pmod{p_i^{e_i}}$ for $1 \le i \le k$. If $e = 2$, then $a \equiv 1 \pmod 4$. In this case, we may take $x = 1$. Suppose $e \ge 3$. We show by induction on m that if $a \equiv 1 \pmod 8$, then $x^2 \equiv a \pmod{2^m}$ is solvable for all $m \ge 3$. The base case, $m = 3$, is clear. Now assume x_0 is a solution to $x^2 \equiv a \pmod{2^m}$; we will show it how to determine a solution x_1 to $x^2 \equiv a \pmod{2^{m+1}}$. We do this by setting $x_1 = x_0 + b2^{m-1}$. Substituting, we find $x_1^2 \equiv x_0^2 + 2^m b x_0 \pmod{2^{m+1}}$; here is where $m \ge 3$ is used. In order that $x_1^2 \equiv a \pmod{2^{m+1}}$, we must have $(x_0^2 - a)/2^m \equiv bx_0 \pmod 2$. We can now solve for b. A similar technique can be used to show by induction on e_i that if $x^2 \equiv a \pmod{p_i^{e_i}}$ is solvable, then so is $x^2 \equiv a \pmod{p_i^{e_i+1}}$. (Hint: set $x_1 = x_0 + bp_i^{e_i}$.)

Our solution is based on Ireland and Rosen [1990, Prop. 5.1.1].

36. See Ireland and Rosen [1990, p. 57].

37. Using the binomial theorem, we have

$$(a+b)^p = a^p + \binom{p}{1}a^{p-1}b + \binom{p}{2}a^{p-2}b^2 + \cdots + \binom{p}{p-1}ab^{p-1} + b^p.$$

Now use Exercise 12. The result for more than two summands is easily proved by induction.

38. Using the hint, we want 2^a3^b to be a cube, $2^{a+1}3^b$ to be a fourth power, and 2^a3^{b+1} to be a fifth power. Hence we must solve the systems of congruences $a \equiv 0 \pmod 3$, $a+1 \equiv 0 \pmod 4$, $a \equiv 0 \pmod 5$, and $b \equiv 0 \pmod 3$, $b \equiv 0 \pmod 4$, and $b+1 \equiv 0 \pmod 5$. We get $a \equiv 15 \pmod{60}$ and $b \equiv 24 \pmod{60}$.

39. See Artin and Whaples [1945]; Watrous [1995].

40. See Leung and Whitehead [1982]; Baruah, Rosier, and Howell [1990]; Baruah, Howell, and Rosier [1993].

41. It is clear that if $M_A = M_B$, then $M_A \equiv M_B \pmod{p_i}$. We now show that if $M_A \ne M_B$, then $M_A \equiv M_B \pmod{p_i}$ with probability $t < 1/2$. Let P be the set of the first $2n$ primes. Let $S = \{p \in P : M_A \equiv M_B \pmod p\}$. Assume $|S| \ge n$. Then $M_A \equiv M_B \pmod g$, where $g = \prod_{p\in S} p \ge 2^{|S|} \ge 2^n$. It follows that $M_A = M_B$, a contradiction. Hence $|S| < n$, and $t = |S|/|P| < 1/2$, as desired.

42. See Karp and Rabin [1987] and Tompa [1991].

43. It suffices to prove the results for q prime. We have $2^p \equiv 1 \pmod q$ and by Fermat's theorem, $2^{q-1} \equiv 1 \pmod q$. Hence, by Exercise 1, we have $2^{\gcd(p,q-1)} \equiv 1 \pmod q$. Since p is a prime, $\gcd(p, q-1)$ equals 1 or p. It cannot equal 1, since then we would have $2^1 \equiv 1 \pmod q$. Hence $p \mid q-1$, or $q \equiv 1 \pmod p$.

To prove (b), note that

$$1 = \left(\frac{1}{q}\right) = \left(\frac{2^p}{q}\right) = \left(\frac{2}{q}\right)^p = \left(\frac{2}{q}\right),$$

so $q \equiv \pm 1 \pmod 8$.

To prove $n = 2^{13} - 1$ is prime, we need only trial divide up to $\sqrt{n} \doteq 90.5$. But the exercise says any such divisor must be congruent to 1 or 79 (mod 104); since $79 \nmid n$, we conclude n is prime. A similar argument works for $n = 2^{17} - 1$, where we only need to trial divide by primes congruent to 1 or 103 (mod 136).

We also can use the information to quickly find the following prime divisors: 47 of $2^{23} - 1$; 233 of $2^{29} - 1$; 223 of $2^{37} - 1$; and 431 of $2^{43} - 1$.

44. It suffices to prove the result for q prime. We have $2^{2^k} \equiv -1 \pmod{q}$, so $2^{2^{k+1}} \equiv 1 \pmod{q}$. Hence 2 is of order 2^{k+1}, modulo q, and thus $2^{k+1} \mid q - 1$. Thus $q \equiv 1 \pmod{2^{k+1}}$. For $k \geq 2$, this implies that 2 is a quadratic residue, modulo q. Hence there exists $a \in (\mathbb{Z}/(q))^*$ such that $a^2 \equiv 2 \pmod{q}$. Now a is an element of order 2^{k+2}, so $2^{k+2} \mid q - 1$.

45. This exercise is essentially a combination of the proofs of Zolotarev [1872] and Swan [1962]. It was shown to one of the authors by J. Tunnell.

46. This exercise is based on the algorithm of Eisenstein [1844a].

47. This exercise is based on the algorithm of Lebesgue [1847].

48. This exercise is based on the algorithm of H. Williams [1980].

For the solution to exercises 46–48, see Shallit [1990].

49. See van Lint [1973].

50. See Blakley and Borosh [1983].

51. One possible implementation is as follows:

```
BINARY JACOBI(a, n)
{a ≥ 0; n > 0; n odd}
t ← 1
while (a ≠ 0) do
        while (a mod 2 = 0) do
            a ← a/2
            if (n mod 8 = 3) or (n mod 8 = 5) then t ← −t
        if (a < n) then
            interchange(a, n)
            if (a mod 4 = 3) and (n mod 4 = 3) then t ← −t
(*)     a ← (a − n)/2
        if (n mod 8 = 3) or (n mod 8 = 5) then t ← −t
if (n = 1) then return(t) else return(0)
```

See, for example, Shallit and Sorenson [1993].

52. Define $b' = \lfloor b/2 \rfloor$ for any integer b. Define $\psi(m, n) = \sum_{1 \leq k \leq n'} \lfloor km/n \rfloor$, and let $\mu(m, n)$ be the function defined by

$$\left(\frac{m}{n}\right) = (-1)^{\mu(m,n)}.$$

First show that, modulo 2, we have

$$\mu(m,n) = \begin{cases} \psi(m,n), & \text{if } m \text{ is odd;} \\ \psi(m,n), & \text{if } m \text{ is even and } n \equiv 1, 7 \pmod 8; \\ \psi(m,n)+1, & \text{if } m \text{ is even and } n \equiv 3, 5 \pmod 8). \end{cases}$$

For example, this can be found in Bachmann [1902].

Next, show that

$$\psi(m,n) + \psi(n,m) = m'n'.$$

This can be done by counting the lattice points (x, y) in $m\mathbb{Z} + n\mathbb{Z}$ contained in the rectangle defined by $0 < x \leq n'm$, $0 < y \leq m'n$.

Now suppose $m = qn + r$, with $0 \leq r < n$. Then $m/n = q + r/n$; $2m/n = 2q + 2r/n; \ldots,$ $n'm/n = n'q + n'r/n$. Adding these up, and applying the result above, we get

$$\psi(m,n) = \binom{n'+1}{2} q + n'r' - \psi(n,r). \tag{A.6}$$

Now assume $u, v > 0$, with v odd. Expand u/v as a continued fraction, obtaining $u_0 = a_0 u_1 + u_2$; $u_1 = a_1 u_2 + u_3; \ldots, u_{n-1} = a_{n-1} u_n + u_{n+1}$, where $u_0 = u$, $u_1 = v$, and $u_{n+1} = 0$. If $u_n \neq 1$, then $\gcd(u, v) > 1$, and so $\left(\frac{u}{v}\right) = 0$. Otherwise, use Eq. (A.6) and compute

$$\psi = \sum_{0 \leq i \leq n-1} a_i \binom{u'_{i+1}+1}{2} + \sum_{0 \leq i \leq n-1} (u'_{i+1} u'_{i+2} \bmod 2).$$

Then $\left(\frac{u}{v}\right) = (-1)^{\psi}$ if u is odd or $v \equiv \pm 1 \pmod 8$, and $\left(\frac{u}{v}\right) = (-1)^{\psi+1}$ otherwise.

Using Schönhage's rapid method for computing continued fractions, it follows that $\left(\frac{u}{v}\right)$ can be computed in the stated time bound.

This complexity bound is part of the "folklore" and apparently has never appeared in print. The basic idea can be found in Gauss [1876]. Our presentation is based on that in Bachmann [1902]. H. W. Lenstra, Jr. also informed us of this idea; he attributes it to A. Schönhage.

53. See Nagasaka, Shiue and Ho [1991].

Chapter 6

1. See, e.g., Lang [1965].

2. Choose a nonzero element a. By the pigeonhole principle, there must be positive numbers $i < j$ with $a^i = a^j$. Therefore $a^{j-i} = 1$, so a is invertible.

3. See Albert [1961, p. 124].

4. See van der Waerden [1970, pp. 54–60].

5. See Bach, Driscoll, and Shallit [1990, 1993]. The latter paper gives references to previous work on gcd-free bases for polynomials.

6. **(a)** This is similar to the integer multiplication algorithm of Karatsuba and Ofman [1962]. We may as well assume n is a power of 2. Write $f = f_0 + f_1X^{n/2}$ and $g = g_0 + g_1X^{n/2}$, where f_0, f_1, g_0, g_1 have degree $< n/2$. Then recursively compute $a = (f_0 + g_0)(f_1 + g_1)$, $b = f_0g_0$, and $c = f_1g_1$. The product is

$$fg = b + (a - b - c)X^{n/2} + cX^n.$$

Let $M(n)$ and $A(n)$ denote the number of multiplications and addition/subtractions in k used by this method. We have

$$M(n) = 3M(n/2); \qquad M(1) = 1;$$

$$A(n) = 2M(n/2) + O(n); \qquad A(1) = 0.$$

From this, we find $M(n) = 3^{\log_2 n} = n^{\log_2 3}$ and $A(n) = O(n \lg n)$.

(b) Let h be a polynomial in $\mathbb{F}_p[Z]$, of degree $< d$ and divisible by Z. Using the Newton iteration $\xi' = 2\xi - (1 - h)\xi^2$ and the result of (a), it is possible to find $(1 - h)^{-1} \bmod Z^d$ with $O(d^{\log_2 3})$ bit operations. (Start with $1 + h \bmod Z^2$, and compute further iterates mod Z^4, Z^8, and so on.) Now, assume that $n = \deg f \geq m = \deg g$. To divide f by g within our time bound it is enough to find $\lfloor f/g \rfloor$. We compute an approximate inverse of g (in the power series ring $k((1/X))$), as follows. Let

$$g(X) = X^m - c_1X^{m-1} - \cdots - c_m = X^m(1 - h(Z)),$$

where $Z = 1/X$. Find a polynomial ℓ for which $\ell(Z)(1 - h(Z)) \equiv 1 \pmod{Z^{n-m+1}}$, as outlined above; then $\lfloor f/g \rfloor = f(X)X^{-m}\ell(X^{-1})$.

(c) If $q = p$ (a prime), the result follows from (a) and Exercise 3.10. Otherwise, we use (a) and (b) to implement multiplication in $\mathbb{F}_q$.

Under mild assumptions, the proof of (b) can be extended to show that the complexity of polynomial division (measured in field operations) is within a constant factor of the complexity of polynomial multiplication. (See Aho, Hopcroft, and Ullman [1974].) No such result is known for the greatest common divisor.

7. We suppose that $\mathbb{F}_q = \mathbb{F}_p(\alpha)$, where α is a zero of the irreducible polynomial $X^2 - AX + B$. Then for $u, v \in \mathbb{F}_p$, $\overline{u\alpha + v} = -u\alpha + (uA + v)$, $N(u\alpha + v) = u^2B + v^2 + uvA$, and (as in high school) $x^{-1} = \bar{x}/N(x)$. This reduces inversion in $\mathbb{F}_q$ to $O(1)$ operations in $\mathbb{F}_p$.

8. The hard part is showing that every nonzero element has an inverse. Evidently X is invertible, as are nonzero constants, so we are reduced to finding the inverse of $1 + a_1/X + a_2/X^2 + \cdots$. This has the form $1 + \sum_{i \geq 1} b_i/X^i$, and we can obtain $b_1, \ldots, b_n$ using $O(n^2)$ field operations with the formulas

$$b_1 = -a_1,$$

$$b_2 = -(a_2 + a_1b_1),$$

$$b_3 = -(a_3 + a_2b_1 + a_1b_2),$$

and so on. We note that $k((1/X)) \cong k((X))$.

9. See Artin [1924a].

10. See van der Waerden [1970, pp. 131–138].

11. This algorithm is from Ranum [1911]. The running time analysis was pointed out to us by Gary Shute. For related methods and more references, see Cerlienco and Piras [1983].

12. **(a)** The upper bound is given by Theorem 6.5.1. We can verify the lower bound directly for $n \le 3$, by explicitly calculating $I_q(n)$. If $n \ge 4$, then

$$I_q(n) \ge (1 - 2q^{-n/2})/n \ge (1 - 2q^{-2})/n \ge 1/2n.$$

So the bounds hold for all n. (We note that that $I_q(n) = \Theta(q^n/n)$, uniformly in q and n.) The second result follows by taking logarithms.

(b) The length of $I_q(n)$ is about $n \log_2 q$, which is not bounded by a polynomial in $\lg n$ and $\lg q$. We sketch an algorithm that will compute $I_q(n)$ using $O(n^2(\lg q)^2)$ bit operations. First, use trial division to find the prime divisors of n. Combine these to get n's squarefree divisors (there are $2^{\omega(n)}$ of these, which is $o(n)$ by the prime number theorem). Then, compute $q^{n/d}$ for every squarefree divisor d of n and form $I_q(n)$ by adding and subtracting these powers and dividing by n, as specified by Theorem 6.5.1. The bulk of the work is in computing the powers of q. By Exercise 3.3, this uses $\sum_{d \mid n} \mu^2(d)(n/d)^2(\lg q)^2$ bit operations. Since

$$\sum_{d \mid n} \frac{\mu^2(d)}{d^2} = \prod_{p \mid n} \left(1 + \frac{1}{p^2}\right) \le \prod_{p} \left(1 - \frac{1}{p^2}\right)^{-1} = \zeta(2),$$

this is $O(n^2(\lg q)^2)$.

Our analysis assumed an arbitrary base. We note that $I_q(n)$ is much easier to compute in base q. For example, when $q = 2$, the problem is essentially one of converting out of a redundant base 2 number system that allows the digits $\{1, 0, -1\}$. (Such systems were studied by Avizienis [1961].) This conversion can be done in $O(n)$ steps as follows. We can assume that the first digit is 1 and the last is $\{\pm 1\}$. Read the digits from left to right, successively replacing each block of the form $10^k\{\pm 1\}$ by itself (if the sign is plus) or by 01^k1 (if the sign is minus). This is easy to do with a stack, and we find that that $I_2(n)$ can be computed using $O(n \lg \lg n)$ bit operations.

Prabhu and Bose [1979] give an algorithm for counting bivariate irreducible polynomials in $\mathbb{F}_q[X, Y]$.

13. By Theorem 6.5.2 and the lower bound of Exercise 12, the probability that a monic degree n polynomial is primitive, given that it is irreducible, is at least

$$\frac{\varphi(q^n - 1)}{q^n} \ge \frac{\varphi(q^n - 1)}{2(q^n - 1)}.$$

We may as well assume $q^n \ge 4$. Landau [1903a] proved that $m/\varphi(m) = \Omega(\log \log m)$. Since $\log \log m$ is increasing and positive for $m \ge 3$, this probability is at least a constant times

$$\frac{1}{\log \log(q^n - 1)} \ge \frac{1}{\log \log q^n} = \frac{1}{\log n + \log \log q}.$$

(See Theorem 8.8.7 for an explicit version of Landau's result.)

14. Consider

$$F(f) - \lfloor f \rfloor = (F(f) - f) + (f - \lfloor f \rfloor).$$

The left hand side is a polynomial, whereas the right hand side belongs to $X^{-1}k[[X^{-1}]]$. Therefore both must be zero. This result states that the Euclidean algorithm for $k[X]$ is essentially unique.

Lazard [1977] proved the following generalization: suppose that f_0 and f_1 are polynomials, and there is a chain of n "division steps" $f_{i+1} = f_{i-1} - q_i f_i$, $i = 1, \dots, n$ (with quotients $q_i \in k[X]$) making $f_n = \gcd(f_0, f_1)$ and $f_{n+1} = 0$. Then n is at least as large as the number of division steps taken by the Euclidean algorithm on f_0, f_1.

15. Our definition of the resultant follows Euler [1750]; nowadays it is usually defined as a determinant, following Sylvester [1840]. To prove (a) it suffices to count changes of sign. The "division rule" for resultants—part (b)—is due to O. Biermann [1891]. (See Berlekamp [1968] for an elegant proof.) This leads to a procedure similar to the Euclidean algorithm or the Jacobi symbol algorithm for computing the resultant. The running time was analyzed by G. Collins [1971].

Asymptotically, resultants can be computed in nearly linear time. Let $f, g \in k[X]$ have degrees m and n respectively, and assume that $m \geq n$. Using the algorithm of Moenck [1973], it is possible to obtain the degrees and leading coefficients of the Euclidean algorithm's remainder sequence using $O(m(\lg n)^2(\lg\lg n))$ field operations. Using (b), we obtain $R(f, g)$ within the same time bound. This shows that the norm from $\mathbb{F}_q$ to $\mathbb{F}_p$ can be found using $(\lg q)^{1+o(1)}$ bit operations.

16. If C is a circulant matrix of complex numbers whose first row is $(a_0, \dots, a_{n-1})$, then

$$\det(C) = \prod_{\zeta}(a_0 + a_1\zeta + \cdots + a_{n-1}\zeta^{(n-1)}),$$

where the product is taken over all n-th roots of unity. This is due to Spottiswoode [1853]; a proof can be found in Aitken [1949]. Therefore,

$$\det(C) = R(a_0 + a_1X + \cdots + a_{n-1}X^{n-1}, X^n - 1).$$

This is a polynomial identity among the a_i's, with integer coefficients, so it must continue to hold when elements of $\mathbb{F}_q$ are substituted for the a_i's. Applying Exercise 15 yields the running time.

17. Assume that $\mathbb{F}_q = \mathbb{F}_p(\alpha)$, where α is a root of the irreducible equation

$$f(X) = X^d - \sigma_1X^{d-1} + \sigma_2X^{d-2} - \cdots + (-1)^d\sigma_d = 0.$$

Let $t_i = \mathrm{Tr}(\alpha^i)$. Since $t_0 = d$ and

$$\mathrm{Tr}\left(\sum_{i=0}^{d-1} c_i\alpha^i\right) = \sum_{i=0}^{d-1} c_i t_i,$$

it will suffice to compute $t_1, \dots, t_{d-1}$. Von zur Gathen and Shoup [1992b] give the following algorithm for computing $t_1, \dots, t_{d-1}$. Let $g(X) = \prod_{1\leq i\leq d}(1 - \alpha^{p^i}X)$ be the reverse of $f(X)$. Then

$$-(g'/g) \bmod X^{d-1} = \sum_{0\leq i<d-1} t_{i+1}X^i$$

has the desired traces as coefficients. Thus, by Theorem 6.4.3, the trace of any element can be found using $O((\lg q)^2)$ bit operations, and (once the t_i's are known) additional traces cost

$O(d(\lg p)^2)$. Using quadratically convergent Newton iteration to invert g and fast arithmetic in $\mathbb{F}_p$, traces can be found using $(\lg q)^{1+o(1)}$ bit operations, asymptotically.

18. Relations of this type go back at least to Dirichlet's work on prime numbers in arithmetic progressions. For a proof see Ireland and Rosen [1990, p. 254].

19. The equation is either $x^2 = a$ (which is always solvable) or it can be put in the standard form $x^2 + x + a = 0$, which is solvable iff $\mathrm{Tr}(a) = 0$. (This is an example of an Artin-Schreier equation.)

This problem was discussed by Berlekamp, Rumsey, and Solomon [1967]. See also Shayan and Le-Ngoc [1988].

20. Let $n = n_1 n_2 \cdots n_k$, and define $q_0 = p$, $q_i = q_{i-1}^{n_i}$ for $i = 1, \ldots, k$. We suppose that for $i = 1, \ldots, k$, $\mathbb{F}_{q_i}$ is implemented by adjoining α_i, a zero of f_i, a degree n_i irreducible monic polynomial, to $\mathbb{F}_{q_{i-1}}$. Thus, elements of $\mathbb{F}_{q_i}$ are represented as polynomials in α_i, whose coefficients are elements of $\mathbb{F}_{q_{i-1}}$. Let $q = p^n$, and assume all $n_i \geq 2$.

With this representation, we can add and subtract elements of $\mathbb{F}_q$ using n additions or subtractions in $\mathbb{F}_p$. So these operations use $O(\lg q)$ bit operations.

Surprisingly, a naive recursive implementation of multiplication will not use $O((\lg q)^2)$ bit operations. To see this, let us carefully analyze a "shift-and-add" algorithm for multiplying polynomials modulo f, when f is a monic polynomial of degree m. Let the multiplicands be $g = a_{m-1}X^{m-1} + \cdots + a_0$ and be $h = b_{m-1}X^{m-1} + \cdots + b_0$. We compute their product modulo f as

$$a_0 h + a_1(Xh \bmod f) + a_2(X^2 h \bmod f) + \cdots + a_{m-1}(X^{m-1}h \bmod f),$$

where $X^j h \bmod f$ is obtained from $X^{j-1} \bmod f$ using m multiplications of coefficients. We form $m-1$ of these powers. and the multiplications by a_j, $0 \leq j < m$, use m^2 further multiplications. Therefore, $(m-1)m + m^2 = 2m^2 - m$ multiplications are used to compute $gh \bmod f$. Let $M(n)$ denote the number of $\mathbb{F}_p$-multiplications used to multiply two elements of $\mathbb{F}_{p^n}$ in this representation. Abusing notation slightly, we see that $M(n)$ must satisfy

$$M(n) = (2n_k^2 - n_k)M(n/n_k).$$

Since $M(1) = 1$, we have $M(n) = \prod_{1 \leq i \leq k}(2n_i^2 - n_i)$. Now, suppose that for $m = 2, \ldots, r$, k_m of the n_i's have the value m. Further let $n'_m = m^{k_m}$. We can rewrite $M(n)$ as

$$\prod_{2 \leq m \leq r} (2m^2 - m)^{k_m} = n \prod_{2 \leq m \leq r} (2m-1)^{\log_m n'_m} = n \prod_{2 \leq m \leq r} (n'_m)^{\log(2m-1)/\log m}.$$

Some calculus shows that $\frac{\log 2x-1}{\log x}$ decreases for $x \geq 2$, so

$$M(n) \leq n \left(\prod_{2 \leq m \leq r} n'_m \right)^{\log_2 3} = n^{1+\log_2 3}.$$

This bound is attainable, because we could have a tower of degree 2 extensions, making $n_i = 2$ for $i = 1, \ldots, \log_2 n$. In this case the recursive algorithm uses $n^{2.58\ldots}$ multiplications in $\mathbb{F}_p$, which is not $O(n^2)$.

This suggests, but does not prove, that multiplication in this representation of $\mathbb{F}_{p^n}$ uses $O(n^{1+\log_2 3}(\lg p)^2)$ bit operations. (Note that we did not consider the *additions* used.) This is true, however, and can be proved by analyzing $A(n)$, the number of $\mathbb{F}_p$-additions used for multiplication in $\mathbb{F}_{p^n}$. Since the "shift-and-add" polynomial algorithm above uses $(2m-1)(m-1)$ additions of coefficients, we have

$$A(n) = (2n_k^2 - n_k)A(n/n_k) + (2n_k - 1)(n_k - 1)n/n_k$$

when $n > 1$. This implies that $A(n) = O(n^{1+\log_2 3})$.

Similar remarks could be made about the complexity of division. We will be content to observe that inversion can be done in polynomial time (since $\alpha^{-1} = \alpha^{q-2}$ in $\mathbb{F}_q$).

It is well-known that constant factors in the running time of recursive steps of an algorithm affect the exponent in the ultimate running time. For finite fields, this issue was apparently first raised by Chor [1982]. Iterated field extensions play a role in the nim-multiplication procedure of Conway [1976, Chapter 6].

21. For properties of tensor products, see Lang [1965, Chapter 16.]

(a) It is enough to consider the case $k = 2$. Consider the $\mathbb{F}_p$-linear map from R into $\mathbb{F}_p(\alpha_1, \alpha_2)$ determined by $X_i \longmapsto \alpha_i$, $i = 1, 2$. If this is not 1-1, then α_2 satisfies an polynomial equation of degree $< n_2$ over $\mathbb{F}_{q^{n_1}}$. This is impossible since the degree of $\mathbb{F}_p(\alpha_1, \alpha_2)$ is $n_1 n_2$. We note that the hypothesis on the n_i's is necessary: for example, if q and r are quadratic nonresidues mod p, $\mathbb{F}_p(\sqrt{q}, \sqrt{r}) \cong \mathbb{F}_{p^2}$, whereas $\mathbb{F}_p(\sqrt{q}) \otimes \mathbb{F}_p(\sqrt{r})$ has dimension 4 as a vector space over $\mathbb{F}_p$.

(b) Let f_i be an irreducible monic polynomial in $\mathbb{F}_p[X]$ of degree $n_i \geq 2$, with the zero α_i. We use the relation $f_i(\alpha_i) = 0$ to replace powers of α higher than n_i by smaller ones. As in the solution for Exercise 20, we let $q_i = p^{n_1 \cdots n_i}$, and implement $\mathbb{F}_{q_i}$ as $\mathbb{F}_{q_{i-1}}[X]/(f_i)$. (Therefore, $q_0 = p$ and $q_k = q = p^n$.) As before, addition and subtraction use $O(\lg q)$ bit operations. Let $M(n)$ be the number of $\mathbb{F}_p$-multiplications used to multiply elements of $\mathbb{F}_q$, by the recursive "shift-and-add" procedure. We find that

$$M(n) = n_k^2 M(n/n_k) + (n_k - 1)n; \qquad M(1) = 1.$$

By induction on n, we have $M(n) \leq 2n^2 - n$. Similarly, $A(n)$, the number of $\mathbb{F}_p$-additions/subtractions used, satisfies

$$A(n) \leq n_k^2 A(n/n_k) + 2(n_k - 1)n; \qquad A(1) = 0.$$

By induction, $A(n) \leq 4n^2 - 2n$. Therefore, multiplication in this representation of $\mathbb{F}_q$ uses $O((\lg q)^2)$ bit operations.

22. Let $\{\beta_1, \ldots, \beta_n\}$ be a basis for $\mathbb{F}_q$, as a vector space over $\mathbb{F}_p$. We represent elements of $\mathbb{F}_q$ by their coefficients relative to this basis, and are given a multiplication table of n^3 elements $c_k^{(ij)}$, for which

$$\beta_i \beta_j = \sum_{1 \leq k \leq n} c_k^{(ij)} \beta_k.$$

In this representation, addition or subtraction reduces to addition or subtraction of basis elements, which uses $O(\lg q)$ bit operations. To multiply, we use the relation

$$\left(\sum_i a_i\beta_i\right)\left(\sum_j b_j\beta_j\right) = \sum_{1\le k\le n}\left(\sum_{i,j} a_i b_j c_k^{(ij)}\right)\beta_k.$$

Thus, multiplication can be done using $O(n^3(\lg p)^2)$ bit operations. Division can be effected by solving a system of linear equations, and has the same complexity. It is interesting that a representation of 1 can be found with the same amount of work, as a solution to the linear equation $\beta x = \beta$.

23. Part (a) is due to Eisenstein [1850]. For a proof, see, e.g., Albert [1961, p. 119]. For normal bases with special properties, see H. W. Lenstra and Schoof [1987]; Lempel and Weinberger [1988]; Hachenberger [1992]; Akbik [1992]; and references therein.

Hensel [1888] showed the fraction of elements whose conjugates form a normal basis is

$$\prod_{1\le i\le r}\left(1 - p^{-d_i}\right),$$

where $d_1, \ldots, d_r$ are the degrees of the distinct irreducible factors of $X^n - 1 \bmod p$. Von zur Gathen and Giesbrecht [1990] showed this fraction to be $\Omega(1)$ when $p \le n^4$, and $\Omega((\log_p n)^{-1})$ otherwise. (See also Frandsen [1991b].)

For references to algorithms for constructing normal bases, see Exercise 24.

By the results of Exercise 7.24, addition and subtraction in the normal basis representation use $O(\lg q)$ bit operations, and multiplication and division can be done using $O(n(\lg q)^2)$. Raising an element to the p-th power can be done with a cyclic shift of coefficients, since we have

$$\left(\sum_{1\le i\le n} a_i\beta_i\right)^p = \sum_{1\le i\le n} a_{i-1}\beta_i,$$

taking indices mod n. This uses $O(\lg q)$ bit operations, the time to copy n elements of $\mathbb{F}_p$.

Turning now to exponentiation, let $e < p^n - 1$. Represent this in base p, so that

$$e = e_0 + e_1 p + e_2 p^2 + \cdots + e_k p^k,$$

with $0 \le e_i < p$. If $x \ne 0$, the following routine computes $y = x^e$:

```
NORMAL-POWER(x, e)
y ← 1
for i = k, ..., 0 do
        y ← x^{e_i} y
        if (i = 0) exit the loop
        y ← y^p
return(y)
```

Since $k < n$, this uses at most n exponentiations to powers less than p, n multiplications, and n shifts. The total bit complexity (including the cost of writing e in base p) is $O((\lg e)n(\lg q)^2)$.

In certain situations, though, exponentiation using a normal basis will outperform the binary method (Algorithm POWER). Suppose that p is small, and we are willing to precompute x^j for $2 \le j < p$. With this modification, we can compute x^e using about $n + p$ multiplications in $\mathbb{F}_q$ (worst case), whereas the binary method uses $2n \lg p$. This is especially attractive when $p = 2$ and the $\mathbb{F}_q$ operations are realized in hardware.

Normal bases for finite field computation were popularized by Omura and Massey [1986], who used a symmetry property of the multiplication table to design a small multiplier for $\mathbb{F}_{2^n}$. For other hardware designs using normal bases, and related theoretical results, see Agnew, Mullin, and Vanstone [1988]; C. Wang et al. [1985]; Beth, Cook, and Gollmann [1986]; Itoh and Tsujii [1988a, 1988b, 1989b]; Mullin, Onyszchuk, Vanstone, and Wilson [1988]; Onyszchuk, Mullin, and Vanstone [1988]; Scott, Simmons, Tavares, and Peppard [1988]; Asano, Itoh, and Tsujii [1989]; Ash, Blake, and Vanstone [1989]; Pincin [1989]; Geiselmann and Gollmann [1990]; Séguin [1990]; Stinson [1990]; C. Wang and Pei [1990]; Beth, Geiselmann, and Meyer [1991]; von zur Gathen [1991a, 1991b, 1992]; Gao [1992]; Gao and Lenstra [1992]; Hasan, Wang, and Bhargava [1992]; Agnew, Beth, Mullin, and Vanstone [1993]; Blake, Gao, and Mullin [1994]; and Gao and Vanstone [1995].

Regarding (c), it is known that the quality lies between n^2 and $2n - 1$. See Mullin, Onyszchuk, Vanstone, and Wilson [1988].

24. It suffices to prove that $m(\lg p - 1) < \lg q \le m \lg p$. Let $k = \lg p$. Then $p < 2^k$, so $p^m < 2^{mk}$, which implies that $\lg q \le mk$. Also, $p \ge 2^{k-1}$, so $p^m \ge 2^{m(k-1)}$, implying $\lg q > m(k-1)$.

25. This was proved by Kühne [1902]. Note that Theorem 6.5.3 is a special case of this result. More proofs and related results can be found by consulting the references for Theorem 6.5.3.

26. Suppose a is a quadratic nonresidue mod p. Then we have $\left(\frac{a}{X}\right) = -1$ in $\mathbb{F}_p[X]$, whereas $\left(\frac{a}{X}\right) = +1$ in $\mathbb{F}_{p^2}[X]$.

27. We may as well assume the characteristic is odd (otherwise everything is a square). Suppose $k = \mathbb{F}_q$ and the extension has degree n. If x is a nonzero element of K,

$$x^{(q^n-1)/2} = (x^{q^{n-1}+\cdots+q+1})^{(q-1)/2} = N(x)^{(q-1)/2}.$$

Hence $N(x)$ is a square iff x is one. Dirichlet [1832] proved this for extensions of the form $\mathbb{F}_p[\sqrt{-1}]$. (See also Hardman and Jordan [1969].)

To get the running time bound, combine part c) of Exercise 15 with Theorem 5.9.3. This gives an alternative proof for the last running time bound in Theorem 6.7.2.

A corresponding result holds for e-th powers when $q \equiv 1 \bmod e$. In this case, we can determine if x is an e-th power in $\mathbb{F}_{q^n}$ using $O((n^2 + \lg q)(\lg q)^2)$ bit operations.

28. **(a)** Since f is monic, we can use the division algorithm as when A is a field.

(b) Let $d = \deg f$. Since R is a free module of rank d over A, addition reduces to d additions of elements in A; the running time bound follows since $\lg f \ge d \lg n$. Subtraction is similar.

(c) With the algorithm outlined in the solution to Exercise 20, multiplication of two elements of R uses $\le 2d^2$ multiplications mod n.

(d) Sufficiency is clear. To prove necessity, suppose that $g \in R^*$. Then there are polynomials $\gamma, \delta \in A[X]$ with $\gamma g + \delta f = 1$. Applying (a), we can write $\gamma = qf + \alpha$ where $\deg \alpha < \deg f$. Let $\beta = \delta + qg$. Evidently $\alpha g + \beta f = 1$, and the degree of βf is less than $\deg f + \deg g$. Since f is monic, this implies $\deg \beta < \deg g$. (Compare Sanderson [1911]; McCoy [1948, p. 168].)

(e) Use Gaussian elimination to solve the system of (d). If a non-unit pivot is encountered, use the gcd-free basis construction (Algorithm REFINE, Section 4.8) to obtain a relatively prime factorization $n = n_1 n_2$ and solve the system modulo n_1 and n_2. (It may be necessary to do this recursively.) Recombine the results using the Chinese remainder theorem (Theorem 5.5.3).

For results on the structure of the rings discussed in this exercise, see McDonald [1974].

29. Let g generate the cyclic group $\mathbb{F}_{q^n}^*$. If $a = g^x$, then $\mathrm{N}(a) = 1$ iff $x(1 + q + \cdots + q^{n-1}) \equiv 0 \pmod{q^n - 1}$, that is, iff x is a multiple of $q - 1$.

30. J. Gordon [1976] proved this for $q = 2$. To verify the result, observe that

$$\prod_{1 \le i \le d} \left(Y - a(X)^{q^i}\right) = m(Y)$$

in $k[Y]$, and substitute $Y = X$. This method uses $O(n(\lg q)^3)$ bit operations. We note that a straightforward approach (find a linear relation among the powers of a) would use $O(n^3(\lg q)^2)$ bit operations, and the Berlekamp-Massey algorithm $O(n^2(\lg q)^2)$.

Shoup [1994] showed that the minimal polynomial for any element of $\mathbb{F}_{p^n}$ can be found using $O(n^{(\omega+1)/2})$ arithmetic operations in $\mathbb{F}_p$, assuming that $n \times n$ matrices can be multiplied in $O(n^\omega)$ steps. (According to Coppersmith and Winograd [1987, 1990], $2 \le \omega \le 2.376$.) For other recent minimal polynomial algorithms, see Rifà and Borrell [1991]; Thiong Ly [1989b]. Di Porto, Guida, and Montolivo [1992] gave an efficient construction of minimal polynomials for primitive elements.

31. This is due to Berlekamp, McEliece, and van Tilborg [1978].

32. This goes back at least to Landsberg [1893], who computed the number of $m \times n$ matrices over $\mathbb{F}_p$ with a given rank. For similar results on probabilities associated with random matrices over finite fields, see Reiner [1961]; Berlekamp [1966]; Strong [1988]; and references therein.

33. This result is a special case of a result of Dickson [1905]. Also see Szele [1947]; Szep [1947]; Jungnickel [1992].

34. Assertions (a) and (b) can be found in Lang [1965, pp. 206 ff.] For short proofs of of (c) and (d), see Guerrier [1968].

35. To prove (a), use Galois theory and the fact that η is an algebraic integer. Part (b) is a generalization of the law of quadratic reciprocity. A proof highlighting this viewpoint can be found in Bach and Shallit [1985, 1989].

For expository papers about higher reciprocity laws, see Wyman [1972] and E. Lehmer [1978].

Chapter 7

1. First observe that if $q \equiv 5 \pmod 8$, then q cannot be a square. Hence $2 \notin \mathbb{F}_q^2$, that is, $2^{(q-1)/2} = -1$. Use this and the condition on a to show that the formula is correct, by squaring both sides.

The first case is from Legendre [1798, pp. 231–238]; the complete formula is due to Pocklington [1917]. (See also Cipolla [1903].)

Atkin [1992] suggested a way to avoid the test on a. Let $b = (2a)^{(q-5)/8}$, $i = 2ab^2$. Then $x = ab(i-1)$ is a square root of a.

2. For $k = 1, 2, \ldots$, Linnik's result (Theorem 8.5.6) implies there is a prime $p \equiv 1 \pmod{2^k}$ with $p = 2^{O(k)}$. This gives a sequence of primes p with $\nu_2(p-1) = \Omega(\log p)$.

3. The algorithm maintains the loop invariants $a^{e_1}b^{e_2} = 1$ and $\nu_2(e_1) < \nu_2(e_2)$. Since $a^{e_1+1}b^{e_2} = a$, it will return a correct square root of a as soon as e_1 becomes odd. To prove termination, observe that each pass through the loop reduces $\nu_2(e_1)$ by 1. There are thus $O(\nu_2(q-1))$ passes; as before, the algorithm uses $O(\nu_2(q-1)(\lg q)^3)$ bit operations.

4. This exercise is from Bumby [1989]. Since $\left(\frac{1}{p}\right) = 1$ and $\left(\frac{-1}{p}\right) = -1$, we can find x using binary search. Use the proof of Corollary 7.1.2 to find $u, v \in \mathbb{F}_p$ with $x = u^2$, and $-(x+1) = v^2$. Then $u^2 + v^2 = -1$, so $g = u + vi$ is a non-square in $\mathbb{F}_{p^2}$. Finish off with the algorithm TONELLI, taking $q = p^2$. The algorithm uses $O((\lg p)^4)$ bit operations.

5. This seems to be folklore; see, e.g., Angluin [1982] or Shoup [1988]. Together with Theorem 7.1.3, the result shows that computing square roots and finding non-squares in $\mathbb{F}_q$ are polynomial-time Turing equivalent. It seems to be an open problem to do this with many-one reductions.

6. If a is not a square, then $a^{(q-1)/2} = -1$. Because of this,

$$(\mathrm{Tr}(b))^2 = 2(a + a^{(q+1)/2}) = 0,$$

so b is a nonzero element whose trace vanishes and thus cannot be in $\mathbb{F}_q$.

7. The relation $X = (Y + t)/2$ gives an isomorphism between $\mathbb{F}_q[X]/(f)$ and $\mathbb{F}_q[Y]/(g)$. Thus a square root of a is given by $(\frac{Y+t}{2})^{(q+1)/2} \bmod g(Y)$. We note that multiplication and squaring mod f use 6 and 5 $\mathbb{F}_q$-multiplications, respectively; the corresponding counts mod g are 5 and 4. On average, the exponentiation should use half as many multiplications as squarings, so the sparse implementation will take about 4/5 the time of the original.

8. This algorithm is due in essence to Pocklington [1917]. The error bound is from Peralta [1986]; for additional properties of this method see Bach [1990b].

9. The lower bound is part of the prime number theorem for arithmetic progressions in $k[X]$, due to Kornblum [1919]. (See also Artin [1924a].) For an elementary proof of the lower bound, see Bach and Shallit [1984].

The following algorithm is due to D. Cantor (personal communication). Let $f(X) = X^k + \cdots + a$ be irreducible. Then $\mathbb{F}_q[X]/(f)$ is a model of $\mathbb{F}_{q^k}$. If $b = X^{(1+q+q^2+\cdots+q^{k-1})/k} \bmod f$, then $(-b)^k = (-1)^k \mathrm{N}(X) = a$. The running time is dominated by the cost of finding f; as in Theorem 7.6.3, $O((k + \lg q)(k^3(\lg q)^2))$ bit operations are used, on average.

10. After a rational change of variables, we can take the equation to be $x^2 + \beta x = \gamma$; we can solve this by inverting a matrix, since the left hand side is a linear function of x. This solution is from Berlekamp, Rumsey, and Solomon [1967]. See also C. Chen [1982]; Shayan and Le-Ngoc [1988].

11. A random $x \in \mathbb{F}_q^*$ is a solution to $x^r = a$ with probability $r/(q-1)$. Therefore we expect to use $O(\frac{\lg r}{r} q (\lg q)^2)$ bit operations before one is found.

12. **(a)** Let $r^\nu \,||\, q-1$, and $m = (r^\mu)^{-1} \bmod (q-1)/r^\nu$. Consider

$$G_0 \subset G_1 \subset \cdots \subset G_{\nu-1} \subset G_\nu = \mathbb{F}_q^*,$$

where $G_{\nu-i}$ denotes the subgroup of r^i-th powers. (An element belongs to $G_{\nu-i}$ iff it becomes 1 when raised to the power $(q-1)/r^i$.) If $\mu \geq \nu$, then $x^{r^\mu} = a$ is solvable iff $a \in G_0$; in this case a solution is a^m, which can be found using $O((\lg q)^3)$ bit operations. If $\mu < \nu$, let $g \in \mathbb{F}_q^* - (\mathbb{F}_q^*)^r$. Then for $i = \mu, \ldots, \nu$, we find e_i, $0 \leq e_i < r$ so that

$$g^{e_\mu r^\mu + e_{\mu+1} r^{\mu+1} + \cdots + e_\nu r^\nu} h = a$$

for some $h \in G_0$. Then we can take

$$x = g^{e_\mu + e_{\mu+1} r + \cdots + e_\nu r^{\nu-\mu}} h^m.$$

The major cost is in finding the e_i's; counting them, we find the total expected work is $O(r(\nu - \mu + 1)(\lg q)^3)$. (Curiously, this does not increase with μ; in part this is because increasing μ pushes the action into a smaller subgroup.)

(b) Let $g = \gcd(d, q-1)$. We first observe that g can be factored by trial division (i.e. by trying all possible prime divisors up to $\sqrt{g}$) using $O(p(\lg g)^2)$ bit operations, if p is the largest prime factor of g. Expressing $d = r_1^{\mu_1} \cdots r_k^{\mu_k} s$ with $\gcd(r_i, s) = 1$ for all i requires $\nu_{r_i}(d) + 1$ divisions by each r_i, so the work for this is easily bounded by $O(\lg d)^3$. Then $x_0 = a^m$, where $ms \equiv 1 \bmod q-1$, and we get $x_1, \ldots, x_k$ by using the algorithm of part (a) k times. The total work for finding the x_i's is

$$O\left(\sum_{1 \leq i \leq k} r_i \nu_{r_i}(d) (\lg q)^3 \right).$$

To get the result, we use the bound $\nu_1 r_1 + \cdots + \nu_k r_k \leq r_1^{\nu_1} \cdots r_k^{\nu_k}$.

13. We will consider worst case estimates only. Suppose $f \in \mathbb{F}_q[X]$ has degree d. Factoring this completely by Berlekamp's algorithm uses $O((d^2 + dq)(\lg f)^2)$ bit operations. As an alternative, consider the following algorithm. At all times, we maintain a partial factorization $f_1, f_2, \ldots, f_k$ of f. (Note that any factor of degree 1 can be dropped from the list.) Using a basis $\{b_1, \ldots, b_r\}$ of $\mathcal{B}$, we compute each possible $\gcd(b_i - \alpha, f_j)$, updating the f_i's as new factors are found. Since there are $\leq qd$ possible $b_i - \alpha$'s, and $\sum \deg f_i = \Theta(d)$, the total work is $O(dq(\lg f)^2)$. This is certainly preferable when q is small, and no worse in general.

14. If $a^p = a$, then $a^q = a$. By Theorem 7.4.1, we must have $a_i \in k$ for $i = 1, \ldots, r$. But if $a_i^p = a_i$, $a_i \in \mathbb{F}_p$. The other direction is similar to the proof of Theorem 7.4.1.

15. Let $k = \mathbb{F}_p(\alpha)$, where α satisfies a monic polynomial equation of degree n. First compute α^{ip}, $i = 1, \ldots, n-1$, as in the proof of Theorem 7.4.2, using $O((n + \lg p)(\lg q)^2)$ bit operations.

Similarly, compute $X^{jp} \bmod f$, $j = 0, \ldots, d-1$, at a cost of $O((d + \lg p)(\lg f)^2)$. From these results, obtain the nd products $\alpha^{ip}X^{jp}$, at a cost of $O(d(\lg q)^2)$ bit operations per product.

16. Using Exercise 15, find a matrix A for the $\mathbb{F}_p$-linear transformation $a \longmapsto a^p \bmod f$. This will have dimension $nd \times nd$, so the kernel of $A - I$ (which has dimension $\leq d$), can be found using $O(n^2d^3(\lg p)^2) = O(d(\lg f)^2)$ bit operations.

17. Assume that f has at least two distinct irreducible factors. Using Exercise 16, find a basis for the absolute Berlekamp algebra $\mathcal{A}$. Then choose any $a \in \mathcal{A} - \mathbb{F}_p$, and try to split f with $\gcd(a - \alpha, f)$, $\alpha \in \mathbb{F}_p$. If $q = p^n$, the algorithm will use $O(\deg f + n + p)(\lg f)^2)$ bit operations.

Recall that Theorem 7.4.5 gives a bound of $O(\deg f + p^n)(\lg f)^2)$ bit operations for the relative version of Berlekamp's algorithm. Comparing this with the above bound, we see that the absolute version does less work, except when n and p are small.

18. The kernel K of ϕ is a normal subgroup of G. Therefore the chance that $\phi(x)$ takes any particular value y is $|K|/|G|$, which does not depend on y.

19. This algorithm is due to McEliece [1969]. For the correctness, observe that $M(a) = a + a^2 + \cdots + a^{2^n}$ maps $\mathcal{B}$ onto $\mathcal{A} = \mathbb{F}_2 \times \cdots \times \mathbb{F}_2$. The algorithm fails to split iff $M(a) \in \{0, 1\}$, which happens with probability $1/2^{r-1}$. We can evaluate $M(a)$ using $n - 1$ applications of the absolute Frobenius map, by a process similar to Horner's rule. If this is done, then by Theorem 7.4.4 and Exercise 15, a total running time of $O((\mathrm{dg}\, f + n)(\lg f)^2)$ results.

For similar methods, see Shiva and Allard [1970]; Thiong Ly [1989a].

20. Assume that q is odd and f is squarefree, with $s \geq 2$ distinct irreducible factors. If all of these factors have degree m, then $k[X]/(f) \cong \mathbb{F}_{q^m} \times \cdots \times \mathbb{F}_{q^m}$. If $\gcd(r, f) = 1$, then $s = r^{(q^m-1)/2} \equiv \pm 1$ modulo each factor, so the chance that $\gcd(s - 1, f)$ splits f is $1 - 1/2^{s-1}$. (If $\gcd(r, f) \neq 1$, we already have a factor unless $r = 0$.) The running time bound follows from Theorems 7.5.3 and 6.4.4.

21. The proof of correctness is similar to that of Theorem 7.4.6; the probability of failure is at most $1/2^{r-1}$, when f has r distinct irreducible factors. Using Exercise 16, $O((d + n + \lg p)(\lg f)^2)$ bit operations are required. As a general rule, we would prefer this algorithm to CZ whenever $n + \lg p = o(\lg q)$.

22. **(a)** This algorithm is from Rabin [1980a]. One trial uses $O((\lg q)(\lg f)^2)$ bit operations. In discussing the error probability, we may as well assume that f is the product of distinct linear factors in $\mathbb{F}_q[X]$, since we can reduce to this case within the above time bound. (Replace f by $\gcd(f, X^q - X)$.) Ben-Or [1981] proved that one trial of the algorithm fails to find a nontrivial factor of f with probability $1/2^{d-1} + O(d/\sqrt{q})$, where d is the degree of f. For d fixed and $q \longrightarrow \infty$, this is comparable to the error probability of the Cantor-Zassenhaus algorithm (CZ).

(b) The running time for one trial is the same as for (a), so for similar reasons we will assume that f has degree $d > 1$, with distinct nonvanishing zeroes $z_1, \ldots, z_d \in \mathbb{F}_q$. To estimate the error probability, we consider the isomorphism

$$\mathbb{F}_q[X]/(f) \cong \mathbb{F}_q \oplus \cdots \oplus \mathbb{F}_q,$$

and let Y_i be the 0-1 random variable equal to the i-th component of $T(aX)$. We note that

$$\Pr[Y_i = 1] = \Pr[\mathrm{Tr}(az_i) = 0] = 1/2,$$

and for $i \neq j$,

$$\Pr[Y_i = Y_j] = \Pr[\mathrm{Tr}(a(z_i - z_j)) = 0] = 1/2.$$

Therefore, Y_i and Y_j are pairwise independent when $i \neq j$, so the degree of $\gcd(T(aX), f)$, which equals $\sum_{1\leq i\leq d} Y_i$, has expected value $d/2$ and variance $d/4$. The probability of failure is

$$\Pr\left[\deg\left(\gcd(T(aX), f)\right) = 0, d\right] = \Pr\left[\left|\sum_{1\leq i\leq d} Y_i - d/2\right| \geq d/2\right],$$

and this is at most $1/d$, by Chebyshev's inequality. This estimate is due to Ben-Or [1981], who also showed it to be best possible. Earlier, Rabin [1980] proved that the failure probability is at most $1/2$.

23. This exercise is due to Niederreiter [1993a]; the precise formulation is from Göttfert [1994]. Observing that both $(fh)'$ and h^2 are polynomials in X^2, and solving the resulting linear system, one obtains a factorization algorithm with complexity similar to the algorithm BERLEKAMP.

For more about this algorithm, see V. Miller [1992a]; Fleischmann [1993]; Niederreiter and Göttfert [1993, 1995]; Niederreiter [1993d, 1994a, 1994b]; and Gao and von zur Gathen [1994].

24. A deterministic polynomial-time algorithm for constructing a normal basis of $\mathbb{F}_q$ was apparently first given by Lüneburg [1985]. For other such algorithms, see H. W. Lenstra [1991a]; von zur Gathen and Giesbrecht [1990]; and Bach, Driscoll, and Shallit [1990, 1993]. For randomized algorithms for the problem, see the last two references, as well as Frandsen [1991b].

Normal basis construction can also be reduced in polynomial time to factorization of polynomials. For such results, see Sidel´nikov [1987]; Schwarz [1988]; Semaev [1988].

Stepanov and Shparlinski [1991] surveyed efficient algorithms for constructing normal bases, including normal bases generated by a primitive element.

25. This result is often attributed to Stickelberger [1898], but it was apparently first proved by Pellet [1878]. For other references and related formulas see Schwarz [1940].

The time bound on computing $\mu(f)$ follows from Theorem 5.9.3 and Exercise 6.15, after observing that $D = (-1)^{n(n-1)/2}R(f, f')$. (Note that $D \neq 0$ iff f is squarefree.)

There is a corresponding result for fields of characteristic 2, but it requires the evaluation of D mod 8. See Carlitz [1953a] and Swan [1962]. Apparently, these formulas do not lead to a better method than Berlekamp's algorithm unless the degree of f is small. For more on this problem, see Berlekamp [1968].

26. This is from Pellet [1878]. See also Mirimanoff and Hensel [1905].

27. The following algorithm was given (in characteristic zero) by D. Yun [1977]. We have extended it to work in characteristic p.

```
YUN(f)
{returns a list of pairs (f_i, e_i) with f_i squarefree, ∏ f_i^{e_i} = f.}
g ← gcd(f, f')
c_1 ← f/g
d_1 ← f'/g
i ← 1
while c_i ≠ 1 do
        f_i ← gcd(c_i, d_i − c_i')
        c_{i+1} ← c_i/f_i
        d_{i+1} ← (d_i − c_i')/f_i
        output ((f_i, i))
        f ← f/f_i^i
        i ← i + 1
if f ≠ 1 then
        find h so that h^p = f
        (h_1, e_1), ..., (h_s, e_s) ← YUN (h)
        output ( (h_1, pe_1), ..., (h_s, pe_s) )
```

We first show that it correctly computes a squarefree factorization.

Suppose $f = \hat{f}h^p$, where no irreducible polynomial divides $\hat{f}$ to a power p or higher. A direct computation shows that when given $\hat{f}$ as input, the algorithm finds the same values of c_1 and d_1 as it would when given f as input. We may therefore take $h = 1$.

Now, let $\pi = X - a$, a divisor of f. (The constant term a may belong to an extension field of $\mathbb{F}_q$.) We want to show that if $e < p$ and $\pi^e \,||\, f$, then π divides f_e exactly once. So let $f = \epsilon\pi^e$, where $\pi \not| \epsilon$. Then $f' = e\epsilon\pi^{e-1} + \epsilon'\pi^e$, so $g = \eta\pi^{e-1}$, where $\pi \not| \eta$.

Now we show by induction that for $i \le e$, there is an α_i not divisible by π, and a β_i for which

$$c_i = \alpha_i \pi,$$

$$d_i = (e - (i-1))\alpha_i + \beta_i \pi.$$

For the base case, we observe that $c_1 = \epsilon\pi/\eta$ and $d_1 = e\epsilon/\eta + \eta'\pi$. For the induction step, if the result is true for i, we find that

$$c_i' = \alpha_i + \alpha_i'\pi,$$

$$d_i - c_i' = (e - i)\alpha_i + (\beta_i - \alpha_i')\pi,$$

so that

$$f_i = \begin{cases} \theta_i & \text{if } i < e; \\ \theta_i \pi & \text{if } i = e, \end{cases}$$

where $\pi \not| \theta_i$. We may therefore take $\alpha_{i+1} = \alpha_i/\theta_i$ and $\beta_{i+1} = (\beta_i - \alpha_i')/\theta_i$.

Now we take $i = e$ in the above expression for f_i, to show that $\pi \,||\, f_e$. But inspection of the algorithm shows that every c_i, and hence every f_i, is a divisor of f. From this the correctness follows.

We now analyze the running time.

We note first that given the inverse of the Frobenius matrix for $\mathbb{F}_q$, p-th roots can be computed in k with $O((\lg q)^2)$ bit operations.

We also note the following fact. Given two polynomials a_1 and a_2, we can compute the three quantities $g = \gcd(a_1, a_2)$, a_1/g, and a_2/g in $O((\lg a_1)(\lg a_2))$ bit operations. For this reason, the first three steps of the algorithm use $O((\lg f)^2)$ bit operations.

We now find an expression for the degrees of the polynomials c_i appearing in the algorithm. In the notation used to prove Theorem 7.5.1, if $\pi = X - a$ divides $\hat{f}$ to the power e, then we have $c_i = \alpha_i \pi$ for $i \leq e$. Each such π contributes 1 to the degree of c_i, and appears only in f_e; we therefore have

$$\deg c_i = \sum_{e \geq i} \deg f_e.$$

Because $\deg d_i \leq \deg \hat{f} \leq \deg f$, and the loop terminates by the d-th iteration, the gcd computations in the loop use at most a constant times

$$(\deg f) \sum_{i \geq 1} \sum_{e \geq i} \deg f_e = (\deg f)(\deg \hat{f})$$

field operations. Therefore $O((\lg f)^2)$ bit operations suffice for the loop.

At the end of the loop f must be a p-th power, and we can find h in time $O((\lg f)^2)$, by using the inverse Frobenius matrix to compute p-th roots of the coefficients of f.

This shows that the computation up to the first recursion uses $O((\lg f)^2)$ bit operations. But because $p \geq 2$, the algorithm works recursively on a polynomial whose degree is at most half that of f. For this reason, this bound holds for the whole algorithm.

We now prove the second assertion. If it is necessary to take p-th roots at all, then $\lg p \leq p \leq d$. By applying Theorem 7.4.2 to the defining polynomial of k, we see that a matrix for the Frobenius automorphism on k can be found using $O((n + d)(\lg q)^2)$ bit operations, and the same bound suffices for inverting this matrix by Gaussian elimination. Because $(n + d)(\lg q)^2 = O((1 + n/d^2)(\lg f)^2)$, and the rest of the algorithm takes $O((\lg f)^2)$ steps, the result follows.

28. Prove that each polynomial is squarefree and vanishes at the roots of the other.

29. See D. H. Lehmer [1972]. For a proof that $\tilde{D}_n = e^{-\gamma}(1 + 1/n) + O((\log n)/n^2)$, see Greene and Knuth [1990]. We note that by Theorem 6.5.1, $D_n \leq \tilde{D}_n$, and for fixed n, $\lim_{q \to \infty} D_n = \tilde{D}_n$.

30. See Carlitz [1932].

31. This algorithm, and a proof of its correctness, appeared in Rabin [1980a]. To analyze the running time, first observe that the time bound is sufficient for us to factor d. For each prime l dividing d, we need to do $(d/l) \lg q$ multiplications mod f; now use the prime number theorem to show that

$$\sum_{\substack{l \mid d \\ l \text{ prime}}} \frac{1}{l} \leq \sum_{m \leq \omega(d)} \frac{1}{p_m} = O(\log \log \log d).$$

The expected number of bit operations to find an irreducible polynomial f using this method is $O(d(\lg\lg\lg d)(\lg f)^3)$.

32. The following algorithm is from Ben-Or [1981]; we assume $d > 1$.

```
BEN-OR(d)
choose f, a random degree d monic polynomial
u ← X
for k = 1, ..., ⌊n/2⌋ do
        u ← u^q mod f
        if gcd(f, u − X) ≠ 1 fail
return(f)
```

Ben-Or gave a heuristic analysis of its running time, based on an analogy between random polynomials and random permutations. We can make his argument rigorous as follows. Let $m(f)$ be the minimum degree of any irreducible factor of f. If we choose a random degree d polynomial in $\mathbb{F}_q[X]$ (with the uniform distribution), then

$$\begin{aligned} \mathrm{E}[m(f)] &= \sum_{k\geq 0} \Pr[m(f) > k] \\ &\leq \sum_{k\geq 0} \Pr[\tilde{m}(\sigma) > k] \\ &= \mathrm{E}[\tilde{m}(\sigma)], \end{aligned}$$

where σ is a random permutation on d letters, with minimum cycle length $\tilde{m}(\sigma)$. (The inequality can be proved as follows. According to Exercise 6.12, the density of degree j irreducible polynomials is at most $1/j$, which is exactly the density of cyclic permutations on j letters. Therefore, the fraction of polynomials, all of whose factors have degree $> k$, is at most the fraction of permutations, all of whose cycles have length $> k$.) Shepp and Lloyd [1966] proved that $E[\tilde{m}(\sigma)] = O(\log d)$. Therefore, the expected number of passes through the loop is $O(\lg d)$. By Theorems 6.4.2–6.4.4, each pass through the loop uses $O((\lg q)(\lg f)^2)$ bit operations, so the expected time for one invocation of the algorithm is $O((\lg d)(\lg q)(\lg f)^2)$. If we run the algorithm repeatedly until it succeeds, the expected number of bit operations used is is $O(d(\lg d)(\lg q)(\lg f)^2)$. This analysis assumed that the binary algorithm is used to compute q-th powers; we note that an implementation using a matrix for the Frobenius map would have an expected cost of $O(d(d + \lg q)(\lg f)^2)$.

33. See Bach and von zur Gathen [1988].

34. To compute a square root of a in $\mathbb{F}_q$, apply Theorem 7.8.1 to $f(X) = X^2 - a$. Roots of the resulting quadratic polynomial h can be found using the quadratic formula and Cipolla's algorithm.

35. This is from Zassenhaus [1969]; the construction is similar to the proof of Theorem 7.8.1. Assume that $f \in \mathbb{F}_q[X]$ is squarefree, with at least two distinct irreducible factors. Find a nonconstant $g \in \mathcal{B}$, the Berlekamp subalgebra of $\mathbb{F}_q[X]/(f)$. Then, use linear algebra to compute the least degree $G \in \mathbb{F}_q[X]$ such that $G(g) \equiv 0 \pmod{f}$. All of G's zeroes lie in $\mathbb{F}_q$, and if $G(a) = 0$, $\gcd(g - a, f)$ splits f. This construction uses $O((\deg f + \lg q)(\lg f)^2)$ bit operations.

36. **(a)** Let $f = 2X^2 + X + 1$ and $g = 2X + 1$ in $\mathbb{Z}/(4)[X]$. Then $fg = 3X + 1$.

(b) In $\mathbb{Z}/(8)[X]$, we have $(X + 2)^2 = (X - 2)^2$. We first verify that the polynomials $X \pm 2$ are irreducible. (We say that a non-unit f is *irreducible* if whenever $f = gh$, then g or h must be a unit.) Suppose that $X \pm 2 = gh$. Reducing this mod 2 and applying unique factorization in $\mathbb{F}_2[X]$, we see that one of the factors, say g, has the form $\epsilon + 2g'$, with $\epsilon \in (\mathbb{Z}/(8))^*$. This is a unit, since $g(\epsilon - 2g')(\epsilon^2 - 4(g')^2) = \epsilon^4$. Now, we must rule out the possibility that $X + 2$ is a unit multiple of $X - 2$. Suppose that $(X + 2)u(X) = X - 2$, with u a unit. Substituting $X = Y - 2$, we obtain $Yu(Y - 2) = Y - 4$, which is impossible since the right-hand side has a constant term.

(c) Continuing with this example, let $f = X + 2$ and $g = X - 2$. Any common divisor d of f and g must be a unit. (If not, we would have $X + 2 = ad$ and $X - 2 = bd$; since f and g are irreducible, a and b are units. But then we would have $X + 2 = (X - 2)b^{-1}a$, which was ruled out in (b).) This shows that 1 is a gcd of f and g. On the other hand, since 4 belongs to the ideal I generated by f and g, we have $\mathbb{Z}/(8)[X]/I \cong \mathbb{Z}/(4)$. Therefore, 1 does not belong to I and cannot be expressed as a linear combination of f and g.

Phenomena related to (a) are discussed more fully by Boos [1948]. Examples (b) and (c) come from Zassenhaus [1978], who points out that greatest common divisors need not even exist in $\mathbb{Z}/(p^n)[X]$. Chapter XIII of McDonald [1974] discusses the general theory of polynomials over finite local rings (of which $\mathbb{Z}/(p^n)[X]$ is an example).

37. This is proved in Lang [1986, p. 42].

For other "starting criteria," see Cauchy [1847]; Thurston [1943]; and Carlitz [1953b].

38. **(a)** To show $((\mathbb{Z}/(2^n))^*)^2 = \{a \in (\mathbb{Z}/(2^n))^* : a \equiv 1 \pmod 8\}$, prove that both subgroups have index 4 and the first is contained in the second.

(b) By the proof of (a), each square has exactly 4 square roots; the ones given are all distinct.

(c) Suppose $x^2 \equiv a \pmod{2^k}$. Then

$$x' = x - \left(\frac{x^2 - a}{2}\right) \cdot x^{-1}$$

is a square root of $a \pmod{2^{2k-2}}$. Starting with a square root of $a \pmod 8$, and using this $O(\lg n)$ times, we can obtain a square root of $a \pmod{2^n}$. The total work is within a constant factor of the work required for the last stage: $O(n^2)$ bit operations.

39. The following algorithm is from Cipolla [1907b]. Consider the formal power series

$$(1 - Z)^{1/p} = 1 - c_1 Z - c_2 Z^2 - \cdots .$$

We observe that

$$c_i = (-1)^{i+1} \frac{\prod_{0 \le j \le i-1}(1 - jp)}{p^i \, i!},$$

so that $\hat{c}_i = p^{2i} c_i$ is a p-adic integer. (That is, it can be written as a fraction whose denominator is not divisible by p.) To find an $x \in \mathbb{Z}/(p^n)$ with $x^p = a$, write

$$a = a^p(1 - p^2 w),$$

where $w = (1 - a^{1-p})/p^2$. Then

$$x = a\left(1 - \sum_{1 \le i \le k} \hat{c}_i w^i\right),$$

with $k = n+1+\lfloor (n-2)/(p-2) \rfloor$. To get the running time bound, observe that we can compute $\hat{c}_i$ from $\hat{c}_{i-1}$ using $O(1)$ operations in $\mathbb{Z}/(p^n)$, and that $k = O(n)$.

For another algorithm (which uses $O((\lg q)^3)$ bit operations), see Volpi [1995].

40. The essential idea is to find new polynomials λ, μ (satisfying the same degree constraints) making

$$\lambda g + \mu h = 1 \bmod p^n.$$

Once this is done, we can go from a factorization mod p^n to a factorization mod p^{2n} with one loop iteration. (The same equations work, provided we replace p^{i-1} by p^n, and compute q mod p^n.)

To get λ, we use Newton's method. Let $A = \mathbb{Z}[X]/(h)$. Suppose that $\lambda g \equiv 1 \bmod p^{n/2}$ in A. If

$$\lambda' = 2\lambda - g\lambda^2,$$

(this is just the usual Newton iteration for computing g^{-1}), then

$$\lambda' g = 1 - (1 - \lambda g)^2,$$

which shows that $\lambda' g \equiv 1 \bmod p^n$. Using a similar computation for μ, we obtain polynomials λ', μ' with $\deg \lambda' < \deg h$, $\deg \mu' < \deg g$, and

$$\lambda' g + \mu' h \equiv 1 \pmod{p^n}.$$

(This relation can be proved by applying the Chinese remainder theorem to the ring $\mathbb{Z}/(p^n)[X]/(gh)$.)

There are many variants of this idea. A particularly clear exposition is given by Davenport, Siret, and Tournier [1988]. The quadratically convergent Hensel method for polynomial factorization is apparently due to Rychlik [1924]. For other such algorithms, see Zassenhaus [1969, 1978]; von zur Gathen [1984c]; Lugiez [1985]; and Böffgen and Reichert [1987].

41. The first assertion is clear, since $f \equiv X^2 \pmod{p}$. To prove the second, let a be a square root of q mod p. Then $\nu_p(f(\pm ap)) \ge 3$, whereas $\nu_p(f'(\pm ap)) = 1$. Applying Exercise 37, we have the factorization

$$f(X) \equiv (X - pa + \cdots)(X + pa + \cdots) \bmod p^n$$

whenever $n \ge 2$.

42. Since p-adic numbers are not, in general, finitely representable, we first have to decide what we mean by "algorithm." One interpretation is the following. We will design processes that operate on infinite sequences of p-adic digits, such that: i) the n-th output digit depends only on the first n digits of the inputs, ii) once an output digit is produced it is never changed, and iii) the n-th output digit appears after a finite amount of time.

For simplicity we discuss only p-adic integers. Assume the digits are chosen from $\{0, \ldots, p-1\}$, and let the inputs be

$$x = \cdots + x_2 p^2 + x_1 p + x_0$$

and

$$y = \cdots + y_2p^2 + y_1p + y_0$$

The digits of

$$x + y = \cdots + z_2p^2 + z_1p + z_0$$

can be obtained with the following scheme. Let $c_{-1} = 0$, and compute for $i = 0, 1, 2, \ldots$

$$z_i = x_i + y_i + c_{i-1} \bmod p;$$

$$c_i = \left\lfloor \frac{x_i + y_i + c_{i-1}}{p} \right\rfloor.$$

Subtraction is similar, but with borrowing instead of carrying. To obtain

$$xy = \cdots + w_2p^2 + w_1p + w_0,$$

we set $c_{i-1} = 0$, then compute for $i = 0, 1, 2, \ldots$

$$w_i = x_0y_i + \cdots + x_iy_0 + c_{i-1} \bmod p;$$

$$c_i = \left\lfloor \frac{x_0y_i + \cdots + x_iy_0 + c_{i-1}}{p} \right\rfloor.$$

To implement division, it is enough to give a procedure for inversion and combine this with multiplication. If $x_0 = 0$, then the inverse cannot exist (if it did, we could reduce mod p and get a contradiction), so we assume $x_0 \neq 0$. Let

$$x^{-1} = u = \cdots + u_2p^2 + u_1p + u_0.$$

We can find the digits of u with the following scheme. Let $u_0 = x_0^{-1} \bmod p$ and $c_0 = \lfloor (x_0u_0 - 1)/p \rfloor$. Then, for $i = 1, 2, 3, \ldots$, we find u_i such that

$$x_0u_i + x_1u_{i-1} + \cdots + x_iu_0 + c_{i-1} \equiv 0 \bmod p$$

and set

$$c_i = \left\lfloor \frac{x_0u_i + x_1u_{i-1} + \cdots + x_iu_0 + c_{i-1}}{p} \right\rfloor.$$

These algorithms were given by Hensel [1904].

We note that the first n digits of a sum or difference are obtained using $O(n \lg p)$ bit operations, whereas computing the first n digits of a product or inverse takes time $O((n \lg p)^2)$. These procedures are efficient, in the sense that the time required to produce m bits of output is bounded by a polynomial in m. It is interesting that addition and subtraction can be done by finite-state machines (using $O(1)$ states), but the other operations cannot be.

43. This result was stated without proof by Pellet [1870]. For the proof, see, e.g. Lidl and Niederreiter [1983, p. 127]. Equations of the form $X^p - X - \alpha = 0$ in characteristic p are often called *Artin-Schreier equations*, after a well-known paper by Artin and Schreier [1927].

44. Using Lagrange multipliers, we see that

$$\max \left\{ \sum_{1 \le i \le r} x_i^m : x_1 \cdots x_r = n, x_i > 0 \right\}$$

is attained when $x_1 = \cdots = x_r$. Therefore the sum is bounded by $kn^{m/k}$. This equals n^m when $k = 1$, and, since $k \le n$, is bounded by $n^{m/2+1}$ otherwise.

45. **(a)** See Gauss [1801, §356].

(b) Gauss [1801, §358] first determined the explicit form of the cubic period equation. To find L and M within the stated time bound, form the values of $27m^2$ and $4q - \ell^2$ for $\ell, m \le 2\sqrt{q}$, sort these, and look for a match.

Similar explicit forms are known for other small values of e, but the history is too complicated to trace here. For references and formulas, see Berndt and Evans [1981] ($e \le 5$); D. H. Lehmer and E. Lehmer [1984] ($e = 6$); and Evans [1983] ($e = 8$). (According to H. J. S. Smith [1860], formulas for $e = 8, 12$ were known to Jacobi.)

46. For $n = 2$, this is Theorem 7.8.2. (An algorithm using Ψ_2 will have the same complexity.) For $n = 3$, we find the least prime $q \equiv 1 \pmod 3$ with $q \ne p$ and $p^{(q-1)/3} \not\equiv 1 \pmod q$. Using Euler's criterion and the prime number theorem, the cost is $O(q(\lg p + (\lg q)^2))$, and this is $O((\lg p)^3)$ by Theorem 8.7.13. We can get an irreducible polynomial of degree 6 within this time bound by combining degree 2 and degree 3 polynomials, as indicated in the proof of Theorem 7.8.3.

47. Cardano's formula is discussed in detail by van der Waerden [1970, v. 1]. Using the results presented therein, it is not difficult to verify that $3^2 v_1 v_2 = -3B$, from which it follows that $u_1 u_2 = -B^3/27$. Also, $u_1 + u_2 = v_1^3 + v_2^3 = -C$. The discriminant of $h = 0$ is $D = ((y_1 - y_2)(y_2 - y_3)(y_1 - y_2))^2$, which is a square in $\mathbb{F}_{p^m}$, as is -3 by quadratic reciprocity. Therefore, the

$$v_i^3 = 27\left(\frac{-27}{2}C \pm \frac{3}{2}\sqrt{-3D}\right)$$

are cubes in $\mathbb{F}_{p^m}$. (They are distinct because $D \ne 0$.)

48. Let $q = p^n$. We first use Theorem 7.8.1, as in the proof of Theorem 7.8.6. This produces an equation of degree at most 4, with distinct roots in $\mathbb{F}_p$, and previous results suffice to handle all cases but degree 4. Therefore, we may assume that

$$f(X) = X^4 + a_3X^3 + a_2X^2 + a_1X + a_0$$

is a polynomial in $\mathbb{F}_p[X]$ with four distinct zeroes in $\mathbb{F}_p$. (Necessarily, then, $p \ge 5$.) The following algorithm from Cauchy [1829] reduces the factorization of f to the solution of three lower-degree equations. Put $X = Y - a_3/4$, and the equation becomes

$$g(Y) = Y^4 + BY^2 + CY + D = 0.$$

We may assume $D \ne 0$. If C vanishes mod p, we take the square root of each zero of $U^2 + BU + D$. Otherwise, we form

$$h(U) = U^3 + 2BU^2 + (B^2 - 4D)U - C^2;$$

this has three distinct roots, all squares, in $\mathbb{F}_p$. Let u_1 and u_2 be two of the roots (which we can find via Theorem 7.8.6), then put $v_1 = \sqrt{u_1}$, $v_2 = \sqrt{u_2}$, and $v_3 = -C/(v_1v_2)$. The roots $y_1, \ldots, y_4$ of $g(Y) = 0$ are obtained by solving the linear equations

$$\begin{aligned} y_1 + y_2 + y_3 + y_4 &= 0 \\ y_1 - y_2 + y_3 - y_4 &= 2v_1 \\ y_1 - y_2 - y_3 + y_4 &= 2v_2 \\ y_1 + y_2 - y_3 - y_4 &= 2v_3. \end{aligned}$$

For the theory behind this algorithm, we refer the reader to van der Waerden [1970, v. 1]. (It helps to observe that $(y_1 - y_2 + y_3 - y_4)^2$ is the same as $4(y_1 + y_3)(y_2 + y_4)$ whenever $\sum y_i = 0$.)

The running time is a consequence of Theorem 7.8.4 and Theorem 7.8.6.

49. Let p be the characteristic of $\mathbb{F}_q$. We may as well assume that p is odd (otherwise we can use linear algebra), f is monic, and $\deg a < \deg f$. Combining a gcd-free basis construction with squarefree decomposition, we first find a factorization

$$f = f_1^{e_1} \cdots f_r^{e_r},$$

in which the f_i are squarefree and pairwise relatively prime. (See, e.g. Bach, Driscoll, and Shallit [1990, 1993].) After applying distinct degree factorization to each f_i, we can assume that the irreducible factors of f_i in $\mathbb{F}_q[X]$ all have the same degree. By the Chinese remainder theorem, a is a square mod f iff a is a square mod $f_i^{e_i}$, for $1 \le i \le r$. Thus, it is enough to consider the case $r = 1$. To simplify the notation, assume $f = g^e$, where g is squarefree and factors over $\mathbb{F}_q$ into s irreducible factors of degree d. Let $t = \lceil \log_p e \rceil$ and $R = \mathbb{F}_q[X]/(g^e)$. There are now four cases: i) if $a = 0$, it is a square; ii) if $a \in R^*$, then a is a square iff $a^{p^t(q^d-1)/2} \equiv 1 \pmod{f}$; iii) if $a \notin R^* \cup \{0\}$ and $s > 1$, use $\gcd(a, g)$ to split g, and apply the test recursively; iv) if $a \notin R^* \cup \{0\}$ and $s = 1$, write $a = bf^\nu$, where $f^\nu \,\|\, a$; necessarily $b \in R^*$, and a is a square iff ν is even and $b^{p^t(q^d-1)/2} = 1$. The correctness of this test relies on the Chinese remainder theorem and Theorem 6.6.6.

50. Consider the irreducible factorization

$$f = \prod_{i \le i \le r} f_i^{e_i}.$$

Let d_i be the degree of f_i. As in the solution to Exercise 49, we can compute $r, d_1, \ldots, d_r$, and $e_1, \ldots, e_r$. Let $t_i = \lceil \log_p e_i \rceil$. Now use the following formulas.

(a) $\varphi(f) = \prod_{1 \le i \le r} q^{e_i}(q^{d_i} - 1)$;

(b) $\omega(f) = r$;

(c) $\omega'(f) = \sum_{1 \le i \le r} e_i$;

(d) $d(f) = \prod_{1 \le i \le r}(e_i + 1)$;

(e) $\sigma(f) = \prod_{1 \le i \le r}(1 + q^{d_i} + \cdots + q^{d_i e_i})$.

(f) $\lambda(f) = \mathrm{lcm}_{1 \le i \le r}\{(q^{d_i} - 1)p^{t_i}\}$.

Formulas (a)–(e) can be found in Carlitz [1930, 1932]. Formula (f) is a consequence of Theorem 6.6.6.

Chapter 8

1. Nielsen [1906] devoted a book to the logarithmic integral, including methods by which it could be calculated. Nowadays, it is more common to express it in terms of the *exponential integral*, given by the Cauchy principal value of

$$\mathrm{Ei}(x) = \int_{-\infty}^{x} t^{-1} e^{t}\, dt.$$

Thus $\mathrm{li}(x) = \mathrm{Ei}(\log x) - \mathrm{Ei}(\log 2)$. Cody and Thatcher [1969] gave accurate approximations for the exponential integral. Riemann's approximation, in the form

$$R(x) = \sum_{n \geq 1} \frac{\mu(n)}{n} \mathrm{li}_0(x^{1/n}),$$

can be computed easily using a result of Gram [1884]:

$$R(x) = 1 + \sum_{n \geq 1} \frac{(\log x)^n}{n\zeta(n+1)n!}.$$

(See Cody, Hillstrom, and Thatcher [1971] for accurate numerical approximations to the zeta function.)

Tables listing the error in these approximations can be found in D. N. Lehmer [1914]; Ingham [1932, p. 7]; D. H. Lehmer [1959]; Bohman [1972]; Lagarias, Miller, and Odlyzko [1985]; Shiu [1987]; and Deleglise and Rivat [1995]. Riesel and Göhl [1970] studied a more precise version of Riemann's formula, which uses complex zeroes of the zeta function.

2. This proof is from Kronecker [1901]. Sylvester [1888] gave the following elegant variation of Euler's argument. Since the prime factors of any number $\leq n$ are themselves primes $\leq n$, we have

$$\prod_{p \leq n} (1 - 1/p)^{-1} \geq 1 + 1/2 + \cdots + 1/n.$$

The right-hand side is the n-th harmonic number, which goes to infinity with n. (See Theorem 2.5.3.) Therefore, there are infinitely many primes.

3. Wright [1955] proved that every solution to the differential equation for $\Delta(x)$ is asymptotic to $1/\log x$. (See also Wright [1961].)

Legendre [1798, p. 18] believed that the number of primes $\leq x$ was well approximated by $\prod_{p<\sqrt{x}}(1 - p^{-1})$, and provided tables for this purpose. However, if n is a random integer from $[1, x]$, the events that $p \mid n$ are only approximately independent, and this approximation fails badly even when a few primes are considered. (Note that $m \mid n$ with probability $1/m + O(1/x)$. Taking $m = p_1 \cdots p_r$, the two terms are comparable when $r \approx (\log x)/(\log\log x)$, by the prime number theorem.) The deviation from independence was studied numerically by Furry [1942], who noted that "the last and largest of the trial divisors finds roughly twice as large a proportion of victims among the survivors of previous trials as it would in a virgin population."

This heuristic argument has been given by many authors, including Cherwell [1941] and Hoffman de Visme [1961] (who identified the error); and Sispánov [1942] and Schroeder [1984] (who did not). Sutton [1937] and Pólya [1959] observed that the problem is "fixed" if one sieves up to x^c, where $c = \exp(-\gamma)$. See also Sylvester [1888]; and G. Hardy and Littlewood [1922].

4. Proofs of (a)–(c) can be found in Ingham [1932].

To prove (d), we follow G. Hardy and Wright [1985, p. 10]. We have $n = \Theta(p_n/(\log p_n))$ which implies

$$\log n = \log p_n - \log\log p_n + O(1) = (1 + o(1))\log p_n.$$

Therefore, $p_n = \Theta(n \log p_n) = \Theta(n \log n)$.

To prove (e), observe that if $\omega(n) = r$, then $p_1 \cdots p_r \leq n$. But $p_1 \cdots p_r = \exp(\vartheta(p_r)) \geq \exp(Cr \log r)$ for some $C > 0$. Thus, $r \log r = O(\log n)$, and as above, this implies $r = O((\log n)/(\log\log n))$.

Result (f) is due to S. Wigert [1906]. See also E. Landau [1909, Vol. 1, p. 219.]

Result (g) is due to E. Landau [1903a].

5. For $s > 1$, we have

$$\Gamma(s) = \int_0^\infty y^{s-1} e^{-y}\, dy,$$

so

$$\Gamma'(s) = \int_0^\infty (\log y) y^{s-1} e^{-y}\, dy.$$

Letting $s \to 1^+$, using the special value $\Gamma'(1) = -\gamma$, and substituting $y = \alpha x$ yields (a).

To prove (b), we use Euler's summation formula (Corollary 2.5.2) with $f(x) = x^{-s}$ to find

$$\sum_{1 \leq n \leq x} \frac{1}{n^s} = 1 - \{x\}x^{-s} + \frac{1}{s-1}\left(1 - \frac{1}{x^{s-1}}\right) - s\int_1^x \{t\}t^{-s-1}\, dt,$$

where $\{x\} = x - \lfloor x \rfloor$. If $s > 1$, we let $x \to \infty$ and conclude

$$\zeta(s) = \frac{1}{s-1} + O(1).$$

Proofs of (c) and (d) can be found in Ingham [1932, p. 23].

6. **(a)** By Theorem 8.2.5, it suffices to prove $\sum_{p^k \leq x} 1 = \pi(x) + \Theta(\sqrt{x}/(\log x))$. We have

$$\sum_{p^k \leq x} 1 = \sum_{1 \leq k \leq \log_2 x} \pi(x^{1/k}) \leq \pi(x) + \pi(\sqrt{x}) + (\log_2 x)\pi(\sqrt[3]{x}) = \pi(x) + \Theta\left(\frac{\sqrt{x}}{\log x}\right).$$

(b) Using part (b) of Exercise 4, it is enough to prove that $\sum_{\substack{p^k \leq x \\ k \geq 2}} 1/p^k = O(1)$. We have

$$\sum_{k \geq 2} \sum_{p^k \leq x} \frac{1}{p^k} \leq \sum_{k \geq 2} \sum_{n \geq 2} \frac{1}{n^k} = \sum_{n \geq 2} \frac{1}{n(n-1)} = 1.$$

7. See Apostol [1976, p. 279].

8. See Ingham [1932, p. 35].

9. We have

$$\psi(x) - \theta(x) \le \sum_{2\le k\le \log_2 x} \sum_{p\le x^{1/k}} \log p \le (\log x) \sum_{2\le k\le \log_2 x} \sum_{p\le x^{1/k}} 1 \le \frac{\sqrt{x}(\log x)^2}{\log 2}.$$

Once we know Theorem 8.2.5, however, we can sharpen this to $\psi(x) = \vartheta(x) + O(\sqrt{x})$. (See the solution to Exercise 11.) The improved estimate is due to Chebyshev [1852].

10. This exercise is from Cesàro [1894], who corrected a similar (but erroneous) formula found by Pervushin [1898]. The series for $\mathrm{li}(x)$ is easily obtained using integration by parts. (See, e.g. E. Landau [1909, Vol. 1].) It can be shown that

$$\mathrm{li}^{-1}(y) = y \log y(1 + f(y)),$$

where $f(y)$ is a power series in $\log\log y$ and $1/\log y$. The coefficients of f can be computed as far as desired; recurrence relations for this purpose were given by Cipolla [1902].

11. Using our bounds from Section 8.2, we prove this for $c = 6$. From the proof of Theorem 8.2.5, we have $\vartheta(x) \le (4\log 2)x$ and $\psi(x) \ge x/2 - 3/2$. Observing that $(\log_2 x)\sqrt[3]{x} < 3\sqrt{x}$ for $x > 4096$, we have

$$\psi(x) - \vartheta(x) \le \vartheta(\sqrt{x}) + (\log_2 x)\vartheta(\sqrt[3]{x}) \le 16\log 2\sqrt{x}$$

in this range. Using this estimate, we find

$$\begin{aligned}\vartheta(cx) - \vartheta(x) &= \psi(cx) - \vartheta(x) - (\psi(cx) - \vartheta(cx))\\ &\ge \left(\frac{1}{2} - \frac{4\log 2}{c}\right) y^2 - 16\log 2 y - \frac{3}{2}\end{aligned}$$

when $y = \sqrt{cx}$. Substituting $c = 6$ and solving this for y, we find that this is positive when $y > 292.742\ldots$, i.e. for $x > 14283$. This proves the result except for a finite range of x. To prove it completely, it suffices to check that there is a prime between n and $6(n-1)$ for all integers n, $2 \le n \le 14283$. This is easily done by a computer program.

For $c = 2$, this result was conjectured by Bertrand [1845] and proved by Chebyshev [1852]. See Moree [1993b] for extensions to arithmetic progressions.

12. Let q be the smallest prime such that n is not a qth power. Considering the exponents in n's prime factorization, we have

$$n \ge 2^{\prod_{p<q} p} = 2^{e^{\vartheta(q-1)}}.$$

Since $\vartheta(q-1) = \Theta(q)$, the result follows. This is best possible, since when $n = 2^{\mathrm{lcm}(2,\ldots,k)}$, $\log_2 \log n = \psi(k) \sim k$.

13. Let $x \ge 1$ be an integer. We have

$$\sum_{1\le d\le x} \psi\left(\frac{x}{d}\right) = \sum_{1\le d\le \frac{x}{\log x}} \psi\left(\frac{x}{d}\right) + \sum_{\frac{x}{\log x}<d\le x} \psi\left(\frac{x}{d}\right)$$

By Theorem 8.2.5, the second sum is bounded by a constant times

$$x \sum_{\frac{x}{\log x} < d \le x} \frac{1}{d},$$

which is $O(x \log\log x)$, by Theorem 2.5.3. On the other hand, if $\psi(x) \sim cx$, then the first sum is asymptotic to

$$cx \sum_{1 \le d \le \frac{x}{\log x}} \frac{1}{d} \sim cx \log x.$$

Since $\log(x!) \sim x \log x$ by Stirling's approximation, we must have $c = 1$. This result was proved by Chebyshev [1851].

14. See G. Hardy and Wright [1985, p. 267].

15. This exercise is due to E. Landau [1903c]. For recent work on this problem and more references, see W. Miller [1987]; and Massias, Nicolas, and Robin [1988].

16. This proof comes from Ingham [1932, pp. 34–35]. The idea of using the Riemann-Lebesgue theorem is apparently due to E. Landau [1908]. The bound on ζ'/ζ is due to E. Landau [1903b]. A complete proof along these lines may be found in Apostol [1976, Chapter 13], with the exception of the Riemann-Lebesgue theorem (part (g)), for which see Royden [1968].

17. The result is due to E. Cahen [1894]. For a proof, see Rademacher [1973, p. 115].

18. Assertions (b) and (c) are equivalent by Exercise 9. For the rest, see Ingham [1932, p. 83].

19. For proofs that (a) implies (b), see Titchmarsh [1930] (for $\sigma = 1/2$), and Prachar [1957, p. 235] (for $\sigma \ge 1/2$).

Below, we prove that (b) implies (a). Define $\psi(x, n, a)$ and $\vartheta(x, n, a)$ as in Theorem 8.8.13. As in the proof of Theorem 2.7.1, we have

$$\vartheta(x, n, a) = \frac{1}{\varphi(n)} \int_2^x dt + \eta(x) \log x + \int_2^x \frac{\eta(t)\, dt}{t},$$

where $\eta(x) = \pi(x, n, a) - \mathrm{li}(x)/\varphi(n)$. Using (b) with ϵ replaced by $\epsilon/2$, we find that

$$\vartheta(x, n, a) = \frac{x}{\varphi(n)} + O(x^{\sigma+\epsilon/2} \log x).$$

By Exercise 9, we have

$$0 \le \psi(x, n, a) - \vartheta(x, n, a) \le \psi(x) - \vartheta(x) = O(x^{1/2+\epsilon}(\log x)^2),$$

so

$$\psi(x, n, a) = \frac{x}{\varphi(n)} + O(x^{\sigma+\epsilon}).$$

Let χ be a Dirichlet character mod n, and define 1_χ as in Theorem 8.5.2. Multiplying the last estimate by $\chi(a)$, summing over $a \in (\mathbb{Z}/(n))^*$ and using Exercise 6.18, we find

$$\sum_{m \le x} \Lambda(m)\chi(m) = 1_\chi x + O(x^{\sigma+\epsilon}),$$

which implies

$$\sum_{m \le x} \left(\Lambda(m)\chi(m) - 1_\chi \right) = O(x^{\sigma+\epsilon}).$$

Now, we observe that

$$\sum_{m \ge 1} \frac{\Lambda(m)\chi(m) - 1_\chi}{m^s} = -\left(\frac{L'}{L}(s, \chi) + 1_\chi \zeta(s) \right).$$

By Exercise 17, this function is analytic for $\Re(s) > \sigma + \epsilon$. Since the poles at $s = 1$ cancel, $L(s, \chi) \ne 0$ in this half-plane as well. Since ϵ is arbitrary, we get (a). This result seems to be folklore. We learned of the above proof from R. Roy (personal communication).

For a more precise version of this exercise, see see Bach, Giesbrecht, and McInnes [1991]. Presumably, an analogous result can be proved for Dedekind zeta functions.

20–23. For solutions to these exercises, see Edwards [1974]. Exercise 20 is from Backlund [1918]. Exercise 21 is due to Stieltjes [1889]. The functional equation of Exercise 22 was proved by Riemann [1860].

We remark that the "naive" algorithm to evaluate $\zeta(s)$ for real $s > 1$ (sum the series) does not lead to a polynomial-time algorithm, whereas the asymptotic expansion does. See H. Cohen and Olivier [1992]. It is not known how to compute $\zeta(1/2 + it)$ to relative accuracy 2^{-k} in polynomial time (i.e. using $(k(\log t))^{O(1)}$ bit operations).

The coefficients in the Laurent expansion of the zeta function around $s = 1$ are called *Stieltjes constants*. For information about these, see Kluyver [1924]; Wilton [1927]; Briggs [1955]; Briggs and Chowla [1955]; Liang and Todd [1972]; Israilov [1983]; and Bohman and Fröberg [1988].

24. This is due to van der Pol [1947], who built an electromechanical device to evaluate $\zeta(1/2 + it)$ in this manner. The reader may wish to experiment with the use of this formula, together with the fast Fourier transform (FFT), in locating zeroes of the zeta function.

25. Abel [1823] evaluated the integral in (a) when s was an even integer. For (b), see Sommerfeld [1956, p. 151].

26. For any n, we have $P(n) \ge p_{\varphi(n)}$. If n is prime, $\varphi(n) = n - 1$. Therefore, for such n,

$$P(n) > (n-1)\log(n-1) = n \log n + O(\log n),$$

by Theorem 8.8.4. This result is not best possible, as Prachar [1961] proved that

$$P(n) = \Omega_\infty \left(n(\log n) \frac{(\log\log n)(\log\log\log\log n)}{(\log\log\log n)^2} \right).$$

The first nontrivial result on this problem seems to be that of E. Landau [1901], who proved that $P(n) > n + 1$ for all $n \ge 2$.

27. Let $C > 0$ be such that $P(n) \le C\varphi(n)(\log n)^2$. For each i with $1 \le i \le Cn(\log n)^2$, let $A(i)$ be 1 or 0 depending on whether i is prime or composite. (Store these values in an array.) Make a list of all the elements of $(\mathbb{Z}/(n))^*$ (say, by computing $\gcd(a, n)$ for every positive $a < n$). Then, for each $a \in (\mathbb{Z}/(n))^*$, find the least k for which $A(a + kn) = 1$. $P(n)$ is then the maximum of the $P(n, a)$ found in this manner.

28. We follow Bach, Giesbrecht, and McInnes [1991]. Fix $n \geq 1$ and $a \in (\mathbb{Z}/(n))^*$. To avoid problems later, we assume, unconventionally, that $2 \leq a \leq n+1$. Consider the probability that a random integer k, with $1 \leq k \leq x$ and congruent to a mod n, is prime. By Theorem 8.4.2, this is asymptotic to

$$\frac{\mathrm{li}(x)/\varphi(n)}{x/n} \sim \frac{n}{\varphi(n)\log x}$$

as $x \to \infty$. (We get an extra constant factor of $n/\varphi(n)$ from conditioning; a number is more likely to be prime is increased if it is relatively prime to n.)

Now, consider the following experiment: for each $k \geq 2$, congruent to a mod n, independently label k "prime" with probability $n/(\varphi(n)\log k)$. Let T_k be a 0-1 random variable indicating whether k was labelled prime, and let $X_k = T_k - \mathrm{E}[T_k]$. Choose $b_k = \sqrt{k}\log k$. We have

$$\sum_{k\geq 1} \frac{\mathrm{E}[X_k^2]}{b_k^2} \leq \sum_{k\geq 1} \frac{\mathrm{E}[T_k^2]}{b_k^2} = \sum_{k\geq 1} \frac{\mathrm{E}[T_k]}{b_k^2} = \sum_{k\geq 1} \frac{O(\log\log n)}{k(\log k)^3} < \infty.$$

Let $\Pi(x, n, a)$ be the number of integers $\leq x$ labelled "prime," when x is a positive integer congruent to a mod n. By Euler's summation formula (Corollary 2.5.2), we have

$$\begin{aligned}\mathrm{E}[\Pi(x,n,a)] &= \sum_{\substack{1\leq k\leq x \\ k\equiv a \pmod{n}}} \frac{n}{\varphi(n)\log k} \\ &= \frac{n}{\varphi(n)}\left[\int_0^{\frac{x-a}{n}} \frac{dt}{\log(a+nt)} + \frac{1}{\log a} - \int_0^{\frac{x-a}{n}} \frac{n(t-\lfloor t\rfloor)\,dt}{(a+nt)(\log(a+nt))^2}\right].\end{aligned}$$

Setting $u = a + nt$, we find

$$\frac{n}{\varphi(n)}\int_0^{\frac{x-a}{n}} \frac{dt}{\log(a+nt)} = \frac{1}{\varphi(n)}\int_a^x \frac{du}{\log u} = \frac{1}{\varphi(n)}[\mathrm{li}(x) - \mathrm{li}(a)]$$

and

$$\left|\int_0^{\frac{x-a}{n}} \frac{n(t-\lfloor t\rfloor)\,dt}{(a+nt)(\log(a+nt))^2}\right| \leq \int_a^x \frac{du}{u(\log u)^2} \leq \frac{1}{\log a}.$$

Therefore,

$$\mathrm{E}[\Pi(x,n,a)] = \frac{\mathrm{li}(x)}{\varphi(n)} + O(1)$$

as $x \to \infty$. By Lemma 8.3.2, the following holds with probability 1:

$$\Pi(x,n,a) = \frac{\mathrm{li}(x)}{\varphi(n)} + O(x^{1/2}\log x).$$

29. Let $H_k = 1 + \cdots + 1/k$ denote the k-th harmonic number. Let

$$\eta(j,n,a) = \begin{cases} p, & \text{if } j = p^k \text{ (a prime power) and } j \equiv a \pmod{n}; \\ 1, & \text{otherwise.}\end{cases}$$

Define

$$\delta(x, n, a) = \prod_{1 \le m < x} \prod_{1 \le j \le m} \eta(j, n, a).$$

We claim that the ERH holds iff

$$\left| H_{\delta(x,n,a)} - \frac{x^2}{2\varphi(m)} \right| \le 9x^{3/2}(H_n + 5) + 1 \tag{A.7}$$

for every $n \ge 1$, $a \in (\mathbb{Z}/(n))^*$, and integer $x \ge 2$. (Given particular values for n, a, and x, we can clearly check the inequality in a finite number of steps.)

To prove this, we first observe that for integer $x \ge 1$,

$$\log \delta(x, n, a) = \sum_{m<x} \psi(m, n, a) = \int_0^x \psi(t, n, a)\,dt = \sum_{\substack{m \le x \\ m \equiv a \pmod{n}}} \Lambda(m)(x - m).$$

(Here $\psi(m, n, a) = \sum_{\substack{m \le x \\ m \equiv a \pmod{n}}} \Lambda(m)$.) Therefore, if the inequality holds, we have

$$\sum_{\substack{m \le x \\ m \equiv a \pmod{n}}} \Lambda(m)(x - m) = \frac{x^2}{2\varphi(n)} + O(x^{3/2} \log n)$$

If χ is a Dirichlet character mod n, we can multiply this by $\chi(a)$ and sum over all $a \in (\mathbb{Z}/(n))^*$ to get

$$\psi_1(x, \chi) := \sum_{m \le x} \Lambda(m)\chi(m)(x - m) = 1_\chi \frac{x^2}{2} + O(x^{3/2} \log n),$$

where 1_χ is as defined in Theorem 8.5.2. Applying this bound to the formula

$$-\frac{L'}{L}(s, \chi) - 1_\chi \frac{s(s+1)}{2(s-1)} = s(s+1) \int_1^\infty \frac{\psi_1(t, \chi) - 1_\chi t^2/2}{t^{s+2}}\,dt,$$

we see that the function on the left is analytic for $\Re(s) > 1/2$, so the ERH is true. Conversely, if the ERH holds, we have the estimate

$$\left| \sum_{m \le x} \Lambda(m)\chi(m)(1 - m/x) - 1_\chi \frac{x}{2} \right| \le 9\sqrt{x}(\log n + 5).$$

(The constants 9 and 5, which are surely not optimal, can be obtained from some inequalities appearing in Bach [1982].) Multiplying this by $x\chi(a)^{-1}$ and summing over χ, we easily find

$$\left| \log \delta(x, n, a) - \frac{x^2}{2\varphi(n)} \right| \le 9x^{3/2}(\log n + 5).$$

whenever $a \in (\mathbb{Z}/(n))^*$. Since $H_n - 1 \le \log n \le H_n$, this implies (A.7).

The idea behind this proof is due to Shapiro, who found a corresponding formula equivalent to the Riemann hypothesis. (See M. Davis, Matiyasevich, and Robinson [1976].) This exercise has the following interesting consequence: if the ERH is independent of Peano arithmetic, it must be true.

30. Let $x = 65n^2(\log n)^4$, and choose a random integer k with $1 \le k \le (x-a)/n$. Then test $p = nk + a$ for primality using any of the methods in Chapter 9. By Theorem 8.8.18, the chance that p is prime is

$$\ge \frac{\mathrm{li}(x)/\varphi(n) - \sqrt{x}(\log x + 2\log n)}{x/n + O(1)} = \Omega\left(\frac{n/\varphi(n)}{\log n}\right),$$

and since $n \ge \varphi(n)$, this is $\Omega(1/(\log n))$. This exercise (with the larger value $x = n^3$) is from Rabin and Shallit [1986]. See also Boyar, Friedl, and Lund [1990].

31. Let $p_1, \ldots, p_k$ be the first k primes. Suppose q is a prime, congruent to 1 mod 8, such that

$$\left(\frac{q}{p_2}\right) = \cdots = \left(\frac{q}{p_k}\right) = 1.$$

By quadratic reciprocity,

$$\left(\frac{2}{q}\right) = \left(\frac{3}{q}\right) = \cdots = \left(\frac{p_k}{q}\right) = 1.$$

Let $m = 8p_2 \cdots p_k$. By Linnik's theorem, there is such a q,·no larger than m^C, so

$$\log q \le C(\vartheta(p_k) + \log 4) \le (C + o(1))p_k.$$

The first result follows, after noting that the least quadratic nonresidue mod q is larger than p_k. This implies the first result.

Fridlender [1949] and Salié [1949] proved this without considering the dependence on C. The extension to primitive roots is due to Turán [1950], who noted that the least primitive root cannot be smaller than the least quadratic nonresidue.

This result is not best possible, as Graham and Ringrose [1990] showed that the least quadratic nonresidue mod p is $\Omega_\infty((\log p)(\log\log\log p))$. (More is true if the ERH holds; see Exercise 56.) Elliott [1968] showed that the least k-th power nonresidue mod p is also $\Omega_\infty(\log p)$; here the implied constant depends on k.

32. Let χ denote the quadratic character mod p. By Dirichlet's theorem, there are prime powers $q_1 < q_2 < q_3 < \cdots$ with $\chi(q_i) = -1$. Let $x = q_k$. We have

$$\sum_{n \le x} \Lambda(n)\left(1 - \frac{n}{x}\right) = \sum_{n \le x} \Lambda(n)\chi(n)\left(1 - \frac{n}{x}\right) + 2\sum_{1 \le i < k} \Lambda(q_i)\left(1 - \frac{q_i}{x}\right)$$
$$+ \sum_{p^m \le x} \Lambda(p^m)\left(1 - \frac{p^m}{x}\right).$$

Assuming ERH, this implies

$$x/2 + O(\sqrt{x}) \le O(\sqrt{x}\log p) + O(k\log x) + O(\log x).$$

Therefore,

$$\sqrt{x} = O\left(\log p + k\frac{\log x}{\sqrt{x}}\right) = O(\log p + k),$$

from which the result follows.

33. Burgess [1967] showed this is true when a is fixed. This result, applied to Cipolla's algorithm, would give an alternative proof that square roots can be computed in $\mathbb{F}_p$ in deterministic polynomial time. (See Section 7.2.)

For an unconditional bound of $p^{1/(2\sqrt{e})+o(1)}$, see Friedlander [1973].

34. See H. W. Lenstra [1979].

35. Find a prime $p \equiv 1 \pmod{k}$. To see whether m is a k-th power, first compute $m^{(p-1)/k} \bmod p$. If this differs from 0 or 1, m cannot be a k-th power. Otherwise, compute $\lfloor m^{1/k} \rfloor$ using, say, Newton's method. We will think of the search for p as a "precomputation," and analyze the running time assuming that m is chosen uniformly from $\{1, \ldots, n\}$. Evidently, we can assume $k \leq \log_2 n$, so the chance that m passes the first test is $\Theta(1/k)$. By Linnik's theorem, $\lg p = O(\lg k) = O(\lg \lg n)$. Applying Theorem 5.4.1 and part (c) of Exercise 3.1, we get an expected running time of $O((\lg n)(\lg \lg n) + (\lg n)^2(\lg \lg n)/k)$. When $k = \Theta(\lg n)$, this is $O((\lg n)(\lg \lg n))$.

For more on this algorithm, and a generalization of Wagstaff's conjecture to the problem of finding several primes in an arithmetic progression, see Bach and Sorenson [1993].

The solutions to Exercises 36–39 rely on the following result from probability theory, called the *Borel-Cantelli lemma*. Suppose that $E_1, E_2, \ldots$ are events in some probability space. If $\sum_{i \geq 1} \Pr[E_i] < \infty$, then with probability 1, only finitely many E_i occur. For a proof, see Feller [1968, p. 201].

36. **(a)** This is proved in Feller [1968, p. 224].

(b) We have

$$\Pr[W(m) \geq t] \leq \sum_{1 \leq i \leq m} \Pr[i \text{ is omitted from the samples}] = m\left(1 - \frac{1}{m}\right)^t.$$

Let $t(m) = (2 + \epsilon)m \log m$. Because $(1 - 1/m)^m \leq e^{-1}$, the estimate above implies

$$\sum_{m \geq 1} \Pr[W(m) \geq t(m)] \leq \sum_{m \geq 1} \frac{1}{m^{1+\epsilon}} < \infty.$$

The result follows from the Borel-Cantelli lemma.

(c) Assume heuristically that the primes $p = 2, 3, 5, \ldots$ represent random elements of $(\mathbb{Z}/(n))^*$. Apply (b) with $m = \varphi(m)$, observing that $\log \varphi(m) \sim \log m$. This conjecture is from McCurley [1986], who derived it in a different way. Earlier, Wagstaff [1979] conjectured that $P(n) \sim \varphi(n) \log n \log \varphi(n)$, except on a set of density 0.

37. A proof of (b) appears in Bach [1995]. Part (c) is from Bach and Huelsbergen [1993]. Part (a) is a consequence of (c).

We note that if these bounds held for the actual elements $2, 3, 5, \ldots \in (\mathbb{Z}/(n))^*$, we could improve Theorem 7.8.2 (and many of its consequences) by a factor $\lg p$.

38. We use the following result from probability theory: if $Y_1, \ldots, Y_n$ are independent random variables with an exponential distribution and mean 1 (i.e., with $Y_i \geq 0$ and $\Pr[Y_i \geq y] = e^{-y}$), and S is their sum, then $\{Y_1/S, \ldots, Y_n/S\}$ are independent of S and have the same distribution as $X_1, \ldots, X_n$. Furthermore, the maximum Y_i has expected value H_n. (See Feller [1971, p. 76].)

To prove (a), observe that

$$H_n = \mathrm{E}[\max\{Y_i\}] = \mathrm{E}[\max\{X_i \cdot S\}] = \mathrm{E}[\max\{X_i\}] \cdot \mathrm{E}[S] = n\mathrm{E}[\max\{X_i\}].$$

To prove (b), let A_n be the event that one of $X_1, \ldots, X_n$ exceeds $(2 + \epsilon)\frac{\log n}{n}$, and let B_n be the event that one of $Y_1, \ldots, Y_n$ exceeds $(2 + 2\epsilon)\log n$. By the law of large numbers, $S \sim n$ with probability 1, so the chance that A_n occurs infinitely often is at most the chance that B_n occurs infinitely often. We also have

$$\Pr[B_n] \leq n \int_{(2+2\epsilon)\log n}^{\infty} e^{-t}\, dt = \frac{1}{n^{1+2\epsilon}}.$$

Applying the Borel-Cantelli lemma, we get (b).

This problem is well-known in probability theory. Result (a) goes back to Laplace [1812, p. 272]. We do not know a reference for (b).

39. For $n \geq 2$, label n "prime" with probability $1/\log n$, and let P_i be the i-th number labelled "prime." Let $n-k+1, \ldots, n-1, n$ be k consecutive numbers. The probability that none of them are labelled "prime" is at most $(1 - 1/(\log n))^k$. Let $k(n) = (1 + \epsilon)(\log n)^2$; we have

$$\sum_{n\geq 2}\left(1 - \frac{1}{\log n}\right)^{k(n)} \leq \sum_{n\geq 2}\frac{1}{n^{1+\epsilon}} < \infty.$$

By the Borel-Cantelli lemma, we have, almost surely,

$$P_i - P_{i-1} \leq (1 + \epsilon)(\log P_i)^2.$$

Since $P_i \sim i \log i$ with probability 1, the result follows.

40. Let $p_1, \ldots, p_k$ be the first k primes, and form $A = \prod_{1\leq i\leq k} p_i$. Using the Chinese remainder theorem, find a positive number $t < p_{k+1}p_{k+2}$, satisfying

$$tA \equiv \begin{cases} +1 \pmod{p_{k+2}}; \\ -1 \pmod{p_{k+1}}. \end{cases}$$

Then

$$\{tA, tA \pm 1, tA \pm 2, \ldots, tA \pm p_k\}$$

is a set of $2p_k + 1$ consecutive composite numbers. The complexity is dominated by the cost to compute A, which is $O(k^2(\lg k)^2)$ by the prime number theorem and Exercise 3.3. Taking $m = 2p_k + 1$ and observing that $m = \Theta(k \lg k)$ yields the result.

This algorithm was found by Backlund [1929] and S. Johnson [1965]. Neither author discussed the running time.

41. This method is due to Lander and Parkin [1967b], who did not discuss its running time. We will be content with a rough, necessarily heuristic, analysis of this matter. Using the Cramèr model, we should have $\Delta \approx (\log p)^2$. The Δ numbers $\leq q$ are all composite with probability $(1 - 1/(\log p))^{\Delta}$, so we expect to update p about

$$\left(1 - \frac{1}{\log p}\right)^{-(\log p)^2} \approx p$$

times before a record is found. On average, each update increases p by $\Delta - \log p \approx (\log p)^2$, so a new record value of Δ should occur around $p(\log p)^2$. The average number of prime tests needed to find this place should be about $p \log p$, which beats direct search by the factor $\log p$. Using randomized primality tests, the program uses $O(\log p)$ space.

The main advantage of this method seems to be its small space usage. In practice, it is outperformed by the segmented sieve of Eratosthenes, which lists all primes up to a given bound in nearly linear time. (See Section 9.8.)

42. See Rubinstein [1993] for $k = 2$, and Bateman and Horn [1962] for the general case.

43. This exercise is from Riesel [1985a, p. 67] (who did not discuss its running time). Observe that if $\nu(p) = p$, we must have $p \leq k$. Therefore, the following algorithm does the job. Make a list of all primes $p \leq k$. For each such p, compute $a_1 \bmod p, \ldots, a_k \bmod p$, and use these values to determine $\nu(p)$. If $|a_i| \leq A$, we use $O(k(\lg A)(\lg p))$ bit operations to find the residues mod p, and $O(k(\lg k)(\lg p))$ to determine $\nu(p)$, say, by sorting. By the prime number theorem, the total work is $O(k^2(\lg k + \lg A))$. (We can safely ignore the cost to find the primes, since it is much less than this.)

44. Let us examine the consequences of the (admittedly naive) idea that n is prime with "probability" $1/\log n$. The "expected" number of Mersenne primes $\leq x$ is about

$$\sum_{p \leq \log_2 x} \frac{1}{p \log 2} = \Theta(\log \log \log x),$$

whereas the "expected" number of Fermat primes is

$$\sum_{k \geq 1} \frac{1}{2^k \log 2} < \infty.$$

(Compare G. Hardy and Wright [1985, p. 15].)

Both of these calculations suffer from the defect that the prime divisors of Mersenne and Fermat numbers are restricted to certain congruence classes, which increases the chance (in some sense) that numbers having these forms are prime. For example, if $q \mid 2^p - 1$, then $q \equiv \pm 1 \pmod 8$ and $q \equiv 1 \pmod p$, and if $q \mid 2^{2^k} + 1$ with $k \geq 2$, then $q \equiv 1 \pmod{2^{k+2}}$. These results are due to Euler; see Exercises 5.43 and 5.44.)

Wagstaff [1983], using a more sophisticated probabilistic argument, conjectured that there are about $(e^{\gamma}/\log 2) \log \log x$ Mersenne primes $\leq x$. This is supported by numerical data. See also Good [1955]; Gillies [1964]; Ehrman [1967]; Shanks and Kravitz [1967]; Schroeder [1983]; Wagstaff [1983]; and Lenstra [1986a].

Schinzel and Sierpiński [1958] noted that if $4n - 1$, $8n - 1$ are simultaneously prime infinitely often, there are infinitely many Mersenne composites.

45. Using polar coordinates and the easily-proved inequality $(x^n + y^n)^{1/n} \geq (\frac{x^2+y^2}{2})^{1/2}$, we have

$$\int_B^\infty \int_B^\infty \frac{dx\,dy}{n(x^n + y^n)^{1-1/n}} \leq \frac{2^{(n+1)/2}}{n} \int_{\sqrt{2}B}^\infty \frac{dr}{r^{n-2}} \int_0^{\pi/4} d\theta = \frac{\pi B^{3-n}}{n(n-3)}.$$

If x, y, z are positive integers with $x^n + y^n = z^n$, and $n \geq 3$, then $x, y \geq 2$. Hence the "expected" number of solutions is $O(n^{-2}2^{-n})$, which decreases rapidly with n. This argument comes from an unpublished manuscript of R. Feynman. Knowing that Fermat's last theorem is true for $n \leq 100$, and using another estimate for the integral, Feynman concluded "for my money Fermat's theorem is true." See Schweber [1986]. Skeptics may wish to note that a similar argument applies to $x^{n-1} + y^{n-1} = z^n$, which has the solution $x = y = z = 2$ for every positive integer n.

For other probability arguments in favor of Fermat's last theorem, see Erdős and Ulam [1971]; Kofler [1972].

46. By the prime number theorem, we have

$$\sum_{\substack{\mathbf{Np}\leq x \\ \deg \mathbf{p}>1}} 1 = \sum_{p\leq\sqrt{x}} \sum_{\substack{\mathbf{p}|p \\ \deg \mathbf{p}>1}} 1 \leq [K:\mathbb{Q}] \sum_{\mathbf{p}\leq\sqrt{x}} 1 = O(\sqrt{x}),$$

and from this, the result follows.

47. The powers $1, \alpha, \ldots, \alpha^{n-1}$ form a basis for $K/\mathbb{Q}$. Relative to this basis, A is a companion matrix and (a) and (b) are immediate. See, e.g., H. Cohn [1978, p. 74]. To prove (c), we use the fact that linear equations over $\mathbb{Q}$ can be solved in deterministic polynomial time. (See, e.g., Schrijver [1986].) Suppose that $\beta = \sum_{0\leq i<n} b_i\alpha^i$ is an algebraic number, with $c_i \in \mathbb{Q}$. Using linear algebra, we find the minimal polynomial $g(X) = X^m + c_{m-1}X^{m-1} + \cdots + c_0$ for β. (Note that $m \mid n$.) Then

$$\mathrm{Tr}(\beta) = \frac{n}{m}(-c_{m-1})$$

and

$$\mathrm{N}(\beta) = ((-1)^m c_0)^{n/m}.$$

48. **(a)** Let B be the $n \times n$ matrix $(\sigma^i(\beta_j))$. Now observe that

$$\det(B)^2 = \det(B)\det(B^T) = \det(BB^T) = \det(\mathrm{Tr}(\beta_j\beta_j)).$$

(b) See Ireland and Rosen [1990, p. 174].

(c) Let $\gamma_1, \ldots, \gamma_n$ be the zeroes of g, and $\delta_1, \ldots, \delta_{n-1}$ be the zeroes of g'. We have

$$N(g'(\gamma)) = \prod_{1\leq i\leq n} g'(\gamma_i) = \prod_{1\leq i\leq n}\left(n \prod_{1\leq j\leq n-1}(\gamma_i - \delta_j)\right) = R(g, g').$$

(d) This is from Dedekind [1878, §5].

49. **(a)** Let $\beta_1, \ldots, \beta_n$ be an integral basis for K. Since the powers of α are algebraic integers, there is an $n \times n$ matrix M with entries in $\mathbb{Z}$ for which

$$(1, \alpha, \ldots, \alpha^{n-1}) = (\beta_1, \ldots, \beta_n)M.$$

Taking the conjugates of this relation, we see that $(\sigma_i(\alpha^j)) = (\sigma_i(\beta_j))M$, and the result follows by taking determinants.

(b) Let $f = gh$, where $g, h \in \mathbb{Z}[X]$ and g is the minimal polynomial of α. Then $f'(\alpha) = g'(\alpha)h(\alpha)$, so that

$$\operatorname{disc}(\alpha) = \mathrm{N}(g'(\alpha)) \mid \mathrm{N}(f'(\alpha)).$$

(c) Define g and h as in (b). We have

$$R(f, f') = R(gh, f') = R(g, f')R(h, f').$$

But since $f' = g'h + gh'$, we also have

$$R(g, f') = R(g, f' \bmod g) = R(g, g'h \bmod g) = R(g, g'h) = R(g, g')R(g, h).$$

50. Let $\Delta = R(f, f')$. This can be expressed as a determinant involving the coefficients of f and f'. (See, e.g., Lang [1965, p. 135]). From this, we infer that $|\Delta| < (2n)!(nB)^{2n}/2$. (This bound is certainly not best possible.) Then, by Stirling's approximation,

$$C := \log\left((2n)!(nB)^{2n}\right) = O(n \log(nB)).$$

To compute Δ, we make a list of all primes $p \le C$. For each such p, we reduce f mod p and compute $\Delta_p = R(f, f')$ mod p. Then, Δ is the unique integer satisfying $-C/2 < \Delta < C/2$ and $\Delta \equiv \Delta_p \pmod p$ for $p \le C$, which can be found using the Chinese remainder theorem. We estimate the running time as follows. Obtaining the values of f mod $p \le C$ costs $n \sum_{p \le C} O((\lg B)(\lg p))$, which is $O(nC(\lg B))$ by the prime number theorem. Similarly, the values of f' and $R(f, f')$ mod p can be found at a cost of

$$O\left(n^2 \sum_{p \le C} (\lg p)^2\right) = O(n^2 C(\lg C)).$$

Finally, recovering Δ from Δ_p uses $O(C^2)$ bit operations, by Corollary 5.5.3. The total work is easily seen to be is $O(n^3(\lg(nB))^2)$.

This algorithm and its running time estimate are due to G. Collins [1971]. This paper gives a similar bound for the complexity of computing the resultant of two polynomials in $\mathbb{Z}[X]$.

51. Let $f(X) = X^n - a$. We use the Euclidean algorithm, as suggested in Exercise 6.15, to compute the resultant of f and f'.

$$R(X^n - a, nX^{n-1}) = \pm n^n R(X^n - a, X^{n-1}) = \pm n^n R(X^{n-1}, -a) = \pm n^n a^{n-1} R(X^{n-1}, 1) = \pm n^n a^{n-1}.$$

In light of Theorem 8.7.1, this proves (b). Assertion (a) is a special case of this, for $n = 2$ and $d = a$.

To prove (c), we use Exercise 49. Let ζ be a primitive n-th root of unity and $f(X) = X^n - 1$. We have

$$\mathrm{N}\left(f'(\zeta)\right) = n^{\varphi(n)} \prod_{i \in (\mathbb{Z}/(n))^*} \zeta^i.$$

Since the product has absolute value 1, the result follows.

52. We first estimate the discriminant $\boldsymbol{\Delta}$ of K/E. Theorem 8.7.1 holds for relative extensions, and from this we find

$$\boldsymbol{\Delta} \mid \left(R(X^p - n, pX^{p-1})\right) = \left(p^p n^{p-1}\right) = (1-\zeta)^{p^e(p-1)} n^{p-1}.$$

On the other hand, there is a conductor $\mathbf{f}$ of L/K for which $\mathbf{f}^{p-1} \mid \boldsymbol{\Delta}$, by the conductor-discriminant formula. Therefore $\mathbf{f} \mid (1-\zeta)^{p^e} n$, so we can take the latter ideal as a conductor for L/K.

We can check this result by verifying that it implies a weak form of the quadratic reciprocity law. Let q be an odd prime and $p = 2$. Then $\zeta_2 = -1$, and we can take $f = 4q$ as a conductor for $\mathbb{Q}(\sqrt{q})/\mathbb{Q}$. If r is a prime $\equiv 1 \pmod{4q}$, then r splits in this extension, so $\left(\frac{q}{r}\right) = +1$.

53. For this exercise, we let ζ be a primitive n-th root of unity, and $L = \mathbb{Q}(\zeta)$. The Galois group of $L/\mathbb{Q}$ is isomorphic to $(\mathbb{Z}/(n))^*$. Let σ_a denote the automorphism taking taking ζ to ζ^a, when $a \in (\mathbb{Z}/(n))^*$. We have $\left(\frac{L/\mathbb{Q}}{p}\right) = \sigma_p$.

Assume the conclusion of Theorem 8.7.7. Let G be a nontrivial subgroup of $(\mathbb{Z}/(n))^*$. By Galois theory, L has a subfield K whose Galois group is isomorphic to $(\mathbb{Z}/(n))^*/G$. Furthermore, if

$$\eta = \sum_{x \in G} \zeta^x$$

(this is a Gaussian period), then $K = \mathbb{Q}(\eta)$. Since primes congruent to 1 mod n split in L, we see that n is a conductor for $K/\mathbb{Q}$. The discriminant Δ of $\mathbb{Q}$ equals 1. Applying Theorem 8.7.7 with $E = \mathbb{Q}$, there is a prime $p = O((\log n)^2)$, not dividing n, and not splitting in $K/\mathbb{Q}$. Therefore, $\left(\frac{K/\mathbb{Q}}{p}\right) \neq 1$, so that η is not fixed by σ_p. This implies $p \notin G$. Since G was arbitrary, $(\mathbb{Z}/(n))^*$ is generated by primes that are $O((\log n)^2)$. This is the conclusion of Ankeny's theorem (Theorem 8.5.3).

Next, assume the conclusion of Theorem 8.7.8. Let $a \in (\mathbb{Z}/(n))^*$. The discriminant of L divides n^n, by Exercise 51. Applying Theorem 8.7.8 with $E = \mathbb{Q}$, there is a prime p, not dividing n, with $p = O(n^2(\log n)^2)$ such that $\left(\frac{L/\mathbb{Q}}{p}\right) = \sigma_a$, i.e. $p \equiv a \pmod{n}$. Thus, Linnik's theorem holds with an exponent of $C > 2$. (Compare Theorem 8.5.8.)

Finally, we apply Theorem 8.7.9 with $E = \mathbb{Q}$. For $a \in (\mathbb{Z}/(n))^*$, the number of primes $p \leq x$ with $\left(\frac{K/\mathbb{Q}}{p}\right) = \sigma_a$, i.e. with $p \equiv a \pmod{n}$, is asymptotic to $\mathrm{li}(x)/\varphi(n)$. This is the prime number theorem for arithmetic progressions (Theorem 8.4.2). (Similarly, the conclusion of Theorem 8.7.10 implies the conclusion of Theorem 8.4.5.)

54. Result (a) is proved in Stark [1987, pp. 265 ff.]. To prove (b), evaluate $\mathrm{N}(2\beta)$, or $\mathrm{N}(2\beta - 1)$, depending on cases.

55. Let $n = 2^e r$, where r is odd. Clearly it is enough to compute r', the squarefree part of r. Let $\Delta = \operatorname{disc}(\mathbb{Q}(\sqrt{r}))$. We have

$$r' = \begin{cases} \Delta, & \text{if } \Delta \equiv 1 \pmod 4; \\ \Delta/4, & \text{if } \Delta \equiv 0 \pmod 4. \end{cases}$$

(These are the only two possible cases.)

This exercise is due to Chistov [1989], who proved a similar result for integral bases. See also H. W. Lenstra [1992]. We note that a corresponding result could be proved for conductors, since the conductor of a quadratic field equals, up to sign, its discriminant.

56. This result is due to H. Montgomery [1971]. For a short proof using algebraic numbers see Bach and Huelsbergen [1993].

57. This exercise is from Bach and Shallit [1985, 1989].

58. We may as well assume d is not a square (for otherwise Theorem 8.5.8 suffices). Let $E = \mathbb{Q}(\sqrt{d})$ and $K = E(\zeta)$, where ζ is a primitive n-th root of unity. There is an automorphism σ_a of K/E taking ζ to ζ_a. Use Theorem 8.7.10 to estimate the smallest degree 1 prime $\mathbf{p}$ of $\mathbb{Q}(\sqrt{d})$ with $\left(\frac{K/E}{\mathbf{p}}\right) = \sigma_a$, observing that the discriminant of K is bounded by $(nd^2)^{\varphi(n)}$.

Chapter 9

1. If $n = 2^{p-1}(2^p - 1)$, and $2^p - 1$ is prime, then by Exercise 9 we have $\sigma(n) = (2^p - 1)2^p = 2n$. On the other hand, suppose n is an even and $\sigma(n) = 2n$. Write $n = 2^k \cdot r$, where $k \geq 1$ and r is odd. Then

$$2^{k+1}r = 2n = \sigma(n) = \sigma(2^k)\sigma(r) = (2^{k+1} - 1)\sigma(r),$$

so $(2^{k+1} - 1) \mid r$. Write $r = (2^{k+1} - 1)s$; then $\sigma(r) = 2^{k+1}s$. Now r and s are distinct divisors of r, and their sum is $2^{k+1}s = \sigma(r)$; hence r cannot have any other divisors and must be prime. Thus $s = 1$, $r = 2^{k+1} - 1$. Thus $n = 2^k(2^{k+1} - 1)$, where this second factor is prime.

2. The result is true for $p = 2$. Assume p odd. In $(p - 1)!$, pair each element with its inverse. The elements 1 and $p - 1$ are their own inverse, so $(p - 1)! \equiv 1(p - 1) \equiv -1 \pmod p$.

This exercise was apparently known to Leibniz; see Vacca [1899] and Mahnke [1912]. Lagrange was the first to publish a proof; see Lagrange [1771]. For a primality criterion similar to Wilson's theorem, see Mann and Shanks [1972]; W. Adams, Liverance, and Shanks [1991].

3. It suffices to show that $(n - 1)! \equiv 0 \pmod{p^e}$ for every prime power $p^e \,\|\, n$. If n is divisible by at least two distinct primes, say p and q, then $n - 1 \geq 2p^e - 1 \geq p^e$, so $p^e \mid (n - 1)!$. Thus we may assume n is a prime power p^e, with $e \geq 2$. By Exercise 2.15, we know that $\nu_p(p^e!) \geq p^{e-1} - 1$. Now if $p \geq 3$, then it is easy to prove by induction that $p^{e-1} - 1 \geq e$ for $e \geq 2$, while $2^{e-1} - 1 \geq e$ for $e \geq 3$. Hence for n composite and > 4, we have $(n - 1)! \equiv 0 \pmod{p^e}$.

4. Using Theorem 8.5.7, it follows that there exists a constant c such that the least prime p congruent to $1 \pmod{2^k}$ is $\leq c2^{11k/2}$. For these p, we have $T(p) \leq 2 + 2\log_2(c2^{9k/2})$. It follows that $\liminf_{p\to\infty} T(p)/\log_2 p \leq 18/11$. As C. Pomerance has noted (personal communication), the constant of $18/11$ can be improved to $4/3$ by using the result of Pintz, Steiger, and Szemerédi [1989] mentioned in the notes to section 9.1.

5. This exercise is from Brillhart, Lehmer, and Selfridge [1975]. Also see H. W. Lenstra [1986a].

6. Let $r = (6n+1)(12n+1)(18n+1)$; if all these factors are prime, then $\lambda(r) = \mathrm{lcm}(6n, 12n, 18n) = 36n$. On the other hand, $r - 1 = 36n(36n^2 + 11n + 1)$. Hence $\lambda(r) \mid r - 1$.

This exercise is from Chernick [1939].

7. First, show that $\sigma(n)$ is a power of 2 if and only if n is the product of distinct Mersenne primes; see, for example, Sierpiński [1959]; Koster and Shapiro [1990]. Next, trial divide by $r = 2^p - 1$ for prime p. Test if r is prime using the Lucas-Lehmer test, and if so, test if $r^2 \mid n$. All this can be done in deterministic polynomial time.

8. Show that if $\varphi(n) = 2^k$, then $n = 2^j p_1 p_2 \cdots p_j$, where the p_i are distinct odd primes of the form $2^{2^i} + 1$. Then follow the solution of the previous exercise, using Pepin's test in place of the Lucas-Lehmer test.

9. We have $S(65) = \{1, 8, 18, 47, 57, 64\}$; for example, $8^2 \equiv -1 \pmod{65}$ and $18^2 \equiv -1 \pmod{65}$. However $14 \equiv 8 \cdot 18 \pmod{65}$, and $14 \notin S(65)$.

10. See Gerlach [1994].

11. The computation of the gcd is unnecessary because $\left(\frac{a}{n}\right) = 0$ iff $\gcd(a, n) \neq 1$.

12. We modify our routine PrimeTest(n, k) as follows: before invoking the Miller-Rabin test, we trial divide by "small" primes. More formally, we choose a bound B; we will see in a minute that B can be taken to be $B = O((\lg x)^2)$. Compute a table of the primes $\leq B$ by any reasonable method; see Section 9.8. (In practice, this table would be precomputed.) Trial divide n by the primes $\leq B$. If for some $p \leq B$ we find $p \mid n$ and $p \neq n$, then return "composite". If $p = n$, return "prime". Otherwise, n has no prime divisor $\leq B$, and apply the Miller-Rabin test at most k times, as before.

To analyze this modification, we will use a theorem of Jurkat and Richert [1965], which states that

$$|\{n \leq y : \text{ no } p \leq x \text{ divides } n\}| = y \prod_{p \leq x} \left(1 - \frac{1}{p}\right) \left(1 + O\left(\frac{1}{\log y}\right)\right),$$

for $\log x \leq (\log y)/(2 \log \log 3y)$.

The expected running time of this modified version of PrimeTest(n, k) (with n randomly chosen in $[1, x]$) can be computed as follows. First, as in Theorem 9.8.1, the primes $\leq B$ can be computed in $O(B(\lg B)^2(\lg \lg B))$ bit operations. The trial division can be done in $O(\pi(B)(\lg B)(\lg x)) = O(B \lg x)$ bit operations. The probability that n survives the trial division and goes on to experience the Miller-Rabin test is, using Theorem 8.9.6 and the theorem of Jurkat and Richert:

$$\frac{e^{-\gamma}}{\log B} \left(1 + O\left(\frac{1}{\log x}\right)\right) \left(1 + O\left(\frac{1}{(\log B)^2}\right)\right) = \Omega\left(\frac{1}{\log B}\right),$$

provided $x, B \geq 3$ and $\log B \leq (\log x)/(2 \log \log 3x)$. The expected number of bit operations used is therefore $O(B(\lg B)^2(\lg \lg B)) + O(B \lg x) + O(k(\lg x)^3/(\lg B))$.

It remains to compute the error probability of the modified algorithm, that is, the probability that the algorithm returns a composite number which it falsely asserts is prime. Following the analysis in Lemma 9.7, let S be the set of all numbers in $[1, x]$ that are not divisible by any prime $\leq B$. Then, again using the result of Jurkat and Richert, we see that $|S| = \Omega(\frac{x}{\log B})$. It follows that $\Pr[X']$, the probability that a randomly chosen member of S is prime is

$$\frac{\frac{x}{\log x} - \frac{B}{\log B}}{\frac{x}{\log B}},$$

up to a constant factor. Provided B is sufficiently small, this is $\Omega((\log B)/(\log x))$. Thus, using Lemma 9.7, the error probability is $O(4^{-k}(\log x)/(\log B))$.

To get an error probability of any fixed bound $\epsilon > 0$, we can now take $k = \log\log x - \log\epsilon$, and $B = (\log x)^2$. The expected running time of the modified algorithm is $O((\lg x)^3)$, and the expected running time of RANDOMPRIME is $O((\lg x)^4)$.

15. The number $30^{32} + 1$ is prime; use Theorem 9.1.1. The number $2^{127} - 1$ is also a prime; for this, use Theorem 9.2.4. However, $n = 2^{101} - 1$ is composite, as can be readily seen by computing $3^{n-1} \pmod n$. For a solution to Exercise 16, see Wilkins [1950].

17. See Rotkiewicz [1964a].

18. See Malo [1903]; Sierpiński [1947].

19. See Rotkiewicz [1964c].

20. See Rotkiewicz [1980]. Pomerance, Selfridge, and Wagstaff [1980, Thm. 1] proved that there are infinitely many strong pseudoprimes to the base a.

21. One direction follows from Theorem 9.3.10. For the other, observe that since $n = 3 \pmod 4$, $n - 1 = 2d$, where d is odd. If $a^{(n-1)/2} \equiv \left(\frac{a}{n}\right) \pmod n$, then $a^d \equiv \pm 1 \pmod n$. Hence n is a strong pseudoprime to the base a.

This exercise is due to Malm [1977].

22. Let $p = 12k + 1$, $q = 24k + 1$, and assume both of these are primes. Let $n = pq$. Then $a^{n-1} = a^{12k(24k+3)} = (a^{p-1})^{(24k+3)} \equiv 1 \pmod p$ for all a relatively prime to p; in particular, it is true for $a = 2$ and $a = 3$. Also, both 2 and 3 are quadratic residues of the prime q, so $b^{n-1} = b^{12k(24k+3)} = (b^{(q-1)/2})^{24k+3} \equiv 1 \pmod q$ for $b = 2$ and $b = 3$.

This exercise is due to Shanks [1985, p. 228].

23. It is easy to verify that $L_n = \alpha^n + \beta^n$, where α and β are the roots of the equation $X^2 - X - 1 = 0$. Then, using Exercise 5.37, we have $1^n = (\alpha + \beta)^n \equiv \alpha^n + \beta^n = L_n$, where the congruence is $(\text{mod } n)$. Thus $L_n \equiv 1 \pmod n$. The converse is not true, as $L_n \equiv 1 \pmod n$ for $n = 705 = 3 \cdot 5 \cdot 47$.

See Lind [1967]; Hoggatt and Bicknell [1974]; Pollin and Schoenberg [1980].

24. It is easy to verify that $u_n = \alpha^n + \beta^n + \gamma^n$, where α, β, γ are the roots of the equation $X^3 - X - 1 = 0$. Since $(X - \alpha)(X - \beta)(X - \gamma) = X^3 - X - 1$, we have $\alpha + \beta + \gamma = 0$. Then, as in the solution given above, $0^n = (\alpha + \beta + \gamma)^n \equiv \alpha^n + \beta^n + \gamma^n = u_n$, where the congruence is $(\text{mod } n)$. Thus $u_n \equiv 0 \pmod n$. The converse is not true; the smallest counterexamples are $n = 271441 = 521^2$ and $n = 904631 = 7 \cdot 13 \cdot 9941$.

See Perrin [1899]; Malo [1900]; Escott [1901]; Dickson [1908]; Neumann and Wilson [1979]; W. Adams and Shanks [1982]; W. Adams [1987]; Arno [1991].

25. See Baillie and Wagstaff [1980].

26. We have $n+1 = 1+bp$ for some $b \geq 1$, and $(1+bp)c = p^2 - 1$. Thus $c \equiv -1 \pmod{p}$, so $c \geq p-1$. Therefore $1+bp \leq p+1$, so $b = 1$.

27. See Lehmann [1982].

28. See Gandhi [1971]; Vanden Eynden [1972]; and Golomb [1974].

29. See Mills [1947].

30. Suppose that there exists an integer r such that $|f(r)| = p$, a prime number. Since the polynomial is nonconstant, there exists an integer m such that $|f(k)| > p$ for all $k \geq m$. Now $f(r+jp) - f(r)$ is clearly divisible by p for all j, so $|f(r+jp)|$ cannot be a prime for all $j \geq (m-r)/p$.

31. By Lemma 8.8.4, we have $p_{n+1} < (n+1)(\log(n+1) + \log\log(n+1)) < 2(n+1)\log(n+1)$ for $n \geq 5$. Now it is easy to see that $n+1 < n^2$ for $n \geq 2$, $\log n > 3$ for $n \geq 21$, and $n^3 > (n+1)^2$ for $n \geq 3$; hence $n^{\log n} > n^3 > (n+1)^2$ for $n \geq 21$. Therefore $2\log(n+1) < (\log n)^2$ for $n \geq 21$. Hence for $n \geq 21$, $p_{n+1} < 2(n+1)\log(n+1) < n^2(\log n)^2 < p_n$, where we have again used Lemma 8.8.4. It now suffices to verify the result for $1 \leq n \leq 20$.

32. Replace line (5) of the algorithm with

```
(5a) j ← p
(5b) while (j ≤ n) do
(5c)        a[j] ← 0
(5d)        j ← j + p
```

33. Consider the following modified algorithm:

```
SEGMENT2(n, Δ)
a ← FIND PRIMES(√n)
r ← Length(a)
for i ← 1 to r do c[i] ← a[i]^2
m ← 0
while (m < n) do
        for j ← 1 to Δ do
              d[j] ← 1
        if m = 0 then d[1] ← 0
        for i ← 1 to r do
              while (c[i] ≤ Δ) do
                    d[c[i]] ← 0
                    c[i] ← c[i] + a[i]
              c[i] ← c[i] − Δ
        for j ← 1 to Δ do
              if (d[i] = 1) and (m + i ≤ n) then output(m + i)
        m ← m + Δ
```

Here, FIND PRIMES(n) is a routine that returns a list of the primes $\leq n$.

34. As in the solution to Exercise 2.7, use the principle of inclusion-exclusion. Let $p_1, p_2, \ldots, p_k$ be the primes $\le \sqrt{x}$.

Now the number of positive integers $\le x$ that are not divisible by any of $p_1, p_2, \ldots, p_k$ is just $\pi(x) - \pi(\sqrt{x}) + 1$. But from the principle, we know that the number of positive integers $\le x$ that are not divisible by any of the primes $p_1, p_2, \ldots, p_k$ is

$$\lfloor x \rfloor - \sum_{p \le \sqrt{x}} \left\lfloor \frac{x}{p} \right\rfloor + \sum_{p<q\le\sqrt{x}} \left\lfloor \frac{x}{pq} \right\rfloor - \cdots.$$

Hence these two sums are equal.

35. Call a prime p *large* if $p > \sqrt{x}$. Using the hint, we see that there are exactly $\lfloor x/p \rfloor$ integers divisible by large prime. Since none of these integers have any other large prime divisors it follows that the total number of integers $\le x$ with a large prime divisor is

$$\begin{aligned}\sum_{\sqrt{x}<p\le x} \left\lfloor \frac{x}{p} \right\rfloor &= \left(x \sum_{\sqrt{x}<p\le x} \frac{1}{p} \right) + O(\pi(x)) \\ &= x(\log\log x + B + o(1) - (\log\log\sqrt{x} + B + o(1))) + O(\pi(x)) \\ &= x \log 2 + o(x).\end{aligned}$$

36. First, note that every integer $n \le x^2$ can be written uniquely as $n = a^2 b$, where $1 \le a \le x$, and b is squarefree. It follows that

$$\lfloor x^2 \rfloor = \sum_{1\le k\le x} S_2\left(\frac{x^2}{k^2}\right).$$

Now let $f(x) = S_2(x^2)$, and apply Exercise 2.19. Then $g(x) = \lfloor x^2 \rfloor$, and

$$\begin{aligned} S_2(x^2) &= \sum_{1\le k\le x} \mu(k) \left\lfloor \frac{x^2}{k^2} \right\rfloor \\ &= x^2 \sum_{1\le k\le x} \frac{\mu(k)}{k^2} + O(x) \\ &= x^2 \frac{6}{\pi^2} + O(x),\end{aligned}$$

where we have used Exercise 2.21. This exercise is due to Gegenbauer [1885].

37. A term $\lfloor x/n \rfloor$ in the sum is nonzero if and only if n is squarefree and divisible only by primes $\le \sqrt{x}$. Let $T(x)$ be the number of squarefree integers $\le x$ that are divisible by a "large" prime (one $> \sqrt{x}$). Then

$$T(x) = \sum_{\sqrt{x}<p\le x} S_2(x/p)$$

$$= \sum_{\sqrt{x}<p\le x} \left(\frac{6}{\pi^2}\frac{x}{p} + O\left(\sqrt{\frac{x}{p}}\right)\right)$$

$$= \frac{6x}{\pi^2}\left(\sum_{\sqrt{x}<p\le x} \frac{1}{p}\right) + O(\sqrt{x})\left(\sum_{\sqrt{x}<p\le x} p^{-1/2}\right)$$

$$= \frac{6x}{\pi^2}(\log 2 + o(1)) + O(x/\log x),$$

where we have used Exercise 2.36. Since the total number of squarefree numbers $\le x$ is $6x/\pi^2 + O(\sqrt{x})$ by Exercise 36, the total number of squarefree numbers not divisible by a large prime is asymptotically $(6/\pi^2)(1 - \log 2)x$.

38. As in Exercise 9.36, we have

$$S_2(x) = \sum_{1\le k\le \sqrt{x}} \mu(k) \left\lfloor \frac{x}{k^2} \right\rfloor . \tag{A.8}$$

Using this, we can compute $S_2(x)$ as follows: first determine the primes $\le \sqrt{x}$ using a segmented sieve (as in Theorem 9.8.2) using $O(x^{1/2+\epsilon})$ bit operations and $O(x^{1/4+\epsilon})$ space. Next, sieve by all powers of these primes to determine $\mu(k)$ for $1 \le k \le \sqrt{n}$; it will be necessary to make a final pass through the list to handle the case where k has a prime factor $> \sqrt{x}$. This can also be done in a "segmented" fashion in the same time bound. Use this information to compute the sum (A.8).

39. For correctness, observe the number of divisors of a number r is equal to twice the number of divisors of r that are $< \sqrt{r}$, plus 1 if r is a square.

The number of bit operations used can be computed as follows: the outer loop is performed $O(\sqrt{n})$ times. During loop j, at most $\lfloor n/j \rfloor$ multiplications are performed (to compute the index kj); each multiplication costs $O((\lg n)^2)$. Thus the total cost is, up to a constant factor, bounded by

$$\sum_{1\le j\le\sqrt{n}} \left\lfloor \frac{n}{j} \right\rfloor (\lg n)^2 = O(n(\lg n)^3),$$

where we have used the estimate in Theorem 2.5.3. This can be improved to $O(n(\lg n)^2)$ by the same sort of trick employed in Exercise 9.32.

The space used is

$$\sum_{1\le k\le n} \lg d(k) \le 1 + 2 \sum_{1\le k\le n} \log_2 d(k)$$

$$\le 1 + 2n \log_2 \left(\frac{1}{n} \sum_{1\le k\le n} d(k)\right)$$

$$\leq 1 + 2n \log_2 \left(\frac{1}{n} (n \log n + O(n)) \right)$$
$$= O(n \lg \lg n).$$

In the last step, we used Theorem 2.5.4.

40. The idea is to use Exercise 2.8. Initialize each entry of the array x of size n to 1; then sieve by each prime power $p^k \leq n$, multiplying $x[r]$ by $k + 1$ if $p^k \,\|\, r$. (This can be done by running through the multiples $s \cdot p^k$, and not changing those entries where $p \mid s$. To get the desired running time, one must use the bound on the size of $d(n)$ given in Theorem 8.8.9.

41. One way to do this is simply to change line (4) of algorithm Number of Divisors Table to read $x[j] \leftarrow x[j] + j$, and line (6) to read $x[kj] \leftarrow x[kj] + k + j$. This gives an algorithm that uses $O(n(\lg n)^3)$ bit operations.

An even better algorithm can be obtained following the idea in the solution to the previous exercise: use the formula in Exercise 2.9 and sieve by all prime powers $p^k \leq n$. This gives an algorithm that uses $O(n(\lg n)^2(\lg \lg n))$ bit operations. This same method can be adapted to tabulate $\varphi(n)$, in the same time bound.

To tabulate $\mu(k)$, initialize an array of size n to all 1's, then sieve by all primes $p \leq n$, multiplying each entry by -1. Then sieve by all $p^2 \leq n$, multiplying each entry by 0.

42. The necessary data structure is discussed in Lagarias, Miller and Odlyzko [1985]. For more on the problem of maintaining an array and its partial sums, see the papers of Fredman [1982]; Yao [1985]; Fredman and Saks [1989]; and Dietz [1989].

43. See Brillhart and Selfridge [1967].

44. If n is an odd prime power, say $n = p^e$, then $\lambda(n) = p^{e-1}(p-1)$. Now $(\mathbb{Z}/(p^e))^*$ is a cyclic group of order $\lambda(n)$ (which is even), so half its members satisfy $x^{\lambda(n)} \equiv 1 \pmod{n}$, and the other half satisfy $x^{\lambda(n)} \equiv -1 \pmod{n}$.

On the other hand, if n is odd and composite, then choose a prime divisor p of n such that $\nu_2(p-1) = \nu_2(\lambda(n))$. Write $n = p^e \cdot r$, where $p^e \,\|\, n$. Let t be a quadratic nonresidue (mod p^e) and let x satisfy the congruences $x \equiv t \pmod{p^e}$ and $x \equiv 1 \pmod{r}$. Now $\lambda(n) = p^{e-1}(p-1)u$, where u is odd. Then $x^{\lambda(n)/2} \equiv (x^{p^{e-1}(p-1)/2})^u \equiv (-1)^u \equiv -1 \pmod{p^e}$, and $x^{\lambda(n)/2} \equiv 1 \pmod{r}$. Hence $x^{\lambda(n)/2} \not\equiv \pm 1 \pmod{n}$, as desired.

46. It suffices to consider prime divisors r of n. Let w be the order of a in $(\mathbb{Z}/(r))^*$. Since $a^{(n-1)/2} \equiv -1 \pmod{n}$, we have $a^{(n-1)/2} \equiv -1 \pmod{r}$ and $a^{n-1} \equiv 1 \pmod{r}$. Hence $\nu_2(w) = \nu_2(n-1)$. On the other hand, by Exercise 5.1 we know that $w \mid r - 1$, so $\nu_2(w) \leq \nu_2(r-1)$. Combining these gives us the desired inequality. The inequality is strict if and only if $w \mid (r-1)/2$, which occurs if and only if $a^{(r-1)/2} \equiv -1 \pmod{r}$. See H. Cohen and H. W. Lenstra [1984c, p. 311].

47. If we pick a random $q \equiv 1 \pmod{p}$, the chance that it is prime is $p/((p-1)\log q)$. On the other hand, given that q is a prime, congruent to 1 modulo p, n is a p-th power modulo q with probability $1/p$. So the chance that a random q is "good" is $(1 - 1/p)(p/(p-1))/(\log q) = 1/(\log q)$.

Now, suppose we try q's up to m. The probability that no q is "good" should be at most $(1 - 1/\log m)^{m/p} \leq e^{-m/(p \log m)}$. This bound is close to 1 when m is small, and decreases rapidly when m is large. Thus, a good guess for the expected value should be the value of m that makes $e^{-m/(p \log m)} = e^{-c}$, where c is a positive constant, say 1. Solving this for m gives $m \approx cp \log p$, i.e. $m = O(p \log p)$.

48. To see that FACTOR $\in \mathcal{NP}$, it suffices to observe that a nondeterministic machine, on input n, can simply guess a nontrivial factor d, and then verify that $1 < d \leq k$, $d < n$, and $k \mid n$.

To see that FACTOR $\in$ co-$\mathcal{NP}$, on input n the nondeterministic machine guesses the complete factorization of n, as well as proofs in the style of Theorem 9.1.4 that p is prime, for each of the given factors p^e. For each p^e listed, the machine then verifies these Pratt-style proofs, and checks to make sure that $p^e \,\|\, n$. Next, the machine verifies that n is the product of the given prime powers. Finally, the machine verifies that all the listed primes are $> k$. All these verifications can be done in time polynomial in $\log n$.

This exercise has the following interesting consequence: assuming ERH, there exists a deterministic algorithm that one can write down explicitly, that always succeeds in factoring n as a product of prime powers, and runs in polynomial time if *any* polynomial-time algorithm for factoring exists. The idea is simply to enumerate all Turing machines, and dovetail the computations, running Turing machine 1 for 1 step, then machine 1 for 2 steps and machine 2 for 1 step, etc. If there is a polynomial-time factoring algorithm, one will eventually reach a polynomial-time Turing machine in the enumeration that given n, will correctly produce the complete factorization of n, along with a Pratt-style proof that that the given factors are truly prime. (We need ERH so the primitive roots of each prime p can be determined in polynomial time.) If this Turing machine uses $f(n)$ steps on input n, and is numbered j in the enumeration, the total cost for the simulation is $O((j + f(n))^2)$.

49. Since $E(n)$ is a subgroup of $(\mathbb{Z}/(n))^*$, the least unit outside $E(n)$ must be prime. The idea behind this exercise appears in G. Miller [1976].

50. This is the original proof, by G. Miller [1976], that PRIMES belongs to $\mathcal{P}$ if ERH is true.

51. Prove the following algebraic result first. Let $\psi(x) = x^{(n-1)/2}$; this is a homomorphism from $(\mathbb{Z}/(n))^*$ to itself. Then for odd n, $\psi((\mathbb{Z}/(n))^*) = \{\pm 1\}$ iff n is prime. Now use Ankeny's theorem. If n is prime, there is a small quadratic nonresidue, so $\psi(x) = -1$ will occur for some $x \leq 2(\log n)^2$; also $\psi(1) = 1$. If n is composite, we may as well assume that ψ does not annihilate $(\mathbb{Z}/(n))^*$. Apply Ankeny's theorem to the proper subgroup $\psi^{-1}(\pm 1)$ to conclude that some $x \leq 2(\log n)^2$ has $\psi(x) \neq \pm 1$. This exercise is a deterministic version of a randomized algorithm due to D. Lehmann [1982].

52. Use Exercise 8.34 for an appropriate character mod p^2.

For part (b), see H. W. Lenstra [1979]; Bach [1991b].

For part (c), use appropriate results from Chapter 8. See H. W. Lenstra [1979] and Pajunen [1980].

53. Here is a heuristic argument, suggesting that the expected time to test a random $x \leq n$ for primality is $O((\log n)^4)$. By the prime number theorem, the primes $\leq n$ contribute this much to the average. Now we consider composite integers. Suppose that for each prime $a = 2, 3, 5, \ldots,$ the chance that the algorithm does not halt on that a is $\leq 1/2$. Then, we should expect to use $O(1)$ primes on average, so the average run time contributed by composite n should be $O((\log n)^3)$.

54. One such number is 5689 71935 26942 02437 03269 72321. See J. Davenport [1992].

Bibliography

[Abbott et al. 1980]
W. L. Abbott et al. Solution to problem E 2766. *Amer. Math. Monthly* **87** (1980), 406.

[Abel 1823]
N. H. Abel. Oplosning af et par opgaver ved hjelp af bestemte Integraler. *Magazin for Naturvidenskaberne* **2** (1823), 55–68; 205–216. French translation in *Oeuvres,* Vol. 1, pp. 11–27.

[Abramov 1979]
S. A. Abramov. Some estimates related to the Euclidean algorithm. *Ž. Vyčisl. Mat. i Mat. Fiz.* **19** (1979), 756–760. English translation in *U.S.S.R. Comput. Maths. Math. Phys.* **19** (1979) 207–212.

[W. Adams 1987]
W. W. Adams. Characterizing pseudoprimes for third-order linear recurrences. *Math. Comp.* **48** (1987), 1–15.

[W. Adams, Liverance, and Shanks 1991]
W. W. Adams, E. Liverance, and D. Shanks. Infinitely many necessary and sufficient conditions for primality. *Bull. Inst. Combin. Appl.* **3** (1991), 69–76.

[W. Adams and Shanks 1982]
W. W. Adams and D. Shanks. Strong primality tests that are not sufficient. *Math. Comp.* **39** (1982), 255–300.

[Adleman 1978]
L. M. Adleman. Two theorems on random polynomial time. In *Proc. 19th Ann. Symp. Found. Comput. Sci.*, pp. 75–83. IEEE Press, 1978.

[Adleman 1980]
L. M. Adleman. On distinguishing prime numbers from composite numbers. In *Proc. 21st Ann. Symp. Found. Comput. Sci.*, pp. 387–406, 1980.

[Adleman and Huang 1994]
L. M. Adleman and M.-D. Huang, editors. *Algorithmic Number Theory, First International Symposium, ANTS-I*, Vol. 877 of *Lecture Notes in Computer Science*. Springer-Verlag, 1994.

[Adleman and Kompella 1988]
L. M. Adleman and K. Kompella. Using smoothness to achieve parallelism. In *Proc. Twentieth ACM Symp. Theor. Comput.*, pp. 528–538, 1988.

[Adleman and Leighton 1981]
L. M. Adleman and F. T. Leighton. An $O(n^{1/10.89})$ primality testing algorithm. *Math. Comp.* **36** (1981), 261–266.

[Adleman and H. W. Lenstra 1986]
L. M. Adleman and H. W. Lenstra, Jr. Finding irreducible polynomials over finite fields. In *Proc. Eighteenth Ann. ACM Symp. Theor. Comput.*, pp. 350–355. ACM, 1986.

[Adleman, Manders, and Miller 1977]
L. M. Adleman, K. Manders, and G. L. Miller. On taking roots in finite fields. In *Proc. 18th Ann. Symp. Found. Comput. Sci.*, pp. 175–178, 1977.

[Adleman, Pomerance, and Rumely 1983]
L. M. Adleman, C. Pomerance, and R. S. Rumely. On distinguishing prime numbers from composite numbers. *Ann. Math.* **117** (1983), 173–206.

[Agnew, Beth, Mullin, and Vanstone 1993]
G. B. Agnew, T. Beth, R. C. Mullin, and S. A. Vanstone. Arithmetic operations in $GF(2^m)$. *J. Cryptology* **6** (1993), 3–13.

[Agnew, Mullin, and Vanstone 1988]
G. B. Agnew, R. C. Mullin, and S. A. Vanstone. Fast exponentiation in $GF(2^n)$. In C. G. Günther, editor, *Advances in Cryptology–EUROCRYPT '88 Proceedings*, Vol. 330 of *Lecture Notes in Computer Science*, pp. 251–255. Springer-Verlag, 1988.

[Agou 1976]
S. Agou. Factorisation des polynômes à coefficients dans un corps fini. *Publications du Dept. Math. Lyon* **13** (1976), 63–71.

[Aho, Hopcroft, and Ullman 1974]
A. Aho, J. Hopcroft, and J. Ullman. *The Design and Analysis of Computer Algorithms.* Addison-Wesley, Reading, Massachusetts, 1974.

[Aiello and Subbarao 1993]
W. Aiello and M. V. Subbarao. A conjecture in addition chains related to Scholz's conjecture. *Math. Comp.* **61** (1993), 17–23; S1–S6.

[Aitken 1949]
A. C. Aitken. *Determinants and Matrices*. Oliver and Boyd, Edinburgh, 1949.

[Akbik 1992]
S. Akbik. Normal generators of finite fields. *J. Number Theory* **41** (1992), 146–149.

[Akl 1980]
S. G. Akl. Professor Jevons and his logical machine. *Austral. Comput. Bull.* **5** (1981), 28–30.

[Akushskiĭ and Burtsev 1986]
I. Ya. Akushskiĭ and V. M. Burtsev. Realization of primality tests for Mersenne and Fermat numbers. *Vestnik Akademii Nauk Kazakhskoi SSR* (1) (1986), 52–59.

[Alagić and Arbib 1978]
S. Alagić and M. A. Arbib. *The Design of Well-Structured and Correct Programs*. Springer-Verlag, New York, 1978.

[Albert 1941]
A. A. Albert. Some mathematical aspects of cryptography. In R. E. Block, N. Jacobson, J. M. Osborn, D. J. Saltman, and D. Zelinsky, editors, *A. Adrian Albert: Collected Mathematical Papers*, pp. 903–920. Amer. Math. Soc., 1993. Invited address to the Manhattan, Kansas meeting of the American Mathematical Society, November 22, 1941.

[Albert 1961]
A. A. Albert. *Fundamental Concepts of Higher Algebra*. University of Chicago Press, 1961.

[Alford, Granville, and Pomerance 1994a]
W. R. Alford, A. Granville, and C. Pomerance. There are infinitely many Carmichael numbers. *Ann. Math.* **140** (1994), 703–722.

[Alia and Martinelli 1991]
G. Alia and E. Martinelli. A VLSI modulo m multiplier. *IEEE Trans. Comput.* **40** (1991), 873–878.

[Allouche 1987]
J.-P. Allouche. Automates finis en théorie des nombres. *Exposition. Math.* **5** (1987), 239–226.

[Allouche and Shallit 1992]
J.-P. Allouche and J. O. Shallit. The ring of k–regular sequences. *Theoret. Comput. Sci.* **98** (1992), 163–187.

[Anderson, Earle, Goldschmidt, and Powers 1967]
S. F. Anderson, J. G. Earle, R. E. Goldschmidt, and D. M. Powers. The IBM System/360 Model 91: floating-point execution unit. *IBM J. Res. Develop.* **11** (1967), 34–53.

[Angluin 1982]
D. Angluin. Lecture notes on the complexity of some problems in number theory. Technical Report 243, Yale University, Department of Computer Science, August 1982.

[Ankeny 1952]
N. C. Ankeny. The least quadratic non residue. *Ann. Math.* **55** (1952), 65–72.

[Ansari 1951]
A. R. Ansari. On prime representing function. *Ganita* **2** (1951), 81–82.

[Apostol 1967]
T. M. Apostol. *Mathematical Analysis*. Addison-Wesley, 1967.

[Apostol 1976]
T. M. Apostol. *Introduction to Analytic Number Theory*. Springer-Verlag, New York, 1976.

[Apostol and Ford 1991]
T. M. Apostol and K. Ford. Solution to problem 6615. *Amer. Math. Monthly* **98** (1991), 562–565.

[Archibald 1914]
R. C. Archibald. Remarks on Klein's "Famous Problems of Elementary Geometry". *Amer. Math. Monthly* **21** (1914), 247–259.

[Archibald 1935]
R. C. Archibald. Mersenne's numbers. *Scripta Math.* **3** (1935), 112–119.

[Arnault 1991]
F. Arnault. Le test de primalité de Rabin-Miller: un nombre composé qui le "passe". Technical Report 61, Université de Poitiers, November 1991.

[Arnault 1993]
F. Arnault. Carmichaels fortement pseudo-premiers, pseudo-premiers de Lucas. Technical Report 73, Université de Poitiers, January 1993.

[Arnault 1995]
F. Arnault. Rabin-Miller primality test: composite numbers which pass it. *Math. Comp.* **64** (1995), 355–361.

[Arno 1991]
S. Arno. A note on Perrin pseudoprimes. *Math. Comp.* **56** (1991), 371–376.

[Artin 1924a]
E. Artin. Quadratische Körper im Gebiete der höheren Kongruenzen. *Math. Zeitschrift* **19** (1924), 153–246. Reprinted in *Collected Papers,* pp. 1–94.

[Artin 1924b]
E. Artin. Über eine neue Art von L-Reihen. *Mat. Sem. Univ. Hamburg* **3** (1924), 89–108. Reprinted in *Collected Papers*, pp. 105–124.

[Artin 1927]
E. Artin. Beweis des allgemeinen Reciprocitätsgesetzes. *Mat. Sem. Univ. Hamburg* **5** (1927), 353–363. Reprinted in *Collected Papers*, pp. 131–141.

[Artin 1942]
E. Artin. *Galois Theory*, Vol. 2 of *Notre Dame Mathematical Lectures*. University of Notre Dame Press, 1942.

[Artin 1965]
E. Artin. *The Collected Papers of Emil Artin*. Addison-Wesley, 1965.

[Artin and Schreier 1927]
E. Artin and O. Schreier. Eine Kennseichnung der reell abgeschlossenen Körper. *Mat. Sem. Univ. Hamburg* **5** (1927), 225–231. Reprinted in *Collected Papers*, pp. 289–295.

[Artin and Whaples 1945]
E. Artin and G. Whaples. Axiomatic characterization of fields by the product formula for valuations. *Bull. Amer. Math. Soc.* **51** (1945), 469–492. Reprinted in *Collected Papers*, pp. 202–225.

[Arwin 1920]
A. Arwin. Über Kongruenzen von dem fünften und höheren Graden nach einem Primzahlmodulus. *Arkiv för Matematik, Astronomi och Fisik* **14** (1920), 1–46.

[Asano, Itoh and Tsujii 1989]
Y. Asano, T. Itoh, and S. Tsujii. Generalised fast algorithm for computing multiplicative inverses in $GF(2^m)$. *Electronics Letters* **25** (1989), 664–665.

[Ash, Blake, and Vanstone 1989]
D. W. Ash, I. F. Blake, and S. A. Vanstone. Low complexity normal bases. *Disc. Appl. Math.* **25** (1989), 191–210.

[Atkin 1992]
A. O. L. Atkin. Square roots and cognate matters modulo $p = 8n + 5$. Unpublished manuscript, 1992.

[Atkin and Birch 1971]
A. O. L. Atkin and B. J. Birch, editors. *Computers in Number Theory*. Academic Press, New York, 1971.

[Atkin and Larson 1982]
A. O. L. Atkin and R. G. Larson. On a primality test of Solovay and Strassen. *SIAM J. Comput.* **11** (1982), 789–791.

[Averbuch, Bshouty, and Kaminski 1992]
A. Averbuch, N. H. Bshouty, and M. Kaminski. A classification of algorithms for multiplying polynomials of small degree over finite fields. *J. Algorithms* **13** (1992), 577–588.

[Avizienis 1961]
A. Avizienis. Signed-digit number representations for fast parallel arithmetic. *IRE Trans. Elect. Comput.* **10** (1961), 389–400.

[Ayoub and Chowla 1981]
R. G. Ayoub and S. Chowla. On Euler's polynomial. *J. Number Theory* **13** (1981), 443–445.

[Babbage 1889]
H. P. Babbage. *Babbage's Calculating Engines*. Spon & Company, London, 1889. Reprinted by Tomash Publishers, Los Angeles, 1982.

[Bach 1982]
E. Bach. Fast algorithms under the extended Riemann hypothesis: a concrete estimate. In *Proc. Fourteenth Ann. ACM Symp. Theor. Comput.*, pp. 290–295. ACM Press, 1982.

[Bach 1985]
E. Bach. *Analytic Methods in the Analysis and Design of Number-theoretic Algorithms*. The MIT Press, 1985.

[Bach 1987]
E. Bach. Realistic analysis of some randomized algorithms. In *Proc. Nineteenth Ann. ACM Symp. Theor. Comput.*, pp. 453–461. ACM, 1987.

[Bach 1988]
E. Bach. How to generate factored random numbers. *SIAM J. Comput.* **17** (1988), 179–193.

[Bach 1990a]
E. Bach. Explicit bounds for primality testing and related problems. *Math. Comp.* **55** (1990), 355–380.

[Bach 1990b]
E. Bach. A note on square roots in finite fields. *IEEE Trans. Inform. Theory* **36** (1990), 1494–1498.

[Bach 1991b]
E. Bach. Realistic analysis of some randomized algorithms. *J. Comput. System Sci.* **42** (1991), 30–53.

[Bach 1994]
E. Bach. Polynomial-time algorithms for some number-theoretic constants. Unpublished manuscript, 1994.

[Bach 1995]
E. Bach. Comments on search procedures for primitive roots. Unpublished manuscript, 1995. To appear, *Math. Comp.*

[Bach, Driscoll, and Shallit 1990]
E. Bach, J. Driscoll, and J. O. Shallit. Factor refinement. In *Proc. 1st ACM-SIAM Symp. Discrete Algorithms (SODA)*, pp. 201–211, 1990.

[Bach, Driscoll, and Shallit 1993]
E. Bach, J. Driscoll, and J. O. Shallit. Factor refinement. *J. Algorithms* **15** (1993), 199–222.

[Bach and von zur Gathen 1988]
E. Bach and J. von zur Gathen. Deterministic factorization of polynomials over special finite fields. Technical Report 799, University of Wisconsin – Madison, Computer Sciences Department, October 1988.

[Bach, von zur Gathen, and Lenstra 1994]
E. Bach, J. von zur Gathen, and H. W. Lenstra, Jr. Deterministic factorization of polynomials over special finite fields. Unpublished manuscript, dated June, 1994.

[Bach, Giesbrecht, and McInnes 1991]
E. Bach, M. Giesbrecht, and J. McInnes. The complexity of number theoretic problems. Technical Report No. 247/91, Dept. Computer Science, Univ. Toronto, January 1991.

[Bach and Huelsbergen 1993]
E. Bach and L. Huelsbergen. Statistical evidence for small generating sets. *Math. Comp.* **61** (1993), 69–82.

[Bach, Miller, and Shallit 1984]
E. Bach, G. Miller, and J. O. Shallit. Sums of divisors, perfect numbers, and factoring. In *Proc. Sixteenth Ann. ACM Symp. Theor. Comput.*, pp. 183–190. ACM, 1984.

[Bach, Miller, and Shallit 1986]
E. Bach, G. Miller, and J. O. Shallit. Sums of divisors, perfect numbers and factoring. *SIAM J. Comput.* **15** (1986), 1143–1154.

[Bach and Shallit 1984]
E. Bach and J. O. Shallit. A class of functions equivalent to factoring. Technical Report 84-008, University of Chicago, Department of Computer Science, 1984.

[Bach and Shallit 1985]
E. Bach and J. O. Shallit. Factoring with cyclotomic polynomials. In *Proc. 26th Ann. Symp. Found. Comput. Sci.*, pp. 443–450, 1985.

[Bach and Shallit 1989]
E. Bach and J. O. Shallit. Factoring with cyclotomic polynomials. *Math. Comp.* **52** (1989), 201–219.

[Bach and Shoup 1990]
E. Bach and V. Shoup. Factoring polynomials using fewer random bits. *J. Symbolic Comput.* **9** (1990), 229–239.

[Bach and Sorenson 1993]
E. Bach and J. Sorenson. Sieve algorithms for perfect power testing. *Algorithmica* **9** (1993), 313–328.

[Bach and Sorenson 1994]
E. Bach and J. Sorenson. Explicit bounds for primes in residue classes. *Proc. Symp. Appl. Math.* **48** (1994), 535–539.

[Bachmann 1894]
P. Bachmann. *Die Analytische Zahlentheorie*. Teubner, Leipzig, 1894.

[Bachmann 1902]
P. Bachmann. *Niedere Zahlentheorie*, Vol. I. Teubner, Leipzig, 1902. Reprinted by Chelsea, New York, 1968.

[Backlund 1918]
R. J. Backlund. Über die Nullstellen der Riemannschen Zetafunktion. *Acta Math.* **41** (1918), 345–375.

[Backlund 1929]
R. J. Backlund. Über die Differenzen zwischen den Zahlen, die zu den n ersten Primzahlen teilerfremnd sind. In *Commentationes in Honorem Ernesti Leonardi Lindelöf*, pp. 3–9. Suomalainen Tiedeakatemia, 1929. (= *Ann. Acad. Sci. Fenn.* (A) **32**.).

[Baillie and Wagstaff 1980]
R. Baillie and S. S. Wagstaff, Jr. Lucas pseudoprimes. *Math. Comp.* **35** (1980), 1391–1417.

[Baker and Gruenberger 1959]
C. L. Baker and F. J. Gruenberger. *The First Six Million Prime Numbers*. Microcard Foundation, Madison, Wisconsin, 1959.

[Baker 1987]
P. W. Baker. Fast computation of $A * B$ modulo N. *Electronics Letters* **23** (1987), 794–795.

[Balasubramanian 1986]
R. Balasubramanian. Number theory and primality testing. In K. S. Rao, editor, *Proc. Workshop on Mathematics of Computer Algorithms*, pp. A.5.1–A.5.29. Inst. Math. Sci., Madras, India, 1986.

[Balasubramanian, Deshouillers, and Dress 1986a]
R. Balasubramanian, J.-M. Deshouillers, and F. Dress. Problème de Waring pour les bicarrés. I. Schéma de la solution. *C. R. Acad. Sci. Paris* **303** (1986), 85–88.

[Balasubramanian, Deshouillers, and Dress 1986b]
R. Balasubramanian, J.-M. Deshouillers, and F. Dress. Problème de Waring pour les bicarrés. II. Résultats auxiliaires pour le théorème asymptotique. *C. R. Acad. Sci. Paris* **303** (1986), 161–163.

[Bang 1952]
T. Bang. A function representing prime numbers. *Norsk Mat. Tidskrift* **34** (1952), 117–118.

[Baratz 1978]
A. E. Baratz. An analysis of the Solovay and Strassen test for primality. Technical Report MIT/LCS/TM-108, MIT Laboratory for Computer Science, July 1978.

[Barlow 1811]
P. Barlow. *An Elementary Investigation of the Theory of Numbers, With its Application to the Indeterminate and Diophantine Analysis, the Analytical and Geometrical Division of the Circle, and Several Other Curious Algebraical and Arithmetical Problems*. J. Johnson and Co., London, 1811.

[Bartee and Schneider 1963]
T. C. Bartee and D. I. Schneider. Computation with finite fields. *Inform. Control* **6** (1963), 79–98.

[Baruah, Rosier, and Howell 1990]
S. K. Baruah, R. R. Howell, and L. E. Rosier. Algorithms and complexity concerning the preemptive scheduling of periodic, real-time tasks on one processor. *Real-Time Systems* **2** (1990), 301–324.

[Baruah, Howell, and Rosier 1993]
S. K. Baruah, R. R. Howell, and L. E. Rosier. Feasibility problems for recurring tasks on one processor. *Theoret. Comput. Sci.* **118** (1993), 3–20.

[Bassalygo 1978]
L. A. Bassalygo. A remark on fast multiplication of polynomials over Galois fields. *Problemy Peredachi Informatsii* **14** (1978), 101–102. In Russian. English translation in *Problems Inform. Transmission*, **14**, 1978, 71–72.

[Bateman and Horn 1962]
P. T. Bateman and R. A. Horn. A heuristic asymptotic formula concerning the distribution of prime numbers. *Math. Comp.* **16** (1962), 363–367.

[Bateman, Selfridge, and Wagstaff 1989]
P. T. Bateman, J. L. Selfridge, and S. S. Wagstaff, Jr. The new Mersenne conjecture. *Amer. Math. Monthly* **96** (1989), 125–128.

[Bateman and Stemmler 1962]
P. T. Bateman and R. M. Stemmler. Waring's problem for algebraic number fields and primes of the form $(p^r - 1)/(p^d - 1)$. *Illinois J. Math.* **6** (1962), 142–156.

[Baum and Shokrollahi 1991]
U. Baum and M. A. Shokrollahi. An optimal algorithm for multiplication in F_{256}/F_4. *Appl. Alg. Eng. Comm. Comp.* **2** (1991), 15–20.

[Bays and Hudson 1977]
C. Bays and R. H. Hudson. The segmented sieve of Eratosthenes and primes in arithmetic progressions to 10^{12}. *BIT* **17** (1977), 121–127.

[Beame, Cook, and Hoover 1986]
P. W. Beame, S. A. Cook, and H. J. Hoover. Log depth circuits for division and related problems. *SIAM J. Comput.* **15** (1986), 994–1003.

[Beard 1974]
J. T. B. Beard, Jr. Computing in $GF(q)$. *Math. Comp.* **28** (1974), 1159–1166.

[Beauchemin, Brassard, Crépeau, and Goutier 1987]
P. Beauchemin, G. Brassard, C. Crépeau, and C. Goutier. Two observations on probabilistic primality testing. In A. M. Odlyzko, editor, *Advances in Cryptology–CRYPTO '86 Proceedings*, Vol. 263 of *Lecture Notes in Computer Science*, pp. 443–450. Springer-Verlag, Berlin, 1987.

[Beauchemin, Brassard, Crepeau, Goutier, and Pomerance 1988]
P. Beauchemin, G. Brassard, C. Crépeau, C. Goutier, and C. Pomerance. The generation of random numbers that are probably prime. *J. Cryptology* **1** (1988), 53–64.

[Beeger 1950]
N. G. W. H. Beeger. On composite numbers n for which $a^{n-1} \equiv 1 \pmod{n}$ for every a prime to n. *Scripta Math.* **16** (1950), 133–135.

[Beeger 1951]
N. G. W. H. Beeger. On even numbers m dividing $2^m - 2$. *Amer. Math. Monthly* **58** (1951), 553–555.

[Beiler 1964]
A. H. Beiler. *Recreations in the Theory of Numbers*. Dover, New York, 1964.

[Bell 1951]
E. T. Bell. *Mathematics: Queen and Servant of Science*. G. Bell & Sons, 1951.

[Ben-Or 1981]
M. Ben-Or. Probabilistic algorithms in finite fields. In *Proc. 22nd Ann. Symp. Found. Comput. Sci.*, pp. 394–398. IEEE Press, 1981.

[Bengelloun 1986]
S. A. Bengelloun. An incremental primal sieve. *Acta Infor.* **23** (1986), 119–125.

[Bentley 1982]
J. L. Bentley. *Writing Efficient Programs*. Prentice-Hall, 1982.

[Bergeron, Berstel, and Brlek 1994]
F. Bergeron, J. Berstel, and S. Brlek. Efficient computation of addition chains. *J. Théorie Nombres Bordeaux* **6** (1994), 21–38.

[Bergeron, Berstel, Brlek, and Duboc 1989]
F. Bergeron, J. Berstel, S. Brlek, and C. Duboc. Addition chains using continued fractions. *J. Algorithms* **10** (1989), 403–412.

[Berlekamp 1966]
E. R. Berlekamp. Distribution of cyclic matrices in a finite field. *Duke Math. J.* **33** (1966), 45–48.

[Berlekamp 1967]
E. R. Berlekamp. Factoring polynomials over finite fields. *Bell System Tech. J.* **46** (1967), 1853–1859.

[Berlekamp 1968]
E. R. Berlekamp. *Algebraic Coding Theory*. McGraw-Hill, 1968.

[Berlekamp 1970]
E. R. Berlekamp. Factoring polynomials over large finite fields. *Math. Comp.* **24** (1970), 713–735.

[Berlekamp 1972]
E. R. Berlekamp. Factoring polynomials. In *Proc. 3rd Southeastern Conference on Combinatorics, Graph Theory, and Computing*, pp. 1–7, Winnipeg, 1972. Utilitas Mathematica. (= *Congr. Numer.* **VII**).

[Berlekamp, McEliece, and van Tilborg 1978]
E. R. Berlekamp, R. J. McEliece, and H. C. A. van Tilborg. On the inherent intractability of certain coding problems. *IEEE Trans. Inform. Theory* **IT-24** (1978), 384–386.

[Berlekamp, Rumsey, and Solomon 1967]
E. R. Berlekamp, H. Rumsey, and G. Solomon. On the solution of algebraic equations over finite fields. *Inform. Control* **10** (1967), 553–564.

[Berndt and Evans 1981]
B. C. Berndt and R. J. Evans. The determination of Gauss sums. *Bull. Amer. Math. Soc.* **5** (1981), 107–129.

[D. Bernstein 1992]
D. Bernstein. The coprime base algorithm. Unpublished manuscript, 1992.

[D. Bernstein 1995]
D. Bernstein. Detecting perfect powers in essentially linear time. Unpublished manuscript, 1995.

[L. Bernstein 1971]
L. Bernstein. *The Jacobi-Perron Algorithm, its Theory and Applications*, Vol. 207 of *Lecture Notes in Mathematics*. Springer-Verlag, Berlin, 1971.

[Bertrand 1845]
J. Bertrand. Mémoire sur le nombre de valeurs que peut prendre une function quand on y permute les lettres qu'elle renferme. *J. École Polytechnique* **28** (1845), 123–140.

[Beth 1986]
T. Beth. On the arithmetics of Galoisfields [sic] and the like. In J. Calmet, editor, *Algebraic Algorithms and Error-Correcting Codes: AAEEC-3*, pp. 2–16. Springer-Verlag, 1986.

[Beth, Cook, and Gollmann 1986]
T. Beth, B. M. Cook, and D. Gollmann. Architectures for exponentiation in $GF(2^n)$. In A. M. Odlyzko,

editor, *Advances in Cryptology–CRYPTO '86 Proceedings*, Vol. 263 of *Lecture Notes in Computer Science*, pp. 302–310. Springer-Verlag, 1986.

[Beth, Geiselmann, and Meyer 1991]
T. Beth, W. Geiselmann, and F. Meyer. Finding (good) normal bases in finite fields. In S. M. Watt, editor, *Proc. Int'l. Symp. on Symbolic and Algebraic Computation: ISSAC '91*, pp. 173–178. ACM Press, 1991.

[K. Biermann 1964]
K.-R. Biermann. Thomas Clausen, Mathematiker und Astronom. *J. Reine Angew. Math.* **216** (1964), 159–198.

[O. Biermann 1891]
O. Biermann. Über die Resultante ganzer Functionen. *Monatsh. Math. Phys.* **2** (1891), 143–146.

[Bilharz 1937]
H. Bilharz. Primdivisoren mit vorgegebener Primitivwurzel. *Math. Ann.* **114** (1937), 476–492.

[Billingsley 1969]
P. Billingsley. On the central limit theorem for the prime divisor function. *Amer. Math. Monthly* **76** (1969), 132–139.

[Billingsley 1973]
P. Billingsley. Prime numbers and Brownian motion. *Amer. Math. Monthly* **80** (1973), 1099–1115.

[Binet 1841]
J. P. M. Binet. Recherches sur la théorie des nombres entiers et sur la résolution de l'équation indéterminée du premier degré qui n'admet que des solutions entières. *J. Math. Pures Appl.* **6** (1841), 449–494.

[Binet 1843]
J. P. M. Binet. Mémoire sur l'intégration des équations linéaires aux différences finies, d'un ordre quelconque, à coefficients variables. *C. R. Acad. Sci. Paris* **17** (1843), 559–567.

[Birch and Swinnerton-Dyer 1963]
B. J. Birch and H. P. F. Swinnerton-Dyer. Notes on elliptic curves I. *J. Reine Angew. Math.* **212** (1963), 7–25.

[Birch and Swinnerton-Dyer 1965]
B. J. Birch and H. P. F. Swinnerton-Dyer. Notes on elliptic curves II. *J. Reine Angew. Math.* **218** (1965), 79–108.

[Birkhoff and Hall 1971]
G. Birkhoff and M. Hall, Jr., editors. *Computers in Algebra and Number Theory*. Amer. Math. Soc., New York, 1971. Proc. SIAM-AMS Sympos. Appl. Math. 1970, Vol. IV.

[Birkhoff and Mac Lane 1977]
G. Birkhoff and S. Mac Lane. *A Survey of Modern Algebra*. Macmillan, 4th edition, 1977.

[Blake, Gao, and Mullin 1994]
I. F. Blake, S. Gao, and R. C. Mullin. Factorization of $cx^{q+1} + dx^q - ax - b$ and normal bases over $GF(q)$. *SIAM J. Disc. Math.* **7** (1994), 499–512.

[Blakley 1983]
G. R. Blakley. A computer algorithm for calculating the product AB modulo M. *IEEE Trans. Comput.* **C-32** (1983), 497–500.

[Blakley and Borosh 1983]
G. R. Blakley and I. Borosh. Modular arithmetic of iterated powers. *Comput. Math. with Appl.* **9** (1983), 567–581.

[Blankinship 1963]
W. A. Blankinship. A new version of the Euclidean algorithm. *Amer. Math. Monthly* **70** (1963), 742–745.

[Bleichenbacher and Maurer 1994]
D. Bleichenbacher and U. Maurer. Finding all strong pseudoprimes $\leq x$. Unpublished manuscript, 1994.

[Blokh 1964]
E. L. Blokh. Method of decoding Bose-Chaudhuri triple error correcting codes. *Techn. Kibern.* **3** (1964), 30–37. In Russian. English translation in *Engineering Cybernetics* **3** (1964), 22–32.

[Blum and Kannan 1989]
M. Blum and S. Kannan. Designing programs that check their work. In *Proc. Twenty-first Ann. ACM Symp. Theor. Comput.*, pp. 86–97, 1989.

[Blum and Kannan 1995]
M. Blum and S. Kannan. Designing programs that check their work. *J. Assoc. Comput. Mach.* **42** (1995), 269–291.

[Böffgen and Reichert 1987]
R. Böffgen and M. A. Reichert. Computing the decomposition of primes p and p-adic absolute values in semi-simple algebras over $\mathbb{Q}$. *J. Symbolic Comput.* **4** (1987), 3–10.

[Bohman 1972]
J. Bohman. On the number of primes less than a given limit. *BIT* **12** (1972), 576–577.

[Bohman 1973]
J. Bohman. Some computational results regarding the prime numbers below 2,000,000,000. *BIT* **13** (1973), 242–244. Errata in **14** (1974), 127.

[Bohman and Fröberg 1988]
J. Bohman and C.-E. Fröberg. The Stieltjes function – definition and properties. *Math. Comp.* **51** (1988), 281–289.

[Bohr and Cramér 1923]
H. Bohr and H. Cramér. Die neuere Entwicklung der analytischen Zahlentheorie. *Encyklopädie der Mathematischen Wissenschaften* **II C 8** (1923), 722–849. Reprinted in Cramér, *Collected Works*, Vol. I, pp. 289–416.

[du Bois-Reymond 1870]
P. du Bois-Reymond. Sur la grandeur relative des infinis des fonctions. *Ann. Mat. Pura Appl.* **4** (1870), 338–353.

[du Bois-Reymond 1872]
P. du Bois-Reymond. Théorème général concernant la grandeur relative des infinis des fonctions et de leurs derivées. *J. Reine Angew. Math.* **74** (1872), 294–304.

[du Bois-Reymond 1879]
P. du Bois-Reymond. Ueber Integration und Differentiation infinitärer Relationen. *Math. Ann.* **14** (1879), 498–506.

[Bojanczyk and Brent 1987]
A. W. Bojanczyk and R. P. Brent. A systolic algorithm for extended GCD computation. *Comput. Math. with Appl.* **14** (1987), 233–238.

[Bokhari 1987]
S. H. Bokhari. Multiprocessing the sieve of Eratosthenes. *IEEE Computer* **20** (1987), 50–58.

[Bombieri 1965]
E. Bombieri. On the large sieve. *Mathematika* **12** (1965), 201–225.

[D. Bond 1984]
D. J. Bond. Practical primality testing. In *Proc. Int'l. Conf. Secure Communication Systems*, pp. 50–53. IEE, 1984.

[Boos 1948]
P. Boos. Divisibilité des polynômes relativement aux puissances d'un nombre entier. *Bull. Soc. Math. France* **76** (1948), 65–78.

[Borho, Buhl, Hoffmann, Mertens, Nebgen, and Reckow 1983]
W. Borho, J. Buhl, H. Hoffmann, S. Mertens, E. Nebgen, and R. Reckow. Große Primzahlen und befreundete Zahlen: Über den Lucas-Test und Thabit-Regeln. *Mitt. Math. Ges. Hamburg* **11** (1983), 232–256.

[Borning 1972]
A. Borning. Some results for $k! \pm 1$ and $2 \cdot 3 \cdot 5 \cdots p \pm 1$. *Math. Comp.* **26** (1972), 567–570.

[Borodin 1973]
A. Borodin. Computational complexity: theory and practice. In A. V. Aho, editor, *Currents in the Theory of Computing*, pp. 35–89. Prentice-Hall, 1973.

[Borodin, von zur Gathen, and Hopcroft 1982]
A. Borodin, J. von zur Gathen, and J. Hopcroft. Fast parallel matrix and gcd computations. *Inform. Control* **52** (1982), 241–256.

[Borodin and Munro 1975]
A. Borodin and I. Munro. *The Computational Complexity of Algebraic and Numeric Problems*. American Elsevier, 1975.

[Borwein 1985]
P. B. Borwein. On the complexity of calculating factorials. *J. Algorithms* **6** (1985), 376–380.

[Bos and Coster 1990]
J. Bos and M. Coster. Addition chain heuristics. In G. Brassard, editor, *Advances in Cryptology–CRYPTO '89 Proceedings*, Vol. 435 of *Lecture Notes in Computer Science*, pp. 400–407. Springer-Verlag, 1990.

[Bosma 1993]
W. Bosma. Explicit primality criteria for $h \cdot 2^k \pm 1$. *Math. Comp.* **61** (1993), 97–109.

[Bosma and van der Hulst 1990a]
W. Bosma and M.-P. van der Hulst. Faster primality testing. In J.-J. Quisquater and J. Vandewalle, editors, *Advances in Cryptology–EUROCRYPT '89 Proceedings*, pp. 652–656. Springer-Verlag, 1990.

[Bosma and van der Hulst 1990b]
W. Bosma and M.-P. van der Hulst. *Primality proving with cyclotomy*. PhD thesis, Faculteit Wiskunde en Informatica, Universiteit van Amsterdam, 1990.

[Bouniakowsky 1857]
V. Bouniakowsky. Sur les diviseurs numériques invariables des fonctions rationnelles entières. *Mem. Imp. Acad. Sci. St.-Pétersbourg* **6** (1857), 305–329.

[Bouniakowsky 1884]
V. Bouniakowsky. Protokol'i fiziko-matematicheskago otddjleniya. *Zapiski Imperatorskoj Akademii Nauk* **48** (1884), 112.

[Bourbaki 1951]
N. Bourbaki. *Fonctions d'une Variable Réelle*, Vol. 12. Hermann, 1951.

[Boute 1992]
R. T. Boute. The Euclidean definition of the functions div and mod. *ACM Trans. Prog. Lang. Syst.* **14** (1992), 127–144.

[Boyar, Frandsen, and Sturtivant 1992]
J. Boyar, G. Frandsen, and C. Sturtivant. An arithmetic model of computation equivalent to threshold circuits. *Theoret. Comput. Sci.* **93** (1992), 303–319.

[Boyar, Friedl, and Lund 1990]
J. Boyar, K. Friedl, and C. Lund. Practical zero-knowledge proofs: giving hints and using deficiencies. In J.-J. Quisquater and J. Vendewalle, editors, *Advances in Cryptology–EUROCRYPT '89 Proceedings*, pp. 155–172. Springer-Verlag, 1990.

[Boyd 1988]
C. Boyd. Probabilistic prime tests. In N. M. Stephens and M. P. Thorne, editors, *Computers in Mathematical Research*, pp. 57–68. Clarendon Press, Oxford, 1988.

[Boys 1901]
C. V. Boys. The comptometer. *Nature* **64** (1901), 265–268.

[Bradley 1970]
G. H. Bradley. Algorithm and bound for the greatest common divisor of n integers. *Comm. ACM* **13** (1970), 433–436.

[Brassard, Monet, and Zuffellato 1986]
G. Brassard, S. Monet, and D. Zuffellato. Algorithmes pour l'arithmétique des très grands entiers. *Techniques et Science Informatique* **5** (1986), 89–102.

[Brauer 1939]
A. Brauer. On addition chains. *Bull. Amer. Math. Soc.* **45** (1939), 736–739.

[Brent 1973]
R. P. Brent. The first occurrence of large gaps between successive primes. *Math. Comp.* **27** (1973), 959–963.

[Brent 1974]
R. P. Brent. The distribution of small gaps between successive primes. *Math. Comp.* **28** (1974), 315–324.

[Brent 1975]
R. P. Brent. Irregularities in the distribution of primes and twin primes. *Math. Comp.* **29** (1975), 43–56. Corrigenda in **30** (1976), 198.

[Brent 1976a]
R. P. Brent. Analysis of the binary Euclidean algorithm. In J. F. Traub, editor, *Algorithms and Complexity: New Directions and Recent Results*, pp. 321–355. Academic Press, New York, 1976.

[Brent 1979]
R. P. Brent. On the zeros of the Riemann zeta function in the critical strip. *Math. Comp.* **33** (1979), 1361–1372.

[Brent 1980a]
R. P. Brent. The first occurrence of certain large prime gaps. *Math. Comp.* **35** (1980), 1435–1436.

[Brent and Cohen 1989]
R. P. Brent and G. L. Cohen. A new lower bound for odd perfect numbers. *Math. Comp.* **53** (1989), 431–437.

[Brent, Cohen, and te Riele 1989]
R. P. Brent, G. L. Cohen, and H. J. J. te Riele. Improved techniques for lower bounds for odd perfect numbers. Technical Report CMA-R50-89, Centre for Mathematical Analysis, Australian National University, Canberra, October 1989.

[Brent and Kung 1983]
R. P. Brent and H. T. Kung. Systolic VLSI arrays for linear-time GCD computation. In F. Anceau and E. J. Aas, editors, *VLSI '83*, pp. 145–154. Elsevier Science Publishers B. V., 1983.

[Brent and Kung 1985]
R. P. Brent and H. T. Kung. A systolic algorithm for integer GCD computation. In K. Hwang, editor, *Proc. 7th Symp. Comp. Arith.*, pp. 118–125. IEEE Press, 1985.

[Brent, Kung, and Luk 1983]
R. P. Brent, H. T. Kung, and F. T. Luk. Some linear time algorithms for systolic arrays. In *Information Processing 83: Proc. IFIP 9th World Computer Congress*, pp. 865–876. North-Holland, 1983.

[Brent and McMillan 1980]
R. P. Brent and E. M. McMillan. Some new algorithms for high-precision computation of Euler's constant. *Math. Comp.* **34** (1980), 305–312.

[Brent, van de Lune, and te Riele 1982]
R. P. Brent, J. van de Lune, H. J. J. te Riele, and D. T. Winter. On the roots of the Riemann zeta function in the critical strip. II. *Math. Comp.* **39** (1982), 681–688.

[Brentjes 1981]
A. J. Brentjes. *Multi-Dimensional Continued Fraction Algorithms*, Vol. 145 of *Mathematical Centre Tracts*. Mathematisch Centrum, Amsterdam, 1981.

[Bressoud 1989]
D. M. Bressoud. *Factorization and Primality Testing*. Springer-Verlag, 1989.

[Breusch 1979]
R. Breusch. Solution to elementary problem E2686. *Amer. Math. Monthly* **86** (1979), 131.

[Brewer 1951]
B. W. Brewer. Tests for primality. *Duke Math. J.* **18** (1951), 757–763.

[Brezinski 1991]
C. Brezinski. *History of Continued Fractions and Padé Approximants*, Vol. 12 of *Springer Series in Computational Mathematics*. Springer-Verlag, 1991.

[Brickell 1983]
E. Brickell. A fast modular multiplication algorithm with application to two key cryptography. In D. Chaum, R. L. Rivest, and A. T. Sherman, editors, *Advances in Cryptology–Proceedings of CRYPTO 82*, pp. 51–60, New York, 1983. Plenum Press.

[Brickell, Gordon, McCurley, and Wilson 1993]
E. F. Brickell, D. M. Gordon, K. S. McCurley, and D. B. Wilson. Fast exponentiation with precomputation. In R. A. Rueppel, editor, *Advances in Cryptology–EUROCRYPT '92 Proceedings*, Vol. 658 of *Lecture Notes in Computer Science*, pp. 200–207. Springer-Verlag, 1993.

[Briggs 1955]
W. E. Briggs. Some constants associated with the Riemann zeta-function. *Michigan Math. J.* **3** (1955), 117–121.

[Briggs and Chowla 1955]
W. E. Briggs and S. Chowla. The power series coefficients of $\zeta(s)$. *Amer. Math. Monthly* **62** (1955), 323–325.

[Brillhart, Lehmer, and Selfridge 1975]
J. Brillhart, D. H. Lehmer, and J. L. Selfridge. New primality criteria and factorizations of $2^m \pm 1$. *Math. Comp.* **29** (1975), 620–647.

[Brillhart, Lehmer, Selfridge, Tuckerman, and Wagstaff 1983]
J. Brillhart, D. H. Lehmer, J. L. Selfridge, B. Tuckerman, and S. S. Wagstaff, Jr. *Factorizations of $b^n \pm 1$, $b = 2, 3, 5, 6, 7, 10, 11, 12$ Up to High Powers*, Vol. 22 of *Contemporary Mathematics*. American Mathematical Society, 1983.

[Brillhart, Lehmer, Selfridge, Tuckerman, and Wagstaff 1988]
J. Brillhart, D. H. Lehmer, J. L. Selfridge, B. Tuckerman, and S. S. Wagstaff, Jr. *Factorizations of $b^n \pm 1$, $b = 2, 3, 5, 6, 7, 10, 11, 12$ Up to High Powers*, Vol. 22 of *Contemporary Mathematics*. American Mathematical Society, 1988. 2nd edition.

[Brillhart and Selfridge 1967]
J. Brillhart and J. L. Selfridge. Some factorizations of $2^n \pm 1$ and related results. *Math. Comp.* **21** (1967), 87–96. Corrigendum, *Math. Comp.* **21** (1967) 751.

[Brlek, Castéran, Habsieger and Mallette 1995]
S. Brlek, P. Castéran, L. Habsieger, and R. Mallette. On-line evaluation of powers using Euclid's algorithm. *RAIRO Inform. Théor.* **29** (1995) 431–450.

[Brlek and Mallette 1992]
S. Brlek and R. Mallette. Sur le calcul des chaines d'additions optimales. Technical Report 173, Département de mathématiques et d'informatique, Université du Québec à Montréal, February 1992. Abbreviated version in J. Labelle and J.-G. Penaud, eds., *Atelier de Combinatoire Franco-Québecois, 6-7 mai 1991*, Publications du LaCIM **10** (1992), 71–85.

[Brooks et al. 1995]
A. S. Brooks et al. Dating and context of three middle stone age sites with bone points in the Upper Semliki Valley, Zaire. *Science* **268** (1995), 548–553.

[Brooks and Smith 1987]
A. S. Brooks and C. C. Smith. Ishango revisited: new age determinations and cultural interpretations. *African Archaeological Review* **5** (1987), 65–78.

[Browder 1976]
F. E. Browder, editor. *Mathematical Developments Arising from Hilbert Problems*, Vol. 28 of *Symposium in pure mathematics*. Amer. Math. Soc., 1976.

[Brown, Noll, Parady, Smith, Smith, and Zarantonello 1990]
J. Brown, L. C. Noll, B. K. Parady, J. F. Smith, G. W. Smith, and S. E. Zarantonello. Letter to the editor. *Amer. Math. Monthly* **97** (1990), 214.

[Brown and Duncan 1971]
J. L. Brown, Jr. and R. L. Duncan. The least remainder algorithm. *Fibonacci Quart.* **9** (1971), 347–350,401.

[Bruckman 1994a]
P. S. Bruckman. On the infinitude of Lucas pseudoprimes. *Fibonacci Quart.* **32** (1994), 153–154.

[Bruckman 1994b]
P. S. Bruckman. Lucas pseudoprimes are odd. *Fibonacci Quart.* **32** (1994), 155–157.

[Bruckman 1994c]
P. S. Bruckman. On a conjecture of Di Porto and Filipponi. *Fibonacci Quart.* **32** (1994), 158–159.

[Brun 1920]
V. Brun. Le crible d'Eratosthène et le théorème de Goldbach. *Videnskapsselskaptes Skrifter, Mat. Naturv. Klasse. (Kristiania)* **1**(3) (1920), 1–36.

[Brun 1931]
V. Brun. Algorithme pour calculer le $n$ième nombre premier. *Det Kongelige Norske Videnskabers Selskab* **4** (1931), 66–69.

[Brun 1964]
V. Brun. Euclidean algorithms and musical theory. *Enseign. Math.* **10** (1964), 125–137.

[Bshouty 1992]
N. H. Bshouty. A lower bound for the multiplication of polynomials modulo a polynomial. *Inform. Process. Lett.* **41** (1992), 321–326.

[Bshouty and Kaminski 1990]
N. H. Bshouty and M. Kaminski. Multiplication of polynomials over finite fields. *SIAM J. Comput.* **19** (1990), 452–456.

[Buchmann 1990]
J. Buchmann. Complexity of algorithms in algebraic number theory. In R. A. Mollin, editor, *Number Theory*, pp. 37–53. Walter de Gruyter, Berlin, 1990.

[Buchmann and Shoup 1991]
J. Buchmann and V. Shoup. Constructing nonresidues in finite fields and the extended Riemann hypothesis. In *Proc. Twenty-third Ann. ACM Symp. Theor. Comput.*, pp. 72–79, 1991.

[Buchmann and Lenstra 1994]
J. A. Buchmann and H. W. Lenstra, Jr. Approximating rings of integers in number fields. *J. Théorie Nombres Bordeaux* **6** (1994), 221–260.

[Buck 1946]
R. C. Buck. Prime-representing functions. *Amer. Math. Monthly* **53** (1946), 265.

[Buell and Ward 1989]
D. A. Buell and R. L. Ward. A multiprecise integer arithmetic package. *J. Supercomputing* **3** (1989), 89–107.

[Buhler, Crandall, Ernvall, and Metsänkylä 1993]
J. P. Buhler, R. E. Crandall, R. Ernvall, and T. Metsänkylä. Irregular primes and cyclotomic invariants to four million. *Math. Comp.* **61** (1993), 151–153.

[Buhler, Crandall, and Penk 1982]
J. P. Buhler, R. E. Crandall, and M. A. Penk. Primes of the form $n! \pm 1$ and $2 \cdot 3 \cdot 5 \cdots p \pm 1$. *Math. Comp.* **38** (1982), 639–643. Corrigendum in *Math. Comp.* **40** (1983), 727.

[Buhler, Crandall, and Sompolski 1992]
J. P. Buhler, R. E. Crandall, and R. W. Sompolski. Irregular primes to one million. *Math. Comp.* **59** (1992), 717–722.

[Bumby 1989]
R. T. Bumby. How not to compute square roots mod p with applications. Unpublished manuscript, 1989.

[Burgess 1957]
D. A. Burgess. The distribution of quadratic residues and non-residues. *Mathematika* **4** (1957), 106–112.

[Burgess 1962a]
D. A. Burgess. On character sums and primitive roots. *Proc. Lond. Math. Soc.* **12** (1962), 179–192.

[Burgess 1963a]
D. A. Burgess. A note on the distribution of residues and non-residues. *J. London Math. Soc.* **38** (1963), 253–256.

[Burgess 1967]
D. A. Burgess. On the quadratic character of a polynomial. *J. London Math. Soc.* **42** (1967), 73–80.

[Burgess and Elliott 1968]
D. A. Burgess and P. D. T. A. Elliott. The average of the least primitive root. *Mathematika* **15** (1968), 39–50.

[Burthe 1995]
R. J. Burthe, Jr. *The Average Witness is* 2. PhD thesis, University of Georgia, 1995.

[Butler 1954]
M. C. R. Butler. On the reducibility of polynomials over a finite field. *Quart. J. Math. Oxford* **(2) 5** (1954), 102–107.

[Cadwell 1971]
H. Cadwell. Large intervals between consecutive primes. *Math. Comp.* **25** (1971), 909–913.

[E. Cahen 1894]
E. Cahen. Sur la fonction $\zeta(s)$ de Riemann et sur des fonctions analogues. *Ann. Sci. École Normale Supérieure* **11** (1894), 75–164.

[Caldwell 1995]
C. K. Caldwell. On the primality of $n! \pm 1$ and $2 \cdot 3 \cdot 5 \cdots p \pm 1$. *Math. Comp.* **64** (1995), 889–890.

[Callan 1991]
D. Callan. Solution to problem 1344. *Math. Mag.* **64** (1991), 134.

[Calmet 1985]
J. Calmet. Algebraic algorithms in $GF(q)$. *Discrete Math.* **56** (1985), 101–109.

[Calmet and Loos 1980]
J. Calmet and R. Loos. An improvement of Rabin's probabilistic algorithm for generating irreducible polynomials over $GF(p)$. *Inform. Process. Lett.* **11** (1980), 94–95.

[Calude and Zimand 1984]
C. Calude and M. Zimand. A relation between correctness and randomness in the computation of probabilistic algorithms. *Internat. J. Comput. Math.* **16** (1984), 47–53.

[Camion 1980]
P. F. Camion. Un algorithme de construction des idempotents primitifs d'idéaux d'algèbres sur $\mathbb{F}_q$. *C. R. Acad. Sci. Paris* **291** (1980), A479–A482.

[Camion 1982]
P. F. Camion. Un algorithme de construction des idempotents primitifs d'idéaux d'algèbres sur $\mathbb{F}_q$. *Ann. Disc. Math.* **12** (1982), 55–63.

[Camion 1983a]
P. F. Camion. Improving an algorithm for factoring polynomials over a finite field and constructing large irreducible polynomials. *IEEE Trans. Inform. Theory* **IT-29** (1983), 378–385.

[Camion 1983b]
P. F. Camion. A deterministic algorithm for factoring polynomials of $\mathbb{F}_q[X]$. *Ann. Disc. Math.* **17** (1983), 149–157.

[Camion 1989]
P. F. Camion. An iterative Euclidean algorithm. In L. Huguet and A. Poli, editors, *Applied Algebra, Algebraic Algorithms and Error-Correcting Codes: AAECC-5*, Vol. 356 of *Lecture Notes in Computer Science*, pp. 88–128. Springer-Verlag, 1989.

[Campbell-Kelly 1980]
M. Campbell-Kelly. Programming on the Mark I: early programming activity at the University of Manchester. *Ann. Hist. Comput.* **2** (1980), 130–168.

[Cantor 1989]
D. G. Cantor. On arithmetical algorithms over finite fields. *J. Combin. Theory. Ser. A* **50** (1989), 285–300.

[Cantor and Zassenhaus 1981]
D. G. Cantor and H. Zassenhaus. A new algorithm for factoring polynomials over finite fields. *Math. Comp.* **36** (1981), 587–592.

[Capocelli, Cerbone, Cull, and Holloway 1990]
R. M. Capocelli, G. Cerbone, P. Cull, and J. L. Holloway. Fibonacci facts and formulas. In R. M. Capocelli, editor, *Sequences*, pp. 123–137. Springer-Verlag, 1990.

[Carissan 1920]
E. Carissan. Machine à résoudre les congruences. *Bulletin de la Société d'Encouragement pour L'Industrie Nationale* **119** (1920), 600–607.

[Carlitz 1930]
L. Carlitz. The arithmetic of polynomials in a Galois field. *Proc. Nat. Acad. Sci. U. S. A.* **17** (1930), 120–122.

[Carlitz 1932]
L. Carlitz. The arithmetic of polynomials in a Galois field. *Amer. J. Math.* **54** (1932), 39–50.

[Carlitz 1953a]
L. Carlitz. A theorem of Stickelberger. *Math. Scand.* **1** (1953), 82–84.

[Carlitz 1953b]
L. Carlitz. A theorem on congruences. *J. Indian Math. Soc.* **17** (1953), 43–45.

[Carmichael 1910]
R. D. Carmichael. Note on a new number theory function. *Bull. Amer. Math. Soc.* **16** (1910), 232–238.

[Carmichael 1912]
R. D. Carmichael. On composite numbers P which satisfy the Fermat congurence $a^{P-1} \equiv 1 \bmod P$. *Amer. Math. Monthly* **19** (1912), 22–27.

[Carmichael 1913]
R. D. Carmichael. On the numerical factors of the arithmetic forms $\alpha^n \pm \beta^n$. *Ann. Math.* **15** (1913-14), 30–70.

[Carmichael 1914]
R. D. Carmichael. *The Theory of Numbers*. John Wiley & Sons, 1914. Reprinted by Dover, 1959.

[Carmichael 1919]
R. D. Carmichael. Fermat numbers $F_n = 2^{2^n} + 1$. *Amer. Math. Monthly* **26** (1919), 137–146.

[Caron and Silverman 1988]
T. R. Caron and R. D. Silverman. Parallel implementation of the quadratic sieve. *J. Supercomputing* **1** (1988), 273–290.

[Cartier 1970]
P. Cartier. Sur une généralisation des symboles de Legendre-Jacobi. *Enseign. Math.* **16** (1970), 31–48.

[Cassels and Fröhlich 1967]
J. W. S. Cassels and A. Fröhlich, editors. *Algebraic Number Theory*. Thompson Book Co., Washington, 1967. Reprinted by Academic Press, 1986.

[Cauchy 1829]
A. Cauchy. Sur la résolution des équivalences dont les modules se réduisent à des nombres premiers. *Exercices Math.* **4** (1829), 253–292. Reprinted in *Oeuvres (2)*, Vol. 9, pp. 298–341.

[Cauchy 1840]
A. Cauchy. Mémoire sur l'élimination d'une variable entre deux équations algébriques. In *Exercices d'Analyse et de Physique Mathématique*, Vol. I. Bachelier, 1840. Reprinted in *Oeuvres (2)*, Vol. 11, pp. 466-509.

[Cauchy 1841]
A. Cauchy. Mémoire sur diverses formules relatives à l'algèbre et à la théorie des nombres (suite). *C. R. Acad. Sci. Paris* **12** (1841), 813–846. Reprinted in *Oeuvres (1)*, Vol. 6, pp. 113–146.

[Cauchy 1847]
A. Cauchy. Mémoire sur les racines des équivalences correspondantes à des modules quelconques premiers ou non premiers, et sur les avantages que présente l'emploi de ces racines dans la théorie des nombres. *C. R. Acad. Sci. Paris* **25** (1847), 37–54. Reprinted in *Oeuvres (1)*, Vol. 10, pp. 324–333.

[Caviness and Collins 1976]
B. F. Caviness and G. E. Collins. Algorithms for Gaussian integer arithmetic. In R. D. Jenks, editor, *Symsac '76: Proceedings of the 1976 ACM Symposium on Symbolic and Algebraic Computation*, pp. 36–45, 1976.

[Cayley 1848]
A. Cayley. Nouvelles recherches sur les fonctions de M. Sturm. *J. Math. Pures Appl.* **13** (1848), 269–274. Reprinted in *Collected Papers,* Vol. 1, pp. 392–401.

[Cerlienco and Piras 1983]
L. Cerlienco and F. Piras. Powers of a matrix. *Boll. Unione Mat. Italiana* (6) **2-B** (1983), 681–690.

[Cesàro 1881a]
E. Cesàro. Démonstration élémentaire et généralisation de quelques théorèmes de M. Berger. *Mathesis* **1** (1881), 99–102.

[Cesàro 1881b]
E. Cesàro. Question 75. *Mathesis* **1** (1881), 184.

[Cesàro 1883]
E. Cesàro. Sur divers questions d'arithmétique. *Mem. Soc. Roy. Sci. Liège* **10** (1883), 1–350. Reprinted in *Opere Scelte I*, Vol. 1, pp. 10–362.

[Cesàro 1885]
E. Cesàro. Étude moyenne du plus grand commun diviseur de deux nombres. *Ann. Mat. Pura Appl.* **13** (1885), 235–250. Reprinted in *Opere Scelte I*, Vol. 2, pp. 2–18.

[Cesàro 1894]
E. Cesàro. Sur une formule empirique de M. Pervouchine. *C. R. Acad. Sci. Paris* **119** (1894), 848–849. Reprinted in *Opere Scelte I*, Vol. 2, pp. 417–418.

[Chaitin and Schwartz 1978]
G. J. Chaitin and J. T. Schwartz. A note on Monte Carlo primality tests and algorithmic information theory. *Comm. Pure Appl. Math.* **31** (1978), 521–527.

[Chang 1960]
T.-H. Chang. Lösung der Kongruenz $x^2 \equiv a \pmod{p}$ nach einem Primzahlmodul $p = 4n+1$. *Math. Nachrichten* **22** (1960), 136–142.

[Chartres 1967a]
B. A. Chartres. Algorithm 310: prime number generator 1. *Comm. ACM* **10** (1967), 569.

[Chartres 1967b]
B. A. Chartres. Algorithm 311: prime number generator 2. *Comm. ACM* **10** (1967), 570.

[Chatland 1949]
H. Chatland. On the Euclidean algorithm in quadratic number fields. *Bull. Amer. Math. Soc.* **55** (1949), 948–953.

[Chatland and Davenport 1950]
H. Chatland and H. Davenport. Euclid's algorithm in real quadratic fields. *Canad. J. Math.* **2** (1950), 289–296.

[Chebotarev 1926]
N. G. Chebotarev. Die Bestimmung der Dichtigkeit einer Menge von Primzahlen, welche zu einer gegebenen Substitutionsklasse gehören. *Math. Ann.* **95** (1926), 191–228.

[Chebyshev 1851]
P. L. Chebyshev. Sur la fonction qui détermine la totalité des nombres premiers inférieurs à une limite donnée. *Mem. Imp. Acad. Sci. St.-Pétersbourg* **6** (1851), 141–157. Reprinted in *Oeuvres,* Vol. I, pp. 29–48.

[Chebyshev 1852]
P. L. Chebyshev. Mémoire sur les nombres premiers. *J. Math. Pures Appl.* **17** (1852), 366–390. Reprinted in *Oeuvres*, Vol. 1, pp. 51–70.

[C. Chen 1982]
C.-L. Chen. Formulas for the solutions of quadratic equations over $GF(2^m)$. *IEEE Trans. Inform. Theory* **IT-28** (1982), 792–794.

[J. Chen and Li 1977]
J. M. Chen and X. M. Li. The structure of the polynomials over the finite field defined by $Q[f]$-matrix. *Acta Math. Sinica* **20** (1977), 294–297. In Chinese. Reviewed in *Math. Reviews* **80** (1980), #12021.

[Cheng 1984]
U. Cheng. On the continued fraction and Berlekamp's algorithm. *IEEE Trans. Inform. Theory* **IT-30** (1984), 541–544.

[Chernick 1939]
J. Chernick. On Fermat's simple theorem. *Bull. Amer. Math. Soc.* **45** (1939), 269–274.

[Chernoff 1952]
H. Chernoff. A measure of asymptotic efficiency for tests of a hypothesis based on the sum of observations. *Ann. Math. Stat.* **23** (1952), 493–507.

[Cherwell 1941]
Lord Cherwell. Number of primes and probability considerations. *Nature* **148** (1941), 436.

[Cherwell 1946]
Lord Cherwell. Note on the distribution of the intervals between prime numbers. *Quart. J. Math. Oxford* **17** (1946), 46–62.

[Cherwell and Wright 1960]
Lord Cherwell and E. M. Wright. The frequency of prime-patterns. *Quart. J. Math. Oxford* **11** (1960), 60–63.

[Chien, Cunningham, and Oldman 1969]
R. T. Chien, B. D. Cunningham, and I. B. Oldham. Hybrid methods for finding roots of a polynomial – with application to BCH decoding. *IEEE Trans. Inform. Theory* **IT-15** (1969), 329–335.

[Chistov 1984a]
A. L. Chistov. Polynomial time construction of a finite field. In *Abstracts of Lectures at 7th All-Union Conference in Mathematical Logic, Novosibirsk*, p. 196, 1984. Not available to us. Cited in Shoup [1993a].

[Chistov 1984b]
A. L. Chistov. An algorithm of polynomial complexity for factoring polynomials, and determination of the components of a variety in subexponential time. *Zap. Nauchn. Sem. Leningrad Otdel. Mat. Inst. Steklov* **137** (1984), 124–188. In Russian with English summary. English translation in *J. Soviet Math.* **34** (1986), 1838-1882.

[Chistov 1987]
A. L. Chistov. Efficient factorization of polynomials over local fields. *Dokl. Akad. Nauk SSSR* **293** (1987), 1073–1077. In Russian. English translation in *Soviet Math. Doklady* **35** (1987), 430–433.

[Chistov 1989]
A. L. Chistov, The complexity of constructing the ring of integers of a global field. *Dokl. Akad. Nauk SSSR* **306** (1989), 1063–1067. In Russian. English translation in *Soviet Math. Doklady* **39** (1989), 597–600.

[Chistov 1991]
A. L. Chistov. Efficient factoring polynomials [sic] over local fields and its applications. In I. Satake, editor, *Proc. 1990 International Congress of Mathematicians*, pp. 1509–1519. Springer-Verlag, 1991.

[Chor 1982]
B. Chor. Arithmetic of finite fields. *Inform. Process. Lett.* **14** (1982), 4–6.

[Chor and Goldreich 1990]
B. Chor and O. Goldreich. An improved parallel algorithm for integer gcd. *Algorithmica* **5** (1990), 1–10.

[Chudnovsky, Chudnovsky, Denneau, and Younis 1988]
D. V. Chudnovsky, G. V. Chudnovsky, M. M. Denneau, and S. G. Younis. A design of general purpose number-theoretic computer. In L. P. Kartashev and S. I. Kartashev, editors, *Third International Conference on Supercomputing: Proceedings, Supercomputing '88*, Vol. II, pp. 498–499. International Supercomputing Institute, Inc., 1988.

[Church 1932]
A. Church. A set of postulates for the foundation of logic. *Ann. Math.* **33** (1932), 346–366.

[Church 1933]
A. Church. A set of postulates for the foundation of logic (second paper). *Ann. Math.* **34** (1933), 839–864.

[Churchhouse 1988a]
R. F. Churchhouse. Computers in number theory. In C. Verkerk, editor, *Proceedings, 1987 CERN School of Computing*, pp. 357–377. CERN, Geneva, Switzerland, 1988.

[Churchhouse 1988b]
R. F. Churchhouse. Some recent discoveries in number theory and analysis made by the use of a computer. In N. M. Stephens and M. P. Thorne, editors, *Computers in Mathematical Research*, pp. 1–14. Clarendon Press, Oxford, 1988.

[Cipolla 1902]
M. Cipolla. La determinazione assintotica dell'n^{imo} numero primo. *Rend. Accad. Sci. Fis. Mat. Napoli* **8** (1902), 132–166.

[Cipolla 1903]
M. Cipolla. Un metodo per la risoluzione della congruenza di secondo grado. *Rend. Accad. Sci. Fis. Mat. Napoli* **9** (1903), 154–163.

[Cipolla 1904a]
M. Cipolla. Applicazione della teoria delle funzioni numeriche del second'ordine alla risoluzione della congruenza di secondo grado. *Rend. Accad. Sci. Fis. Mat. Napoli* **10** (1904), 135–150.

[Cipolla 1904b]
M. Cipolla. Sui numeri composti P, che verificano la congruenza di Fermat $a^{P-1} \equiv 1 \pmod{P}$. *Ann. Mat. Pura Appl.* **9** (1904), 139–160.

[Cipolla 1907a]
M. Cipolla. Sulla risoluzione apiristica delle congruenze binomie secondo un modulo primo. *Math. Ann.* **63** (1907), 54–61.

[Cipolla 1907b]
M. Cipolla. Sulla risoluzione apiristica delle congruenze binomie. *Atti R. Accad. Lincei* **16** (1907), 603–608.

[Cipolla 1930]
M. Cipolla. Formule di risoluzione apiristica delle equazioni di grado qualunque in un corpo finito. *Rend. Circ. Mat. Palermo* **54** (1930), 199–206.

[Claasen 1977]
H. L. Claasen. The group of units in $GF(q)[x]/(a(x))$. *Proc. Konin. Neder. Akad. Wet.* **80** (1977), 245–255. (= *Indag. Math.* **39**).

[Clark, Bannon, and Keller 1988]
D. W. Clark, P. J. Bannon, and J. B. Keller. Measuring VAX 8800 performance with a histogram hardware monitor. In *Proc. 15th Ann. Intern. Symp. on Computer Architecture*, pp. 176–185, 1988.

[Clarke and Shannon 1983]
J. H. Clarke and A. G. Shannon. Some aspects of Carmichael numbers. *New Zealand Math. Mag.* **20** (1983), 99–101.

[Cobham 1964]
A. Cobham. The intrinsic computational difficulty of functions. In Y. Bar-Hillel, editor, *Logic, Methodology, and Philosophy of Science*, pp. 24–30. North-Holland, 1964.

[Cobham 1966]
A. Cobham. The recognition problem for the set of perfect squares. In *Proc. 7th Ann. Symp. Switching and Automata Theory*, pp. 78–87. IEEE Press, 1966.

[Cody, Hillstrom, and Thatcher 1971]
W. J. Cody, K. E. Hillstrom, and H. C. Thatcher, Jr. Chebyshev approximations for the Riemann zeta function. *Math. Comp.* **25** (1971), 537–547.

[Cody and Thatcher 1969]
W. J. Cody and H. C. Thatcher, Jr. Chebyshev approximations for the exponential integral $\mathrm{Ei}(x)$. *Math. Comp.* **23** (1969), 289–338.

[A. Cohen 1988]
A. M. Cohen. Some computations relating to the Riemann zeta function. In N. M. Stephens and M. P. Thorne, editors, *Computers in Mathematical Research*, pp. 15–29. Oxford University Press, 1988.

[H. Cohen 1979]
H. Cohen. Arithmétique et informatique. In *Journées Arithmétiques de Luminy*, Vol. 61 of *Astérisque*, pp. 57–61. Société Mathématique de France, 1979.

[H. Cohen 1993]
H. Cohen. *A Course in Computational Algebraic Number Theory*. Springer-Verlag, New York, 1993.

[H. Cohen and A. K. Lenstra 1987]
H. Cohen and A. K. Lenstra. Implementation of a new primality test. *Math. Comp.* **48** (1987), 103–121.

[H. Cohen and H. W. Lenstra 1984a]
H. Cohen and H. W. Lenstra, Jr. Heuristics on class groups. In D. V. Chudnovsky, editor, *Number Theory (New York, 1982)*, Vol. 1052 of *Lecture Notes in Mathematics*, pp. 26–36. Springer-Verlag, 1984.

[H. Cohen and H. W. Lenstra 1984b]
H. Cohen and H. W. Lenstra, Jr. Heuristics on class groups of number fields. In H. Jager, editor, *Number Theory, Noordwijkerhout 1983*, Vol. 1068 of *Lecture Notes in Mathematics*, pp. 33–62. Springer-Verlag, 1984.

[H. Cohen and H. W. Lenstra 1984c]
H. Cohen and H. W. Lenstra, Jr. Primality testing and Jacobi sums. *Math. Comp.* **42** (1984), 297–330.

[H. Cohen and Olivier 1992]
H. Cohen and M. Olivier. Calcul des valeurs de la fonction zêta de Riemann en multiprécision. *C. R. Acad. Sci. Paris* **314** (1992), 427–430.

[S. Cohen 1968]
S. D. Cohen. The distribution of irreducible polynomials in several indeterminates over a finite field. *Proc. Edinburgh Math. Soc.* **16** (1968-9), 1–17.

[S. Cohen 1992]
S. D. Cohen. The explicit contruction of irreducible polynomials over finite fields. *Designs, Codes and Cryptography* **2** (1992), 169–174.

[S. Cohen, Odoni and Strothers 1974]
S. D. Cohen, R. W. Odoni, and W. W. Strothers. On the least primitive root modulo p^2. *Bull. Lond. Math. Soc.* **6** (1974), 42–46.

[H. Cohn 1978]
H. Cohn. *A Classical Invitation to Algebraic Numbers and Class Fields*. Springer-Verlag, 1978.

[H. Cohn 1980]
H. Cohn. *Advanced Number Theory*. Dover, 1980.

[H. Cohn 1985]
H. Cohn. *Introduction to the Construction of Class Fields*. Cambridge University Press, 1985.

[P. Cohn 1961]
P. M. Cohn. On a generalization of the Euclidean algorithm. *Proc. Cambridge Phil. Soc.* **57** (1961), 18–30.

[P. Cohn 1991]
P. M. Cohn. Euclid's algorithm – since Euclid. *Math. Medley* **19**(2) (1991), 65–72.

[Colburn 1833]
Z. Colburn. *A Memoir of Zerah Colburn*. G. and C. Merriam, Springfield, Mass., 1833.

[Cole 1903]
F. N. Cole. On the factoring of large numbers. *Bull. Amer. Math. Soc.* **10** (1903), 134–137.

[G. Collins 1968]
G. E. Collins. Computing time analyses for some arithmetic and algebraic algorithms. Technical Report 36, University of Wisconsin, Madison; Computer Sciences, July 1968. Appeared in *Proc. 1968 Summer Institute on Symbolic Mathematical Computations*, IBM Federal Systems Center, 1968, pp. 195–231.

[G. Collins 1969]
G. E. Collins. Computing multiplicative inverses in $GF(p)$. *Math. Comp.* **23** (1969), 197–200.

[G. Collins 1971]
G. E. Collins. The calculation of multivariate polynomial resultants. *J. Assoc. Comput. Mach.* **18** (1971), 515–532.

[G. Collins 1974a]
G. E. Collins. The computing time of the Euclidean algorithm. *SIAM J. Comput.* **3** (1974), 1–10.

[G. Collins 1974b]
G. E. Collins. Quantifier elimination for real closed fields by cylindrical algebraic decomposition–preliminary report. *ACM SIGSAM Bull.* **8** (1974), 80–90.

[G. Collins and Loos 1982]
G. E. Collins and R. G. K. Loos. The Jacobi symbol algorithm. *ACM SIGSAM Bull.* **16**(1) (1982), 12–16.

[G. Collins, Mignotte, and Winkler 1982]
G. E. Collins, M. Mignotte, and F. Winkler. Arithmetic in basic algebraic domains. In B. Buchberger, G. E.

Collins, and R. Loos, editors, *Computer Algebra: Symbolic and Algebraic Computation*, pp. 189–220. Springer-Verlag, 1982.

[S. Collins, Reddy, and Sloane 1978]
S. Collins, S. M. Reddy, and N. J. A. Sloane. Elementary problem 2704. *Amer. Math. Monthly* **85** (1978), 198.

[Colquitt and Welsh 1991]
W. N. Colquitt and L. Welsh, Jr. A new Mersenne prime. *Math. Comp.* **56** (1991), 867–870.

[Comba 1989]
P. G. Comba. Experiments in fast multiplication of large integers. Technical Report G320-2158, IBM, Cambridge Scientific Center, February 1989.

[Comrie 1946]
L. J. Comrie. The application of commercial calculating machines to scientific computing. *Math. Tables Aids Comput.* **2** (1946-7), 149–159.

[Condie 1988]
L. Condie. Primality tests. Technical Report CS 88/24, University College, University of New South Wales, Department of Computer Science, October 1988.

[Condie 1993]
L. Condie. Prime generation with the Demytko-Miller-Trbovich algorithm. In J. Seberry and Y. Zheng, editors, *Advances in Cryptology–AUSCRYPT '92*, Vol. 718 of *Lecture Notes in Computer Science*, pp. 413–421. Springer-Verlag, 1993.

[Conrey 1989]
J. B. Conrey. At least two fifths of the zeroes of the Riemann zeta function are on the critical line. *Bull. Amer. Math. Soc.* **20** (1989), 79–81.

[Conway 1968]
J. H. Conway. A tabulation of some information concerning finite fields. In R. F. Churchhouse and J.-C. Herz, editors, *Computers in Mathematical Research*, pp. 37–50. North-Holland, 1968.

[Conway 1976]
J. H. Conway. *On Numbers and Games*. Academic Press, 1976.

[Cook 1971]
S. A. Cook. The complexity of theorem-proving procedures. In *Proc. Third Ann. ACM Symp. Theor. Comput.*, pp. 151–158. ACM Press, 1971.

[Cook 1973]
S. A. Cook. An observation on time-storage tradeoff. In *Proc. Fifth Ann. ACM Symp. Theor. Comput.*, pp. 29–33. ACM Press, 1973.

[Cook 1985]
S. A. Cook. A taxonomy of problems with fast parallel algorithms. *Inform. Control* **64** (1985), 2–22.

[Cook and Reckhow 1973]
S. A. Cook and R. A. Reckhow. Time bounded random access machines. *J. Comput. System Sci.* **7** (1973), 354–375.

[A. Cooper 1926]
A. E. Cooper. *A Topical History of the Theory of Quadratic Residues*. PhD thesis, University of Chicago, June 1926.

[Coppersmith 1990]
D. Coppersmith. Fermat's last theorem (case 1) and the Wieferich criterion. *Math. Comp.* **54** (1990), 895–902.

[Coppersmith and Winograd 1987]
D. Coppersmith and S. Winograd. Matrix multiplication via arithmetic progressions. In *Proc. Nineteenth Ann. ACM Symp. Theor. Comput.*, pp. 1–6. ACM, 1987.

[Coppersmith and Winograd 1990]
D. Coppersmith and S. Winograd. Matrix multiplication via arithmetic progressions. *J. Symbolic Comput.* **9** (1990), 23–52.

[Cormack and Williams 1980]
G. V. Cormack and H. C. Williams. Some very large primes of the form $k \cdot 2^m + 1$. *Math. Comp.* **35** (1980), 1419–1421.

[Cormen, Leiserson, and Rivest 1990]
T. H. Cormen, C. E. Leiserson, and R. L. Rivest. *Introduction to Algorithms*. The MIT Press, 1990.

[Cosnard and Philippe 1989]
M. Cosnard and J.-L. Philippe. Discovering new parallel algorithms: the sieve of Eratosthenes revisited. In J. Della Dora and J. Fitch, editors, *Computer Algebra and Parallelism*, pp. 1–18. Academic Press, 1989.

[Costa Pereira 1985]
N. Costa Pereira. Estimates for the Chebyshev function $\psi(x) - \theta(x)$. *Math. Comp.* **44** (1985), 211–221.

[Cottrell 1973]
A. Cottrell. A lower bound for the Scholz-Brauer problem. *Notices Amer. Math. Soc.* **20** (1973), A–476. Abstract 73T-A200.

[Couffignal 1933]
L. Couffignal. *Les Machines à Calculer*. Gauthier-Villars, Paris, 1933.

[Couvreur and Quisquater 1982]
C. Couvreur and J. J. Quisquater. An introduction to fast generation of large prime numbers. *Philips J. Res.* **37** (1982), 231–264. Errata in *Philips J. Res.* **38** (1983), 77.

[Cramér 1921]
H. Cramér. Some theorems concerning prime numbers. *Arkiv för Matematik, Astronomi och Fisik* **15** (1921), 1–33. Reprinted in *Collected Works*, Vol. I, pp. 138–170.

[Cramér 1935]
H. Cramér. Prime numbers and probability. In *Åttonde Skandinaviska Matematikerkongressen*, pp. 107–115. Håkan Ohlsson, Lund, 1935. Reprinted in *Collected Works*, Vol. II, pp. 827–835.

[Cramér 1937]
H. Cramér. On the order of magnitude of the difference between consecutive prime numbers. *Acta Arith.* **2** (1937), 23–46. Reprinted in *Collected Works*, Vol. II, pp. 871–894.

[Crandall, Doenias, Norrie, and Young 1995]
R. E. Crandall, J. Doenias, C. Norrie, and J. Young. The twenty-second Fermat number is composite. *Math. Comp.* **64** (1995), 863–868.

[Crandall and Penk 1979]
R. E. Crandall and M. A. Penk. A search for large twin prime pairs. *Math. Comp.* **33** (1979), 383–388.

[Cunningham 1908]
A. Cunningham. Solution to problem 16189. *Mathematical Questions and Solutions from the Educational Times* **13** (1908), 19–20.

[Cunningham and Woodall 1925]
A. J. C. Cunningham and H. J. Woodall. *Factorisation of* $y^n \mp 1$, $y = 2, 3, 5, 6, 7, 10, 11, 12$ *up to High Powers* (n). Hodgson, London, 1925.

[Curtze 1895]
M. Curtze. Mathematisch-historische Miscellen. *Bibliotheca Math.* **9** (1895), 33–42.

[Curtze 1899]
M. Curtze. Eine Studienreise. *Centralblatt für Bibliothekswesen* **16** (1899), 288–289.

[Dai and Zeng 1990]
Z. Dai and K. Zeng. Continued fractions and the Berlekamp-Massey algorithm. In J. Seberry and J. Pieprzyk, editors, *Advances in Cryptology–AUSCRYPT '90*, Vol. 453 of *Lecture Notes in Computer Science*, pp. 24–31. Springer-Verlag, 1990.

[Damgård, Landrock, and Pomerance 1993]
I. Damgård, P. Landrock, and C. Pomerance. Average case error estimates for the strong probable prime test. *Math. Comp.* **61** (1993), 177–194.

[Damgård and Landrock 1993]
I. Damgård and P. Landrock. Improved bounds for the Rabin primality test. In M. J. Ganley, editor, *Cryptography and Coding III*, pp. 117–128. Clarendon Press, Oxford, 1991.

[Damgård 1990]
I. B. Damgård. On the randomness of Legendre and Jacobi sequences. In S. Goldwasser, editor, *Advances in Cryptology–CRYPTO '88 Proceedings*, number 403 in Lecture Notes in Computer Science, pp. 162–173. Springer-Verlag, 1990.

[Datta and Singh 1935]
B. Datta and A. N. Singh. *History of Hindu Mathematics, Part I*. Motilal Banarsi Das, Lahore, 1935.

[Datta and Singh 1938]
B. Datta and A. N. Singh. *History of Hindu Mathematics, Part II*. Motilal Banarsi Das, Lahore, 1938.

[H. Davenport 1980]
H. Davenport. *Multiplicative Number Theory*. Springer-Verlag, Berlin, 1980.

[H. Davenport 1983]
H. Davenport. *The Higher Arithmetic*. Dover, New York, 1983.

[J. Davenport 1992]
J. H. Davenport. Primality testing revisited. In P. S. Wang, editor, *Proc. Int'l. Symp. on Symbolic and Algebraic Computation: ISSAC '92*, pp. 123–129, 1992.

[J. Davenport, Siret, and Tournier 1988]
J. H. Davenport, Y. Siret, and E. Tournier. *Computer Algebra*. Academic Press, 1988.

[J. Davenport and Smith 1987]
J. H. Davenport and G. C. Smith. Rabin's primality testing algorithm—a group theory view. Technical Report 87-04, University of Bath, Department of Computer Science, 1987.

[Davida 1972]
G. I. Davida. Inverse of elements of a Galois field. *Electronics Letters* **8** (1972), 518–520.

[Davida and Litow 1985]
G. I. Davida and B. E. Litow. Fast parallel inversion in finite fields. In *Proceedings of the 19th Annual Conference on Information Sciences and Systems*, pp. 305–308. Dept. of EECS, Johns Hopkins University, 1985.

[Davies 1961]
D. Davies. The computation of the zeroes of Hecke zeta functions in the Gaussian field. *Proc. Roy. Soc. London* **264** (1961), 496–502.

[Davies 1965]
D. Davies. An approximate functional equation for Dirichlet L-functions. *Proc. Roy. Soc. London* **264** (1965), 224–236.

[Davies and Haselgrove 1961]
D. Davies and C. B. Haselgrove. The evaluation of Dirichlet L-functions. *Proc. Roy. Soc. London* **264** (1961), 122–132.

[J. Davis and Holdridge 1983]
J. A. Davis and D. B. Holdridge. Factorization using the quadratic sieve algorithm. In D. Chaum, editor, *Advances in Cryptology–Proceedings of Crypto 83*, pp. 103–113. Plenum Press, 1983.

[J. Davis and Holdridge 1988]
J. A. Davis and D. B. Holdridge. Factorization of large integers on a massively parallel computer. In C. G. Günther, editor, *Advances in Cryptology–EUROCRYPT '88 Proceedings*, Vol. 330 of *Lecture Notes in Computer Science*, pp. 235–243. Springer-Verlag, 1988.

[M. Davis 1965]
M. Davis. *The Undecidable*. Raven Press, New York, 1965.

[M. Davis 1973]
M. Davis. Hilbert's tenth problem is unsolvable. *Amer. Math. Monthly* **80** (1973), 233–269.

[M. Davis, Matiyasevich, and Robinson 1976]
M. Davis, Y. Matijasevič, and J. Robinson. Hilbert's tenth problem. Diophantine equations: positive aspects of

a negative solution. In *Mathematical Developments Arising from Hilbert Problems*, Vol. 28 of *Proceedings of Symposia in Pure Mathematics*, pp. 323–378. Amer. Math. Soc., 1976.

[P. Davis 1981]
P. J. Davis. Are there coincidences in mathematics? *Amer. Math. Monthly* **88** (1981), 311–320.

[P. Davis and Hersh 1980]
P. J. Davis and R. Hersh. *The Mathematical Experience*. Birkhäuser, 1980.

[Daykin 1970]
D. E. Daykin. An addition algorithm for greatest common divisor. *Fibonacci Quart.* **8** (1970), 347–349.

[Debnath 1982]
L. Debnath. Some recent results of number theory discovered with electronic computers. *Int. J. Math. Educ. Sci. Technol.* **13** (1982), 603–617.

[Dedekind 1857]
J. W. R. Dedekind. Abriß einer Theorie der höhern Congruenzen in Bezug auf einen reellen Primzahl-Modulus. *J. Reine Angew. Math.* **54** (1857), 1–26. Reprinted in *Werke,* Vol. 1, pp. 40–66.

[Dedekind 1878]
J. W. R. Dedekind. Über den Zussamenhang zwischen der Theorie der Ideale und der Theorie der höheren Congruenzen. *Abhandlungen der Königlichen Gesellschaft der Wissenschaften zu Göttingen* **23** (1878), 3–37. Reprinted in *Werke*, Vol. I, pp. 202–232.

[Dedekind 1879]
J. W. R. Dedekind. *Vorlesungen über Zahlentheorie von P. G. Lejeune Dirichlet*. Viewig, Braunschweig, 3rd edition, 1879.

[Dedekind 1882]
J. W. R. Dedekind. Über die Diskriminanten endlicher Körper. *Abhandlungen der Königlichen Gesellschaft der Wissenschaften zu Göttingen* **29** (1882), 1–56. Reprinted in *Werke,* Vol. 1, pp. 351–396.

[Dedekind 1897]
J. W. R. Dedekind. Über Zerlegungen von Zahlen durch ihre größten gemeinsamen Teiler. In *Festschrift der Technischen Hochschule zu Braunschweig bei Gelegenheit der 69. Versammlung Deutscher Naturforscher und Ärzte*, pp. 1–40. 1897. Reprinted in *Werke*, Vol. 2, pp. 104–147.

[Deleglise and Rivat 1995]
M. Deleglise and J. Rivat. Computing $\pi(x)$: the Meissel, Lehmer, Lagarias, Miller, Odlyzko method. To appear, *Math. Comp.*, 1995.

[Demeczky 1891]
M. Demeczky. Ein Beitrag zur Theorie der Congruenzen höhern Grades. *Mathematische und Naturwissenschaftliche Berichte aus Ungarn* **8** (1891), 51–59.

[Demytko 1989]
N. Demytko. Generating multiprecision integers with guaranteed primality. In W. J. Caelli, editor, *Computer Security in the Age of Information (Proceedings of the Fifth IFIP International Conference on Computer Security)*, pp. 1–8. North-Holland, Amsterdam, 1989.

[Deng 1989]
X. Deng. On the parallel complexity of integer programming. In *Proc. 1st Ann. ACM Symp. Parallel Algorithms and Architectures*, pp. 110–116. ACM, 1989.

[Denjoy 1931]
A. Denjoy. L'hypothèse de Riemann sur la distribution des zéros de $\zeta(s)$, reliée à la théorie des probabilités. *C. R. Acad. Sci. Paris* **192** (1931), 656–658.

[Denjoy 1964]
A. Denjoy. Probabilités confirmant l'hypothèse de Riemann sur les zéros de $\zeta(s)$. *C. R. Acad. Sci. Paris* **259** (1964), 3143–3145.

[Deshouillers 1979]
J.-M. Deshouillers. Progrès récents des petits cribles arithmétiques [d'après Chen, Iwaniec, ...]. In *Séminaire Bourbaki vol. 1977/78: Exposés 507-524*, number 710 in Lecture Notes in Mathematics, pp. 248–262. Springer-Verlag, 1979.

[Deshouillers and Dress 1992]
J.-M. Deshouillers and F. Dress. Sums of 19 biquadrates: on the representation of large integers. *Ann. Sc. Norm. Super. Pisa* **19** (1992), 113–153.

[Diab 1991]
M. Diab. Systolic architectures for multiplication over finite field $GF(2^m)$. In S. Sakata, editor, *Applied Algebra, Algebraic Algorithms and Error-Correcting Codes: AAECC-8*, Vol. 508 of *Lecture Notes in Computer Science*, pp. 329–340. Springer-Verlag, 1991.

[Diaconis and Erdős 1977]
P. Diaconis and P. Erdős. On the distribution of the greatest common divisor. Technical Report 12, Stanford University, Department of Statistics, 1977.

[Diamond 1982]
H. G. Diamond. Elementary methods in the study of the distribution of prime numbers. *Bull. Amer. Math. Soc.* **7** (1982), 553–589.

[Dickson 1904]
L. E. Dickson. A new extension of Dirichlet's theorem on prime numbers. *Messenger Math.* **33** (1904), 155–161.

[Dickson 1905]
L. E. Dickson. Definitions of a group and a field by independent postulates. *Trans. Amer. Math. Soc.* **6** (1905), 198–204.

[Dickson 1908]
L. E. Dickson. Solution to problem 151. *Amer. Math. Monthly* **15** (1908), 209.

[Dickson 1919]
L. E. Dickson. *History of the Theory of Numbers*, Vol. I: Divisibility and Primality. Carnegie Institute of Washington, Publication No. 256, 1919. Reprinted by Chelsea, New York, 1971.

[Dickson 1926]
L. E. Dickson. *Modern Algebraic Theories*. Sanborn, Chicago, 1926. Reprinted as *Algebraic Theories* by Dover, New York, 1959.

[Dickson, Mitchell, Vandiver, and Wahlin 1923]
L. E. Dickson, H. H. Mitchell, H. S. Vandiver, and G. E. Wahlin. *Algebraic Numbers*. National Research Council, Washington, 1923. (= *Bull. Nat. Res. Council* **28**). Reprinted by Chelsea, New York, 1967.

[Dietz 1989]
P. F. Dietz. Optimal algorithms for list indexing and subset rank. In F. Dehne, J.-R. Sack, and N. Santoro, editors, *Algorithms and Data Structures. WADS '89. Proceedings*, Vol. 382 of *Lecture Notes in Computer Science*, pp. 39–46. Springer-Verlag, 1989.

[Dieudonné 1971]
J. Dieudonné. *Infinitesimal Calculus*. Hermann, Paris, 1971.

[Dijkstra 1957]
E. W. Dijkstra. A method to investigate primality. *Math. Tables Aids Comput.* **11** (1957), 195–196.

[Dirichlet 1832]
P. G. L. Dirichlet. Démonstration d'une propriété analogue à la loi de réciprocité qui existe entre deux nombres premiers quelconques. *J. Reine Angew. Math.* **9** (1832), 379–389. Reprinted in *Werke*, Vol. 1, pp. 173–188.

[Dirichlet 1837a]
P. G. L. Dirichlet. Beweis eines Satzes über die arithmetische Progression. *Bericht Ak. Wiss. Berlin* (1837), 108–110. Reprinted in *Werke*, Vol. 1, pp. 307–312.

[Dirichlet 1837b]
P. G. L. Dirichlet. Beweis des Satzes, daß jede unbegrenzte arithmetische Progression, deren erstes Glied und Differenz ganze Zahlen ohne gemeinschaftlichen Factor sind, unendlich viele Primzahlen enthält. *Abhand. Ak. Wiss. Berlin* (1837-9), 45–81. Reprinted in *Werke*, Vol. 1, pp. 313–342.

[Dirichlet 1849]
P. G. L. Dirichlet. Über die Bestimmung der mittleren Werthe in der Zahlentheorie. *Abhand. Ak. Wiss. Berlin* (1849), 69–83. Reprinted in *Werke*, Vol. 2, pp. 49–66.

[Dirichlet 1889]
P. G. L. Dirichlet. *G. Lejeune Dirichlet's Werke*, Vol. 1. Georg Reimer, Berlin, 1889. Reprinted by Chelsea, New York, 1969.

[Dixon 1970]
J. D. Dixon. The number of steps in the Euclidean algorithm. *J. Number Theory* **2** (1970), 414–422.

[Dixon 1971]
J. D. Dixon. A simple estimate for the number of steps in the Euclidean algorithm. *Amer. Math. Monthly* **78** (1971), 374–376.

[Dixon 1984]
J. D. Dixon. Factorization and primality tests. *Amer. Math. Monthly* **91** (1984), 333–352.

[Dobkin and Lipton 1980]
D. Dobkin and R. J. Lipton. Addition chain methods for the evaluation of specific polynomials. *SIAM J. Comput.* **9** (1980), 121–125.

[D'ooge, Robbins, and Karpinski 1926]
M. L. D'ooge, F. E. Robbins, and L. C. Karpinksi. *Nicomachus of Gerasa*. Macmillan, London, 1926.

[Dornsetter 1987]
J. L. Dornsetter. On the equivalence between Berlekamp's and Euclid's algorithms. *IEEE Trans. Inform. Theory* **IT-33** (1987), 428–431.

[Downey, Leong, and Sethi 1981]
P. Downey, B. Leong, and R. Sethi. Computing sequences with addition chains. *SIAM J. Comput.* **10** (1981), 638–646.

[Drake 1971]
S. Drake. The rule behind "Mersenne's numbers". *Physis-Riv. Internaz. Storia Sci.* **13** (1971), 421–424.

[Dubner 1989]
H. Dubner. A new method for producing large Carmichael numbers. *Math. Comp.* **53** (1989), 411–414.

[Dubner 1993]
H. Dubner. Generalized repunit primes. *Math. Comp.* **61** (1993), 927–930.

[Dubner and Dubner 1986]
H. Dubner and R. Dubner. The development of a powerful, low-cost computer for number theory applications. *J. Recreational Math.* **18** (1986), 81–86.

[Dubois 1971]
R. Dubois. *Utilisation d'un théorème de Fermat à la découverte des nombres premiers et notes sur les nombres de Fibonacci*. A. Blanchard, Paris, 1971.

[Ducray 1965]
S. Ducray. Probability and prime numbers. *Proc. Indian Acad. Sci.* **40** (1965), 159–164.

[Dudley 1969]
U. Dudley. History of a formula for primes. *Amer. Math. Monthly* **76** (1969), 23–28.

[Dunten, Jones, and Sorenson 1994]
B. Dunten, J. Jones, and J. Sorenson. Prime number sieves: new and old. Unpublished manuscript, dated 12 July, 1994.

[Duparc 1951]
H. J. A. Duparc. On Carmichael numbers. *Simon Stevin* **29** (1951-52), 21–24.

[Dupré 1846]
A. Dupré. Sur le nombre de divisions à effectuer pour obtenir le plus grand commun diviseur entre deux nombres entiers. *J. Math. Pures Appl.* **11** (1846), 41–64.

[Eastman 1990]
W. L. Eastman. Inside Euclid's algorithm. In D. Ray-Chaudhuri, editor, *Coding Theory and Design Theory, Part I*, Vol. 20 of *IMA Volumes in Mathematics and Its Applications*, pp. 113–127. Springer-Verlag, 1990.

[Eberly 1989]
W. Eberly. Very fast parallel polynomial arithmetic. *SIAM J. Comput.* **18** (1989), 955–976.

[Ecker 1979]
M. W. Ecker. Elementary problem 2773. *Amer. Math. Monthly* **86** (1979), 393.

[Edmonds 1965]
J. Edmonds. Paths, trees, and flowers. *Canadian J. Math.* **17** (1965), 449–467.

[Edwards 1974]
H. M. Edwards. *Riemann's Zeta Function*. Academic Press, New York, 1974.

[Eğecioğlu and Koç 1990]
O. Eğecioğlu and C. K. Koç. Fast modular exponentiation. In *Proc. 1990 Bilkent Int'l. Conf. New Trends in Communication, Control, and Signal Processing*, Vol. 1, pp. 188–194. Elsevier, 1990.

[Ehrman 1967]
J. R. Ehrman. The number of prime divisors of certain Mersenne numbers. *Math. Comp.* **21** (1967), 700–704.

[Eisenstein 1844a]
G. Eisenstein. Einfacher Algorithmus zur Bestimmung des Werthes von $\left(\frac{a}{b}\right)$. *J. Reine Angew. Math.* **27** (1844), 317–318. Reprinted in *Werke,* Vol. 1, pp. 95–96.

[Eisenstein 1850]
G. Eisenstein. Lehrsätze. *J. Reine Angew. Math.* **39** (1850), 180–182. Reprinted in *Werke,* Vol. 2, pp. 620–622.

[Elder 1937]
J. D. Elder. Errata in the Lehmer factor stencils. *Bull. Amer. Math. Soc.* **43** (1937), 253–255.

[Eldridge 1991]
S. E. Eldridge. A faster modular multiplication algorithm. *Internat. J. Comput. Math.* **40** (1991), 63–68.

[Elia 1993]
M. Elia. Some comments on the computation of n'th roots in $\mathbb{Z}_N^*$. In R. Capocelli, A. de Santis, and U. Vaccaro, editors, *Sequences II: Methods in Communication, Security, and Computer Science*, pp. 379–391. Springer-Verlag, 1993.

[Elkies 1988]
N. D. Elkies. On $A^4 + B^4 + C^4 = D^4$. *Math. Comp.* **51** (1988), 825–835.

[Elliott 1967]
P. D. T. A. Elliott. A problem of Erdős concerning power residue sums. *Acta Arith.* **13** (1967), 131–149. Corrigendum in **14** (1968), 437.

[Elliott 1968]
P. D. T. A. Elliott. Some notes on k-th power residues. *Acta Arith.* **14** (1968), 153–162.

[Elliott 1970]
P. D. T. A. Elliott. The distribution of power residues and certain related results. *Acta Arith.* **17** (1970), 141–159.

[Elliott 1972]
P. D. T. A. Elliott. The Riemann zeta function and coin tossing. *J. Reine Angew. Math.* **254** (1972), 100–109.

[Elliott 1979]
P. D. T. A. Elliott. *Probabilistic Number Theory I: Mean-Value Theorems*. Number 239 in Grundlehren der mathematischen Wissenschaften. Springer-Verlag, 1979.

[Elliott 1980]
P. D. T. A. Elliott. *Probabilistic Number Theory II: Central Limit Theorems*. Number 240 in Grundlehren der mathematischen Wissenschaften. Springer-Verlag, 1980.

[Ellison 1975]
W. J. Ellison. *Les Nombres Premiers*. Hermann, 1975.

[Epasinghe 1985]
P. W. Epasinghe. Euclid's algorithm and the Fibonacci numbers. *Fibonacci Quart.* **23** (1985), 177–179.

[Eppstein 1987]
D. Eppstein. On the NP-completeness of cryptarithms. *SIGACT News* **18**(3) (1987), 38–40.

[Er 1983]
M. C. Er. A fast algorithm for computing order-K Fibonacci numbers. *Computer J.* **26** (1983), 224–227.

[Erdős 1932]
P. Erdős. Beweis eines Satzes von Tschebyschef. *Acta Lit. Sci. (Szeged)* **5** (1932), 194–198.

[Erdős 1935a]
P. Erdős. On the difference of consecutive primes. *Quart. J. Math. Oxford* **6** (1935), 124–128.

[Erdős 1949]
P. Erdős. On the converse of Fermat's theorem. *Amer. Math. Monthly* **56** (1949), 623–624.

[Erdős 1950]
P. Erdős. On almost primes. *Amer. Math. Monthly* **57** (1950), 404–407.

[Erdős 1956]
P. Erdős. On pseudoprimes and Carmichael numbers. *Publ. Math. Debrecen* **4** (1956), 201–206.

[Erdős 1960]
P. Erdős. Remarks on number theory III: on addition chains. *Acta Arith.* **6** (1960), 77–81.

[Erdős 1961]
P. Erdős. Remarks on number theory I. *Mat. Lapok* **12** (1961), 10–17. In Hungarian, with English summary.

[Erdős and Jabotinsky 1958]
P. Erdős and E. Jabotinsky. On sequences of integers generated by a sieving process. *Proc. Konin. Neder. Akad. Wet.* **41** (1958), 115–128. (= *Indag. Math.* **20**).

[Erdős and Kac 1940]
P. Erdős and M. Kac. The Gaussian law of errors in the theory of additive number theoretic functions. *Amer. J. Math.* **62** (1940), 738–742.

[Erdős and Pomerance 1986]
P. Erdős and C. Pomerance. On the number of false witnesses for a composite number. *Math. Comp.* **46** (1986), 259–279.

[Erdős, Pomerance, and Schmutz 1991]
P. Erdős, C. Pomerance, and E. Schmutz. Carmichael's lambda function. *Acta Arith.* **58** (1991), 363–385.

[Erdős and Shallit 1991]
P. Erdős and J. O. Shallit. New bounds on the length of finite Pierce and Engel series. *Séminaire de Théorie des Nombres de Bordeaux* **3** (1991), 43–53.

[Erdős and Ulam 1971]
P. Erdős and S. Ulam. Some probabilistic remarks on Fermat's last theorem. *Rocky Mountain J. Math.* **1** (1971), 613–616.

[Escott 1900]
E. B. Escott. Procédé expéditif pour calculer un terme très éloigné dans la série de Fibonacci. *L'Intermédiaire Math.* **7** (1900), 172–175.

[Escott 1901]
E. B. Escott. Question 1484. *L'Intermédiaire Math.* **8** (1901), 63–64.

[Escott 1906]
E. B. Escott. The converse of Fermat's theorem. *Messenger Math.* **36** (1906), 175–176.

[Estes and Pall 1973]
D. R. Estes and G. Pall. A reconsideration of Legendre-Jacobi symbols. *J. Number Theory* **5** (1973), 433–434.

[Euler 1732]
L. Euler. Methodus generalis summandi progressiones. *Comm. Acad. Sci. Petrop.* **6** (1732), 68–97. Reprinted in *Opera Omnia*, Series 1, Vol. 14, pp. 42–72.

[Euler 1741]
L. Euler. Inventio summae cuiusque seriei ex dato termino generali. *Comm. Acad. Sci. Petrop.* **8** (1736) (1741), 9–22. Reprinted in *Opera Omnia*, Series 1, Vol. 14, pp. 108–123.

[Euler 1744]
L. Euler. Variae observationes circa series infinitas. *Comm. Acad. Sci. Petrop.* **9** (1744), 160–188. Reprinted in *Opera Omnia*, Series 1, Vol. 14, pp. 216–244.

[Euler 1750]
L. Euler. Demonstration sur le nombre des point ou deux lignes des ordres quelconques peuvent se couper. *Mémoires de l'Académie des sciences de Berlin* **4** (1748) (1750), 234–248. Reprinted in *Opera Omnia*, Ser. 1, Vol. 26, pp. 46–59.

[Euler 1760]
L. Euler. Theoremata arithmetica nova methodo demonstrata. *Novi Comm. Acad. Sci. Petrop.* **8** (1760), 74–104. Reprinted in Euler [1849], Vol. 1, pp. 274–286; also reprinted in *Opera Omnia*, Ser. 1, Vol. 2, pp. 531–555.

[Euler 1761]
L. Euler. Theoremata circa residua ex divisione potestatum relicta. *Novi Comm. Acad. Sci. Petrop.* **7** (1758/9) (1761), 49–82. Reprinted in *Opera Omnia*, Ser. 1, Vol. 2, pp. 493–518.

[Euler 1762]
L. Euler. Specimen algorithmi singularis. *Novi Comm. Acad. Sci. Petrop.* **9** (1762), 53–69. Reprinted in *Opera Omnia*, Ser. 1, Vol. 15, pp. 31–49.

[Euler 1849]
L. Euler. *Commentationes Arithmeticae Collectae*, Vol. 1. 1849.

[Evans 1983]
R. J. Evans. The octic period polynomial. *Proc. Amer. Math. Soc.* **87** (1983), 389–393.

[Evdokimov 1989]
S. A. Evdokimov. Factoring a solvable polynomial over a finite field and generalized Riemann hypothesis. *Zapiski Nauchnyck Seminarov Leningradskogo Otdeleniya Mat. Inst. V.A. Steklova Akad. Nauk SSSR (LOMI)* **176** (1989), 104–117. In Russian, with English summary.

[Evdokimov 1994]
S. A. Evdokimov. Factorization of polynomials over finite fields in subexponential time under GRH. In L. M. Adleman and M.-D. Huang, editors, *Algorithmic Number Theory, First International Symposium, ANTS-I*, Vol. 877 of *Lecture Notes in Computer Science*, pp. 209–219. Springer-Verlag, 1994.

[Even 1991]
S. Even. Systolic modular multiplication. In A. J. Menezes and S. A. Vanstone, editors, *Advances in Cryptology–CRYPTO '90 Proceedings*, Vol. 537 of *Lecture Notes in Computer Science*, pp. 619–624. Springer-Verlag, 1991.

[Ewing 1983]
J. Ewing. $2^{86243} - 1$ is prime. *Math. Intelligencer* **5**(1) (1983), 60.

[Ewing 1992]
J. Ewing. The latest Mersenne prime. *Amer. Math. Monthly* **99** (1992), 360.

[Ewing 1994]
J. Ewing. $2^{858433} - 1$ [sic; exponent should be 859433] is prime. *Amer. Math. Monthly* **101** (1994), 338.

[Fagin 1990]
B. S. Fagin. Large integer multiplication on massively parallel processors. In *Proc. Frontiers '90: Third Symposium on the Frontiers of Massively Parallel Computation*, pp. 38–42. IEEE Press, 1990.

[Fateman 1974a]
R. J. Fateman. Polynomial multiplication, powers, and asymptotic analysis: some comments. *SIAM J. Comput.* **3** (1974), 196–213.

[Fateman 1974b]
R. J. Fateman. On the computation of powers of sparse polynomials. *Stud. Appl. Math.* **53** (1974), 145–155.

[Feit and Rees 1978]
W. Feit and E. Rees. A criterion for a polynomial to factor completely over the integers. *Bull. Lond. Math. Soc.* **10** (1978), 191–192.

[Feller 1968]
W. Feller. *An Introduction to Probability Theory and its Applications*, Vol. 1. Wiley, New York, 3rd edition, 1968.

[Feller 1971]
W. Feller. *An Introduction to Probability Theory and its Applications*, Vol. 2. Wiley, New York, 2nd edition, 1971.

[Fellows and Koblitz 1992]
M. R. Fellows and N. Koblitz. Self-witnessing polynomial-time complexity and prime factorization. *Designs, Codes and Cryptography* **2** (1992), 231–235.

[Fendel 1985]
D. Fendel. Prime-producing polynomials and principal ideal domains. *Math. Mag.* **58** (1985), 204–210.

[Feng 1989]
G.-L. Feng. A VLSI architecture for fast inversion in $GF(2^m)$. *IEEE Trans. Comput.* **38** (1989), 1383–1386.

[Fenn, Benaissa, and Taylor 1993]
S. T. J. Fenn, M. Benaissa, and D. Taylor. Improved algorithm for division over $GF(2^m)$. *Electronics Letters* **29** (1993), 469–470.

[Ferguson 1986]
H. R. P. Ferguson. A short proof of the existence of vector Euclidean algorithms. *Proc. Amer. Math. Soc.* **97** (1986), 8–10.

[Ferguson 1987]
H. R. P. Ferguson. A non-inductive GF($n, \mathbb{Z}$) algorithm that constructs integral linear relations for n $\mathbb{Z}$-linearly dependent real numbers. *J. Algorithms* **8** (1987), 131–145.

[Ferguson and Forcade 1979]
H. R. P. Ferguson and R. W. Forcade. Generalization of the Euclidean algorithm for real numbers to all dimensions higher than two. *Bull. Amer. Math. Soc.* **1** (1979), 912–914.

[Ferguson and Forcade 1982]
H. R. P. Ferguson and R. W. Forcade. Multidimensional Euclidean algorithms. *J. Reine Angew. Math.* **334** (1982), 171–181.

[Fermat 1894]
P. Fermat. *Oeuvres*. Gauthier-Villars, Paris, 1894.

[Ferrier 1952]
A. Ferrier. The determination of a large prime. *Math. Tables Aids Comput.* **6** (1952), 256.

[Fich and Tompa 1985]
F. E. Fich and M. Tompa. The parallel complexity of exponentiating polynomials over finite fields. In *Proc. Seventeenth Ann. ACM Symp. Theor. Comput.*, pp. 38–47, 1985.

[Fich and Tompa 1988]
F. E. Fich and M. Tompa. The parallel complexity of exponentiating polynomials over finite fields. *J. Assoc. Comput. Mach.* **35** (1988), 651–667.

[Ficken 1943]
F. A. Ficken. Rosser's generalization of the Euclid algorithm. *Duke Math. J.* **10** (1943), 355–379.

[Fiduccia 1985]
C. M. Fiduccia. An efficient formula for linear recurrences. *SIAM J. Comput.* **14** (1985), 106–112.

[Filipponi 1990]
P. Filipponi. Fibonacci pseudoprimes to $4.3 \cdot 10^9$. *Abstracts Amer. Math. Soc.* **11**(2) (1990), 231. Abstract 90T-11-28.

[Finck 1841]
P. J. E. Finck. *Traité Élémentaire d'Arithmétique à l'Usage des Candidats aux Écoles Spéciales*. Derivaux, Strasbourg, 1841.

[Finck 1842]
P. J. E. Finck. Lettre. *Nouvelles Annales de Mathématiques* **1** (1842), 353–355.

[Finck 1843]
P. J. E. Finck. *Traité Élémentaire d'Arithmétique à l'Usage des Candidats aux Écoles Spéciales*. Derivaux, Strasbourg, 1843. 2nd edition.

[Finck 1845]
P. J. E. Finck. Observations sur le théorème de M. Lamé, relativement au plus grand commun diviseur, et nouvelle démonstration de ce théorème. *Nouvelles Annales de Mathématiques* **4** (1845), 71–74.

[Findlay and Johnson 1990]
P. A. Findlay and B. A. Johnson. Modular exponentiation using recursive sums of residues. In G. Brassard, editor, *Advances in Cryptology–CRYPTO '89 Proceedings*, Vol. 435 of *Lecture Notes in Computer Science*, pp. 371–386. Springer-Verlag, 1990.

[Finn 1982a]
J. Finn. Comparison of probabilistic tests for primality. Unpublished manuscript, undated, c. 1982.

[Finn 1982b]
J. Finn. Finding square roots modulo a prime, primality testing, and factoring. Technical Report 295, Princeton University, Dept. of EE & CS, February 1982.

[Finn 1982c]
J. Finn. Comparison of probabilistic tests for primality. Technical Report 297, Princeton University, Dept. of EE & CS, February 1982.

[Finn and Lieberherr 1983]
J. Finn and K. Lieberherr. Primality testing and factoring. *Theoret. Comput. Sci.* **23** (1983), 211–215.

[Fischer 1965c]
P. C. Fischer. Generation of primes by a one-dimensional real-time iterative array. *J. Assoc. Comput. Mach.* **12** (1965), 388–394.

[Fitch 1974]
J. Fitch. A simple method of taking nth roots of integers. *ACM SIGSAM Bull.* **8** (1974), 26.

[Fleischmann 1993]
P. Fleischmann. Connections between the algorithms of Berlekamp and Niederreiter for factoring polynomials over $\mathbb{F}_q$. *Lin. Alg. Appls.* **192** (1993), 101–108.

[Fletcher 1991]
C. R. Fletcher. A reconstruction of the Frenicle-Fermat correspondence of 1640. *Historia Mathematica* **18** (1991), 344–351.

[Floyd and Knuth 1990]
R. W. Floyd and D. E. Knuth. Addition machines. *SIAM J. Comput.* **19** (1990), 329–340.

[Fogels 1961]
È. K. Fogels. On the distribution of prime ideals. *Dokl. Akad. Nauk SSSR* **140** (1961), 1029–1032. In Russian. English translation in *Sov. Math. Doklady* **2**, 1961, 1322–1325.

[Fraleigh 1976]
J. B. Fraleigh. *A First Course in Abstract Algebra*. Addison-Wesley, 1976.

[Frame 1949]
J. S. Frame. Continued fractions and matrices. *Amer. Math. Monthly* **56** (1949), 98–103.

[Frame 1978]
J. S. Frame. A short proof of quadratic reciprocity. *Amer. Math. Monthly* **85** (1978), 818–819.

[Frandsen 1991a]
G. Frandsen. Parallel construction of irreducible polynomials. Unpublished manuscript, 1991.

[Frandsen 1991b]
G. S. Frandsen. Probabilistic construction of normal basis. Technical report 361, Computer Science Dept., Aarhus Univ., August 1991.

[Fredman 1972]
M. L. Fredman. The distribution of absolutely irreducible polynomials in several indeterminates. *Proc. Amer. Math. Soc.* **31** (1972), 387–390.

[Fredman 1982]
M. L. Fredman. The complexity of maintaining an array and computing its partial sums. *J. Assoc. Comput. Mach.* **29** (1982), 250–260.

[Fredman and Saks 1989]
M. L. Fredman and M. E. Saks. The cell probe complexity of dynamic data structures. In *Proc. Twenty-first Ann. ACM Symp. Theor. Comput.*, pp. 345–354. ACM, 1989.

[Fridlender 1949]
V. R. Fridlender. On the least n-th power non-residue. *Dokl. Akad. Nauk SSSR* **66** (1949), 351–352. In Russian.

[Friedlander 1973]
J. B. Friedlander. On characters and polynomials. *Acta Arith.* **25** (1973), 31–37.

[Friedlander 1980]
J. B. Friedlander. Estimates for prime ideals. *J. Number Theory* **12** (1980), 101–105.

[Frobenius 1912]
F. G. Frobenius. Über quadratische Formen, die viele Primzahlen darstellen. *Sitzungsber. d. Königl. Akad. d. Wiss. zu Berlin* (1912), 966–980. Reprinted in *Gesammelte Abhandlungen*, Vol. III, pp. 573–587.

[Frobenius and Stickelberger 1879]
F. G. Frobenius and L. Stickelberger. Über Gruppen von vertauschbaren Elementen. *J. Reine Angew. Math.* **86** (1879), 217–262. Reprinted in *Gesammelte Abhandlungen,* Vol. I, pp. 545–590.

[Fröberg 1968]
C. E. Fröberg. On some number-theoretical problems treated with computers. In R. F. Churchhouse and J.-C. Herz, editors, *Computers in Mathematical Research*, pp. 84–88. North-Holland, 1968.

[Fröhlich and Shepherdson 1955]
A. Fröhlich and J. C. Shepherdson. Effective procedures in field theory. *Phil. Trans. Roy. Soc. London (A)* **248** (1955), 407–432.

[Früchtl 1950]
K. Früchtl. Statistische Untersuchung über die Verteilung von Primzahl-Zwillingen. *Anz. Österreich. Akad. Wiss. (Math.-Natur. Kl.)* **87** (1950), 226–232.

[Fung and Williams 1990]
G. W. Fung and H. C. Williams. Quadratic polynomials which have a high density of prime values. *Math. Comp.* **55** (1990), 345–353.

[Fürer 1985]
M. Fürer. Deterministic and Las Vegas primality testing algorithms. In W. Brauer, editor, *Automata, Languages, and Programming: 12th Colloquium*, Vol. 194 of *Lecture Notes in Computer Science*, pp. 199–209. Springer-Verlag, Berlin, 1985.

[Furry 1942]
W. H. Furry. Number of primes and probability considerations. *Nature* **150** (1942), 120–121.

[Gaal 1971]
L. Gaal. *Classical Galois Theory*. Markham, 1971. Reprinted by Chelsea, New York, 1979.

[Gallagher 1968]
P. X. Gallagher. Bombieri's mean value theorem. *Mathematika* **15** (1968), 1–6.

[Galois 1830]
E. Galois. Sur la théorie des nombres. *Bulletin des sciences mathématiques physiques et chimiques* **13**(218) (1830), 428–435. Reprinted in *Écrits et Mémoires Mathématiques d'Évariste Galois,* pp. 112–128.

[Gandhi 1971]
J. M. Gandhi. Formulae for the Nth prime. In J. H. Jordan and W. A. Webb, editors, *Proc. Washington State University Conference on Number Theory*, pp. 96–106. Department of Mathematics, Washington State University, Pullman, Washington, 1971.

[Gao 1992]
S. Gao. The determination of optimal normal bases of finite fields. Technical Report CORR-92-01, University of Waterloo, Dept. of Combinatorics and Optimization, January 1992.

[Gao and von zur Gathen 1994]
S. Gao and J. von zur Gathen. Berlekamp's and Niederreiter's polynomial factorization algorithms. In G. L. Mullen and P. J.-S. Shiue, editors, *Finite Fields: Theory, Applications, and Algorithms*, Vol. 168 of *Contemporary Mathematics*, pp. 101–116. Amer. Math. Soc., 1994.

[Gao and Lenstra 1992]
S. Gao and H. W. Lenstra, Jr. Optimal normal bases. *Designs, Codes and Cryptography* **2** (1992), 315–323.

[Gao and Vanstone 1995]
S. Gao and S. A. Vanstone. On orders of optimal normal basis generators. *Math. Comp.* **64** (1994), 1227–1233.

[Gardiner, Lazarus, Metropolis, and Ulam 1955]
V. Gardiner, R. Lazarus, N. Metropolis, and S. Ulam. On certain sequences of integers defined by sieves. *Math. Mag.* **29** (1955), 117–122. Reprinted in *Stanislaw Ulam: Sets, Numbers, and Universes*, pp. 340–345.

[Gardner 1971]
M. Gardner. *The Sixth Book of Mathematical Games from Scientific American*. Scribner, New York, 1971.

[Gardner 1982]
M. Gardner. *Logic Machines and Diagrams*. University of Chicago Press, 1982.

[Garey and Johnson 1979]
M. R. Garey and D. S. Johnson. *Computers and Intractability: A Guide to the Theory of NP-Completeness*. Freeman, 1979.

[von zur Gathen 1984a]
J. von zur Gathen. Parallel powering. In *Proc. 25th Ann. Symp. Found. Comput. Sci.*, pp. 31–36, 1984.

[von zur Gathen 1984b]
J. von zur Gathen. Parallel algorithms for algebraic problems. *SIAM J. Comput.* **13** (1984), 802–824.

[von zur Gathen 1984c]
J. von zur Gathen. Hensel and Newton methods in valuation rings. *Math. Comp.* **42** (1984), 637–661.

[von zur Gathen 1986a]
J. von zur Gathen. Irreducible polynomials over finite fields. In K. V. Nori, editor, *Foundations of Software Technology and Theoretical Computer Science '86*, Vol. 241 of *Lecture Notes in Computer Science*, pp. 252–262, 1986.

[von zur Gathen 1986b]
J. von zur Gathen. Representations and parallel computations for rational functions. *SIAM J. Comput.* **15** (1986), 432–452.

[von zur Gathen 1987a]
J. von zur Gathen. Factoring polynomials and primitive elements for special primes. *Theoret. Comput. Sci.* **52** (1987), 77–89.

[von zur Gathen 1987b]
J. von zur Gathen. Computing powers in parallel. *SIAM J. Comput.* **16** (1987), 930–945.

[von zur Gathen 1990]
J. von zur Gathen. Inversion in finite fields using logarithmic depth. *J. Symbolic Comput.* **9** (1990), 175–183.

[von zur Gathen 1991a]
J. von zur Gathen. Efficient exponentiation in finite fields. In *Proc. 32nd Ann. Symp. Found. Comput. Sci.*, pp. 384–391, 1991.

[von zur Gathen 1991b]
J. von zur Gathen. Efficient and optimal exponentiation in finite fields. *Computational Complexity* **1** (1991), 360–394.

[von zur Gathen 1992]
J. von zur Gathen. Processor-efficient exponentiation in finite fields. *Inform. Process. Lett.* **41** (1992), 81–86.

[von zur Gathen and Giesbrecht 1990]
J. von zur Gathen and M. Giesbrecht. Constructing normal bases in finite fields. *J. Symbolic Comput.* **10** (1990), 547–570.

[von zur Gathen and Kaltofen 1985]
J. von zur Gathen and E. Kaltofen. Factorization of multivariate polynomials over finite fields. *Math. Comp.* **45** (1985), 251–261.

[von zur Gathen and Seroussi 1991]
J. von zur Gathen and G. Seroussi. Boolean circuits versus arithmetic circuits. *Inform. Comput.* **91** (1991), 142–154.

[von zur Gathen and Shoup 1992a]
J. von zur Gathen and V. Shoup. Computing Frobenius maps and factoring polynomials. In *Proc. Twenty-fourth Ann. ACM Symp. Theor. Comput.*, pp. 97–105. ACM, 1992.

[von zur Gathen and Shoup 1992b]
J. von zur Gathen and V. Shoup. Computing Frobenius maps and factoring polynomials. *Computational Complexity* **2** (1992), 187–224.

[Gauss 1801]
C. F. Gauss. *Disquisitiones Arithmeticae*. G. Fleischer, Leipzig, 1801. English translation by A. A. Clarke, Springer-Verlag, 1986.

[Gauss 1832]
C. F. Gauss. Theoria residuorum biquadraticorum (commentatio secunda). *Commentationes Societatis Regiae Scientarum Gottingensis Recentiores* **7** (1832), 89–148. Reprinted in *Werke,* Vol. 2, pp. 93–148. German translation in *Untersuchungen über Höhere Arithmetik von Carl Fredrich Gauss*, ed. H. Maser, J. Springer, 1889. Reprinted by Chelsea, New York, 1981.

[Gauss 1849]
C. F. Gauss. Letter to J. F. Encke, December 24, 1849. In *Werke*, Vol. 2, pp. 444–447. K. Gesellschaft der Wissenschaften zu Göttingen, 1863. English translation in L. Goldstein [1973].

[Gauss 1876]
C. F. Gauss. Theorematis fundamentalis in doctrina de residuis quadraticis; demonstrationes et ampliationes novae. In *Werke*, Vol. 2, pp. 49–64. Königlichen Gesellschaft der Wissenschaften, Göttingen, 1876.

[Gauss 1889]
C. F. Gauss. Die Lehre von den Resten. II. In H. Maser, editor, *Untersuchungen über Höhere Arithmetik von Carl Friedrich Gauss*, pp. 602–629. J. Springer, 1889. Reprinted by Chelsea, New York, 1981.

[Ge 1993]
G. Ge. *Algorithms related to multiplicative representations of algebraic numbers*. PhD thesis, University of California, Berkeley, 1993.

[Gegenbauer 1885]
L. Gegenbauer. Asymptotische Gesetze der Zahlentheorie. *Denkschriften der kaiserlichen Akademie der Wissenschaften* [Wien] **49** (1885), 37–80.

[Geiselmann and Gollmann 1990]
W. Geiselmann and D. Gollmann. VLSI design for exponentiation in $GF(2^n)$. In J. Seberry and J. Pieprzyk, editors, *Advances in Cryptology–AUSCRYPT '90*, Vol. 453 of *Lecture Notes in Computer Science*, pp. 398–405. Springer-Verlag, 1990.

[Gel´fond 1946]
A. O. Gel´fond. Commentary [on Chebyshev's number-theory papers]. In *Complete Collected Works of P. L. Chebyshev*, Vol. 1, pp. 285–288. Akademiia Nauk SSSR, Moscow, 1946. In Russian.

[Genaille 1891]
H. Genaille. Piano arithmétique pour la vérification des grands nombres premiers. *Assoc. Française pour l'Avancement des Sciences; Comptes Rendus* **20**(1) (1891), 159.

[Gentleman 1972]
W. M. Gentleman. Optimal multiplication chains for computing a power of a symbolic polynomial. *Math. Comp.* **26** (1972), 935–939.

[Gérardin 1912]
A. Gérardin. Rapport sur diverses méthodes de solutions employées en théorie pour la décomposition des nombres en facteurs. *Assoc. Française pour l'Avancement des Sciences; Comptes Rendus* **41** (1912), 54–57. IIe partie.

[Gerlach 1992]
H. W. Gerlach. On the number of witnesses (a mod n) providing a proper divisor of n. Technical Report 276, University of Auckland, Department of Mathematics and Statistics, November 1992.

[Gerlach 1994]
H. W. Gerlach. A note on closure properties of sets of false witnesses. *New Zealand J. Math.* **23** (1994), 57–64.

[Gibson 1988]
J. K. Gibson. A generalisation of Brickell's algorithm for fast modular multiplication. *BIT* **28** (1988), 755–763.

[Giesbrecht 1992a]
M. Giesbrecht. Factoring in skew-polynomial rings. In I. Simon, editor, *Proc. LATIN '92*, Vol. 583 of *Lecture Notes in Computer Science*, pp. 191–203. Springer-Verlag, 1992.

[Gill 1977]
J. Gill. Computational complexity of probabilistic Turing machines. *SIAM J. Comput.* **6** (1977), 675–695.

[Gillies 1964]
D. B. Gillies. Three new Mersenne primes and a statistical theory. *Math. Comp.* **18** (1964), 93–95. Corrigendum in *Math. Comp.* **31** (1977), 1051.

[Gioia and Subbarao 1978]
A. A. Gioia and M. V. Subbarao. The Scholz-Brauer problem in addition chains II. In D. McCarthy and H. C. Williams, editors, *Proc. Eight Manitoba Conference on Numerical Math. and Computing*, pp. 251–274, 1978. (= *Congr. Numer.* **XXII**).

[Gioia, Subbarao, and Sugunama 1962]
A. A. Gioia, M. V. Subbarao, and M. Sugunama. The Scholz-Brauer problem in addition chains. *Duke Math. J.* **29** (1962), 481–487.

[Glaisher 1878a]
J. W. L. Glaisher. On factor tables, with an account of the mode of formation of the factor table for the fourth million. *Proc. Cambridge Phil. Soc.* **3** (1878), 99–138.

[Glaisher 1878b]
J. W. L. Glaisher. On long successions of composite numbers. *Messenger Math.* **7** (1878), 102–106,171–176.

[Glaisher 1878c]
J. W. L. Glaisher. An enumeration of prime-pairs. *Math. Mag.* **8** (1878), 28–33.

[Gödel 1931]
K. Gödel. Über formal unentscheidbare Sätze der Principia Mathematica und verwandter Systeme I. *Monatsh. Math. Phys.* **38** (1931), 173–198. English translation in M. Davis [1965], pp. 4–38.

[C. Goldstein 1992]
C. Goldstein. On a seventeenth century version of the "fundamental theorem of arithmetic". *Historia Mathematica* **19** (1992), 177–187.

[L. Goldstein 1970a]
L. J. Goldstein. A generalization of the Siegel-Walfisz theorem. *Trans. Amer. Math. Soc.* **149** (1970), 417–429.

[L. Goldstein 1970b]
L. J. Goldstein. Analogues of Artin's conjecture. *Trans. Amer. Math. Soc.* **149** (1970), 431–442.

[L. Goldstein 1971]
L. J. Goldstein. *Analytic Number Theory*. Prentice-Hall, 1971.

[L. Goldstein 1973]
L. J. Goldstein. A history of the prime number theorem. *Amer. Math. Monthly* **80** (1973), 599–615. Corrigendum in **80** (1973), 1115.

[Goldstine 1972]
H. H. Goldstine. *The Computer from Pascal to von Neumann*. Princeton University Press, 1972.

[Golobew 1972a]
W. A. Golobew. Abzählung von 'Vierlingen – Neunlingen' bis 20 000 000. *Anz. Österreich. Akad. Wiss. (Math.-Natur. Kl.)* **108** (1972), 19–22. Also see **93** (1956) 153-157; **94** (1957) 82-87,274-280; **96** (1959) 227-232.

[Golobew 1972b]
W. A. Golobew. Faktorisation der Zahlen der Form $x^4 + x^2 + 41$. Perfekte Magische Quadrate. *Anz. Österreich. Akad. Wiss. (Math.-Natur. Kl.)* **108** (1972), 152–155. For 101 similar papers, see **95** (1958) 9-13,168-172; **96** (1959) 126-129; **97** (1960) 39-44,312-323; **98** (1961) 59-63,165-171; **99** (1962) 33-37,150-160,181-192,270-282; **100** (1963) 55-68,103-121,237-256,292-308; **101** (1964) 117-128,135-143,159-177,289-303, 318-329,374-386,440-450; **102** (1965) 12-26,61-67,113-128,297-300; **103** (1966) 15-22,25-66,169-188,197-

215,289-301; **104** (1967) 6-22,70-83,101-115,160-166, 184-186,235-249,281-306,347-378,387-399; **105** (1968) 17-30,69-85,123-128,182-211,261-283,296-321; **106** (1969) 13-28,46-65,125-195,297-301; **107** (1970) 106-122,223-227; **108** (1971) 93-96.

[Golomb 1967]
S. W. Golomb. *Shift Register Sequences*. Holden-Day, 1967.

[Golomb 1973]
S. W. Golomb. A new arithmetic function of combinatorial significance. *J. Number Theory* **5** (1973), 218–223.

[Golomb 1974]
S. W. Golomb. A direct interpretation of Gandhi's formula. *Amer. Math. Monthly* **81** (1974), 752–754.

[Golomb 1976a]
S. W. Golomb. Formulas for the next prime. *Pacific J. Math.* **63** (1976), 401–404.

[Golomb 1976b]
S. W. Golomb. The "sales tax" theorem. *Math. Mag.* **49** (1976), 187–189.

[Golomb 1981]
S. W. Golomb. The evidence for Fortune's conjecture. *Math. Mag.* **54** (1981), 209–210.

[Golomb, Welch, and Hales 1959]
S. W. Golomb, L. R. Welch, and A. Hales. On the factorization of trinomials over $GF(2)$. Technical Report 20-189, Jet Propulsion Laboratory, July 1959. Reprinted as Chapter 5 of Golomb [1967].

[Golovanov and Solodovnikov 1987]
P. N. Golovanov and V. I. Solodovnikov. Rapid parallel calculation of degrees in a quotient ring of polynomials over a finite field. *Matematicheskie Zametki* **42** (1987), 886–894. English translation in *Mathematical Notes*, **42** (1987), 987–992.

[Good 1955]
I. J. Good. Conjectures concerning the Mersenne numbers. *Math. Tables Aids Comput.* **9** (1955), 120–121.

[Good and Churchhouse 1968]
I. J. Good and R. F. Churchhouse. The Riemann hypothesis and pseudorandom features of the Möbius sequence. *Math. Comp.* **22** (1968), 857–861.

[Goodman and Zaring 1952]
A. W. Goodman and W. M. Zaring. Euclid's algorithm and the least-remainder algorithm. *Amer. Math. Monthly* **59** (1952), 156–159.

[Goodstein and Wormell 1967]
R. L. Goodstein and C. P. Wormell. Formulae for primes. *Math. Gazette* **51** (1967), 35–38. Erratum in **51** (1967), 244.

[D. Gordon 1989]
D. M. Gordon. On the number of elliptic pseudoprimes. *Math. Comp.* **52** (1989), 231–245.

[D. Gordon and Pomerance 1991]
D. M. Gordon and C. Pomerance. The distribution of Lucas and elliptic pseudoprimes. *Math. Comp.* **57** (1991), 825–838. Corrigendum in **60** (1993), 877.

[J. Gordon 1976]
J. A. Gordon. Very simple method to find the minimum polynomial of an arbitrary nonzero element of a finite field. *Electronics Letters* **12** (1976), 663–664.

[J. Gordon 1984a]
J. A. Gordon. Strong primes are easy to find. In T. Beth, N. Cot, and I. Ingemarsson, editors, *Advances in Cryptology–Proceedings of EUROCRYPT 84*, Vol. 209 of *Lecture Notes in Computer Science*, pp. 216–223. Springer-Verlag, 1984.

[J. Gordon 1984b]
J. A. Gordon. Strong RSA keys. *Electronics Letters* **20** (1984), 514–516.

[J. Gordon 1989]
J. A. Gordon. Fast multiplicative inverse in modular arithmetic. In H. J. Beker and F. C. Piper, editors, *Cryptography and Coding*, pp. 269–279. Clarendon Press, Oxford, 1989.

[Göttfert 1994]
R. Göttfert. An acceleration of the Niederreiter factorization algorithm in characteristic 2. *Math. Comp.* **62** (1994), 831–839.

[R. Graham, Knuth, and Patashnik 1989]
R. L. Graham, D. E. Knuth, and O. Patashnik. *Concrete Mathematics*. Addison-Wesley, 1989.

[R. Graham, Yao, and Yao 1978]
R. L. Graham, A. C.-C. Yao, and F.-F. Yao. Addition chains with multiplicative cost. *Discrete Math.* **23** (1978), 115–119.

[S. Graham and Ringrose 1990]
S. Graham and C. J. Ringrose. Lower bounds for least quadratic non-residues. In B. C. Berndt, H.C. Diamond, H. Halberstam, and A. Hildebrand, editors, *Analytic Number Theory: Proceedings of a Conference in Honor of Paul T. Bateman*, pp. 269–309. Birkhauser, Boston, 1990.

[S. Graham 1994]
S. W. Graham. Carmichael numbers with three prime factors. Unpublished manuscript, 1994.

[Gram 1884]
J.-P. Gram. Undersøgelser angaaende Mængden af Primtal under en given Grænse. *Kongelige Danske Videnskabernes Selskabs Skrifter (Naturvidenskabelig og Matematisk Afdeling)* **2** (1884), 183–308. In Danish, with French summary.

[Gram 1895]
J.-P. Gram. Note sur le calcul de la fonction $\zeta(s)$ de Riemann. *Oversigt over det Kongelige Danske Videnskabernes Selskabs Forhandlinger* (1895), 303–308.

[Gram 1902]
J.-P. Gram. Note sur les zéros de la fonction $\zeta(s)$ de Riemann. *Oversigt over det Kongelige Danske Videnskabernes Selskabs Forhandlinger* (1902), 1–15. Reprinted in *Acta Math.* **27** (1903), 289–304.

[Granville 1989a]
A. Granville. Least primes in arithmetic progressions. In J.-M. de Koninck and C. Levesque, editors, *Théorie des Nombres*, pp. 306–321, Walter de Gruyter, Berlin, 1989.

[Granville 1992]
A. Granville. Primality testing and Carmichael numbers. *Notices Amer. Math. Soc.* **39** (1992), 696–700.

[Granville 1994]
A. Granville. Harald Cramér and the distribution of prime numbers. Technical Report 8, University of Georgia, Department of Mathematics, 1994.

[Granville and Monagan 1988]
A. Granville and M. B. Monagan. The first case of Fermat's last theorem is true for all prime exponents up to $714, 591, 416, 091, 389$. *Trans. Amer. Math. Soc.* **306** (1988), 329–359.

[Greene and Knuth 1990]
D. H. Greene and D. E. Knuth. *Mathematics for the Analysis of Algorithms*. Birkhäuser, Boston, 3rd edition, 1990.

[Gregory 1980]
R. T. Gregory. *Error-Free Computation*. Krieger, Huntington, NY, 1980.

[Greibach 1981]
S. A. Greibach. Formal languages: origins and directions. *Ann. Hist. Comput.* **3** (1981), 14–41.

[Gries and Levin 1980]
D. Gries and G. Levin. Computing Fibonacci numbers (and similarly defined functions) in log time. *Inform. Process. Lett.* **11** (1980), 68–69.

[Gries and Misra 1978]
D. Gries and J. Misra. A linear sieve algorithm for finding prime numbers. *Comm. ACM* **21** (1978), 999–1003.

[Grigoriev 1984]
D. Yu. Grigor′ev. Factorization of polynomials over a finite field and the solution of systems of algebraic equations. *Zap. Nauchn. Sem. Leningrad Otdel. Mat. Inst. Steklov* **137** (1984), 20–79. In Russian with English summary. English translation in *J. Soviet Math.* **34** (1986), 1762-1803.

[Grigoriev, Karpinski, and Odlyzko 1992]
D. Yu. Grigoriev, M. Karpinski, and A. M. Odlyzko. Existence of short proofs for nondivisibility of sparse polynomials under the extended Riemann hypothesis. In P. S. Wang, editor, *Proc. Int'l. Symp. on Symbolic and Algebraic Computation: ISSAC '92*, pp. 117–122. ACM, 1992.

[Grimson and Hanson 1977]
W. E. L. Grimson and D. Hanson. Estimates for the product of the primes not exceeding x. In D. McCarthy and H. C. Williams, editors, *Proc. 7th Manitoba Conference on Numerical Mathematics and Computing*, pp. 407–416. Utilitas Mathematica, 1977. (= *Congr. Numer.* **20**).

[von Grosschmid 1911]
L. von Grosschmid. Ueber einen arithmetischen Satz von Lamé. *Mathematisch-Naturwissenschaftliche Blätter* **8** (1911), 125–127.

[Grossman 1924]
H. Grossman. On the number of divisions in finding a G. C. D. *Amer. Math. Monthly* **31** (1924), 443.

[Grosswald 1981]
E. Grosswald. On Burgess's bound for primitive roots modulo primes and an application to $\Gamma(p)$. *Amer. J. Math.* **103** (1981), 1171–1183.

[Grosswald and Hagis 1979]
E. Grosswald and P. Hagis, Jr. Arithmetic progressions consisting only of primes. *Math. Comp.* **33** (1979), 1343–1352.

[Guerrier 1968]
W. J. Guerrier. The factorization of the cyclotomic polynomials mod p. *Amer. Math. Monthly* **75** (1968), 46.

[Guillaume and Morain 1992a]
D. Guillaume and F. Morain. Building Carmichael numbers with a large number of prime factors. Technical Report LIX/RR/92/01, École Polytechnique, Laboratoire d'Informatique, Palaiseau, France, February 1992.

[Guillaume and Morain 1992b]
D. Guillaume and F. Morain. Building Carmichael numbers with a large number of prime factors and generalizations to other numbers. Unpublished manuscript, dated June 30, 1992; submitted to *Math. Comp.*, 1992.

[Gunji and Arnon 1981]
H. Gunji and D. Arnon. On polynomial factorization over finite fields. *Math. Comp.* **36** (1981), 281–287.

[Gupta and Ram Murty 1984]
R. Gupta and M. Ram Murty. A remark on Artin's conjecture. *Inventiones Math.* **78** (1984), 127–130.

[Gurak 1989]
S. Gurak. Cubic and biquadratic pseudoprimes of Lucas type. In *Number Theory: Proceedings of International Number Theory Conference held at Université Laval, July 5-18, 1987*, pp. 330–347. Walter de Gruyter, Berlin, 1989.

[Gurak 1990]
S. Gurak. Pseudoprimes for higher-order linear recurrence sequences. *Math. Comp.* **55** (1990), 783–813.

[Gurari and Ibarra 1977]
E. M. Gurari and O. H. Ibarra. An NP-complete number-theoretic problem. Technical Report 77-12, University of Minnesota, Computer Science Department, August 1977.

[Guthmann 1992]
A. Guthmann. Effective primality tests for integers of the forms $N = k3^n + 1$ and $N = k2^m3^n + 1$. *BIT* **32** (1992), 529–534.

[Guy 1994]
R. K. Guy. *Unsolved Problems in Number Theory*, Vol. I of *Unsolved Problems in Intuitive Mathematics*. Springer-Verlag, 2nd edition, 1994.

[Guy, Lacampagne, and Selfridge 1987]
R. K. Guy, C. B. Lacampagne, and J. L. Selfridge. Primes at a glance. *Math. Comp.* **48** (1987), 183–202.

[Hachenberger 1992]
D. Hachenberger. On primitive and free roots in a finite field. *Appl. Alg. Eng. Comm. Comp.* **3** (1992), 139–150.

[Hadamard 1896]
J. Hadamard. Sur la distribution des zéros de la fonction $\zeta(s)$ et ses conséquences arithmétiques. *Bull. Soc. Math. France* **24** (1896), 199–220.

[Haghighi 1988]
M. Haghighi. Computation of prime numbers by using a probabilistic algorithm. *Comput. Math. with Appl.* **15** (1988), 939–942.

[Hahn 1992]
S. G. Hahn. On Mirimanoff type congruences. *J. Number Theory* **41** (1992), 167–171.

[Halberstam and Richert 1974]
H. Halberstam and H.-E. Richert. *Sieve Methods*. Academic Press, London, 1974.

[Hanson 1972]
D. Hanson. On the product of the primes. *Canad. Math. Bull.* **15** (1972), 33–37.

[Harari 1983]
S. Harari. Primality testing – a deterministic algorithm. In *Secure Digital Communications*, number 279 in Courses and Lectures (International Centre for Mechanical Sciences), pp. 121–125. Springer-Verlag, Vienna, 1983.

[Hardman and Jordan 1969]
N. R. Hardman and J. H. Jordan. The distribution of quadratic residues in fields of order P^2. *Math. Mag.* **42** (1969), 12–17.

[G. Hardy 1910]
G. H. Hardy. *Orders of Infinity*. Cambridge University Press, 1910.

[G. Hardy 1940a]
G. H. Hardy. *A Mathematician's Apology*. Cambridge University Press, 1940.

[G. Hardy 1940b]
G. H. Hardy. *Ramanujan: Twelve Lectures on Subjects Suggested by his Life and Work*. Cambridge University Press, 1940.

[G. Hardy and Littlewood 1914]
G. H. Hardy and J. E. Littlewood. Some problems of Diophantine approximation. *Acta Math.* **37** (1914), 155–238.

[G. Hardy and Littlewood 1922]
G. H. Hardy and J. E. Littlewood. Some problems of 'partitio numerorum': III: on the expression of a number as a sum of primes. *Acta Math.* **44** (1922), 1–70. Reprinted in *Collected Papers of G. H. Hardy*, Vol. I, pp. 561-630.

[G. Hardy and Wright 1985]
G. H. Hardy and E. M. Wright. *An Introduction to the Theory of Numbers*. Oxford University Press, 5th edition, 1985.

[Harkin 1957]
D. Harkin. On the mathematical work of François-Édouard-Anatole Lucas. *Enseign. Math.* **3** (1957), 276–288.

[V. Harris 1969]
V. C. Harris. A test for primality. *Nordisk Mat. Tidskr.* **17** (1969), 82.

[V. Harris 1970a]
V. C. Harris. An algorithm for finding the greatest common divisor. *Fibonacci Quart.* **8** (1970), 102–103.

[V. Harris 1970b]
V. C. Harris. Note on the number of divisions required in finding the greatest common divisor. *Fibonacci Quart.* **8** (1970), 104.

[V. Harris 1974]
V. C. Harris. On Daykin's algorithm for finding the g.c.d. *Fibonacci Quart.* **12** (1974), 80.

[Hartmanis and Shank 1968]
J. Hartmanis and H. Shank. On the recognition of primes by automata. *J. Assoc. Comput. Mach.* **15** (1968), 382–389.

[Hartmanis and Stearns 1965]
J. Hartmanis and R. E. Stearns. On the computational complexity of algorithms. *Trans. Amer. Math. Soc.* **117** (1965), 285–306.

[Hasan and Bhargava 1992a]
M. A. Hasan and V. K. Bhargava. Bit-serial systolic divider and multiplier for finite fields $GF(2^m)$. *IEEE Trans. Comput.* **41** (1992), 972–980.

[Hasan and Bhargava 1992b]
M. A. Hasan and V. K. Bhargava. Low complexity architecture for exponentiation in $GF(2^m)$. *Electronics Letters* **28** (1992), 1984–1986.

[Hasan, Wang, and Bhargava 1992]
M. A. Hasan, M. Wang, and V. K. Bhargava. Modular construction of low complexity parallel multipliers for a class of finite fields $GF(2^m)$. *IEEE Trans. Comput.* **41** (1992), 962–971.

[Hasse 1926]
H. Hasse. Bericht über neuere Untersuchungen und Probleme aus der Theorie der algebraischen Zahlkörper. I. *Jahres. Deutscher Math.-Verein.* **35** (1926), 1–55. Part Ia in **36** (1927) 233–311; Part II in *Erganzungsbände*, v. VI, 1930, 1–201. Reprinted by Physica-Verlag, Würzburg, 1965.

[Hasse 1952]
H. Hasse. Über die Artinsche Vermutung und verwandte Dichtefragen. *Ann. Acad. Sci. Fenn.* (116) (1952), 1–17.

[Hasse 1967]
H. Hasse. *Vorlesungen über Klassenkörpertheorie*. Physica-Verlag, Würzburg, 1967.

[Hastad, Just, Lagarias, and Schnorr 1989]
J. Hastad, B. Just, J. C. Lagarias, and C. P. Schnorr. Polynomial time algorithms for finding integer relations among real numbers. *SIAM J. Comput.* **18** (1989), 859–881.

[Havas, Majewski, and Matthews 1995]
G. Havas, B. S. Majewski, and K. R. Matthews. Extended gcd algorithms. Unpublished manuscript, 1995.

[Hawkins 1958]
D. Hawkins. The random sieve. *Math. Mag.* **31** (1958), 1–3.

[Hawkins 1974]
D. Hawkins. Random sieves, II. *J. Number Theory* **6** (1974), 192–200.

[Haworth 1989]
G. Haworth. Mersenne numbers. Unpublished research bibliography, 1989.

[Hayes 1965]
D. R. Hayes. The distribution of irreducibles in $GF[q, x]$. *Trans. Amer. Math. Soc.* **117** (1965), 101–127.

[Head 1980]
A. K. Head. Multiplication modulo n. *BIT* **20** (1960), 115–116.

[L. Heath 1990]
L. S. Heath. Covering a set with arithmetic progressions is NP-complete. *Inform. Process. Lett.* **34** (1990), 293–298.

[T. Heath 1921]
T. L. Heath. *A History of Greek Mathematics, Volume I: From Thales to Euclid*. Oxford, Clarendon Press, 1921.

[T. Heath 1956]
T. L. Heath. *The Thirteen Books of Euclid's Elements*, Vol. II. Dover, 1956.

[Heath-Brown 1978a]
D. R. Heath-Brown. The differences between consecutive primes. *J. London Math. Soc.* **18** (1978), 7–13.

[Heath-Brown 1982]
D. R. Heath-Brown. Gaps between primes, and the pair correlation of zeros of the zeta-function. *Acta Arith.* **41** (1982), 85–99.

[Heath-Brown 1983]
D. R. Heath-Brown. Prime twins and Siegel zeros. *Proc. Lond. Math. Soc.* **47** (1983), 193–224.

[Heath-Brown 1986]
D. R. Heath-Brown. Artin's conjecture for primitive roots. *Quart. J. Math. Oxford* **37** (1986), 27–38.

[Heath-Brown 1988]
D. R. Heath-Brown. Differences between consecutive primes. *Jahres. Deutscher Math.-Verein.* **90** (1988), 71–89.

[Heath-Brown 1992]
D. R. Heath-Brown. Zero-free regions for Dirichlet L-functions and the least prime in an arithmetic progression. *Proc. Lond. Math. Soc.* **64** (1992), 265–338.

[Heath-Brown and Goldston 1984]
D. R. Heath-Brown and D. A. Goldston. A note on the differences between consecutive primes. *Math. Ann.* **266** (1984), 317–320.

[Hehner and Horspool 1979]
E. C. R. Hehner and R. N. S. Horspool. A new representation of the rational numbers for fast easy arithmetic. *SIAM J. Comput.* **8** (1979), 124–134.

[Heilbronn 1969]
H. Heilbronn. On the average length of a class of finite continued fractions. *Number Theory and Analysis* (1969), 87–96.

[de Heinzelin 1962]
J. de Heinzelin. Ishango. *Scientific American* **206**(6) (1962), 105–116.

[Hejhal 1987]
D. A. Hejhal. Zeroes of Epstein zeta functions and supercomputers. In A. M. Gleason, editor, *Proc. 1986 International Congress of Mathematicians*, pp. 1362–1384. Amer. Math. Soc., 1987.

[Hellegouarch 1986]
Y. Hellegouarch. Loi de réciprocité, critère de primalité dans $\mathbb{F}_q[t]$. *C. R. Math. Rep. Acad. Sci. Canada* **8** (1986), 291–296.

[Hensel 1888]
K. Hensel. Ueber die Darstellung der Zahlen eines Gattungsbereiches für einen beliebigen Primdivisor. *J. Reine Angew. Math.* **103** (1888), 203–207.

[Hensel 1904]
K. Hensel. Neue Grundlagen der Arithmetik. *J. Reine Angew. Math.* **127** (1904), 51–84.

[Hensel 1908]
K. Hensel. *Theorie der Algebraischen Zahlen*. Teubner, Leipzig, 1908.

[Hensley 1992]
D. Hensley. The number of steps in the Euclidean algorithm. Manuscript, dated October, 1992.

[Hensley and Richards 1974]
D. Hensley and I. Richards. Primes in intervals. *Acta Arith.* **25** (1974), 375–391.

[Herlestam 1980]
T. Herlestam. A note on Rabin's probabilistic primality test. *BIT* **20** (1980), 518–521.

[Herstein 1975]
I. N. Herstein. *Topics in Algebra*. Wiley, New York, 2nd edition, 1975.

[Herstein 1987]
I. N. Herstein. A remark on finite fields. *Amer. Math. Monthly* **94** (1987), 290–291.

[Heyde 1978]
C. C. Heyde. A log log improvement to the Riemann hypothesis for the Hawkins random sieve. *Ann. Prob.* **6** (1978), 870–875.

[Heyworth 1982]
M. R. Heyworth. A conjecture on Mersenne's conjecture. *New Zealand Math. Mag.* **19** (1982), 147–151.

[Higgins and Campbell 1994]
J. Higgins and D. Campbell. Mathematical certificates. *Math. Mag.* **67** (1994), 21–28.

[Hikita and Kawai 1980]
T. Hikita and S. Kawai. Parallel sieve methods for generating prime numbers. In S. Lavington, editor, *Information Processing 80*, pp. 257–262. North-Holland, Amsterdam, 1980.

[Hilbert 1897]
D. Hilbert. Die Theorie der algebraischen Zahlkörper. *Jahresbericht der Deutschen Mathematiker-Vereinigung* **4** (1897), 175–546. Reprinted in *Gesammelte Abhandlungen,* Vol. I, pp. 63–363.

[Hilbert 1900]
D. Hilbert. Mathematische Probleme. *Göttinger Nachrichten* (1900), 253–297. Reprinted in *Gesammelte Abhandlungen,* Vol. III, pp. 290–329.

[Hilbert 1902]
D. Hilbert. Mathematical problems. *Bull. Amer. Math. Soc.* **8** (1902), 437–479.

[Hildebrand and Maier 1989]
A. Hildebrand and H. Maier. Irregularities in the distribution of primes in short intervals. *J. Reine Angew. Math.* **397** (1989), 162–193.

[Hill 1979]
J. R. Hill. Large Carmichael numbers with three prime factors. *Notices Amer. Math. Soc.* **26** (1979), A–374. Abstract 79T-A136.

[Hindenburg 1776]
C. F. Hindenburg. *Beschreibung einer ganz neuen Art nach einem bekannten Gesetze fortgehende Zahlen durch Abzählen oder Abmessen bequem u. sicher zu finden. Nebst Anwendung der Methode auf verschiedene Zahlen, besonders auf eine darnach zu fertigende Factorentafel ...,*. Leipzig, 1776.

[Hinz 1980]
J. Hinz. Eine Erweiterung des nullstellenfreien Bereiches der Heckeschen Zetafunktion und Primideale in Idealklassen. *Acta Arith.* **38** (1980), 209–254.

[Hodges 1983]
A. Hodges. *Alan Turing: the Enigma*. Simon & Schuster, 1983.

[Hoffman de Visme 1961]
G. Hoffman de Visme. The density of prime numbers. *Math. Gazette* **45** (1961), 13–14.

[Hoggatt and Bicknell 1974]
V. E. Hoggatt, Jr. and M. Bicknell. Some congruences of the Fibonacci numbers modulo a prime p. *Math. Mag.* **47** (1974), 210–214.

[Hoheisel 1930]
G. Hoheisel. Primzahlproleme in der Analysis. *Situngsberichte der Preussischen Akademie der Wissenschaften* (1930), 573,580–588.

[Hollinger and Serf 1991]
C. Hollinger and P. Serf. Simath—a computer algebra system. In A. Pethö, M. E. Pohst, H. C. Williams, and H. G. Zimmer, editors, *Computational Number Theory*, pp. 331–342. Walter de Gruyter, Berlin, 1991.

[Hooley 1967]
C. Hooley. On Artin's conjecture. *J. Reine Angew. Math.* **225** (1967), 209–220.

[Hooley 1976]
C. Hooley. *Applications of Sieve Methods to the Theory of Numbers*. Cambridge University Press, 1976.

[Hopcroft and Ullman 1979]
J. E. Hopcroft and J. D. Ullman. *Introduction to Automata Theory, Languages, and Computation*. Addison-Wesley, 1979.

[Hsia and Yeh 1973]
P. Hsia and R. T. Yeh. Finite automata with markers. In M. Nivat, editor, *Automata, Languages, and Programming*, pp. 443–451. North-Holland, Amsterdam, 1973.

[Hua 1982]
L. K. Hua. *Introduction to Number Theory*. Springer-Verlag, New York, 1982.

[Huang 1984a]
M.-D. A. Huang. Factorization of polynomials over finite fields and factorization of primes in algebraic number fields. In *Proc. Sixteenth Ann. ACM Symp. Theor. Comput.*, pp. 175–182. ACM, 1984.

[Huang 1984b]
M.-D. A. Huang. On a simple primality testing algorithm. In J. Fitch, editor, *EUROSAM 84*, Vol. 174 of *Lecture Notes in Computer Science*, pp. 321–332. Springer-Verlag, Berlin, 1984.

[Huang 1985]
M.-D. A. Huang. Riemann hypothesis and finding roots over finite fields. In *Proc. Seventeenth Ann. ACM Symp. Theor. Comput.*, pp. 121–130. ACM, 1985.

[Huang 1991a]
M.-D. A. Huang. Generalized Riemann hypothesis and factoring polynomials over finite fields. *J. Algorithms* **12** (1991), 464–481.

[Huang 1991b]
M.-D. A. Huang. Factorization of polynomials over finite fields and decomposition of primes in algebraic number fields. *J. Algorithms* **12** (1991), 482–489.

[Huber 1992]
K. Huber. Solving equations in finite fields and some results concerning the structure of $GF(p^m)$. *IEEE Trans. Inform. Theory* **38** (1992), 1154–1162.

[Hudson 1977]
R. H. Hudson. A formula for the exact number of primes below a given bound in any arithmetic progression. *Bull. Austral. Math. Soc.* **16** (1977), 67–73.

[Hudson and Brauer 1977]
R. H. Hudson and A. Brauer. On the exact number of primes in the arithmetic progressions $4n \pm 1$ and $6n \pm 1$. *J. Reine Angew. Math.* **291** (1977), 23–29.

[Hurwitz 1887]
A. Hurwitz. Über die Entwicklung complexer Grössen in Kettenbrüche. *Acta Math.* **11** (1887), 187–200. Reprinted in *Werke,* Vol. II, pp. 72–83.

[Hurwitz 1896]
A. Hurwitz. Question 801. *L'Intermédiaire Math.* **3** (1896), 214.

[Hurwitz 1962]
A. Hurwitz. New Mersenne primes. *Math. Comp.* **16** (1962), 249–251.

[Hurwitz and Kritikos 1986]
A. Hurwitz and N. Kritikos. *Lectures on Number Theory*. Springer-Verlag, 1986.

[Hurwitz and Selfridge 1961]
A. Hurwitz and J. L. Selfridge. Fermat numbers and perfect numbers. *Notices Amer. Math. Soc.* **8** (1961), 601. Abstract 587-104.

[Hutchinson 1925]
J. I. Hutchinson. On the roots of the Riemann zeta function. *Trans. Amer. Math. Soc.* **27** (1925), 49–60.

[Huxley 1972]
M. N. Huxley. *The Distribution of Prime Numbers*. Oxford University Press, Oxford, 1972.

[Imchenetzki and Bouniakowsky 1887]
V. G. Imchenetzki and V. Bouniakowsky. Sur un nouveau nombre premier, annoncé par le père Pervouchine. *Bull. Acad. Impériale Sci. St. Pétersbourg* **31** (1887), 532–533. Reprinted in *Mélanges Mathématiques et Astronomiques tirés du Bulletin de l'Académie Impériale des Sciences de St. Pétersbourg* **6** (1883), 553–554.

[Ingham 1932]
A. E. Ingham. *The Distribution of Prime Numbers*. Cambridge University Press, Cambridge, 1932.

[Ingham 1945]
A. E. Ingham. Some tauberian theorems connected with the prime number theorem. *J. London Math. Soc.* **20** (1945), 171–180.

[Inkeri 1960]
K. Inkeri. Tests for primality. *Ann. Acad. Sci. Fenn.* (279) (1960), 1–19.

[Ireland and Rosen 1990]
K. Ireland and M. Rosen. *A Classical Introduction to Modern Number Theory*. Springer-Verlag, 2nd edition, 1990.

[Isenkrahe 1900]
C. Isenkrahe. Ueber eine Lösung der Aufgabe, jede Primzahl als Function der vorhergehenden Primzahlen durch einen geschlossenen Ausdruck darzustellen. *Math. Ann.* **53** (1900), 42–44.

[Israilov 1983]
M. I. Israilov. On the Laurent expansion of the Riemann zeta function. *Proc. Steklov Inst. Math.* **4** (1983), 105–112.

[Ito 1989]
H. Ito. An elliptic Fermat test. *Mem. College Ed. Akita Univ. Natur. Sci.* (40) (1989), 5–8.

[Itoh 1987]
T. Itoh. Efficient probabilistic algorithm for solving quadratic equations over finite fields. *Electronics Letters* **23** (1987), 869–870.

[Itoh 1989]
T. Itoh. An efficient probabilistic algorithm for solving quadratic equation over finite fields. *Electronics and Communications in Japan* **72** (1989), 88–96.

[Itoh and Tsujii 1988a]
T. Itoh and S. Tsujii. Effective recursive algorithm for computing multiplicative inverses in $GF(2^m)$. *Electronics Letters* **24** (1988), 334–335.

[Itoh and Tsujii 1988b]
T. Itoh and S. Tsujii. A fast algorithm for computing multiplicative inverses in $GF(2^m)$ using normal bases. *Inform. Comput.* **78** (1988), 171–177.

[Itoh and Tsujii 1989a]
T. Itoh and S. Tsujii. An efficient algorithm for deciding quadratic residuosity in finite fields $GF(p^m)$. *Inform. Process. Lett.* **30** (1989), 111–114.

[Itoh and Tsujii 1989b]
T. Itoh and S. Tsujii. Structure of parallel multipliers for a class of fields $GF(2^m)$. *Inform. Comput.* **83** (1989), 21–40.

[Iverson 1962]
K. E. Iverson. *A Programming Language*. Wiley, 1962.

[Ivić 1985]
A. Ivić. *The Riemann Zeta-Function*. Wiley, 1985.

[Iwamura, Matsumoto, and Imai 1993a]
K. Iwamura, T. Matsumoto, and H. Imai. Systolic-arrays for modular exponentiation using Montgomery method. In R. A. Rueppel, editor, *Advances in Cryptology–EUROCRYPT '92 Proceedings*, Vol. 658 of *Lecture Notes in Computer Science*, pp. 477–481. Springer-Verlag, 1993.

[Iwaniec 1978]
H. Iwaniec. On the problem of Jacobsthal. *Dem. Math.* **11** (1978), 225–231.

[Jacobi 1837]
C. G. J. Jacobi. Über die Kreistheilung und ihre Anwendung auf die Zahlentheorie. *Bericht Ak. Wiss. Berlin* (1837), 127–136. Reprinted in *J. Reine Angew. Math.* **30** (1846), 166–182; *Werke*, Vol. VI, pp. 254–274.

[Jacobson 1974]
N. Jacobson. *Basic Algebra I*. W. H. Freeman, San Francisco, 1974.

[Jaeschke 1990]
G. Jaeschke. The Carmichael numbers to 10^{12}. *Math. Comp.* **55** (1990), 383–389.

[Jaeschke 1993]
G. Jaeschke. On strong pseudoprimes to several bases. *Math. Comp.* **61** (1993), 915–926.

[Jammalamadaka and Uppuluri 1989]
S. R. Jammalamadaka and V. R. R. Uppuluri. Is p a prime number? Some probabilistic tests for primality. *Math. Scientist* **14** (1989), 55–61.

[Jeans 1897]
J. H. Jeans. The converse of Fermat's theorem. *Messenger Math.* **27** (1897-8), 174.

[Jebelean 1993a]
T. Jebelean. Comparing several GCD algorithms. In E. Swartzlander, Jr., M. J. Irwin, and G. Jullien, editors, *Proc. 11th Symp. Comp. Arith.*, pp. 180–185. IEEE Press, 1993.

[Jebelean 1993b]
T. Jebelean. A generalization of the binary GCD algorithm. In M. Bronstein, editor, *ISSAC '93: Proc. 1993 Int'l. Symp. Symb. Alg. Comp.*, pp. 111–116. ACM, 1993.

[Jebelean 1993c]
T. Jebelean. Improving the multiprecision Euclidean algorithm. In A. Miola, editor, *Design and Implementation of Symbolic Computation Systems, (DISCO '93)*, Vol. 722 of *Lecture Notes in Computer Science*, pp. 45–58. Springer-Verlag, 1993.

[Jevons 1870]
W. S. Jevons. On the mechanical performance of logical inference. *Phil. Trans. Roy. Soc. London* **160** (1870), 497–518.

[Jevons 1874]
W. S. Jevons. *Principles of Science*. Macmillan & Co., London, 1874.

[D. Johnson 1984]
D. S. Johnson. The NP-completeness column: an ongoing guide. *J. Algorithms* **5** (1984), 433–447.

[S. Johnson 1965]
S. M. Johnson. An elementary remark on maximal gaps between successive primes. *Math. Comp.* **19** (1965), 675–676.

[J. Jones 1975]
J. P. Jones. Formula for the *n*th prime number. *Canad. Math. Bull.* **18** (1975), 433–434.

[J. Jones and Matiyasevich 1991]
J. P. Jones and Y. V. Matijasevič. Proof of recursive unsolvability of Hilbert's tenth problem. *Amer. Math. Monthly* **98** (1991), 689–709.

[J. Jones, Sato, Wada, and Wiens 1976]
J. P. Jones, D. Sato, H. Wada, and D. Wiens. Diophantine representation of the set of prime numbers. *Amer. Math. Monthly* **83** (1976), 449–464.

[M. Jones, Lal, and Blundon 1967]
M. F. Jones, M. Lal, and W. J. Blundon. Statistics on certain large primes. *Math. Comp.* **21** (1967), 103–107. Corrigenda in **22** (1968) 474,911.

[N. Jones and Laaser 1977]
N. D. Jones and W. T. Laaser. Complete problems for deterministic polynomial time. *Theoret. Comput. Sci.* **3** (1977), 105–117.

[Jönsson 1972]
I. Jönsson. On certain primes of Mersenne-type. *BIT* **12** (1972), 117–118.

[Julia 1990]
B. Julia. Statistical theory of numbers. In J.-M. Luck, P. Moussa, and M. Waldschmidt, editors, *Number Theory and Physics*, number 47 in Proceedings in Physics, pp. 276–293. Springer-Verlag, 1990.

[Jungnickel 1992]
D. Jungnickel. On the uniqueness of the cyclic group of order *n*. *Amer. Math. Monthly* **99** (1992), 545–547.

[Jurkat and Richert 1965]
W. B. Jurkat and H.-E. Richert. An improvement of Selberg's sieve method I. *Acta Arith.* **11** (1965), 217–240.

[Kac 1959]
M. Kac. *Statistical Independence in Probability, Analysis, and Number Theory*. MAA, 1959.

[Kalecki 1964]
M. Kalecki. O pewnych sumach rozciągniętych na liczby pierwsze lub czynniki pierwsze. *Prace Mat.* **8** (1964), 121–129. In Polish; English summary.

[Kaltofen 1985a]
E. Kaltofen. Sparse Hensel lifting. Technical Report 85-12, Department of Computer Science, Rensselaer Polytechnic Institute, 1985.

[Kaltofen 1985b]
E. Kaltofen. Sparse Hensel lifting. In *Proc. EUROCAL '85*, Vol. 204 of *Lecture Notes in Computer Science*, pp. 4–17. Springer-Verlag, 1985.

[Kaltofen and Rolletschek 1989]
E. Kaltofen and H. Rolletschek. Computing greatest common divisors and factorizations in quadratic number fields. *Math. Comp.* **53** (1989), 697–720.

[Kaltofen and Shoup 1995]
E. Kaltofen and V. Shoup. Subquadratic-time factoring of polynomials over finite fields. In *Proc. Twenty-seventh Ann. ACM Symp. Theor. Comput.*, pp. 398–406. ACM, 1995.

[Kaminski and Bshouty 1987]
M. Kaminski and N. H. Bshouty. Multiplicative complexity of polynomial multiplication over finite fields. In *Proc. 28th Ann. Symp. Found. Comput. Sci.*, pp. 138–140. ACM, 1987.

[Kaminski and Bshouty 1989]
M. Kaminski and N. H. Bshouty. Multiplicative complexity of polynomial multiplication over finite fields. *J. Assoc. Comput. Mach.* **36** (1989), 150–170.

[Kangsheng 1988]
S. Kangsheng. Historical development of the Chinese remainder theorem. *Arch. Hist. Exact Sci.* **38** (1988), 285–305.

[Kannan and Lipton 1986]
R. Kannan and R. J. Lipton. Polynomial-time algorithm for the orbit problem. *J. Assoc. Comput. Mach.* **33** (1986), 808–821.

[Kannan, Miller, and Rudolph 1987]
R. Kannan, G. L. Miller, and L. Rudolph. Sublinear parallel algorithm for computing the greatest common divisor of two integers. *SIAM J. Comput.* **16** (1987), 7–16.

[Kaplansky 1945]
I. Kaplansky. Lucas's tests for Mersenne numbers. *Amer. Math. Monthly* **52** (1945), 188–190.

[Karatsuba 1990]
A. A. Karatsuba. The distribution of prime numbers. *Uspekhi. Mat. Nauk* **45** (1990), 81–140. In Russian. English translation in *Russian Math. Surveys* **45** (1990), 99–171.

[Karatsuba and Ofman 1962]
A. A. Karatsuba and Y. Ofman. Multiplication of multidigit numbers by automata. *Dokl. Akad. Nauk SSSR* **145** (1962), 293–294. English translation in *Soviet Physics Doklady* **7** (1963), 595–596.

[Karloff and Ruzzo 1989]
H. J. Karloff and W. L. Ruzzo. The iterated mod problem. *Inform. Comput.* **80** (1989), 193–204.

[Karp 1972]
R. M. Karp. Reducibility among combinatorial problems. In R. E. Miller and J. W. Thatcher, editors, *Complexity of Computer Computations*, pp. 85–103. Plenum Press, 1972.

[Karp 1991]
R. M. Karp. An introduction to randomized algorithms. *Disc. Appl. Math.* **34** (1991), 165–201.

[Karp and Rabin 1987]
R. M. Karp and M. O. Rabin. Efficient randomized pattern-matching algorithms. *IBM J. Res. Develop.* **31** (1987), 249–260.

[Karp and Ramachandran 1990]
R. M. Karp and V. Ramachandran. Parallel algorithms for shared-memory machines. In J. van Leeuwen, editor, *Handbook of Theoretical Computer Science*, Vol. A: Algorithms and Complexity, pp. 869–941. The MIT Press, 1990.

[Karst 1969]
E. Karst. 12 to 16 primes in arithmetical progression. *J. Recreational Math.* **2** (1969), 214–215.

[Karst 1970]
E. Karst. Lists of ten or more primes in arithmetical progression. *Scripta Math.* **28** (1970), 313–317.

[Karst and Root 1973]
E. Karst and S. C. Root. Teilfolgen von Primzahlen in arithmetischer Progression. *Anz. Österreich. Akad. Wiss. (Math.-Natur. Kl.)* **109** (1973), 19–20.

[Katajainen, van Leeuwen, and Penttonen 1988]
J. Katajainen, J. van Leeuwen, and M. Penttonen. Fast simulation of Turing machines by random access machines. *SIAM J. Comput.* **17** (1988), 77–88.

[Kawamura and Hirano 1988]
S.-I. Kawamura and K. Hirano. A fast modular arithmetic algorithm using a residue table. In C. G. Günther, editor, *Advances in Cryptology–EUROCRYPT '88 Proceedings*, Vol. 330 of *Lecture Notes in Computer Science*, pp. 245–250. Springer-Verlag, 1988.

[Keller 1988]
W. Keller. The Carmichael numbers to 10^{13}. *Abstracts Amer. Math. Soc.* **9** (1988), 328–329. Abstract 88T-11-150. Corrigendum, **13** (1992), 505.

[Keller 1991]
W. Keller. Woher kommen die größten derzeit bekannten Primzahlen? *Mitt. Math. Ges. Hamburg* **12** (1991), 211–229.

[Kelvin 1901]
Lord Kelvin. Nineteenth century clouds over the dynamical theory of heat and light. *Philosophical Magazine* **2** (1901), 1–40.

[Kempfert 1969]
H. Kempfert. On the factorization of polynomials. *J. Number Theory* **1** (1969), 116–120.

[Kesava Menon 1969]
P. Kesava Menon. A generalization of the relation $[m, n](m, n) = mn$. *Math. Student* **37** (1969), 194–194.

[Kilian 1983]
H. Kilian. Zur mittleren Anzahl von Schritten beim euklidischen Algorithmus. *Elem. Math.* **38** (1983), 11–15.

[Kim and Pomerance 1989]
S. H. Kim and C. Pomerance. The probability that a random probable prime is composite. *Math. Comp.* **53** (1989), 721–741.

[Kiss, Phong, and Lieuwens 1986]
P. Kiss, B. M. Phong, and E. Lieuwens. On Lucas pseudoprimes which are products of n primes. In A. N. Philippou, G. E. Bergum, and A. F. Horadam, editors, *Fibonacci Numbers and Their Applications*, pp. 131–139. D. Reidel, Dordrecht, 1986.

[Kleene 1981]
S. C. Kleene. Origins of recursive function theory. *Ann. Hist. Comput.* **3** (1981), 52–67.

[Kluyver 1924]
J. C. Kluyver. On certain series of Mr. Hardy. *Quart. J. Pure Appl. Math.* **50** (1924), 185–192.

[Knödel 1953a]
W. Knödel. Carmichaelsche Zahlen. *Math. Nachrichten* **9** (1953), 343–350.

[Knödel 1953b]
W. Knödel. Eine obere Schranke für die Anzahl der Carmichaelschen Zahlen kleiner als x. *Archiv der Mathematik* **4** (1953), 282–284.

[Knopfmacher 1991]
A. Knopfmacher. The length of the continued fraction expansion for a class of rational functions in $\mathbb{F}_q(X)$. *Proc. Edinburgh Math. Soc.* **34** (1991), 7–17.

[Knopfmacher 1992]
A. Knopfmacher. Elementary properties of the subtractive Euclidean algorithm. *Fibonacci Quart.* **30** (1992), 80–83.

[Knopfmacher and Knopfmacher 1988]
A. Knopfmacher and J. Knopfmacher. The exact length of the Euclidean algorithm in $\mathbb{F}_q[X]$. *Mathematika* **35** (1988), 297–304.

[Knopfmacher and Knopfmacher 1990a]
A. Knopfmacher and J. Knopfmacher. Maximum length of the Euclidean algorithm and continued fractions in $\mathbb{F}(x)$. In G. E. Bergum, A. N. Philippou, and A. F. Horadam, editors, *Applications of Fibonacci Numbers*, Vol. 3, pp. 217–222. Kluwer, Boston, 1990.

[Knopfmacher and Knopfmacher 1993]
A. Knopfmacher and J. Knopfmacher. Counting irreducible factors of polynomials over a finite field. *Discrete Math.* **112** (1993), 103–118.

[Knopfmacher, Knopfmacher, and Warlimont 1994]
A. Knopfmacher, J. Knopfmacher, and R. Warlimont. Lengths of factorizations for polynomials over a finite field. In G. L. Mullen and P. J.-S. Shiue, editors, *Finite Fields: Theory, Applications, and Algorithms*, Vol. 168 of *Contemporary Mathematics*, pp. 185–205. Amer. Math. Soc., 1994.

[Knopfmacher and Warlimont 1995a]
A. Knopfmacher and R. Warlimont. Distinct degree factorizations for polynomials over a finite field. *Trans. Amer. Math. Soc.* **347** (1995), 2235–2243.

[Knopfmacher and Warlimont 1995b]
A. Knopfmacher and R. Warlimont. Counting permutations and polynomials with a restricted factorization pattern. To appear, *Australasian J. Combinatorics*, 1995.

[Knuth 1969]
D. E. Knuth. *The Art of Computer Programming. Volume 2: Seminumerical Algorithms*. Addison-Wesley, 1969.

[Knuth 1971]
D. E. Knuth. The analysis of algorithms. In *Actes du Congrès International des Mathématiciens, 1970*, Vol. 3, pp. 269–274. Gauthier-Villars, Paris, 1971.

[Knuth 1975]
D. E. Knuth. *The Art of Computer Programming. Volume 1: Fundamental Algorithms*. Addison-Wesley, 1975. 2nd edition, 2nd printing.

[Knuth 1976a]
D. E. Knuth. Big omicron and big omega and big theta. *SIGACT News* **8**(2) (1976), 18–24.

[Knuth 1976b]
D. E. Knuth. Evaluation of Porter's constant. *Comput. Math. with Appl.* **2** (1976), 137–139.

[Knuth 1981]
D. E. Knuth. *The Art of Computer Programming. Volume 2: Seminumerical Algorithms*. Addison-Wesley, 1981. 2nd edition.

[Knuth and Yao 1976]
D. E. Knuth and A. C. Yao. The complexity of nonuniform random number generation. In J. F. Traub, editor, *Algorithms and Complexity: New Directions and Recent Results*, pp. 357–428. Academic Press, 1976.

[Koblitz 1977]
N. Koblitz. *p-adic Numbers, p-adic Analysis, and Zeta-Functions*. Springer-Verlag, New York, 1977.

[Koblitz 1987a]
N. Koblitz. *A Course in Number Theory and Cryptography*. Springer-Verlag, 1987.

[Koç and Cappello 1989]
C. K. Koç and P. R. Cappello. Systolic arrays for integer Chinese remaindering. In M. D. Ercegovac and E. Swartzlander, editors, *Proc. 9th IEEE Symp. Comp. Arith.*, pp. 216–223, 1989.

[von Koch 1901]
H. von Koch. Sur la distribution des nombres premiers. *Acta Math.* **24** (1901), 159–182.

[Kofler 1972]
J. Kofler. Mit welcher Wahrscheinlichkeit gilt der große FERMAT-Satz? *Praxis Math.* **14** (1972), 281–283.

[Kolden 1949]
K. Kolden. Continued fractions and linear substitutions. *Archiv for Mathematik og Naturvidenskab* **50** (1949), 141–196.

[Kolesnik and Straus 1978]
G. Kolesnik and E. G. Straus. On the first occurrence of values of a character. *Trans. Amer. Math. Soc.* **246** (1978), 385–394.

[Kompella and Adleman 1991]
K. Kompella and L. M. Adleman. Fast checkers for cryptography. In A. J. Menezes and S. A. Vanstone, editors, *Advances in Cryptology–CRYPTO '90 Proceedings*, Vol. 537 of *Lecture Notes in Computer Science*, pp. 515–529. Springer-Verlag, 1991.

[Konyagin and Pomerance 1994]
S. Konyagin and C. Pomerance. On primes recognizable in deterministic polynomial time. To appear, *The Mathematics of Paul Erdős*, R. L. Graham and J. Nesetril, eds., 1994.

[Kornblum 1919]
H. Kornblum. Über die Primfunktionen in einer arithmetischen Progression. *Math. Zeitschrift* **5** (1919), 100–111.

[Kornerup 1993]
P. Kornerup. High-radix modular multiplication for cryptosystems. In E. Swartzlander, Jr., M. J. Irwin, and G. Jullien, editors, *Proc. 11th IEEE Symp. Comp. Arith.*, pp. 277–283. IEEE Press, 1993.

[Korselt 1899]
A. Korselt. Problème chinois. *L'Intermédiaire Math.* **6** (1899), 143.

[Kosambi 1963]
D. D. Kosambi. The sampling distribution of primes. *Proc. Nat. Acad. Sci. U. S. A.* **49** (1963), 20–23.

[Koster and Shapiro 1990]
D. W. Koster and D. B. Shapiro. Solution to problem 1319. *Math. Mag.* **63** (1990), 129.

[Kowol 1992]
G. Kowol. On strong Dickson pseudoprimes. *Appl. Alg. Eng. Comm. Comp.* **3** (1992), 129–138.

[Kraitchik 1926]
M. Kraitchik. *Théorie des Nombres, Tome II*. Gauthier-Villars, Paris, 1926.

[Kraitchik 1929]
M. Kraitchik. *Recherches sur la Théorie des Nombres*, Vol. II. Gauthier-Villars, Paris, 1929.

[Kramer 1981]
E. E. Kramer. *The Nature and Growth of Modern Mathematics*. Princeton University Press, 1981.

[Kranakis 1984a]
E. Kranakis. On the efficiency of probabilistic primality tests. Technical Report 314, Yale University, Department of Computer Science, April 1984.

[Kranakis 1984b]
E. Kranakis. Primality tests. Technical Report 345, Yale University, Department of Computer Science, December 1984.

[Kranakis 1986]
E. Kranakis. *Primality and Cryptography*. Wiley, 1986.

[Kronecker 1854]
L. Kronecker. Mémoire sur les facteurs irréductibles de l'expression $x^n - 1$. *J. Math. Pures Appl.* **19** (1854), 177–192. Reprinted in *Werke*, Vol. I, pp. 77–92.

[Kronecker 1901]
L. Kronecker. *Vorlesungen über Zahlentheorie*, Vol. 1. Teubner, 1901. Reprinted by Springer-Verlag, 1978.

[Ku and Sun 1992]
Y. H. Ku and X. Sun. The Chinese remainder theorem. *J. Franklin Inst.* **329** (1992), 93–97.

[Kuechlin, Lutz, and Nevin 1991]
W. Kuechlin, D. Lutz, and N. Nevin. Integer multiplication in PARSAC-2 on stock microprocessors. In H. F.

Mattson, T. Mora, and T. R. N. Rao, editors, *Applied Algebra, Algebraic Algorithms and Error-Correcting Codes: AAECC-9*, Vol. 539 of *Lecture Notes in Computer Science*, pp. 206–217. Springer-Verlag, 1991.

[Kühne 1902]
H. Kühne. Eine Wechselbeziehung zwischen Functionen mehrerer Unbestimmten, die zu Reciprocitätsgesetzen führt. *J. Reine Angew. Math.* **124** (1902), 121–133.

[Kuipers 1950]
L. Kuipers. Prime-representing functions. *Proc. Konin. Neder. Akad. Wet.* **53** (1950), 309–310. (= *Indag. Math.* **12** (1950), 57–58).

[Kukihara 1985]
K. Kukihara. An algorithm of Euclidean type for multiplication modulo P. *J. Information Processing* **9** (1985), 14–16.

[Kukihara 1989]
K. Kukihara. Euclidean type algorithm for multiplication modulo P II. *J. Information Processing* **12** (1989), 147–153.

[Kurtz, Shanks, and Williams 1986]
G. C. Kurtz, D. Shanks, and H. C. Williams. Fast primality tests for numbers less than $50 \cdot 10^9$. *Math. Comp.* **46** (1986), 691–701.

[Ladner 1975]
R. E. Ladner. The circuit value problem is log space complete for P. *SIGACT News* **7**(1) (1975), 18–20.

[Lagarias 1994]
J. C. Lagarias. Geodesic multidimensional continued fractions. *Proc. Lond. Math. Soc.* **69** (1994), 464–488.

[Lagarias, Miller, and Odlyzko 1985]
J. C. Lagarias, V. S. Miller, and A. M. Odlyzko. Computing $\pi(x)$: the Meissel-Lehmer method. *Math. Comp.* **44** (1985), 537–560.

[Lagarias, Montgomery, and Odlyzko 1979]
J. C. Lagarias, H. L. Montgomery, and A. M. Odlyzko. A bound for the least prime ideal in the Chebotarev density theorem. *Inventiones Math.* **54** (1979), 271–296.

[Lagarias and Odlyzko 1977]
J. C. Lagarias and A. M. Odlyzko. Effective versions of the Chebotarev density theorem. In A. Fröhlich, editor, *Algebraic Number Fields*, pp. 409–464. Academic Press, London, 1977.

[Lagarias and Odlyzko 1979]
J. C. Lagarias and A. M. Odlyzko. On computing Artin L-functions in the critical strip. *Math. Comp.* **33** (1979), 1081–1095.

[Lagarias and Odlyzko 1984]
J. C. Lagarias and A. M. Odlyzko. New algorithms for computing $\pi(x)$. In D. V. Chudnovsky et al., editor, *Number Theory*, Vol. 1052 of *Lecture Notes in Mathematics*, pp. 176–193. Springer-Verlag, 1984.

[Lagarias and Odlyzko 1987]
J. C. Lagarias and A. M. Odlyzko. Computing $\pi(x)$: an analytic method. *J. Algorithms* **8** (1987), 173–191.

[de Lagny 1733]
T. F. de Lagny. Analyse générale, ou méthodes nouvelles pour résoudre les problèmes de tous les genres & de tous les dégrés à l'infini. *Mémoires de l'Académie Royale des Sciences* (Paris) **11** (1733).

[Lagrange 1769]
J. L. Lagrange. Sur la solution des problèmes indéterminés du second degré. *Histoire de l'Académie Royale des Sciences et Belles-Lettres* [Berlin] **23** (1769), 165–310. Reprinted in *Oeuvres,* Vol. 2, pp. 377–535.

[Lagrange 1771]
J. L. Lagrange. Démonstration d'un théorème nouveau concernant les nombres premiers. *Nouveaux Mémoires de l'Académie Royale des Sciences et Belles-Lettres de Berlin* (1771), 125–137. Reprinted in *Oeuvres*, Vol. 3, 1869, pp. 425-438.

[Lal 1967]
M. Lal. Primes of the form $n^4 + 1$. *Math. Comp.* **21** (1967), 245–257.

[Lambek and Moser 1954]
J. Lambek and L. Moser. Inverse and complementary sequences of natural numbers. *Amer. Math. Monthly* **61** (1954), 454–458.

[Lambert 1785]
J. H. Lambert. *Joh. Heinrich Lamberts ehemaligen Königl. Preuss. Oberbaurathes und ordentl. Mitgliedes der Königl. Academie der Wissenschaften zu Berlin &c. deutscher gelehrter Briefwechsel, herausgegeben von Joh. Bernoulli, Volume 5*. Bey dem Herausgeber, Berlin, 1785.

[Lamé 1844]
G. Lamé. Note sur la limite du nombre des divisions dans la recherche du plus grand commun diviseur entre deux nombres entiers. *C. R. Acad. Sci. Paris* **19** (1844), 867–870.

[E. Landau 1899]
E. Landau. Sur la série des inverses des nombres de Fibonacci. *Bull. Soc. Math. France* **27** (1899), 298–300.

[E. Landau 1900]
E. Landau. Sur quelques problèmes relatifs à la distribution des nombres premiers. *Bull. Soc. Math. France* **28** (1900), 25–38. Reprinted in *Collected Works,* Vol. 1, pp. 92–105.

[E. Landau 1903a]
E. Landau. Über den Verlauf der zahlentheoretischer Funktion $\varphi(n)$. *Archiv der Mathematik und Physik* **5** (1903), 86–91. Reprinted in *Collected Works,* Vol. 1, pp. 378–383.

[E. Landau 1903b]
E. Landau. Neuer Beweis des Primzahlsatzes und Beweis des Primidealzatzes. *Math. Ann.* **56** (1903), 645–670. Reprinted in *Collected Works*, Vol. 1, pp. 327–352.

[E. Landau 1903c]
E. Landau. Über die Maximalordnung der Permutationen gegebenen Grades. *Archiv der Mathematik und Physik* **5** (1903), 92–103. Reprinted in *Collected Works,* Vol. 1, pp. 384–395.

[E. Landau 1906]
E. Landau. Über den Zusammenhang einiger neuerer Sätze der analytischen Zahlentheorie. *Akad. Wiss. Wien Sitzungsberichte* **115** (1906), 589–631. Reprinted in *Collected Works*, Vol. 2, pp. 346–388.

[E. Landau 1908]
E. Landau. Zwei neue Herleitungen für die asymptotische Anzahl der Primzahlen unter einer gegebenen Grenze. *Sitzungsberichte der Königlichen Preussischen Akademie der Wissenschaften* (1908), 746–764. Reprinted in *Collected Works,* Vol. 4, pp. 21–39.

[E. Landau 1909]
E. Landau. *Handbuch der Lehre von der Verteilung der Primzahlen*. Teubner, Leipzig, 1909. 2 volumes. Reprinted by Chelsea, New York, 1953.

[E. Landau 1911]
E. Landau. Über die Verteilung der Zahlen, welche aus ν Primfaktoren zusammengesetzt sind. *Göttinger Nachrichten* (1911), 361–381. Reprinted in *Collected Works,* Vol. 4, pp. 443–463.

[E. Landau 1927a]
E. Landau. *Vorlesungen über Zahlentheorie*. Teubner, Leipzig, 2nd edition, 1927. 3 volumes. Reprinted by Chelsea, New York, 1947.

[E. Landau 1927b]
E. Landau. *Einfuhrung in die elementare und analytische Theorie der algebraische Zahlen und der Ideale*. Teubner, Leipzig, 1927. Reprinted by Chelsea, New York, 1949.

[Lander and Parkin 1967a]
L. J. Lander and T. R. Parkin. A counterexample to Euler's sum of powers conjecture. *Math. Comp.* **21** (1967), 101–103.

[Lander and Parkin 1967b]
L. J. Lander and T. R. Parkin. On first appearance of prime differences. *Math. Comp.* **21** (1967), 483–488.

[Landry 1867]
F. Landry. *Aux Mathématiciens de Toutes les Parties du Monde; Communication sur la Décomposition des Nombres en Leurs Facteurs Simples*. Librairie Hachette, Paris, 1867.

[Landsberg 1893]
G. Landsberg. Ueber eine Anzahlbestimmung und eine damit zusammenhängende Reihe. *J. Reine Angew. Math.* **111** (1893), 87–88.

[Lang 1965]
S. Lang. *Algebra*. Addison-Wesley, 1965.

[Lang 1971]
S. Lang. On the zeta function of number fields. *Inventiones Math.* **12** (1971), 337–345.

[Lang 1986]
S. Lang. *Algebraic Number Theory*. Springer-Verlag, New York, 1986.

[Langemyr 1989]
L. Langemyr. Computing the gcd of two polynomials over an algebraic number field. Technical Report TRITA-NA-8804, Royal Institute of Technology (Stockholm), January 1989.

[Lapidus and Maier 1991]
M. L. Lapidus and H. Maier. Hypothèse de Riemann, cordes fractales vibrantes et conjecture de Weyl-Berry modifiée. *C. R. Acad. Sci. Paris* **313** (1991), 19–24.

[Laplace 1812]
P. S. Laplace. *Théorie Analytique des Probabilités*. Courcier, Paris, 1812.

[Lawrence 1896]
F. W. Lawrence. Factorisation of numbers. *Quart. J. Pure Appl. Math.* **28** (1896), 285–311.

[Laws and Rushworth 1971]
B. A. Laws, Jr. and C. K. Rushworth. A cellular-array multiplier for $GF(2^m)$. *IEEE Trans. Comput.* **C-20** (1971), 1573–1578.

[Lazard 1977]
D. Lazard. Le meilleur algorithme d'Euclide pour $K[X]$ et $\mathbb{Z}$. *C. R. Acad. Sci. Paris* **284** (1977), A1–A4.

[Lazard 1982]
D. Lazard. On polynomial factorization. In J. Calmet, editor, *Computer Algebra, EUROCAM '82, European Computer Algebra Conference*, Vol. 144 of *Lecture Notes in Computer Science*, pp. 126–134. Springer-Verlag, 1982.

[Lebesgue 1837]
V.-A. Lebesgue. Recherches sur les nombres. *J. Math. Pures Appl.* **2** (1837), 253–292.

[Lebesgue 1847]
V.-A. Lebesgue. Sur le symbole $\left(\frac{a}{b}\right)$ et quelques-unes de ses applications. *J. Math. Pures Appl.* **12** (1847), 497–517.

[Lebon 1907]
E. Lebon. Pour la recherche rapide des facteurs premiers des grands nombres. *Assoc. Française pour l'Avancement des Sciences; Comptes Rendus* **36** (1907), 49–55.

[Lebon 1908]
E. Lebon. Disposition et étendue d'une table d'éléments donnant les facteurs premiers des nombres jusqu'à cent millions. *Assoc. Française pour l'Avancement des Sciences; Comptes Rendus* **37** (1908), 33–36.

[Lebon 1912]
E. Lebon. Table des facteurs premiers des nombres compris entre 1 et 100 000 000 (début d'une). *Assoc. Française pour l'Avancement des Sciences; Comptes Rendus* **41** (1912), 44–53.

[Lebon 1914]
E. Lebon. Calculs relatifs à la construction d'une nouvelle table de diviseurs des nombres. *Assoc. Française pour l'Avancement des Sciences; Comptes Rendus* **43** (1914), 29–35.

[de Leeuw, Moore, Shannon, and Shapiro 1956]
K. de Leeuw, E. F. Moore, C. E. Shannon, and N. Shapiro. Computability by probabilistic machines. In C. E. Shannon and J. McCarthy, editors, *Automata Studies*, pp. 183–212. Princeton University Press, 1956.

[van Leeuwen 1977]
J. van Leeuwen. An extension of Hansen's theorem for star chains. *J. Reine Angew. Math.* **295** (1977), 202–207.

[Legendre 1785]
A.-M. Legendre. Recherches d'analyse indéterminée. *Histoire de l'Académie Royale des Sciences* [Paris] (1785), 465–559.

[Legendre 1798]
A.-M. Legendre. *Essai sur la Théorie des Nombres*. Duprat, Paris, 1798.

[Legendre 1830]
A.-M. Legendre. *Théorie des Nombres*. Firmin Didot Frères, Paris, 1830.

[Léger 1837]
E. Léger. Note sur le partage d'une droite en moyenne et extrême, et sur un problème d'arithmétique. *Corresp. Math. Phys.* **9** (1837), 483–485.

[Lehman 1960]
R. S. Lehman. On Liouville's function. *Math. Comp.* **14** (1960), 311–320.

[Lehman 1966]
R. S. Lehman. Separation of zeros of the Riemann zeta-function. *Math. Comp.* **20** (1966), 523–541.

[Lehmann 1982]
D. J. Lehmann. On primality tests. *SIAM J. Comput.* **11** (1982), 374–375.

[D. H. Lehmer 1927]
D. H. Lehmer. Tests for primality by the converse of Fermat's theorem. *Bull. Amer. Math. Soc.* **33** (1927), 327–340. Errata in *Math. Comp.* **23** (1969), 217.

[D. H. Lehmer 1928a]
D. H. Lehmer. A further note on the converse of Fermat's theorem. *Bull. Amer. Math. Soc.* **34** (1928), 54–56.

[D. H. Lehmer 1928b]
D. H. Lehmer. The mechanical combination of linear forms. *Amer. Math. Monthly* **35** (1928), 114–121.

[D. H. Lehmer 1929]
D. H. Lehmer. On the number $(10^{23} - 1)/9$. *Bull. Amer. Math. Soc.* **35** (1929), 349–350.

[D. H. Lehmer 1930a]
D. H. Lehmer. An extended theory of Lucas' functions. *Ann. Math.* **31** (1930), 419–448.

[D. H. Lehmer 1932a]
D. H. Lehmer. Note on Mersenne numbers. *Bull. Amer. Math. Soc.* **38** (1932), 383–384.

[D. H. Lehmer 1932b]
D. H. Lehmer. A number theoretic machine. *Bull. Amer. Math. Soc.* **38** (1932), 635.

[D. H. Lehmer 1932c]
D. H. Lehmer. An inversive algorithm. *Bull. Amer. Math. Soc.* **38** (1932), 693–694.

[D. H. Lehmer 1933a]
D. H. Lehmer. A photo-electric number sieve. *Amer. Math. Monthly* **40** (1933), 401–406.

[D. H. Lehmer 1934]
D. H. Lehmer. A machine for combining sets of linear congruences. *Math. Ann.* **109** (1934), 661–667.

[D. H. Lehmer 1935]
D. H. Lehmer. On Lucas's test for the primality of Mersenne's numbers. *J. London Math. Soc.* **10** (1935), 162–165.

[D. H. Lehmer 1936a]
D. H. Lehmer. On the converse of Fermat's theorem. *Amer. Math. Monthly* **43** (1936), 347–354. Errata in *Math. Tables Aids Comput.* **2** (1947), 279; *Math. Comp.* **25** (1971) 943.

[D. H. Lehmer 1936b]
D. H. Lehmer. On the function $x^2 + x + A$. *Sphinx* **6** (1936), 212–214.

[D. H. Lehmer 1938]
D. H. Lehmer. Euclid's algorithm for large numbers. *Amer. Math. Monthly* **45** (1938), 227–233.

[D. H. Lehmer 1939]
D. H. Lehmer. A factorization theorem applied to a test for primality. *Bull. Amer. Math. Soc.* **45** (1939), 132–137.

[D. H. Lehmer 1941]
D. H. Lehmer. *Guide to Tables in the Theory of Numbers*. National Research Council, National Academy of Sciences, Washington, D. C., 1941.

[D. H. Lehmer 1949]
D. H. Lehmer. On the converse of Fermat's theorem II. *Amer. Math. Monthly* **56** (1949), 300–309. Errata in *Math. Comp.* **25** (1971), 943–944.

[D. H. Lehmer 1952a]
D. H. Lehmer. Recent discoveries of large primes. *Math. Tables Aids Comput.* **6** (1952), 61.

[D. H. Lehmer 1952b]
D. H. Lehmer. A new Mersenne prime. *Math. Tables Aids Comput.* **6** (1952), 205.

[D. H. Lehmer 1953a]
D. H. Lehmer. Two new Mersenne primes. *Math. Tables Aids Comput.* **7** (1953), 72.

[D. H. Lehmer 1956a]
D. H. Lehmer. On the roots of the Riemann zeta-function. *Acta Math.* **95** (1956), 291–298.

[D. H. Lehmer 1956b]
D. H. Lehmer. Extended computation of the Riemann zeta-function. *Mathematika* **3** (1956), 102–108.

[D. H. Lehmer 1959]
D. H. Lehmer. On the exact number of primes less than a given limit. *Illinois J. Math.* **3** (1959), 381–388.

[D. H. Lehmer 1963]
D. H. Lehmer. Automation and pure mathematics. In W. F. Freiberger and W. Prager, editors, *Applications of Digital Computers*, pp. 219–231. Ginn and Company, Boston, 1963.

[D. H. Lehmer 1964]
D. H. Lehmer. On a problem of Störmer. *Illinois J. Math.* **8** (1964), 57–79.

[D. H. Lehmer 1968]
D. H. Lehmer. Machines and pure mathematics. In R. F. Churchhouse and J.-C. Herz, editors, *Computers in Mathematical Research*, pp. 1–7. North-Holland, 1968.

[D. H. Lehmer 1969]
D. H. Lehmer. Computer technology applied to the theory of numbers. In W. J. LeVeque, editor, *Studies in Number Theory*, pp. 117–151. Mathematical Association of America, 1969.

[D. H. Lehmer 1971]
D. H. Lehmer. The economics of number theoretic computation. In A. O. L. Atkin and B. J. Birch, editors, *Computers in Number Theory*, pp. 1–9. Academic Press, 1971.

[D. H. Lehmer 1972]
D. H. Lehmer. On reciprocally weighted partitions. *Acta Arith.* **21** (1972), 379–388.

[D. H. Lehmer 1973]
D. H. Lehmer. The economics of quadratic form calculations. *J. Number Theory* **5** (1973), 544–545.

[D. H. Lehmer 1974]
D. H. Lehmer. The influence of computing on research in number theory. In J. P. LaSalle, editor, *The Influence of Computing on Mathematical Research and Education*, Vol. 20 of *Proc. Symp. Appl. Math.*, pp. 3–12. Amer. Math. Soc., 1974.

[D. H. Lehmer 1976a]
D. H. Lehmer. Exploitation of parallelism in number theoretic and combinatorial computation. In *Proc. 5th Manitoba Conf. on Numer. Math.*, pp. 95–111, 1976.

[D. H. Lehmer 1976b]
D. H. Lehmer. Strong Carmichael numbers. *J. Austral. Math. Soc. Ser. A* **21** (1976), 508–510.

[D. H. Lehmer 1980]
D. H. Lehmer. A history of the sieve process. In N. Metropolis, J. Howlett, and Gian-Carlo Rota, editors, *A History of Computing in the Twentieth Century*, pp. 445–456. Academic Press, 1980.

[D. H. Lehmer 1981]
D. H. Lehmer. *Selected Papers of D. H. Lehmer*. Charles Babbage Research Centre, St. Pierre, Manitoba, 1981. 3 volumes.

[D. H. Lehmer and E. Lehmer 1962]
D. H. Lehmer and E. Lehmer. Heuristics, anyone? In G. Szegö, editor, *Studies in Mathematical Analysis and Related Topics*, pp. 202–210. Stanford Univ. Press, Stanford, 1962.

[D. H. Lehmer and E. Lehmer 1984]
D. H. Lehmer and E. Lehmer. The sextic period polynomial. *Pacific J. Math.* **111** (1984), 341–355.

[D. H. Lehmer, Lehmer, and Vandiver 1954]
D. H. Lehmer, E. Lehmer, and H. S. Vandiver. An application of high-speed computing to Fermat's last theorem. *Proc. Nat. Acad. Sci. U. S. A.* **40** (1954), 25–33.

[D. N. Lehmer 1909]
D. N. Lehmer. *Factor Table for the First Ten Millions*. Publication No. 105. Carnegie Institute of Washington, Washington, D. C., 1909. Reprinted by Hafner, New York, 1956.

[D. N. Lehmer 1914]
D. N. Lehmer. *List of Prime Numbers from 1 to 10,006,721*. Publication No. 165. Carnegie Institute of Washington, Washington, D. C., 1914. Reprinted by Hafner, New York, 1956. Errata can be found in *Bull. Amer. Math. Soc.* **38** (1932) 902; *Math. Tables Aids Comput.* **11** (1957) 272-273; *Math. Comp.* **20** (1966) 642; and *Math. Comp.* **29** (1975) 341.

[D. N. Lehmer 1925]
D. N. Lehmer. On a new method of factorization. *Proc. Nat. Acad. Sci. U. S. A.* **11** (1925), 97–98.

[D. N. Lehmer 1929]
D. N. Lehmer. *Factor Stencils*. Carnegie Institute of Washington, Washington, D. C., 1929. Revised and extended by J. D. Elder, 1939. Factor stencil errata can be found in Elder [1937].

[D. N. Lehmer 1932]
D. N. Lehmer. Hunting big game in the theory of numbers. *Scripta Math.* **1** (1932-33), 229–235.

[E. Lehmer 1956]
E. Lehmer. Number theory on the SWAC. In *Numerical Analysis*, Vol. 6 of *Proc. Symp. Appl. Math.*, pp. 103–108. 1956.

[E. Lehmer 1964]
E. Lehmer. On the infinitude of Fibonacci pseudo-primes. *Fibonacci Quart.* **2** (1964), 229–230.

[E. Lehmer 1978]
E. Lehmer. Rational reciprocity laws. *Amer. Math. Monthly* **85** (1978), 467–472.

[Leighton 1992]
F. T. Leighton. *Introduction to Parallel Algorithms and Architectures: Arrays, Trees, Hypercubes*. Morgan Kaufmann, 1992.

[Lemoine 1882]
E. Lemoine. Décomposition d'un nombre entier N en ses puissances $n$ièmes maxima. *C. R. Acad. Sci. Paris* **95** (1882), 719–722.

[Lempel and Weinberger 1988]
A. Lempel and M. J. Weinberger. Self-complementary normal bases in finite fields. *SIAM J. Disc. Math.* **1** (1988), 193–198.

[A. Lenstra 1983]
A. K. Lenstra. Factoring multivariate polynomials over finite fields. In *Proc. Fifteenth Ann. ACM Symp. Theor. Comput.*, pp. 189–192. ACM, 1983.

[A. K. Lenstra, Lenstra, and Lovász 1982]
A. K. Lenstra, H. W. Lenstra, Jr., and L. Lovász. Factoring polynomials with rational coefficients. *Math. Annalen* **261** (1982), 515–534.

[H. W. Lenstra 1977a]
H. W. Lenstra, Jr. Artin's conjecture on primes with prescribed primitive roots. In *Séminaire Delange-Pisot-Poitou*, pp. 14–01 to 14–08. Sécretariat Mathématique, Paris, 1977.

[H. W. Lenstra 1977b]
H. W. Lenstra, Jr. On Artin's conjecture and Euclid's algorithm in global fields. *Inventiones Math.* **42** (1977), 201–224.

[H. W. Lenstra 1979]
H. W. Lenstra, Jr. Miller's primality test. *Inform. Process. Lett.* **8** (1979), 86–88.

[H. W. Lenstra 1981]
H. W. Lenstra, Jr. Primality testing algorithms (after Adleman, Rumely and Williams). In *Séminaire Bourbaki #576*, number 901 in Lecture Notes in Mathematics, pp. 243–257. Springer-Verlag, 1981.

[H. W. Lenstra 1982]
H. W. Lenstra, Jr. Primality testing with Artin symbols. In N. Koblitz, editor, *Number Theory Related to Fermat's Last Theorem*, Vol. 26 of *Progress in Mathematics*, pp. 341–347. Birkhäuser, Boston, 1982.

[H. W. Lenstra 1983]
H. W. Lenstra, Jr. Fast prime number tests. *Nieuw Arch. Wiskunde* **1** (1983), 133–144.

[H. W. Lenstra 1985]
H. W. Lenstra, Jr. Galois theory and primality testing. In I. Reiner and K. W. Roggenkamp, editors, *Orders and their Applications*, Vol. 1142 of *Lecture Notes in Mathematics*, pp. 169–189. Springer-Verlag, 1985.

[H. W. Lenstra 1986a]
H. W. Lenstra, Jr. Primality testing. In J. W. de Bakker, M. Hazewinkel, and J. K. Lenstra, editors, *Mathematics and Computer Science: Proceedings of the CWI Symposium*, pp. 269–287. North-Holland, Amsterdam, 1986.

[H. W. Lenstra 1990]
H. W. Lenstra, Jr. Algorithms for finite fields. In J. H. Loxton, editor, *Number Theory and Cryptography*, Vol. 154 of *London Mathematical Society Lecture Note Series*, pp. 76–85. Cambridge University Press, 1990.

[H. W. Lenstra 1991a]
H. W. Lenstra, Jr. Finding isomorphisms between finite fields. *Math. Comp.* **56** (1991), 329–347.

[H. W. Lenstra 1991b]
H. W. Lenstra, Jr. Algorithms for finite fields. Unpublished lecture notes, 1991.

[H. W. Lenstra 1992]
H. W. Lenstra, Jr. Algorithms in algebraic number theory. *Bull. Amer. Math. Soc.* **26** (1992), 211–244.

[H. W. Lenstra 1993]
H. W. Lenstra, Jr. Generating units modulo an odd integer by addition and subtraction. *Acta Arith.* **64** (1993), 383–388.

[H. W. Lenstra and Schoof 1987]
H. W. Lenstra, Jr. and R. J. Schoof. Primitive normal bases for finite fields. *Math. Comp.* **48** (1987), 217–231.

[Leonard 1974]
P. A. Leonard. Factorizations of general polynomials. *J. Number Theory* **6** (1974), 335–338.

[Lerner 1988]
M. Lerner. Parallel computers, number theory problems, and experimental results. In L. P. Kartashev and S. I. Kartashev, editors, *ICS 88. Third International Conference on Supercomputing. Proceedings, Supercomputing '88*, Vol. 2, pp. 500–509. International Supercomputing Institute, 1988.

[Leung and Whitehead 1982]
J. Y.-T. Leung and J. Whitehead. On the complexity of fixed-priority scheduling of periodic, real-time tasks. *Performance Evaluation* **2** (1982), 237–250.

[Levin 1973]
L. Levin. Universal sequential search problems. *Problemy Peredachi Informatsii* **9**(3) (1973), 115–116. English translation in *Problems of Information Transmission* **9** (1973), 265–266.

[Lévy 1949]
P. Lévy. Arithmétique et calcul des probabilités. In *Congr. Intern. Phil. Sci.*, pp. 125–133, 1949. Reprinted in *Oeuvres*, Vol. VI, pp. 317–325.

[Lewis 1969]
D. J. Lewis. Diophantine equations: p-adic methods. In W. J. LeVeque, editor, *Studies in Number Theory*, pp. 25–75. Mathematical Association of America, 1969.

[Liang and Todd 1972]
J. J. Y. Liang and J. Todd. The Stieltjes constants. *J. Res. Nat. Bur. Standards* **76B** (1972), 161–178.

[Lidl 1991]
R. Lidl. Computational problems in the theory of finite fields. *Appl. Alg. Eng. Comm. Comp.* **2** (1991), 81–89.

[Lidl and Matthews 1988]
R. Lidl and R. W. Matthews. Galois: an algebra microcomputer package. *Congr. Numer.* **66** (1988), 145–156.

[Lidl and Müller 1990]
R. Lidl and W. B. Müller. A note on strong Fibonacci pseudoprimes. In J. Seberry and J. Pieprzyk, editors, *Advances in Cryptology–AUSCRYPT '90*, Vol. 453 of *Lecture Notes in Computer Science*, pp. 311–317. Springer-Verlag, 1990.

[Lidl and Müller 1993]
R. Lidl and W. B. Müller. Primality testing with Lucas functions. In J. Seberry and Y. Zheng, editors, *Advances in Cryptology–AUSCRYPT '92*, Vol. 718 of *Lecture Notes in Computer Science*, pp. 539–542. Springer-Verlag, 1993.

[Lidl, Müller, and Oswald 1990]
R. Lidl, W. B. Müller, and A. Oswald. Some remarks on strong Fibonacci pseudoprimes. *Appl. Alg. Eng. Comm. Comp.* **1** (1990), 59–65.

[Lidl and Niederreiter 1983]
R. Lidl and H. Niederreiter. *Finite Fields*. Cambridge University Press, 1983.

[Lifchitz 1971]
H. Lifchitz. *Table des Nombres Premiers de 0 à 20 millions*. Albert Blanchard, Paris, 1971.

[Lilley 1942]
S. Lilley. Mathematical machines. *Nature* **149** (1942), 462–465.

[Lin-Kriz and Pan 1992]
Y. Lin-Kriz and V. Pan. On parallel complexity of integer linear programming, GCD, and the iterated mod function. In *Proc. 3rd ACM-SIAM Symp. Discrete Algorithms (SODA)*, pp. 124–137, 1992.

[Lind 1967]
D. Lind. Solution to problem B-93. *Fibonacci Quart.* **5** (1967), 111–112.

[Lindemann 1933]
F. A. Lindemann. The unique factorization of a positive integer. *Quart. J. Math.* **4** (1933), 319–320.

[Lindgren 1963]
H. Lindgren. The calculation of $\pi(N)$. *J. Austral. Math. Soc. Ser. A* **3** (1963), 257–266.

[Lindhurst 1995]
S. Lindhurst. An analysis of Shanks' algorithm for computing square roots in finite fields. Unpublished manuscript, 1995.

[Linnik 1944a]
Ju. V. Linnik. On the least prime in an arithmetic progression, I. The basic theorem. *Mat. Sbornik* **15** (1944), 139–178.

[van Lint 1973]
J. H. van Lint. A determinant related to the Jacobi symbol. *Proc. Konin. Neder. Akad. Wet.* **76** (1973), 189–191. (= *Indag. Math.* **35**).

[Lipton and Dobkin 1976]
R. J. Lipton and D. Dobkin. Complexity measures and hierarchies for the evaluation of integers and polynomials. *Theoret. Comput. Sci.* **3** (1976), 349–357.

[Litow and Davida 1988]
B. E. Litow and G. I. Davida. $O(\log(n))$ parallel time finite field inversion. In J. H. Reif, editor, *VLSI Algorithms and Architectures. 3rd Aegean Worskhop on Computing, AWOC 88*, Vol. 319 of *Lecture Notes in Computer Science*, pp. 74–80. Springer-Verlag, 1988.

[Littlewood 1914]
J. E. Littlewood. Sur la distribution des nombres premiers. *C. R. Acad. Sci. Paris* **158** (1914), 1869–1872.

[Littlewood 1928]
J. E. Littlewood. On the class-number of the corpus $P(\sqrt{-k})$. *Proc. Lond. Math. Soc.* **27** (1928), 358–372. Reprinted in *Collected Papers*, v. II, pp. 920–934.

[Littlewood 1965]
J. E. Littlewood. The Riemann Hypothesis. In I. J. Good, editor, *The Scientist Speculates: An Anthology of Partly-Baked Ideas*, pp. 121–122. Capricorn Books, New York, 1965.

[Littlewood 1971]
J. E. Littlewood. The quickest proof of the prime number theorem. *Acta Arith.* **18** (1971), 83–86.

[Livingston 1986]
M. L. Livingston. Explicit estimates for the ψ–function for primes in arithmetic progression. Technical Report No. 69, S. Illinois Univ. at Edwardsville, March 1986.

[Lloyd and Remmers 1967]
D. B. Lloyd and H. Remmers. Polynomial factor tables over finite fields. *Math. Algorithms* **2** (1967), 85–99.

[Löh 1988]
G. Löh. Carmichael numbers with a large number of prime factors. *Abstracts Amer. Math. Soc.* **9** (1988), 329. Abstract 88T-11-151.

[Löh 1989]
G. Löh. Long chains of nearly doubled primes. *Math. Comp.* **53** (1989), 751–759.

[Löh 1993]
G. Löh. On Carmichael numbers whose Carmichael function is squarefree. *Abstracts Amer. Math. Soc.* **14** (1993), 390. Abstract 93T-11-61.

[Löh and Niebuhr 1989]
G. Löh and W. Niebuhr. Carmichael numbers with a large number of prime factors, II. *Abstracts Amer. Math. Soc.* **10** (1989), 305. Abstract 89T-11-131.

[Looff 1851]
W. Looff. Ueber die Periodicität der Decimalbrüche. *Archiv der Mathematik und Physik* **16** (1851), 54–57.

[Low 1968]
M. E. Low. Real zeroes of the Dedekind zeta function of an imaginary quadratic field. *Acta Arith.* **14** (1968), 117–140.

[Lucas 1876a]
E. Lucas. Note sur l'application des séries récurrents à la recherche de la loi de distribution des nombres premiers. *C. R. Acad. Sci. Paris* **89** (1876), 165–167.

[Lucas 1876b]
E. Lucas. Sur la recherche des grands nombres premiers. *Assoc. Française pour l'Avancement des Sciences; Comptes Rendus* **5** (1876), 61–68.

[Lucas 1876c]
E. Lucas. Sur la théorie des nombres premiers. *Atti R. Accad. Sc. Torino* **11** (1875-76), 928–937.

[Lucas 1876d]
E. Lucas. Sur les rapports qui existent entre la théorie des nombres et le calcul intégral. *C. R. Acad. Sci. Paris* **82** (1876), 1303–1305.

[Lucas 1877a]
E. Lucas. Recherches sur plusieurs ouvrages de Léonard de Pise et sur diverses questions d'arithmétique supérieure. *Bull. Biblio. Storia Sci. Mat. Fis.* **10** (1877), 129–193; 239–293.

[Lucas 1878a]
E. Lucas. Théorie des fonctions numériques simplement périodiques. *Amer. J. Math.* **1** (1878), 289–321.

[Lucas 1878b]
E. Lucas. Sur l'emploi de l'arithmomètre Thomas dans l'arithmétique supérieure. *Assoc. Française pour l'Avancement des Sciences; Comptes Rendus* **7** (1878), 94–95.

[Lucas 1878d]
E. Lucas. Théorèmes d'arithmétique. *Atti R. Accad. Sc. Torino* **13** (1877-8), 271–284.

[Lucas 1879]
E. Lucas. Solution to question 453. *Nouv. Corresp. Math.* **5** (1879), 137.

[Lucas 1887]
E. Lucas. Sur le neuvième nombre parfait. *Mathesis* **7** (1887), 45–46.

[Lucas 1891]
E. Lucas. *Théorie des Nombres*. Gauthier-Villars, Paris, 1891.

[Lucas 1896]
E. Lucas. *Récréations Mathématiques*, Vol. II. Gauthiers-Villars, Paris, 1896.

[Lugiez 1985]
D. Lugiez. Fast Hensel's lifting implementation using partial fraction decomposition. *Discrete Math.* **56** (1985), 217–225.

[van de Lune and te Riele 1983]
J. van de Lune and H. J. J. te Riele. On the zeros of the Riemann zeta function in the critical strip. III. *Math. Comp.* **41** (1983), 759–767.

[van de Lune, te Riele, and Winter 1986]
J. van de Lune, H. J. J. te Riele, and D. T. Winter. On the zeros of the Riemann zeta function in the critical strip. IV. *Math. Comp.* **46** (1986), 667–681.

[Lüneburg 1985]
H. Lüneburg. On a little but useful algorithm. In J. Calmet, editor, *AAECC-3*, Vol. 229 of *Lecture Notes in Computer Science*, pp. 296–301, 1985.

[Luo 1989]
X. Luo. A practical sieve algorithm for finding prime numbers. *Comm. ACM* **32** (1989), 344–346.

[Ma and von zur Gathen 1990]
K. Ma and J. von zur Gathen. Analysis of Euclidean algorithms for polynomials over finite fields. *J. Symbolic Comput.* **9** (1990), 429–455.

[MacLagan-Wedderburn 1905]
J. H. MacLagan-Wedderburn. A theorem on finite algebras. *Trans. Amer. Math. Soc.* **6** (1905), 349–352.

[Maeder 1993]
R. E. Maeder. Storage allocation for the Karatsuba integer multiplication algorithm. In A. Miola, editor, *Design and Implementation of Symbolic Computation Systems, (DISCO '93)*, Vol. 722 of *Lecture Notes in Computer Science*, pp. 59–65. Springer-Verlag, 1993.

[Maeder 1991]
R. E. Maeder. Fibonacci on the fast track. *Mathematica Journal* **1**(3) (Winter 1991), 42–46.

[Mahnke 1912]
D. Mahnke. Leibniz auf der Suche nach einer allgemeinen Primzahlgleichung. *Bibliotheca Math.* **13** (1912-3), 29–61.

[Maier 1981]
H. Maier. Chains of large gaps between consecutive primes. *Adv. Math.* **39** (1981), 257–269.

[Maier 1985]
H. Maier. Primes in short intervals. *Michigan Math. J.* **32** (1985), 221–225.

[Maier and Pomerance 1990]
H. Maier and C. Pomerance. Unusually large gaps between consecutive primes. *Trans. Amer. Math. Soc.* **322** (1990), 201–237.

[Mairson 1977]
H. G. Mairson. Some new upper bounds on the generation of prime numbers. *Comm. ACM* **20** (1977), 664–669.

[Majewski and Havas 1994]
B. S. Majewski and G. Havas. The complexity of greatest common divisor computations. In L. M. Adleman and M.-D. Huang, editors, *Algorithmic Number Theory, First International Symposium, ANTS-I*, Vol. 877 of *Lecture Notes in Computer Science*, pp. 184–193. Springer-Verlag, 1994.

[Makowski and Rotkiweicz 1969]
A. Makowski and A. Rotkiewicz. On pseudoprime numbers of special form. *Colloq. Math.* **20** (1969), 269–271.

[Malm 1977]
D. E. G. Malm. On Monte-Carlo primality tests. *Notices Amer. Math. Soc.* **24** (1977), A–529. Abstract 77T-A222.

[Malo 1900]
E. Malo. Question 1484. *L'Intermédiaire Math.* **7** (1900), 280–282,312–314.

[Malo 1903]
E. Malo. Nombres qui, sans être premiers, vérifient exceptionellement une congruence de Fermat. *L'Intermédiaire Math.* **10** (1903), 88.

[Mandelbaum 1988]
D. M. Mandelbaum. New binary Euclidean algorithms. *Electronics Letters* **24** (1988), 857–858.

[Manders and Adleman 1978]
K. L. Manders and L. Adleman. NP-complete decision problems for binary quadratics. *J. Comput. System Sci.* **16** (1978), 168–184.

[von Mangoldt 1905]
H. von Mangoldt. Zur verteilung der Nullstellen der Riemannschen Funktion $\xi(t)$. *Math. Ann.* **60** (1905), 1–19.

[Mann 1974]
H. B. Mann. The solution of equations by radicals. *J. Algebra* **29** (1974), 551–554.

[Mann and Shanks 1972]
H. B. Mann and D. Shanks. A necessary and sufficient condition for primality, and its source. *J. Combin. Theory. Ser. A* **13** (1972), 131–134.

[Mansour, Schieber and Tiwari 1989]
Y. Mansour, B. Schieber, and P. Tiwari. The complexity of approximating the square root. In *Proc. 30th Ann. Symp. Found. Comput. Sci.*, pp. 325–330. IEEE Press, 1989.

[Mansour, Schieber, and Tiwari 1991a]
Y. Mansour, B. Schieber, and P. Tiwari. A lower bound for integer greatest common divisor computations. *J. Assoc. Comput. Mach.* **38** (1991), 453–471.

[Mansour, Schieber, and Tiwari 1991b]
Y. Mansour, B. Schieber, and P. Tiwari. Lower bounds for computations with the floor operation. *SIAM J. Comput.* **20** (1991), 315–327.

[Mapes 1963]
D. C. Mapes. Fast method for computing the number of primes less than a given limit. *Math. Comp.* **17** (1963), 179–183.

[Marcus 1977]
D. A. Marcus. *Number Fields*. Springer-Verlag, New York, 1977.

[Marcus 1981]
D. A. Marcus. An alternative to Euclid's algorithm. *Amer. Math. Monthly* **88** (1981), 280–283.

[Marsaglia and Zaman 1991]
G. Marsaglia and A. Zaman. A new class of random number generators. *Ann. Appl. Prob.* **1** (1991), 462–480.

[Marsh and Edgar 1971]
D. C. B. Marsh and G. A. Edgar. Solution to problem E 2229. *Amer. Math. Monthly* **78** (1971), 299–300.

[Mason 1914a]
T. E. Mason. Mechanical device for testing Mersenne numbers for primes. *Proc. Indiana Acad. Sci.* (1914), 429–431.

[Mason 1914b]
T. E. Mason. Mechanical device for testing Mersenne numbers for primes. *Bull. Amer. Math. Soc.* **21** (1914), 59,68.

[Massey 1969]
J. L. Massey. Shift register synthesis and BCH decoding. *IEEE Trans. Inform. Theory* **IT-15** (1969), 18–27.

[Mathews 1892]
G. B. Mathews. *Theory of Numbers*. G. Bell and Sons, London, 1962. Reprinted by Chelsea, New York, 1962.

[Matiyasevich 1982]
Yu. V. Matiyasevich. Yet another machine experiment in support of Riemann's conjecture. *Kybernetika* **6** (1982), 10–12. In Russian. English translation in *Cybernetics* **18** (1982), 705–707.

[Matiyasevich 1993]
Yu. V. Matiyasevich. *Hilbert's Tenth Problem*. MIT Press, 1993.

[Matthews 1994]
R. Matthews. Strong pseudoprimes and generalized Carmichael numbers. In G. L. Mullen and P. J.-S. Shiue, editors, *Finite Fields: Theory, Applications, and Algorithms*, Vol. 168 of *Contemporary Mathematics*, pp. 227–233. Amer. Math. Soc., 1994.

[U. Maurer 1990]
U. M. Maurer. Fast generation of secure RSA-moduli with almost maximal diversity. In J.-J. Quisquater and J. Vandewalle, editors, *Advances in Cryptology–EUROCRYPT '89 Proceedings*, pp. 636–647. Springer-Verlag, 1990.

[U. Maurer 1992]
U. M. Maurer. Some number-theoretic conjectures and their relation to the generation of cryptographic primes. In C. Mitchell, editor, *Cryptography and Coding II*, Vol. 33 of *Institute of Mathematics and its Applications Conference Series*, pp. 173–191. Clarendon Press, Oxford, 1992.

[W. Maurer 1968]
W. D. Maurer. An improved hash code for scatter storage. *Comm. ACM* **11** (1968), 35–38.

[Mays 1987]
M. E. Mays. Iterating the division algorithm. *Fibonacci Quart.* **25** (1987), 204–213.

[McCarthy 1977]
D. P. McCarthy. The optimal algorithm to evaluate x^n using elementary multiplication methods. *Math. Comp.* **31** (1977), 251–256.

[McCarthy 1986]
D. P. McCarthy. Effect of improved multiplication efficiency on exponentiation algorithms derived from addition chains. *Math. Comp.* **46** (1986), 603–608.

[McCoy 1948]
N. H. McCoy. *Rings and Ideals*. MAA, 1948.

[McCurdy and Wunderlich 1987]
K. J. McCurdy and M. C. Wunderlich. The factorization of large composite numbers on the MPP. In *Frontiers of Massively Parallel Scientific Computation (Proceedings of the First Symposium held at NASA Goddard Space Flight Center, Greenbelt, MD)*, NASA Conference Publication 2478, pp. 265–269. National Aeronautics and Space Administration Scientific and Technical Information Office, 1987.

[McCurley 1984a]
K. S. McCurley. Explicit estimates for the error term in the prime number theorem for arithmetic progressions. *Math. Comp.* **42** (1984), 265–285.

[McCurley 1984b]
K. S. McCurley. Explicit estimates for $\theta(x; 3, l)$ and $\psi(x; 3, l)$. *Math. Comp.* **42** (1984), 287–296.

[McCurley 1984c]
K. S. McCurley. Explicit zero-free regions for Dirichlet L-functions. *J. Number Theory* **19** (1984), 7–32.

[McCurley 1986]
K. S. McCurley. The least r-free number in an arithmetic progression. *Trans. Amer. Math. Soc.* **293** (1986), 467–475.

[McDaniel 1989]
W. L. McDaniel. Some pseudoprimes and related numbers having special forms. *Math. Comp.* **53** (1989), 407–409.

[McDonald 1974]
B. McDonald. *Finite Rings with Identity*. Marcel Dekker, 1974.

[McEliece 1969]
R. J. McEliece. Factorization of polynomials over finite fields. *Math. Comp.* **23** (1969), 861–867.

[Meidanis 1991]
J. Meidanis. Lower bounds for arithmetic problems. *Inform. Process. Lett.* **38** (1991), 83–87.

[Meijer 1992]
A. R. Meijer. How fast is the Euclidean algorithm? *Int. J. Math. Educ. Sci. Technol.* **23** (1992), 324–328.

[Meissel 1870]
E. D. F. Meissel. Über die Bestimmung der Primzahlenmenge innerhalb gegebener Grenzen. *Math. Ann.* **2** (1870), 636–642.

[Meissel 1871]
E. D. F. Meissel. Berechnung der Menge von Primzahlen, welche innerhalb der ersten Hundert Millionen natürlicher Zahlen vorkommen. *Math. Ann.* **3** (1871), 523–525.

[Meissel 1883]
E. D. F. Meissel. Über Primzahlmengen. *Math. Ann.* **21** (1883), 304.

[Meissel 1885]
E. D. F. Meissel. Berechnung der Menge von Primzahlen, welche innerhalb der ersten Milliarde natürlicher Zahlen vorkommen. *Math. Ann.* **25** (1885), 251–257.

[Meller 1958]
N. A. Meller. Computations connected with the check of Riemann's hypothesis. *Dokl. Akad. Nauk SSSR* **123** (1958), 246–248. In Russian.

[de Melo and Svaiter 1995]
W. de Melo and B. F. Svaiter. The cost of computing integers. To appear, *Proc. Amer. Math. Soc.*, 1995.

[Menezes, van Oorschot, and Vanstone 1988]
A. J. Menezes, P. C. van Oorschot, and S. A. Vanstone. Some computational aspects of root finding in $GF(q^m)$. In D. Gianni, editor, *Proc. Int'l. Symp. on Symbolic and Algebraic Computation: ISSAC '88*, pp. 259–270. Springer-Verlag, 1988.

[Menezes, van Oorschot, and Vanstone 1992]
A. J. Menezes, P. C. van Oorschot, and S. A. Vanstone. Subgroup refinement algorithms for root finding in $GF(q)$. *SIAM J. Comput.* **21** (1992), 228–239.

[Mersenne 1644]
M. Mersenne. *Cogitata Physico Mathematica*. Paris, 1644.

[Mertens 1874a]
F. Mertens. Ein Beitrag zur analytischen Zahlentheorie. *J. Reine Angew. Math.* **78** (1874), 46–62.

[Mertens 1874b]
F. Mertens. Ueber einige asymptotische Gesetze der Zahlentheorie. *J. Reine Angew. Math.* **77** (1874), 289–338.

[Michel 1992]
P. Michel. A survey of space complexity. *Theoret. Comput. Sci.* **101** (1992), 99–132.

[Mignotte 1975]
M. Mignotte. Un algorithme sur la décomposition des polynomes dans un corps fini. *C. R. Acad. Sci. Paris* **280** (1969), A137–A139.

[Mignotte 1976a]
M. Mignotte. Factorisation des polynômes sur un corps fini. *Astérisque* **38-39** (1976), 149–157.

[Mignotte 1976b]
M. Mignotte. Algorithmes relatifs à la décomposition des polynômes. *Theoret. Comput. Sci.* **1** (1976), 227–235.

[Mignotte 1980a]
M. Mignotte. Calcul des racines d-ièmes dans un corps fini. *C. R. Acad. Sci. Paris* **290** (1980), A205–A206.

[Mignotte 1980b]
M. Mignotte. Tests de primalité. *Theoret. Comput. Sci.* **12** (1980), 109–117.

[Mignotte 1980c]
M. Mignotte. Factorization of univariate polynomials: a statistical study. *ACM SIGSAM Bull.* **14**(4) (1980), 41–44.

[Mignotte and Schnorr 1988]
M. Mignotte and C. Schnorr. Calcul déterministe des racines d'un polynôme dans un corps fini. *C. R. Acad. Sci. Paris* **306** (1988), 467–472.

[Mihailescu 1988]
P. Mihailescu. A primality test using cyclotomic functions. In T. Mora, editor, *AAECC-6*, Vol. 357 of *Lecture Notes in Computer Science*, pp. 310–323. Springer-Verlag, 1988.

[G. Miller 1976]
G. Miller. Riemann's hypothesis and tests for primality. *J. Comput. System Sci.* **13** (1976), 300–317.

[J. Miller 1951]
J. C. P. Miller. Large primes. *Eureka* (14) (1951), 10–11.

[J. Miller and Spencer Brown 1966]
J. C. P. Miller and D. J. Spencer Brown. An algorithm for evaluation of remote terms in a linear recurrence sequence. *The Computer Journal* **9** (1966), 188–190.

[J. Miller and Wheeler 1951]
J. C. P. Miller and D. J. Wheeler. Large prime numbers. *Nature* **168** (1951), 838.

[V. Miller 1992a]
V. S. Miller. On the factorization method of Niederreiter. Unpublished manuscript, 1992.

[W. Miller 1987]
W. Miller. The maximum order of an element of a finite symmetric group. *Amer. Math. Monthly* **94** (1987), 497–506.

[Mills 1947]
W. H. Mills. A prime-representing function. *Bull. Amer. Math. Soc.* **53** (1947), 604.

[Minkowski 1891]
H. Minkowski. Théorèmes arithmétiques. *C. R. Acad. Sci. Paris* **112** (1891), 209–212. Reprinted in *Gesammelte Abhandlungen*, Vol. I, pp. 261–263.

[Minsky 1967]
M. Minsky. *Computation: Finite and Infinite Machines*. Prentice-Hall, 1967.

[Minsky and Papert 1966]
M. Minsky and S. Papert. Unrecognizable sets of numbers. *J. Assoc. Comput. Mach.* **13** (1966), 281–286.

[Mirimanoff and Hensel 1905]
D. Mirimanoff and K. Hensel. Sur la relation $\left(\frac{D}{p}\right) = (-1)^{n-h}$ et la loi de réciprocité. *J. Reine Angew. Math.* **129** (1905), 86–87.

[Misra 1981]
J. Misra. An exercise in program explanation. *ACM Trans. Prog. Lang. Syst.* **3** (1981), 104–109.

[Mitrinović and Popadić 1978]
D. S. Mitrinović and M. S. Popadić. *Inequalities in Number Theory*. University of Niś, Niś, 1978.

[Mitsui 1968]
T. Mitsui. On the prime ideal theorem. *J. Math. Soc. Japan* **20** (1968), 233–247.

[Miyamoto and Ram Murty 1989]
I. Miyamoto and M. Ram Murty. Elliptic pseudoprimes. *Math. Comp.* **53** (1989), 415–430.

[Mo and Jones 1995]
Z. Mo and J. P. Jones. A new primality test using Lucas sequences. Unpublished manuscript, 1995.

[Möbius 1832]
A. F. Möbius. Über eine besondere Art von Umkehrung der Reihen. *J. Reine Angew. Math.* **9** (1832), 105–123. Reprinted in *Werke*, Vol. 4, pp. 589–612.

[Moenck 1973]
R. T. Moenck. Fast computation of GCDs. In *Proc. Fifth Ann. ACM Symp. Theor. Comput.*, pp. 142–151. ACM, 1973.

[Moenck 1977]
R. T. Moenck. On the efficiency of algorithms for polynomial factoring. *Math. Comp.* **31** (1977), 235–250.

[de Moivre 1730]
A. de Moivre. *Miscellanea Analytica de Seriebus et Quadraturis*. J. Tonson and J. Watts, London, 1730.

[Monier 1980]
L. Monier. Evaluation and comparison of two efficient probabilistic primality testing algorithms. *Theoret. Comput. Sci.* **12** (1980), 97–108.

[H. Montgomery 1975]
H. Montgomery. Distribution of the zeros of the Riemann zeta function. In *Proc. 1974 International Congress of Mathematicians*, Vol. 1, pp. 379–381. 1975.

[H. Montgomery 1971]
H. L. Montgomery. *Topics in Multiplicative Number Theory*, Vol. 227 of *Lecture Notes in Mathematics*. Springer-Verlag, Berlin, 1971.

[H. Montgomery 1973]
H. L. Montgomery. The pair correlation of zeroes of the zeta function. In H. G. Diamond, editor, *Analytic Number Theory*, Vol. 24 of *Proceedings of Symposia in Pure Mathematics*, pp. 181–193. Amer. Math. Soc., 1973.

[H. Montgomery 1980]
H. L. Montgomery. The zeta function and prime numbers. In P. Ribenboim, editor, *Proc. Queen's Number Theory Conf.*, Vol. 54 of *Queen's Papers in Pure and Applied Math.*, pp. 1–31. 1980.

[H. Montgomery and Vaughan 1973]
H. L. Montgomery and R. C. Vaughan. The large sieve. *Mathematika* **20** (1973), 119–134.

[P. Montgomery 1985]
P. L. Montgomery. Modular multiplication without trial division. *Math. Comp.* **44** (1985), 519–521.

[P. Montgomery 1990]
P. L. Montgomery. A problem of Herb Doughty. Unpublished manuscript, 1990.

[E. Moore 1896]
E. H. Moore. A doubly-infinite system of linear groups. In *Mathematical Papers Read at the International Mathematical Congress Held in Connection with the World's Columbian Exposition Chicago 1893*, pp. 208–242. Macmillan, 1896.

[T. Moore 1992]
T. E. Moore. On the least absolute remainder Euclidean algorithm. *Fibonacci Quart.* **30** (1992), 161–165.

[Morain and Olivos 1990]
F. Morain and J. Olivos. Speeding up the computations on an elliptic curve using addition-subtraction chains. *RAIRO Inform. Théor.* **24** (1990), 531–544.

[Moree 1993b]
P. Moree. Bertrand's postulate for primes in arithmetic progressions. *Comput. Math. with Appl.* **26** (1993), 35–43.

[Morehead 1905]
J. C. Morehead. Note on Fermat's numbers. *Bull. Amer. Math. Soc.* **11** (1905), 543–545.

[Morehead and Western 1909]
J. C. Morehead and A. E. Western. Note on Fermat's numbers. *Bull. Amer. Math. Soc.* **16** (1909), 1–6.

[Morii and Takamatsu 1991]
M. Morii and Y. Takamatsu. Exponentiation in finite fields using dual basis multiplier. In S. Sakata, editor, *Applied Algebra, Algebraic Algorithms and Error-Correcting Codes: AAECC-8*, Vol. 508 of *Lecture Notes in Computer Science*, pp. 354–366. Springer-Verlag, 1991.

[Morita 1990a]
H. Morita. A fast modular-multiplication module for smart cards. In J. Seberry and J. Pieprzyk, editors, *Advances*

in Cryptology–AUSCRYPT '90, Vol. 453 of *Lecture Notes in Computer Science*, pp. 406–409. Springer-Verlag, 1990.

[Morita 1990b]
H. Morita. A fast modular-multiplication algorithm based on a higher radix. In G. Brassard, editor, *Advances in Cryptology–CRYPTO '89 Proceedings*, Vol. 435 of *Lecture Notes in Computer Science*, pp. 387–399. Springer-Verlag, 1990.

[Moroz 1961]
B. Z. Moroz. The distribution of power residues and non-residues. *Vestnik Leningrad Univ. Math.* **16** (1961), 164–169. In Russian, with English summary.

[D. Morrison 1968]
D. R. Morrison. Arithmetic in Galois fields of characteristic 2. *Math. Algorithms* **3** (1968), 176–184.

[J. Morrison 1988]
J. F. Morrison. Parallel p-adic computation. *Inform. Process. Lett.* **28** (1988), 137–140.

[M. Morrison 1975]
M. A. Morrison. A note on primality testing using Lucas sequences. *Math. Comp.* **29** (1975), 181–182.

[Moser 1949]
L. Moser. Solution to problem E 848. *Amer. Math. Monthly* **56** (1949), 478.

[Moser 1950]
L. Moser. A prime representing function. *Math. Mag.* **23** (1950), 163–164.

[Moser and Kugelmass 1964]
L. Moser and J. Kugelmass. Solution to problem 518. *Math. Mag.* **37** (1964), 57–59.

[Motwani and Raghavan 1995]
R. Motwani and P. Raghavan. *Randomized Algorithms*. Cambridge University Press, 1995.

[Muir 1874]
T. Muir. Continuants–a new special class of determinants. *Proc. Roy. Soc. Edinburgh* **8** (1872-5), 229–236.

[Muir 1878]
T. Muir. Letter from Mr. Muir to Professor Sylvester on the word continuant. *Amer. J. Math.* **1** (1878), 344.

[Müller and Oswald 1991]
W. B. Müller and A. Oswald. Dickson pseudoprimes and primality testing. In D. W. Davies, editor, *Advances in Cryptology–EUROCRYPT '91 Proceedings*, Vol. 547 of *Lecture Notes in Computer Science*, pp. 512–516. Springer-Verlag, 1991.

[Mullin, Onyszchuk, Vanstone, and Wilson 1988]
R. C. Mullin, I. M. Onyszchuk, S. A. Vanstone, and R. M. Wilson. Optimal normal bases in $GF(p^n)$. *Disc. Appl. Math.* **22** (1988/89), 149–161.

[Murata 1991a]
L. Murata. On the magnitude of the least prime primitive root. *J. Number Theory* **37** (1991), 47–66.

[Murata 1991b]
L. Murata. On the magnitude of the least primitive root. *Astérisque* **198–200** (1991), 253–257.

[Murata 1991c]
L. Murata. A problem analogous to Artin's conjecture for primitive roots and its applications. *Arch. Math.* **57** (1991), 555–565.

[Musser, Rosser, Schoenfeld, and Yohe 1969]
D. R. Musser, J. B. Rosser, L. Schoenfeld, and J. M. Yohe. The Riemann zeta-function magnetic tapes. Technical Report 967, Univ. Wisconsin, Math. Res. Center, 1969.

[Nagasaka, Shiue, and Ho 1991]
K. Nagasaka, J.-S. Shiue, and C.-W. Ho. A fast algorithm of the Chinese remainder theorem and its application to Fibonacci numbers. In G. E. Bergum, A. N. Philippou, and A. F. Horadam, editors, *Applications of Fibonacci Numbers*, Vol. 4, pp. 241–246. Kluwer, Boston, 1991.

[Nair 1982]
M. Nair. On Chebyshev-type inequalities for primes. *Amer. Math. Monthly* **89** (1982), 126–129.

[Narkiewicz 1986]
W. Narkiewicz. *Classical Problems in Number Theory*. PWN – Polish Scientific Publishers, 1986.

[Narkiewicz 1990]
W. Narkiewicz. *Elementary and Analytic Theory of Algebraic Numbers*. Springer-Verlag, 1990.

[Nazarevsky 1904]
Nazarevsky. Question 801. *L'Intermédiaire Math.* **11** (1904), 215.

[Needham 1959]
J. Needham. *Science and Civilisation in China*, Vol. 3 (Mathematics and the Sciences of the Heavens and the Earth). Cambridge University Press, 1959.

[Nelson 1980]
H. L. Nelson. Multi-precise arithmetic on a vector processor, or how we found the 27th Mersenne prime. In *VLSI: New Architectural Horizons (Proc. Twentieth IEEE Computer Society International Conference; Spring CompCon 80)*, pp. 265–269, 1980.

[Netto and le Vavasseur 1907]
E. Netto and R. le Vavasseur. Les fonctions rationnelles. In J. Molk, editor, *Encyclopédie des Sciences Mathématiques*, Vol. I:2, pp. 1–232. Gauthier-Villars, 1907.

[Neudecker 1975]
W. Neudecker. On twin 'primes' and gaps between successive 'primes' for the Hawkins random sieve. *Math. Proc. Cambridge Phil. Soc.* **77** (1975), 365–367.

[Neukirch 1986]
J. Neukirch. *Class Field Theory*. Springer-Verlag, Berlin, 1986.

[Neumann and Wilson 1979]
B. H. Neumann and L. G. Wilson. Some sequences like Fibonacci's. *Fibonacci Quart.* **17** (1979), 80–83.

[von Neumann 1947]
J. von Neumann. Letter to R. D. Richtmyer, March 11, 1947. In A. H. Taub, editor, *John von Neumann: Collected Works, vol. V*, pp. 751–764. Pergamon Press, 1963.

[Newman 1980]
D. J. Newman. Simple analytic proof of the prime number theorem. *Amer. Math. Monthly* **87** (1980), 693–696.

[Nicolas 1981]
J.-L. Nicolas. Tests de primalité (d'après Adleman, Pomerance, Rumely, Lenstra, Cohen). In *Laboratoire Informatique Théorique, Institut de Programmation, Institut Henri Poincaré*, pp. 119–134, 1981.

[Nicolas 1984]
J.-L. Nicolas. Tests de primalité. *Exposition. Math.* **2** (1984), 223–234.

[Nicolas 1985a]
J.-L. Nicolas. Utilisation des ordinateurs en théorie des nombres. *Bull. Soc. Math. Belg. Sér. A* **37** (1985), 11–18.

[Nicolas 1985b]
J.-L. Nicolas. Test de primalité et méthodes de factorisation. In *Colloque d'Algèbre; Publications de L'Institut de Recherche Mathématique de Rennes*, Vol. 4, pp. 148–162, 1985.

[Nicolas and Robin 1983]
J. L. Nicolas and G. Robin. Majorations explicites pour le nombre de diviseurs de N. *Canad. Math. Bull.* **26** (1983), 485–492.

[Niederreiter 1991]
H. Niederreiter. Finite fields and their applications. In D. Dorninger, editor, *Contributions to General Algebra 7*, pp. 251–264. Hölder–Pichler–Tempsky, Vienna, 1991.

[Niederreiter 1993a]
H. Niederreiter. A new efficient factorization algorithm for polynomials over small finite fields. *Appl. Alg. Eng. Comm. Comp.* **4** (1993), 81–87.

[Niederreiter 1993b]
H. Niederreiter. Finite fields and cryptology. In G. L. Mullen and P. J.-S. Shiue, editors, *Finite Fields, Coding Theory, and Advances in Communications and Computing*, Vol. 141 of *Lecture Notes in Pure and Applied Mathematics*, pp. 359–373. Marcel Dekker, New York, 1993.

[Niederreiter 1993c]
H. Niederreiter. Recent advances in the theory of finite fields. In G. L. Mullen and P. J.-S. Shiue, editors, *Finite Fields, Coding Theory, and Advances in Communications and Computing*, Vol. 141 of *Lecture Notes in Pure and Applied Mathematics*, pp. 153–163. Marcel Dekker, New York, 1993.

[Niederreiter 1993d]
H. Niederreiter. Factorization of polynomials and some linear-algebra problems over finite fields. *Lin. Alg. Appls.* **192** (1993), 301–328.

[Niederreiter 1994a]
H. Niederreiter. Factoring polynomials over finite fields using differential equations and normal bases. *Math. Comp.* **62** (1994), 819–830.

[Niederreiter 1994b]
H. Niederreiter. New deterministic factorization algorithms for polynomials over finite fields. In G. L. Mullen and P. J.-S. Shiue, editors, *Finite Fields: Theory, Applications, and Algorithms*, Vol. 168 of *Contemporary Mathematics*, pp. 251–268. Amer. Math. Soc., 1994.

[Niederreiter and Göttfert 1993]
H. Niederreiter and R. Göttfert. Factorization of polynomials over finite fields and characteristic sequences. *J. Symbolic Comput.* **16** (1993), 401–412.

[Niederreiter and Göttfert 1995]
H. Niederreiter and R. Göttfert. On a new factorization algorithm for polynomials over finite fields. *Math. Comp.* **64** (1995), 347–353.

[Nielsen 1906]
N. Nielsen. *Theorie des Integrallogarithmus und verwandter Transzendenten*. Teubner, Leipzig, 1906. Reprinted by Chelsea, New York, 1965.

[Nievengloski 1845]
G.-H. Nievengloski. Sur la limite supérieure du nombre de divisions à faire pour trouver le plus grand commun diviseur de deux nombres. *Nouvelles Annales de Mathématiques* **4** (1845), 568–573.

[Nievengloski 1849]
G.-H. Nievengloski. Note sur une abréviation dans la recherche du plus grand commun diviseur de deux nombres. *Nouvelles Annales de Mathématiques* **8** (1849), 447–448.

[Niven 1951]
I. Niven. Functions which represent prime numbers. *Proc. Amer. Math. Soc.* **2** (1951), 753–755.

[Niven 1984]
I. Niven. Solution to elementary problem 2946. *Amer. Math. Monthly* **91** (1984), 650.

[Noll and Nickel 1980]
C. Noll and L. Nickel. The 25th and 26th Mersenne primes. *Math. Comp.* **35** (1980), 1387–1390.

[Norman 1979]
A. C. Norman. Testing word-sized numbers for primality. *ACM SIGSAM Bull.* **13**(4) (Nov. 1979), 19–20.

[Norris and Simmons 1981]
M. J. Norris and G. J. Simmons. Algorithms for high-speed modular arithmetic. *Congr. Numer.* **31** (1981), 153–163.

[G. Norton 1985]
G. H. Norton. Extending the binary gcd algorithm. In J. Calmet, editor, *AAECC-3*, Vol. 229 of *Lecture Notes in Computer Science*, pp. 363–372, 1985.

[G. Norton 1987]
G. H. Norton. A shift-remainder gcd algorithm. In L. Huguet and A. Poli, editors, *AAECC-5*, Vol. 356 of *Lecture Notes in Computer Science*, pp. 350–356, 1987.

[G. Norton 1989]
G. H. Norton. Precise analyses of the right- and left-shift greatest common divisor algorithms for $GF(q)[x]$. *SIAM J. Comput.* **18** (1989), 608–624.

[G. Norton 1990]
G. H. Norton. On the asymptotic analysis of the Euclidean algorithm. *J. Symbolic Comput.* **10** (1990), 53–58.

[K. Norton 1971]
K. K. Norton. *Numbers With Small Prime Factors and the Least kth Power Non-Residue*, Vol. 106 of *Memoirs Amer. Math. Soc.* Amer. Math. Soc., 1971.

[K. Norton 1973]
K. K. Norton. Bounds for sequences of consecutive power residues. I. In H. G. Diamond, editor, *Analytic Number Theory*, Vol. 24 of *Proceedings of Symposia in Pure Mathematics*, pp. 213–220. Amer. Math. Soc., 1973.

[d'Ocagne 1922]
M. d'Ocagne. Vue d'ensemble sur les machines à calculer. *Bull. Sci. Math.* **46** (1922), 102–144.

[Odlyzko 1987a]
A. M. Odlyzko. On the distribution of spacings between zeros of the zeta function. *Math. Comp.* **48** (1987), 273–308.

[Odlyzko 1987b]
A. M. Odlyzko. New analytic algorithms in number theory. In A. M. Gleason, editor, *Proc. 1986 International Congress of Mathematicians*, pp. 466–475. Amer. Math. Soc., 1987.

[Odlyzko 1990]
A. M. Odlyzko. Bounds for discriminants and related estimates for class numbers, regulators, and zeros of zeta functions: a survey of recent results. *Séminaire de Théorie des Nombres de Bordeaux* **2** (1990), 119–141.

[Odlyzko and Schönhage 1988]
A. M. Odlyzko and A. Schönhage. Fast algorithms for multiple evaluations of the Riemann zeta function. *Trans. Amer. Math. Soc.* **309** (1988), 797–809.

[Odlyzko and te Riele 1985]
A. M. Odlyzko and H. J. J. te Riele. Disproof of the Mertens conjecture. *J. Reine Angew. Math.* **357** (1985), 138–160.

[Oesterlé 1979]
J. Oesterlé. Versions effectives du théorème de Chebotarev sous l'hypothèse de Riemann généralisée. *Astérisque* **61** (1979), 165–167.

[Ofman 1962]
Y. Ofman. On the algorithmic complexity of discrete functions. *Dokl. Akad. Nauk SSSR* **145** (1962), 48–51. English translation in *Soviet Physics Doklady*, **7** (1963), 589–591.

[Ogiwara 1989]
M. Ogiwara. A method for generating cryptographically strong primes. *Res. Rep. Inf. Sci. C, Comput. Sci. (Japan)* (93) (1989), 1–33.

[Olivos 1981]
J. Olivos. On vectorial addition chains. *J. Algorithms* **2** (1981), 13–21.

[Omura and Massey 1986]
J. K. Omura and J. L. Massey. Computational method and apparatus for finite field arithmetic. U. S. patent 4,587,627; May 6, 1986; filed September 14, 1982.

[Onyszchuk, Mullin, and Vanstone 1988]
I. M. Onyszchuk, R. C. Mullin, and S. A. Vanstone. Computational method and apparatus for finite field multiplication. U. S. patent 4,745,568; May 17, 1988; filed May 30, 1985.

[van Oorschot 1988]
P. C. van Oorschot. *Combinatorial and computational issues related to finding roots of polynomials over finite fields*. PhD thesis, Department of Computer Science, University of Waterloo, 1988.

[van Oorschot and Vanstone 1989]
P. C. van Oorschot and S. A. Vanstone. A geometric approach to root finding in $GF(q^m)$. *IEEE Trans. Inform. Theory* **35** (1989), 444–453.

[van Oorschot and Vanstone 1990a]
P. C. van Oorschot and S. A. Vanstone. Some geometric aspects of root finding in $GF(q^m)$. In E. S. Kramer and S. S. Magliveras, editors, *Finite Geometries and Combinatorial Designs*, Vol. 111 of *Contemporary Mathematics*, pp. 303–307. Amer. Math. Soc., 1990.

[van Oorschot and Vanstone 1990b]
P. C. van Oorschot and S. A. Vanstone. On splitting sets in block designs and finding roots of polynomials. *Discrete Math.* **84** (1990), 71–85.

[Ore 1926]
O. Ore. Existenzbeweis für algebraische Körper mit vorgeschriebenden Eigenschaften. *Math. Zeitschrift* **95** (1926), 474–489.

[Ore 1927]
O. Ore. Über den Zusammenhang zwischen den definierenden Gleichungen und der Idealtheorie in algebraischen Körpern. *Math. Ann.* **96** (1927), 313–352. Continued in **97** (1927), 569–598.

[Ore 1934]
O. Ore. Contributions to the theory of finite fields. *Trans. Amer. Math. Soc.* **36** (1934), 243–274.

[Ore 1952]
O. Ore. On the selection of subsequences. *Proc. Amer. Math. Soc.* **3** (1952), 706–712.

[Orton, Peppard, and Tavares 1992]
G. A. Orton, L. E. Peppard, and S. E. Tavares. New fault tolerant techniques for residue number systems. *IEEE Trans. Comput.* **41** (1992), 1453–1464.

[Orup and Kornerup 1991]
H. Orup and P. Kornerup. A high-radix hardware algorithm for calculating the exponential M^E modulo N. In P. Kornerup and D. W. Matula, editors, *Proc. 10th IEEE Symp. Comp. Arith.*, pp. 51–56. IEEE Computer Society Press, 1991.

[Owings 1982]
J. C. Owings. The Lucas sequence as a test for primality. *Abstracts Amer. Math. Soc.* **3**(1) (1982), 131–132. Abstract 82T-10-19.

[Pajunen 1980]
S. Pajunen. On two theorems of Lenstra. *Inform. Process. Lett.* **11** (1980), 224–228.

[Papadimitriou 1975]
M. Papadimitriou. A recursion formula for the sequence of odd primes. *Amer. Math. Monthly* **82** (1975), 289.

[Parady, Smith, and Zarantonello 1990]
B. K. Parady, J. F. Smith, and S. E. Zarantonello. Largest known twin primes. *Math. Comp.* **55** (1990), 381–382.

[E. Parberry 1970]
E. A. Parberry. On primes and pseudo-primes related to the Fibonacci sequence. *Fibonacci Quart.* **8** (1970), 49–60.

[I. Parberry 1981]
I. Parberry. Parallel speedup of sequential prime number sieves. Technical Report 30, Dept. of Computer Science, Univ. of Queensland, Australia, 1981.

[Parikh and Matula 1991]
S. N. Parikh and D. W. Matula. A redundant binary Euclidean GCD algorithm. In P. Kornerup and D. W. Matula, editors, *Proc. 10th IEEE Symp. Comp. Arith.*, pp. 220–225. IEEE Press, 1991.

[Patterson and Williams 1983]
C. D. Patterson and H. C. Williams. A report on the University of Manitoba sieve unit. *Congr. Numer.* **37** (1983), 85–98.

[Paxson 1961]
G. A. Paxson. The compositeness of the the thirteenth Fermat number. *Math. Comp.* **15** (1961), 420.

[Pellet 1870]
A.-E. Pellet. Sur les fonctions irréductibles suivant un module premier et une fonction modulaire. *C. R. Acad. Sci. Paris* **70** (1870), 328–330.

[Pellet 1878]
A.-E. Pellet. Sur la décomposition d'une fonction entière en facteurs irréductibles suivant un module premier *p*. *C. R. Acad. Sci. Paris* **86** (1878), 1071–1072.

[Pepin 1877]
T. Pepin. Sur la formule $2^{2^n} + 1$. *C. R. Acad. Sci. Paris* **85** (1877), 329–331.

[Pepin 1878]
T. Pepin. Sur la formule $2^n - 1$. *C. R. Acad. Sci. Paris* **86** (1878), 307–310.

[Peralta 1986]
R. C. Peralta. A simple and fast probabilistic algorithm for computing square roots modulo a prime number. *IEEE Trans. Inform. Theory* **IT-32** (1986), 846–847.

[Peralta 1992]
R. C. Peralta. On the distribution of quadratic residues and nonresidues modulo a prime number. *Math. Comp.* **58** (1992), 443–440.

[Perrin 1899]
R. Perrin. Question 1484. *L'Intermédiaire Math.* **6** (1899), 76–77.

[Pervushin 1898]
J. Pervouchine. Formules pour la détermination approximative des nombres premiers, de leur somme et de leur différence d'après le numéro de ces nombres. In F. Rudio, editor, *Verhandlungen des ersten internationalen Mathematiker-Kongresses*, pp. 166–167. Teubner, 1898.

[I. Peterson 1985]
I. Peterson. Prime time for supercomputers. *Science News* **128** (1985), 199.

[I. Peterson 1988]
I. Peterson. Priming for a lucky strike. *Science News* **133** (1988), 85.

[W. Peterson and Weldon 1972]
W. W. Peterson and E. J. Weldon, Jr. *Error-correcting Codes*. The MIT Press, 1972.

[Petr 1937]
K. Petr. Über die Redizibilität eines Polynoms mit ganzzahligen Koeffizienten nach einem Primzalhmodul. *Časopis pro Pěstování Matematiky a Fysiky* **66** (1937), 85–94.

[Pettorossi 1980]
A. Pettorossi. Derivation of an $O(k^2 \log n)$ algorithm for computing order-k Fibonacci numbers from the $O(k^3 \log n)$ matrix multiplication method. *Inform. Process. Lett.* **11** (1980), 172–179.

[Phong 1988]
B. M. Phong. On super Lucas and super Lehmer pseudoprimes. *Studia Scientiarum Mathematicarum Hungarica* **23** (1988), 435–442.

[Piarron de Mondesir 1878]
E. S. Piarron de Mondesir. Sur les nombres premiers. Formules pour le calcul exact de la totalité des nombres premiers compris entre 0 et un nombre pair quelconque $2n$. *Assoc. Française pour l'Avancement des Sciences; Comptes Rendus* **6** (1878), 79–92.

[Picutti 1989]
E. Picutti. Pour l'histoire des sept premiers nombres parfaits. *Historia Mathematica* **16** (1989), 123–136.

[Pieper 1978]
H. Pieper. *Variationen über ein zahlentheoretisches Thema von Carl Friedrich Gauss*. Birkhäuser, Basel, 1978.

[Pila 1990]
J. Pila. Frobenius maps of abelian varieties and finding roots of unity in finite fields. *Math. Comp.* **55** (1990), 745–763.

[Piltz 1884]
A. Piltz. *Über die Häufigkeit der Primzahlen in arithmetischen Progressionen und über verwandte Gesetze*. A. Neuenhahn, Jena, 1884.

[Pin 1981]
J.-E. Pin. Test de primalité pour calculatrice de poche, d'après L. Monier. In *Laboratoire Informatique Théorique, Institut de Programmation, Institut Henri Poincaré*, pp. 251–262, 1980/81.

[Pinch 1992]
R. G. E. Pinch. Recognising elements of finite fields. In C. Mitchell, editor, *Cryptography and Coding II*, Vol. 33 of *Institute of Mathematics and its Applications Conference Series*, pp. 193–197. Clarendon Press, Oxford, 1992.

[Pinch 1993a]
R. G. E. Pinch. The Carmichael numbers up to 10^{15}. *Math. Comp.* **61** (1993), 381–391.

[Pinch 1993b]
R. G. E. Pinch. Some primality testing algorithms. *Notices Amer. Math. Soc.* **40** (1993), 1203–1210.

[Pinch and Keller 1992]
R. G. E. Pinch and W. Keller. The Carmichael numbers up to 10^{15}. *Abstracts Amer. Math. Soc.* **13** (1992), 505. Abstract 92T-11-157.

[Pincin 1989]
A. Pincin. A new algorithm for multiplication in finite fields. *IEEE Trans. Comput.* **38** (1989), 1045–1049.

[Pintz, Steiger, and Szemerédi 1989]
J. Pintz, W. L. Steiger, and E. Szemerédi. Infinite sets of primes with fast primality tests and quick generation of large primes. *Math. Comp.* **53** (1989), 399–406.

[Piontas 1989]
M. Piontas. Algorithm for squaring in $GF(2^m)$ in standard basis. *Electronics Letters* **25** (1989), 1262–1263.

[Pippenger 1976]
N. Pippenger. On the evaluation of powers and related problems. In *Proc. 17th Ann. Symp. Found. Comput. Sci.*, pp. 258–263, 1976.

[Pippenger 1979]
N. Pippenger. On simultaneous resource bounds. In *Proc. 20th Ann. Symp. Found. Comput. Sci.*, pp. 307–311. IEEE Press, 1979.

[Pippenger 1980]
N. Pippenger. On the evaluation of powers and monomials. *SIAM J. Comput.* **9** (1980), 230–250.

[Pitteway and Castle 1988]
M. L. V. Pitteway and C. M. A. Castle. On the subtractive version of Euclid's algorithm. *Bull. Inst. Math. Appl.* **24** (1988), 17–21.

[Plaisted 1977]
D. A. Plaisted. Sparse complex polynomials and polynomial reducibility. *J. Comput. System Sci.* **14** (1977), 210–221.

[Plaisted 1978]
D. A. Plaisted. Some polynomial and integer divisibility problems are *NP*-hard. *SIAM J. Comput.* **7** (1978), 458–464.

[Plaisted 1979]
D. A. Plaisted. Fast verification, testing, and generation of large primes. *Theoret. Comput. Sci.* **9** (1979), 1–16. Errata in **14** (1981), 345.

[Plaisted 1984]
D. A. Plaisted. New *NP*-hard and *NP*-complete polynomial and integer divisibility problems. *Theoret. Comput. Sci.* **31** (1984), 125–138.

[Plaisted 1985]
D. A. Plaisted. Complete divisibility problems for slowly utilized oracles. *Theoret. Comput. Sci.* **35** (1985), 245–260.

[Plankensteiner 1970]
B. Plankensteiner. Untersuchung über die Schrittanzahl des euklidischen Algorithmus bei Andwendung auf echte Brüche. *Monatsh. Math.* **74** (1970), 244–257.

[Pocklington 1910]
H. C. Pocklington. The determination of the exponent to which a number belongs, the practical solution of certain congruences, and the law of quadratic reciprocity. *Proc. Cambridge Phil. Soc.* **16** (1910), 1–5.

[Pocklington 1914]
H. C. Pocklington. The determination of the prime or composite nature of large numbers by Fermat's theorem. *Proc. Cambridge Phil. Soc.* **18** (1914-1916), 29–30.

[Pocklington 1917]
H. C. Pocklington. The direct solution of the quadratic and cubic binomial congruences with prime moduli. *Proc. Cambridge Phil. Soc.* **19** (1917), 57–59.

[Pohst and Zassenhaus 1989]
M. Pohst and H. Zassenhaus. *Algorithmic Algebraic Number Theory*. Cambridge University Press, 1989.

[van der Pol 1947]
B. van der Pol. An electro-mechanical investigation of the Riemann zeta function in the critical strip. *Bull. Amer. Math. Soc.* **53** (1947), 976–981.

[Poletti 1929]
L. Poletti. La serie di numeri primi appartenente alle due forme quadratiche (A) $n^2 + n - 1$ (B) $n^2 + n + 1$. *Memorie della Reale Accademia Nazionale dei Lincei, Classe di Scienze Fisiche, Matematiche e Naturali* **3** (1929), 194–218.

[Poli and Gennero 1988]
A. Poli and M. C. Gennero. Fast16: a software program for factorising polynomials over large $GF(p)$. In T. Beth and M. Clausen, editors, *Applicable Algebra, Error-Correcting Codes, Combinatorics, and Computer Algebra: AAEEC-4*, Vol. 307 of *Lecture Notes in Computer Science*, pp. 139–156. Springer-Verlag, 1988.

[de Polignac 1849]
A. de Polignac. Six propositions arithmologiques déduites du crible d'Érathosthène. *Nouvelles Annales de Mathématiques* **8** (1849), 423–429.

[Polkinghorn 1966]
F. Polkinghorn, Jr. Decoding of double and triple error correcting Bose-Chaudhuri codes. *IEEE Trans. Inform. Theory* **IT-12** (1966), 480–481.

[Pollard 1971b]
J. M. Pollard. An algorithm for testing the primality of any integer. *Bull. Lond. Math. Soc.* **3** (1971), 337–340.

[Pollard 1974]
J. M. Pollard. Theorems on factorization and primality testing. *Proc. Cambridge Phil. Soc.* **76** (1974), 521–528.

[Pollin and Schoenberg 1980]
J. M. Pollin and I. J. Schoenberg. On the matrix approach to Fibonacci numbers and the Fibonacci pseudoprimes. *Fibonacci Quart.* **18** (1980), 261–268.

[Polvani 1933]
G. Polvani. Sulla frequenza dei numeri primi. *Rend. Sem. Mat. Fis. Milano* **7** (1933), 303–321.

[Pólya 1959]
G. Pólya. Heuristic reasoning in the theory of numbers. *Amer. Math. Monthly* **66** (1959), 375–384.

[Pomerance 1981a]
C. Pomerance. Recent developments in primality testing. *Math. Intelligencer* **3** (1981), 97–105.

[Pomerance 1981b]
C. Pomerance. On the distribution of pseudoprimes. *Math. Comp.* **37** (1981), 587–593.

[Pomerance 1982b]
C. Pomerance. The search for prime numbers. *Scientific American* **247**(6) (1982), 136–147,178.

[Pomerance 1982c]
C. Pomerance. A new lower bound for the pseudoprime counting function. *Illinois J. Math.* **26** (1982), 4–9.

[Pomerance 1983]
C. Pomerance. References on primality testing. *Congr. Numer.* **39** (1983), 21–23.

[Pomerance 1984a]
C. Pomerance. *Lecture Notes on Primality Testing and Factoring: A Short Course at Kent State University*, Vol. 4 of *MAA Notes*. MAA, 1984.

[Pomerance 1986]
C. Pomerance. Fast, rigorous factorization and discrete logarithm algorithms. In D. S. Johnson et al., editor, *Discrete Algorithms and Complexity*, pp. 119–143. Proc. of the Japan-US Joint Seminar, Academic Press, 1986.

[Pomerance 1987]
C. Pomerance. Very short primality proofs. *Math. Comp.* **48** (1987), 315–322.

[Pomerance 1989]
C. Pomerance. Two methods in elementary analytic number theory. In R. A. Mollin, editor, *Number Theory and Applications*, Vol. 265 of *NATO ASI Series C: Mathematical and Physical Sciences*, pp. 135–161. Kluwer, 1989.

[Pomerance 1993]
C. Pomerance. Carmichael numbers. *Nieuw Arch. Wiskunde* **11** (1993), 199–209.

[Pomerance, Selfridge, and Wagstaff 1980]
C. Pomerance, J. L. Selfridge, and S. S. Wagstaff, Jr. The pseudoprimes to $25 \cdot 10^9$. *Math. Comp.* **35** (1980), 1003–1026.

[Pomerance, Smith, and Tuler 1988]
C. Pomerance, J. W. Smith, and R. Tuler. A pipeline architecture for factoring large integers with the quadratic sieve algorithm. *SIAM J. Comput.* **17** (1988), 387–403.

[van der Poorten and Rotkiewicz 1980]
A. J. van der Poorten and A. Rotkiewicz. On strong pseudoprimes in arithmetic progressions. *J. Austral. Math. Soc. Ser. A* **29** (1980), 316–321.

[Porter 1975]
J. W. Porter. On a theorem of Heilbronn. *Mathematika* **22** (1975), 20–28.

[Di Porto 1993]
A. Di Porto. Nonexistence of even Fibonacci pseudoprimes of the 1st kind. *Fibonacci Quart.* **31** (1993), 173–177.

[Di Porto and Filipponi 1989]
A. Di Porto and P. Filipponi. More on the Fibonacci pseudoprimes. *Fibonacci Quart.* **27** (1989), 232–242.

[Di Porto, Guida, and Montolivo 1992]
A. Di Porto, F. Guida, and E. Montolivo. Fast algorithm for finding primitive polynomials over $GF(q)$. *Electronics Letters* **28** (1992), 118–120.

[Post 1944]
E. L. Post. Recursively enumerable sets of positive integers and their decision problems. *Bull. Amer. Math. Soc.* **50** (1944), 284–316.

[Poulet 1938]
P. Poulet. Table des nombres composés verifiant le théorème de Fermat pour le module 2 jusqu'à 100.000.000. *Sphinx* **8** (1938), 42–52. Errata in *Math. Comp.* **25** (1971), 944–945; **26** (1972), 814.

[Prabhu and Bose 1979]
K. A. Prabhu and N. K. Bose. Number of irreducible q-ary polynomials in several variables with prescribed degrees. *IEEE Trans. Circ. Syst.* **CAS-26** (1979), 973–975.

[Prachar 1957]
K. Prachar. *Primzahlverteilung*, Vol. 91 of *Grundlehren der mathematischen Wissenschaften in Einzeldarstellungen*. Springer-Verlag, Berlin, 1957.

[Prachar 1961]
K. Prachar. Über die kleinste Primzahl einer arithmetischen Reihe. *J. Reine Angew. Math.* **206** (1961), 3–4.

[Prasad 1986]
T. V. S. R. V. Prasad. Primality testing and its implications. In K. S. Rao, editor, *Proc. Workshop on Mathematics of Computer Algorithms*, pp. A.6.1–A.6.6. Inst. Math. Sci., Madras, India, 1986.

[Pratt 1975]
V. R. Pratt. Every prime has a succinct certificate. *SIAM J. Comput.* **4** (1975), 214–220.

[Prešić 1970]
M. D. Prešić. A method for solving equations in finite fields. *Mat. Vesnik* **7** (1970), 507–509.

[Pritchard 1980]
P. Pritchard. On the prime example of programming. In J. M. Tobias, editor, *Proc. Symp. Language Design and*

Programming Methodology, Vol. 79 of *Lecture Notes in Computer Science*, pp. 85–94. Springer-Verlag, Berlin, 1980.

[Pritchard 1981]
P. Pritchard. A sublinear additive sieve for finding prime numbers. *Comm. ACM* **24** (1981), 18–23. Corrigendum in *Comm. ACM*, **24** (1981), 772.

[Pritchard 1982]
P. Pritchard. Explaining the wheel sieve. *Acta Infor.* **17** (1982), 477–485.

[Pritchard 1983a]
P. Pritchard. Fast compact prime number sieves (among others). *J. Algorithms* **4** (1983), 332–344.

[Pritchard 1983b]
P. Pritchard. A case study of number-theoretic computation: searching for primes in arithmetic progression. *Sci. Comput. Programming* **3** (1983), 37–63.

[Pritchard 1983c]
P. Pritchard. Eighteen primes in arithmetic progression. *Math. Comp.* **41** (1983), 697.

[Pritchard 1984]
P. Pritchard. Some negative results concerning prime number generators. *Comm. ACM* **27** (1984), 53–57.

[Pritchard 1985]
P. Pritchard. Long arithmetic progressions of primes: some old, some new. *Math. Comp.* **45** (1985), 263–267.

[Pritchard 1987]
P. Pritchard. Linear prime-number sieves: a family tree. *Sci. Comput. Programming* **9** (1987), 17–35.

[Pritchard 1994]
P. Pritchard. Improved incremental prime number sieves. In L. M. Adleman and M.-D. Huang, editors, *Algorithmic Number Theory, First International Symposium, ANTS-I*, Vol. 877 of *Lecture Notes in Computer Science*, pp. 280–288. Springer-Verlag, 1994.

[Pritchard, Moran, and Thyssen 1995]
P. Pritchard, A. Moran, and A. Thyssen. Twenty-two primes in arithmetic progression. *Math. Comp.* **64** (1995), 1337–1339.

[Prodinger and Tichy 1983]
H. Prodinger and R. F. Tichy. Über ein zahlentheoretisches Problem aus der Informatik. *Öster. Akad. Wiss. Math.-naturw. Klasse* **192** (1983), 385–396.

[Protasi and Talamo 1989]
M. Protasi and M. Talamo. On the number of arithmetical operations for finding Fibonacci numbers. *Theoret. Comput. Sci.* **64** (1989), 119–124.

[Proth 1878]
F. Proth. Théorèmes sur les nombres premiers. *C. R. Acad. Sci. Paris* **87** (1878), 926.

[C. Purdy and Purdy 1990a]
C. N. Purdy and G. B. Purdy. The area-time complexity of the greatest common divisor problem: a lower bound. *Inform. Process. Lett.* **34** (1990), 43–46.

[C. Purdy and Purdy 1990b]
C. N. Purdy and G. B. Purdy. Networks for greatest common divisor computations. *Congr. Numer.* **73** (1990), 125–132.

[G. Purdy, Terras, Terras, and Williams 1979]
G. Purdy, R. Terras, A. Terras, and H. Williams. Graphing L-functions of Kronecker symbols in the real part of the critical strip. *Math. Student* **47** (1979), 101–131.

[G. Purdy 1983]
G. B. Purdy. A carry-free algorithm for finding the greatest common divisor of two integers. *Comput. Math. with Appl.* **9** (1983), 311–316.

[Quesada, Pritchard, and James 1992]
A. R. Quesada, P. Pritchard, and R. E. James III. Technical correspondence. *Comm. ACM* **35** (1992), 11–14.

[Rabin 1960]
M. O. Rabin. Computable algebra, general theory and theory of computable fields. *Trans. Amer. Math. Soc.* **95** (1960), 341–360.

[Rabin 1976]
M. O. Rabin. Probabilistic algorithms. In J. F. Traub, editor, *Algorithms and Complexity: New Directions and Recent Results*, pp. 21–39. Academic Press, New York, 1976.

[Rabin 1980a]
M. O. Rabin. Probabilistic algorithms in finite fields. *SIAM J. Comput.* **9** (1980), 273–280.

[Rabin 1980b]
M. O. Rabin. Probabilistic algorithm for testing primality. *J. Number Theory* **12** (1980), 128–138.

[Rabin and Scott 1959]
M. O. Rabin and D. Scott. Finite automata and their decision problems. *IBM J. Res. Develop.* **3** (1959), 115–125.

[Rabin and Shallit 1986]
M. O. Rabin and J. O. Shallit. Randomized algorithms in number theory. *Comm. Pure Appl. Math.* **39**(S) (1986), 239–256.

[Rabinovitch 1912]
G. Rabinovitch. Eindeutigkeit der Zerlegung in Primzahlfaktoren in quadratischen Zahlkörpern. In E. W. Hobson and A. E. H. Love, editors, *Proc. 5th Int. Cong. of Mathematicians 1912*, Vol. 1, pp. 418–421. 1912.

[Rademacher 1973]
H. Rademacher. *Topics in Analytic Number Theory*, Vol. 169 of *Grundlehren der mathematischen Wissenschaften in Einzeldarstellungen*. Springer-Verlag, Berlin, 1973.

[Radke 1970]
C. E. Radke. The use of quadratic residue research. *Comm. ACM* **13** (1970), 103–105.

[Rados 1906]
G. Rados. Die Diskriminante der allgemeinen Kreisteilungsgleichung. *J. Reine Angew. Math.* **131** (1906), 49–55.

[Raik 1953]
A. E. Raik. Ural'skiĭ matematik Ivan Mikheevich Pervushin. *Istoriko-Matematicheskie Issledovaniya* **6** (1953), 535–572.

[Ram Murty 1988]
M. Ram Murty. Artin's conjecture for primitive roots. *Math. Intelligencer* **10** (1988), 59–67.

[Ramaré and Rumely 1994]
O. Ramaré and R. Rumely. Primes in arithmetic progressions. Technical Report 14, University of Georgia Mathematics Preprint Series, 1994.

[Randell 1982]
B. Randell, editor. *The Origins of Digital Computers*. Springer-Verlag, 1982. 3rd edition.

[Ranum 1911]
A. Ranum. The general term of a recurring series. *Bull. Amer. Math. Soc.* **17** (1911), 457–461.

[Rao and Skavantzos 1992]
P. B. Rao and A. Skavantzos. Efficient computation of the squaring operation in modular rings. *Electronics Letters* **28** (1992), 1628–1630.

[Rédei 1950]
L. Rédei. A short proof of a theorem of Št. Schwartz concerning finite fields. *Časopis pro Pěstování Matematiky a Fysiky* **75** (1950), 211–212.

[Redinbo 1979]
G. R. Redinbo. Finite field arithmetic on an array processor. *IEEE Trans. Comput.* **C-28** (1979), 461–471.

[Ree 1971]
R. Ree. Proof of a conjecture of S. Chowla. *J. Number Theory* **3** (1971), 210–212.

[Regimbal 1975]
S. Regimbal. An explicit formula for the *k*th prime number. *Math. Mag.* **48** (1975), 230–232.

[Reid 1953]
C. Reid. Perfect numbers. *Scientific American* **188**(3) (1953), 84–86.

[Reiner 1961]
I. Reiner. On the number of matrices with given characteristic polynomial. *Illinois J. Math.* **5** (1961), 324–329.

[Rényi and Turán 1958]
A. Rényi and P. Turán. On a theorem of Erdős-Kac. *Acta Arith.* **4** (1958), 71–84.

[Reuschle 1856]
K. G. Reuschle. *Mathematische Abhandlung, enhaltend: Neue Zahlentheoretische Tabellen*. Stuttgart, 1856.

[Ribenboim 1985]
P. Ribenboim. Representation of real numbers by means of Fibonacci numbers. *Enseign. Math.* **31** (1985), 249–259.

[Ribenboim 1988a]
P. Ribenboim. Euler's famous prime generating polynomial and the class number of imaginary quadratic fields. *Enseign. Math.* **34** (1988), 23–42.

[Ribenboim 1988b]
P. Ribenboim. *The Book of Prime Number Records*. Springer-Verlag, 1988.

[Ribenboim 1994]
P. Ribenboim. Prime number records. *Nieuw Arch. Wiskunde* **12** (1994), 53–65.

[Richards 1974]
I. Richards. On the incompatibility of two conjectures concerning primes; a discussion of the use of computers in attacking a theoretical problem. *Bull. Amer. Math. Soc.* **80** (1974), 419–438.

[Richards 1982]
I. Richards. The invisible prime factor. *Amer. Scientist* **70** (1982), 176–179.

[Riemann 1860]
B. Riemann. Ueber die Anzahl der Primzahlen unter einer gegebenen Grosse. *Monatsberichte der Königlichen Preussischen Akademie der Wissenschaften zu Berlin* (1860), 671–680. Reprinted in *Gessammelte Werke*, 2nd edition, pp. 145–153. English translation in Edwards [1974], pp. 299–305.

[Riesel 1958a]
H. Riesel. A new Mersenne prime. *Math. Tables Aids Comput.* **12** (1958), 60.

[Riesel 1958b]
H. Riesel. Mersenne numbers. *Math. Tables Aids Comput.* **12** (1958), 207–213.

[Riesel 1969]
H. Riesel. Lucasian criteria for the primality of $N = h \cdot 2^n - 1$. *Math. Comp.* **23** (1969), 869–875.

[Riesel 1970]
H. Riesel. Primes forming arithmetic series and clusters of large primes. *BIT* **10** (1970), 333–342.

[Riesel 1985a]
H. Riesel. *Prime Numbers and Computer Methods for Factorization*. Birkhäuser, 1985. Revised edition, 1994.

[Riesel and Göhl 1970]
H. Riesel and G. Göhl. Some calculations related to Riemann's prime number formula. *Math. Comp.* **24** (1970), 969–983.

[Rifà and Borrell 1991]
J. Rifà and J. Borrell. Improving the time complexity of the computation of irreducible and primitive polynomials in finite fields. In H. F. Mattson, T. Mora, and T. R. N. Rao, editors, *Applied Algebra, Algebraic Algorithms and Error-Correcting Codes: AAECC-9*, Vol. 539 of *Lecture Notes in Computer Science*, pp. 352–359. Springer-Verlag, 1991.

[Rivest 1991]
R. L. Rivest. Finding four million large random primes. In A. J. Menezes and S. A. Vanstone, editors, *Advances in Cryptology–CRYPTO '90 Proceedings*, Vol. 537 of *Lecture Notes in Computer Science*, pp. 625–626. Springer-Verlag, 1991.

[Rivest, Shamir, and Adleman 1978]
R. L. Rivest, A. Shamir, and L. Adleman. A method for obtaining digital signatures and public-key cryptosystems. *Comm. ACM* **21** (1978), 120–126.

[Robin 1983]
G. Robin. Estimation de la fonction de Tchebychef Θ sur le k-ième nombre premier et grandes valeurs de la fonction $\omega(n)$ nombre de diviseurs premiers de n. *Acta Arith.* **42** (1983), 367–389.

[Robinson 1940]
R. M. Robinson. *Stencils for Solving* $x^2 \equiv a \pmod{m}$. University of California Press, Berkeley, 1940.

[Robinson 1954]
R. M. Robinson. Mersenne and Fermat numbers. *Proc. Amer. Math. Soc.* **5** (1954), 842–846.

[Rogers 1967]
H. Rogers, Jr. *Theory of Recursive Functions and Effective Computability*. McGraw-Hill, New York, 1967. Reprinted by MIT Press, 1987.

[Rolletschek 1986]
H. Rolletschek. On the number of divisions of the Euclidean algorithm applied to Gaussian integers. *J. Symbolic Comput.* **2** (1986), 261–291.

[Rolletschek 1990]
H. Rolletschek. Shortest division chains in imaginary quadratic number fields. *J. Symbolic Comput.* **9** (1990), 321–354.

[Rónyai 1987]
L. Rónyai. Factoring polynomials over finite fields. In *Proc. 28th Ann. Symp. Found. Comput. Sci.*, pp. 132–137. IEEE Press, 1987.

[Rónyai 1989a]
L. Rónyai. Factoring polynomials modulo special primes. *Combinatorica* **9** (1989), 199–206.

[Rónyai 1989b]
L. Rónyai. Galois groups and factoring polynomials over finite fields. In *Proc. 30th Ann. Symp. Found. Comput. Sci.*, pp. 99–104. IEEE Press, 1989.

[Rónyai 1992]
L. Rónyai. Galois groups and factoring polynomials over finite fields. *SIAM J. Disc. Math.* **5** (1992), 345–365.

[Root and Karst 1973]
S. C. Root and E. Karst. Mehr Teilfolgen von Primzahlen in arithmetischer Progression. *Anz. Österreich. Akad. Wiss. (Math.-Natur. Kl.)* **109** (1973), 178–179.

[M. Rosen 1988]
M. I. Rosen. A proof of the Lucas-Lehmer test. *Amer. Math. Monthly* **95** (1988), 855–856.

[Rosser 1939]
J. B. Rosser. The n-th prime is greater than $n \log n$. *Proc. Lond. Math. Soc.* **45** (1939), 21–44.

[Rosser 1941]
J. B. Rosser. Explicit bounds for some functions of prime numbers. *Amer. J. Math.* **63** (1941), 211–232.

[Rosser 1942]
J. B. Rosser. A generalization of the Euclidean algorithm to several dimensions. *Duke Math. J.* **9** (1942), 59–95.

[Rosser 1949]
J. B. Rosser. Real roots of Dirichlet L-series. *Bull. Amer. Math. Soc.* **55** (1949), 906–913.

[Rosser 1950]
J. B. Rosser. Real roots of real Dirichlet L-series. *J. Res. Nat. Bur. Standards* **45** (1950), 505–514.

[Rosser 1963]
J. B. Rosser. Unexpected dividends in the theory of prime numbers. In *Experimental Arithmetic, High Speed Computing, and Mathematics*, Vol. 15 of *Proc. Symp. Appl. Math.*, pp. 259–268. 1963.

[Rosser 1984]
J. B. Rosser. Highlights of the history of the lambda-calculus. *Ann. Hist. Comput.* **6** (1984), 337–349.

[Rosser and Schoenfeld 1962]
J. B. Rosser and L. Schoenfeld. Approximate formulas for some functions of prime numbers. *Ill. J. Math.* **6** (1962), 64–94.

[Rosser and Schoenfeld 1975]
J. B. Rosser and L. Schoenfeld. Sharper bounds for the Chebyshev functions $\theta(x)$ and $\psi(x)$. *Math. Comp.* **29** (1975), 243–269.

[Rosser, Yohe, and Schoenfeld 1969]
J. B. Rosser, J. M. Yohe, and L. Schoenfeld. Rigorous computation and the zeros of the Riemann zeta-function. In *Information Processing 68*, pp. 70–76. North-Holland, 1969.

[Rothstein and Zassenhaus 1994]
M. Rothstein and H. Zassenhaus. Deterministic analysis of aleatoric methods of polynomial factorization over finite fields. *J. Number Theory* **47** (1994), 20–42.

[Rotkiewicz 1962]
A. Rotkiewicz. Sur les nombres premiers p et q tels que $pq \mid 2^{pq} - 2$. *Rend. Circ. Mat. Palermo* **11** (1962), 280–282.

[Rotkiewicz 1963a]
A. Rotkiewicz. Sur les nombres pseudopremiers de la forme $ax + b$. *C. R. Acad. Sci. Paris* **257** (1963), 2601–2604.

[Rotkiewicz 1963b]
A. Rotkiewicz. Sur les nombres composés tels que $n \mid 2^n - 2$ et $n \nmid 3^n - 3$. *Bull. Soc. Mathématiciens et Physiciens de la R. S. de Serbie* **15** (1963), 7–11.

[Rotkiewicz 1964a]
A. Rotkiewicz. Sur les formules donnant des nombres pseudopremiers. *Colloq. Math.* **12** (1964), 69–72.

[Rotkiewicz 1964b]
A. Rotkiewicz. Sur les progressions arithmétiques et géométriques formées de trois nombres pseudopremiers distincts. *Acta Arith.* **10** (1964), 325–328.

[Rotkiewicz 1964c]
A. Rotkiewicz. Une formule explicite pour un nombre pseudopremier par rapport à un nombre entier donné $a > 1$. *Elem. Math.* **19** (1964), 36.

[Rotkiewicz 1964d]
A. Rotkiewicz. Une généralisation d'un théorème de Cipolla. *Glasnik Math.-Fiz. i Astr.* **19** (1964), 187–188.

[Rotkiewicz 1964e]
A. Rotkiewicz. Sur les nombres naturels n et k tels que les nombres n et nk sont à la fois pseudopremiers. *Rendiconta della Classe di Scienze Fisiche, Matematiche e Naturali, Accademia Nazionale dei Lincei* **36** (1964), 816–818.

[Rotkiewicz 1964f]
A. Rotkiewicz. Remarque sur un théorème de F. Proth. *Mat. Vesnik* **1** (1964), 244–245.

[Rotkiewicz 1964g]
A. Rotkiewicz. Sur les polynômes en x qui pour une infinité de nombres naturels x donnent des nombres pseudopremiers. *Rendiconta della Classe di Scienze Fisiche, Matematiche e Naturali, Accademia Nazionale dei Lincei* **36** (1964), 136–140.

[Rotkiewicz 1967]
A. Rotkiewicz. On the pseudoprimes of the form $ax + b$. *Proc. Cambridge Phil. Soc.* **63** (1967), 389–392.

[Rotkiewicz 1969]
A. Rotkiewicz. On arithmetical progressions formed by k different pseudoprimes. *J. Mathematical Sciences* **4** (1969), 5–10.

[Rotkiewicz 1972a]
A. Rotkiewicz. On some problems of W. Sierpiński. *Acta Arith.* **21** (1972), 251–259.

[Rotkiewicz 1972b]
A. Rotkiewicz. On the number of pseudoprimes $\leq x$. *Publ. Elektrotehn. Fak. Univ. Beograd* (389) (1972), 43–45.

[Rotkiewicz 1972c]
A. Rotkiewicz. *Pseudoprime Numbers and Their Generalizations*. University of Novi Sad, Novi Sad, Yugoslavia, 1972.

[Rotkiewicz 1973]
A. Rotkiewicz. On the pseudoprimes with respect to Lucas sequences. *Bull. Acad. Polon. Sci., Sér. Sci. Math. Astr. Phys.* **21** (1973), 793–797.

[Rotkiewicz 1979]
A. Rotkiewicz. The solution of W. Sierpiński's problem. *Rend. Circ. Mat. Palermo* **28** (1979), 62–64.

[Rotkiewicz 1980]
A. Rotkiewicz. Arithmetical progressions formed from three different Euler pseudoprimes for the odd base a. *Rend. Circ. Mat. Palermo* **29** (1980), 420–426.

[Rotkiewicz 1982]
A. Rotkiewicz. On Euler Lehmer pseudoprimes and strong Lehmer pseudoprimes with parameters L, Q in arithmetic progressions. *Math. Comp.* **39** (1982), 239–247.

[Rotkiewicz and Wasén 1979]
A. Rotkiewicz and R. Wasén. On a numbertheoretical series. *Publ. Math. (Debrecen)* **26** (1979), 1–4.

[Rougon and Dussaud 1981]
C. Rougon and R. Dussaud. Sur les nombres de Carmichael. *Publ. du Centre de Recherches en Mathématiques Pures, Neuchâtel* **16** (1981), 1–4.

[Rousseau 1994]
G. Rousseau. On the Jacobi symbol. *J. Number Theory* **48** (1994), 109–111.

[Royden 1968]
H. L. Royden. *Real Analysis*. MacMillan, New York, 1968.

[Rubinstein 1993]
M. Rubinstein. A simple heuristic proof of Hardy and Littlewood's Conjecture B. *Amer. Math. Monthly* **100** (1993), 456–460.

[Rudd, Buell, and Chiarulli 1984]
W. G. Rudd, D. A. Buell, and D. M. Chiarulli. A high performance factoring machine. In *Proc. Eleventh International Symposium on Computer Architecture*, pp. 297–300. IEEE Press, 1984.

[Rudin 1964]
W. Rudin. *Principles of Mathematical Analysis*. McGraw-Hill, 1964.

[Rumely 1983]
R. Rumely. Recent advances in primality testing. *Notices Amer. Math. Soc.* **30** (1983), 475–477.

[Rumely 1993]
R. Rumely. Numerical computations concerning the ERH. *Math. Comp.* **61** (1993), 415–440.

[Rychlik 1924]
K. Rychlik. Zur Bewertungstheorie der algebraischen Körper. *J. Reine Angew. Math.* **153** (1924), 94–107.

[Salié 1949]
H. Salié. Über den kleinsten positiven quadratischen Nichtrest nach einer Primzahl. *Math. Nachrichten* **3** (1949), 7–8.

[Samuel 1971]
P. Samuel. About Euclidean rings. *J. Algebra* **19** (1971), 282–301.

[Sanderson 1911]
M. Sanderson. Generalizations in the theory of numbers and theory of linear groups. *Ann. Math.* **13** (1911), 36–39.

[Santos 1969]
E. S. Santos. Probabilistic Turing machines and computability. *Proc. Amer. Math. Soc.* **22** (1969), 704–710.

[Sarrus 1819]
F. Sarrus. Démonstration de la fausseté du théorème énoncé à la page 320 du *IX.*^{e} volume de ce receuil. *Annales de Math. Pure Appl.* **10** (1819-20), 184–187.

[Sauerbrey 1993]
J. Sauerbrey. A modular exponentiation unit based on systolic arrays. In J. Seberry and Y. Zheng, editors, *Advances in Cryptology–AUSCRYPT '92*, Vol. 718 of *Lecture Notes in Computer Science*, pp. 505–516. Springer-Verlag, 1993.

[Sauerbrey and Dietel 1993]
J. Sauerbrey and A. Dietel. Resource requirements for the application of addition chains in modulo exponentiation. In R. A. Rueppel, editor, *Advances in Cryptology–EUROCRYPT '92 Proceedings*, Vol. 658 of *Lecture Notes in Computer Science*, pp. 174–182. Springer-Verlag, 1993.

[Schaefer 1975]
P. Schaefer. An algorithm for finding the GCD of a finite set of integers. *Delta (Waukesha)* **5** (1975), 19–27.

[Schinzel 1961]
A. Schinzel. Remarks on the paper "Sur certaines hypothèses concernant les nombres premiers". *Acta Arith.* **7** (1961), 1–8.

[Schinzel 1963]
A. Schinzel. A remark on a paper of Bateman and Horn. *Math. Comp.* **17** (1963), 445–447.

[Schinzel and Sierpiński 1958]
A. Schinzel and W. Sierpiński. Sur certaines hypothèses concernant les nombres premiers. *Acta Arith.* **4** (1958), 185–208.

[F. Schmidt 1928]
F. K. Schmidt. Zur Zahlentheorie in Körpern von der Characteristik p. *Sitzungsberichte der Physikalish-medizinischen Societät zu Erlangen* **58-59** (1928), 159–172.

[Schoenfeld 1976]
L. Schoenfeld. Sharper bounds for the Chebyshev functions $\theta(x)$ and $\psi(x)$. II. *Math. Comp.* **30** (1976), 337–360. Corrigenda in *Math. Comp.* **30** (1976), 900.

[Scholz 1937]
A. Scholz. Aufgabe 253. *Jahres. Deutscher Math.-Verein.* **47** (1937), Supplement, pp. 41–42.

[Schönhage 1971]
A. Schönhage. Schnelle Berechnung von Kettenbruchentwicklungen. *Acta Infor.* **1** (1971), 139–144.

[Schönhage 1975]
A. Schönhage. A lower bound for the length of addition chains. *Theoret. Comput. Sci.* **1** (1975), 1–12.

[Schönhage 1977]
A. Schönhage. Schnelle Multiplikation von Polynomen über Körpern der Charakteristik 2. *Acta Infor.* **7** (1977), 395–398.

[Schönhage 1986]
A. Schönhage. Tapes versus pointers, a study in implementing fast algorithms. *Bull. European Assoc. Theor. Comput. Sci.* (30) (1986), 23–32.

[Schönhage, Grotefeld, and Vetter 1994]
A. Schönhage, A. F. W. Grotefeld, and E. Vetter. *Fast Algorithms: A Multitape Turing Machine Implementation*. BI-Wissenschaftsverlag, Mannheim, 1994.

[Schönhage and Strassen 1971]
A. Schönhage and V. Strassen. Schnelle Multiplikation großer Zahlen. *Computing* **7** (1971), 281–292.

[Schoof 1985]
R. Schoof. Elliptic curves over finite fields and the computation of square roots mod p. *Math. Comp.* **44** (1985), 483–494.

[Schrijver 1986]
A. Schrijver. *Theory of Linear and Integer Programming*. Wiley, 1986.

[Schroeder 1983]
M. R. Schroeder. Where is the next Mersenne prime hiding? *Math. Intelligencer* **5**(3) (1983), 31–33.

[Schroeder 1984]
M. R. Schroeder. *Number Theory in Science and Communication*. Springer-Verlag, Berlin, 1984.

[Schützenberger 1968]
M. P. Schützenberger. A remark on acceptable sets of numbers. *J. Assoc. Comput. Mach.* **15** (1968), 300–303.

[Schwarz 1940]
Š. Schwarz. Sur le nombre des racines et des facteurs irréductibles d'une congruence donnée. *Časopis pro Pěstování Matematiky a Fysiky* **69** (1940), 128–145.

[Schwarz 1946]
Š. Schwarz. Příspěvek k reducibilitě binomických kongruencí. *Časopis pro Pěstování Matematiky a Fysiky* **71** (1946), 21–31. In Czech. French summary entitled "Contribution à la reductibilité des congruences binomiques".

[Schwarz 1949]
Š. Schwarz. On the reducibility of binomial congruences and on the bound of the least integer belonging to a given exponent mod *p*. *Časopis pro Pěstování Matematiky a Fysiky* **74** (1949), 1–16.

[Schwarz 1956]
Š. Schwarz. On the reducibility of polynomials over a finite field. *Quart. J. Math. Oxford* **(2) 7** (1956), 110–124.

[Schwarz 1988]
Š. Schwarz. Construction of normal bases in cyclic extensions of a field. *Czech. Math. J.* **38** (1988), 291–312.

[Schweber 1986]
S. Schweber. Feynman and the visualization of space-time processes. *Rev. Mod. Phys.* **58** (1986), 449–508.

[Scott, Simmons, Tavares, and Peppard 1988]
P. A. Scott, S. J. Simmons, S. E. Tavares, and L. E. Peppard. Architectures for exponentiation in $GF(2^m)$. *IEEE J. Selected Areas Commun.* **6** (1988), 578–586.

[Seelhoff 1886]
P. Seelhoff. Die neunte vollkommene Zahl. *Z. Math. Phys.* **31** (1886), 174–178.

[Séguin 1990]
G. E. Séguin. Low complexity normal bases for $F_{2^{mn}}$. *Disc. Appl. Math.* **28** (1990), 309–312.

[Selberg 1943]
A. Selberg. On the normal density of primes in small intervals, and the difference between consecutive primes. *Archiv for Mathematik og Naturvidenskab* **47** (1943), 87–105.

[Selfridge and Guy 1971]
J. L. Selfridge and R. K. Guy. Primality testing with application to small machines. In *Proc. Wash. State Univ. Conf. on Number Theory*, pp. 45–51, 1971.

[Selfridge and Hurwitz 1964]
J. L. Selfridge and A. Hurwitz. Fermat numbers and Mersenne numbers. *Math. Comp.* **18** (1964), 146–148.

[Semaev 1988]
I. A. Semaev. Construction of polynomials irreducible over a finite field with linearly independent roots. *Mat. Sbornik* **135** (1988), 520–532. In Russian. English translation in *Math. USSR Sbornik* **63** (1989), 507–519.

[Semba 1983]
I. Semba. Systematic method for determining the number of multiplications required to compute x^m, where m is a positive integer. *J. Information Processing* **6** (1983), 31–33.

[Serre 1979]
J.-P. Serre. *Local Fields*. Springer-Verlag, 1979.

[Serre 1981]
J.-P. Serre. Quelques applications du théorème de densité de Chebotarev. *Publ. Math. IHES* **54** (1981), 323–401. Reprinted in *Oeuvres*, Vol. III, pp. 563–641.

[Serret 1852]
J.-A. Serret. Théorème d'arithmétique. *Nouvelles Annales de Mathématiques* **11** (1852), 414–416.

[Serret 1885]
J.-A. Serret. *Cours d'Algèbre Supérieure*. Gauthier-Villars, Paris, 1885.

[Sexton 1955]
C. R. Sexton. Abzählung von 'Vierlingen' von 1,000.000 bis 2,000.000. *Anz. Österreich. Akad. Wiss. (Math.-Natur. Kl.)* **92** (1955), 236–239.

[Shallit 1979a]
J. O. Shallit. Simple continued fractions for some irrational numbers. *J. Number Theory* **11** (1979), 209–217.

[Shallit 1979b]
J. O. Shallit. Integer functions and continued fractions. A.B. Thesis, Princeton University, 1979.

[Shallit 1982]
J. O. Shallit. Simple continued fractions for some irrational numbers, II. *J. Number Theory* **14** (1982), 228–231.

[Shallit 1986]
J. O. Shallit. Metric theory of Pierce expansions. *Fibonacci Quart.* **24** (1986), 22–40.

[Shallit 1990]
J. O. Shallit. On the worst case of three algorithms for computing the Jacobi symbol. *J. Symbolic Comput.* **10** (1990), 593–610.

[Shallit 1991]
J. O. Shallit. Problem E 3431. *Amer. Math. Monthly* **98** (1991), 264.

[Shallit 1994]
J. O. Shallit. Origins of the analysis of the Euclidean algorithm. *Historia Mathematica* **21** (1994), 401–419.

[Shallit and Sorenson 1993]
J. O. Shallit and J. Sorenson. A binary algorithm for the Jacobi symbol. *ACM SIGSAM Bull.* **27**(1) (1993), 4–11.

[Shallit and Sorenson 1994]
J. O. Shallit and J. Sorenson. Analysis of a left-shift binary GCD algorithm. *J. Symbolic Comput.* **17** (1994), 473–486.

[Shallit, Williams, and Morain 1995]
J. O. Shallit, H. C. Williams, and F. Morain. Discovery of a lost factoring machine. *Math. Intelligencer* **17**(3) (1995), 41–47.

[Shand, Bertin, and Vuillemin 1990]
M. Shand, P. Bertin, and J. Vuillemin. Hardware speedups in long integer multiplication. In *Proc. 2nd Ann. ACM Symp. Parallel Algorithms and Architectures*, pp. 138–145. ACM, 1990.

[Shanks 1960]
D. Shanks. A note on Gaussian twin primes. *Math. Comp.* **14** (1960), 201–203.

[Shanks 1961]
D. Shanks. On numbers of the form $n^4 + 1$. *Math. Comp.* **15** (1961), 186–189. Corrigendum in **16** (1962), 513.

[Shanks 1962]
D. Shanks. An inductive formulation of the Riemann hypothesis. In *Abstracts of Short Communications, International Congress of Mathematicians*, pp. 51–52. Almqvist and Wiksells, Uppsala, 1962.

[Shanks 1963]
D. Shanks. Supplementary data and remarks concerning a Hardy-Littlewood conjecture. *Math. Comp.* **17** (1963), 188–193.

[Shanks 1964]
D. Shanks. On maximal gaps between successive primes. *Math. Comp.* **18** (1964), 646–651.

[Shanks 1965]
D. Shanks. Review of F. Gruenberger and G. Armerding, Statistics on the First Six Million Prime Numbers. *Math. Comp.* **19** (1965), 503–505.

[Shanks 1969]
D. Shanks. Class number, a theory of factorization, and genera. In *Number Theory Institute, 1969*, Vol. 20 of *Proceedings of Symposia in Pure Mathematics*, pp. 415–440. Amer. Math. Soc., 1969.

[Shanks 1972]
D. Shanks. Five number-theoretic algorithms. In *Proc. 2nd Southeastern Conference on Combinatorics, Graph Theory, and Computing*, pp. 51–70, Winnipeg, 1972. Utilitas Mathematica. (= *Congr. Numer.* **VII**).

[Shanks 1973]
D. Shanks. Systematic examination of Littlewood's bounds on $L(1, \chi)$. In H. G. Diamond, editor, *Analytic Number Theory*, Vol. 24 of *Proc. Symp. Pure Math.*, pp. 267–283. Amer. Math. Soc., 1973.

[Shanks 1978]
D. Shanks. *Solved and Unsolved Problems in Number Theory*. Chelsea Publishing Company, 2nd edition, 1978.

[Shanks 1985]
D. Shanks. *Solved and Unsolved Problems in Number Theory*. Chelsea Publishing Company, 3rd edition, 1985.

[Shanks 1990a]
D. Shanks. Review of S. S. Wagstaff, Jr., Table of all Carmichael numbers $< 25 \cdot 10^9$. *Math. Comp.* **55** (1990), 404–405.

[Shanks 1990b]
D. Shanks. Review of G. Jaeschke, Table of all Carmichael numbers $< 10^{12}$. *Math. Comp.* **55** (1990), 405.

[Shanks and Kravitz 1967]
D. Shanks and S. Kravitz. On the distribution of Mersenne divisors. *Math. Comp.* **21** (1967), 97–101.

[Shannon 1948]
C. E. Shannon. A mathematical theory of communication (part I). *Bell System Tech. J.* **27** (1948), 379–423.

[Shapiro 1946]
H. N. Shapiro. Some assertions equivalent to the prime number theorem for arithmetic progressions. *Comm. Pure Appl. Math.* **2** (1949), 293–308.

[Shawe-Taylor 1986]
J. Shawe-Taylor. Generating strong primes. *Electronics Letters* **22** (1986), 875–877.

[Shawe-Taylor 1992]
J. Shawe-Taylor. Proportion of primes generated by strong prime methods. *Electronics Letters* **28** (1992), 135–137.

[Shayan and Le-Ngoc 1988]
Y. R. Shayan and T. Le-Ngoc. Direct hardware solution to quadratic equation $Z^2 \oplus Z \oplus \beta = 0$ in Galois fields based on normal basis representation. *Electronics Letters* **24** (1988), 847–848.

[Shea 1973]
D. D. Shea. On the number of divisions needed in finding the greatest common divisor. *Fibonacci Quart.* **11** (1973), 508–510.

[Shepp and Lloyd 1966]
L. A. Shepp and S. P. Lloyd. Ordered cycle lengths in a random permutation. *Trans. Amer. Math. Soc.* **121** (1966), 340–357.

[Shiu 1986]
P. Shiu. Counting sums of two squares: the Meissel-Lehmer method. *Math. Comp.* **47** (1986), 351–360.

[Shiu 1987]
P. Shiu. Counting prime numbers on a computer. *Bull. Inst. Math. Appl.* **23** (1987), 89–92.

[Shiva and Allard 1970]
S. G. S. Shiva and P. E. Allard. A few useful details about a known technique for factoring $1 + X^{2q-1}$. *IEEE Trans. Inform. Theory* **IT-16** (1970), 234–235.

[Shlesinger 1986]
M. F. Shlesinger. On the Riemann hypothesis: a fractal random walk approach. *Physica A* **138** (1986), 310–319.

[Shokrollahi 1991]
M. A. Shokrollahi. On the rank of certain finite fields. *Computational Complexity* **1** (1991), 157–181.

[Shokrollahi 1992a]
M. A. Shokrollahi. Efficient randomized generation of optimal algorithms for multiplication in certain finite fields. *Computational Complexity* **2** (1992), 67–96.

[Shokrollahi 1992b]
M. A. Shokrollahi. Optimal algorithms for multiplication in certain finite fields using elliptic curves. *SIAM J. Comput.* **21** (1992), 1193–1198.

[Shortt 1978]
J. Shortt. An iterative program to calculate Fibonacci numbers in $O(\log n)$ arithmetic operations. *Inform. Process. Lett.* **7** (1978), 299–303.

[Shoup 1988]
V. Shoup. New algorithms for finding irreducible polynomials over finite fields. In *Proc. 29th Ann. Symp. Found. Comput. Sci.*, pp. 283–290. IEEE Press, 1988.

[Shoup 1989a]
V. Shoup. *Removing randomness from computational number theory*. PhD thesis, Computer Sciences Department, University of Wisconsin, 1989.

[Shoup 1990a]
V. Shoup. On the deterministic complexity of factoring polynomials over finite fields. *Inform. Process. Lett.* **33** (1990), 261–267.

[Shoup 1990b]
V. Shoup. Searching for primitive roots in finite fields. In *Proc. Twenty-second Ann. ACM Symp. Theor. Comput.*, pp. 546–554. ACM Press, 1990.

[Shoup 1990c]
V. Shoup. New algorithms for finding irreducible polynomials over finite fields. *Math. Comp.* **54** (1990), 435–447.

[Shoup 1991a]
V. Shoup. Smoothness and factoring polynomials over finite fields. *Inform. Process. Lett.* **38** (1991), 39–42.

[Shoup 1991b]
V. Shoup. A fast deterministic algorithm for factoring polynomials over finite fields of small characteristic. In S. M. Watt, editor, *Proc. Int'l. Symp. on Symbolic and Algebraic Computation: ISSAC '91*, pp. 14–21. ACM Press, 1991.

[Shoup 1992]
V. Shoup. Searching for primitive roots in finite fields. *Math. Comp.* **58** (1992), 369–380.

[Shoup 1993a]
V. Shoup. Fast construction of irreducible polynomials over finite fields. In *Proc. 4th ACM-SIAM Symp. Discrete Algorithms (SODA)*, pp. 484–492. ACM, 1993.

[Shoup 1993b]
V. Shoup. Factoring polynomials over finite fields: asymptotic complexity vs. reality. In *Proc. IMACS Symposium (Lille, France)*, 1993.

[Shoup 1994]
V. Shoup. Fast construction of irreducible polynomials over finite fields. *J. Symbolic Comput.* **17** (1994), 371–391.

[Shparlinski 1991]
I. E. Shparlinski. On some problems in the theory of finite fields. *Uspekhi. Mat. Nauk* **46**(1) (1991), 165–200. In Russian. English translation in *Russian Math. Surveys* **46** (1) (1991), 199–240.

[Shparlinski 1992]
I. E. Shparlinski. *Computational and Algorithmic Problems in Finite Fields*. Kluwer Academic, Boston, 1992.

[Shparlinski 1993a]
I. E. Shparlinski. On bivariate polynomial factorization over finite fields. *Math. Comp.* **60** (1993), 787–791.

[Shparlinski 1993b]
I. E. Shparlinski. Finding irreducible and primitive polynomials. *Appl. Alg. Eng. Comm. Comp.* **4** (1993), 263–268.

[Shparlinski, Tsfasman, and Vladut 1992]
I. E. Shparlinski, M. A. Tsfasman, and S. G. Vladut. Curves with many points and multiplication over finite fields. In H. Stichtenoth and M. A. Tsfasman, editors, *Coding Theory and Algebraic Geometry*, pp. 145–169. Springer-Verlag, 1992.

[Sidel´nikov 1987]
V. M. Sidel´nikov. On normal bases of a finite field. *Mat. Sbornik* **133** (1987), 497–507. In Russian. English translation in *Math. USSR Sbornik* **61** (1988), 485–494.

[Siegel 1932]
C. L. Siegel. Über Riemanns Nachlaß zur analytischen Zahlentheorie. *Quel. Stud. Ges. Math. Astron. Phys.* **2** (1932), 45–80.

[Siegel 1935]
C. L. Siegel. Über die Classenzahl quadratischer Zahlkörper. *Acta Arith.* **1** (1935), 83–86. Reprinted in *Gesammelte Abhandlungen,* Vol. 1, pp. 406–409.

[Sierpiński 1947]
W. Sierpiński. Remarque sur une hypothèse des Chinois concernant les nombres $(2^n - 2)/n$. *Colloq. Math.* **1** (1947), 9.

[Sierpiński 1952]
W. Sierpiński. Sur une formule donnant tous les nombres premiers. *C. R. Acad. Sci. Paris* **235** (1952), 1078–1079.

[Sierpiński 1959]
W. Sierpiński. Sur les nombres dont la somme de diviseurs est une puissance du nombre 2. In *Golden Jubilee Commemoration Volume*, pp. 7–9. Calcutta Mathematical Society, Calcutta, India, 1959.

[Sierpiński 1964b]
W. Sierpiński. Sur un théorème de F. Proth. *Mat. Vesnik* **1** (1964), 243–244.

[Silverman 1991a]
R. D. Silverman. A perspective on computational number theory. *Notices Amer. Math. Soc.* **38** (1991), 562–568.

[Singleton 1969a]
R. C. Singleton. Algorithm 356: a prime number generator using the treesort principle. *Comm. ACM* **12** (1969), 563.

[Singleton 1969b]
R. C. Singleton. Algorithm 357: an efficient prime number generator. *Comm. ACM* **12** (1969), 563–564.

[Singmaster 1983]
D. Singmaster. Some Lucas pseudo-primes. *Abstracts Amer. Math. Soc.* **4**(2) (1983), 197. Abstract 83T-10-146.

[Sinisalo 1993]
M. K. Sinisalo. Checking the Goldbach conjecture up to $4 \cdot 10^{11}$. *Math. Comp.* **61** (1993), 931–934.

[Sipser 1992]
M. Sipser. The history and status of the P versus NP question. In *Proc. Twenty-fourth Ann. ACM Symp. Theor. Comput.*, pp. 603–618. ACM Press, 1992.

[Sispánov 1941]
S. Sispánov. Sobre los numeros pseudo-primos. *Boletin Matetmatico* **14** (1941), 99–106.

[Sispánov 1942]
S. Sispánov. La criba de Eratóstenes y la integral logarítmica de Tchebysheff. *Boletin Matematico* **15** (1942), 105–116.

[Skavantzos and Rao 1992]
A. Skavantzos and P. B. Rao. New multipliers modulo $2^N - 1$. *IEEE Trans. Comput.* **41** (1992), 957–961.

[Slisenko 1981]
A. O. Slisenko. Complexity problems in computational theory. *Uspekhi. Mat. Nauk* **36**(6) (1981), 21–103. In Russian. English translation in *Russian Math. Surveys* **36** (1981), 23–125.

[Slot and van Emde Boas 1988]
C. Slot and P. van Emde Boas. The problem of space invariance for sequential machines. *Inform. Comput.* **77** (1988), 93–122.

[Slowinski 1979]
D. Slowinski. Searching for the 27th Mersenne prime. *J. Recreational Math.* **11** (1978-9), 258–261.

[H. F. Smith 1956]
H. F. Smith. On a generalization of the prime pair problem. *Math. Tables Aids Comput.* **11** (1956), 249–254.

[H. J. S. Smith 1859]
H. J. S. Smith. *Report on the Theory of Numbers, Part I.* British Association, 1859. Reprinted by Chelsea, New York, 1965.

[H. J. S. Smith 1860]
H. J. S. Smith. *Report on the Theory of Numbers, Part II.* British Association, 1860. Reprinted by Chelsea, New York, 1965.

[J. Smith and Wagstaff 1983]
J. W. Smith and S. S. Wagstaff, Jr. An extended precision operand computer. In *Proc. 21st Southeast Region ACM Conference*, pp. 209–216, 1983.

[Sokolovskiĭ 1968]
A. V. Sokolovskiĭ. A theorem on the zeroes of Dedekind's zeta-function and the distance between "neighboring" prime ideals. *Acta Arith.* **13** (1968), 321–334. In Russian. Reviewed in *Math. Reviews* **36** (1968), #6380.

[Solovay and Strassen 1977]
R. Solovay and V. Strassen. A fast Monte-Carlo test for primality. *SIAM J. Comput.* **6** (1977), 84–85. Erratum in **7** (1978), 118.

[Somer 1989]
L. Somer. On Fermat d-pseudoprimes. In J.-M. de Koninck and C. Levesque, editors, *Number Theory (Proceedings of the International Number Theory Conference Held at Université Laval, July 5–18, 1987)*, pp. 841–860. Walter de Gruyter, Berlin, 1989.

[Somer 1991]
L. Somer. On even Fibonacci pseudoprimes. In G. E. Bergum, A. N. Philippou, and A. F. Horadam, editors, *Applications of Fibonacci Numbers*, Vol. 4, pp. 277–288. Kluwer, Boston, 1991.

[Sommerfeld 1956]
A. Sommerfeld. *Thermodynamics and Statistical Mechanics.* Academic Press, New York, 1956.

[Sorenson 1990b]
J. Sorenson. An introduction to prime number sieves. Technical Report 909, University of Wisconsin, Computer Sciences Department, January 1990.

[Sorenson 1991a]
J. Sorenson. An analysis of two prime number sieves. Technical Report 1028, University of Wisconsin, Computer Sciences Department, June 1991.

[Sorenson 1994]
J. Sorenson. Two fast GCD algorithms. *J. Algorithms* **16** (1994), 110–144.

[Sorenson 1995]
J. Sorenson. An analysis of Lehmer's Euclidean GCD algorithm. Unpublished manuscript, date January, 1995.

[Sorenson and Parberry 1994]
J. Sorenson and I. Parberry. Two fast parallel prime number sieves. *Inform. Control* **114** (1994), 115–130.

[Spearman and Williams 1975]
B. Spearman and K. S. Williams. Handbook of estimates in the theory of numbers. Carleton Mathematical Lecture Note 14, Carleton University, 1975.

[Spira 1961]
R. Spira. The complex sum of divisors. *Amer. Math. Monthly* **68** (1961), 120–124.

[Spira 1969]
R. Spira. Calculation of Dirichlet L-functions. *Math. Comp.* **23** (1969), 489–497.

[Spottiswoode 1853]
W. Spottiswoode. Elementary theorems relating to determinants. Rewritten and much enlarged by the author. *J. Reine Angew. Math.* **51** (1853), 209–271,328–381.

[Srinivasan 1961]
B. R. Srinivasan. Formulae for the nth prime. *J. Indian Math. Soc.* **25** (1961), 33–39.

[Stäckel 1916]
P. Stäckel. Die Darstellung der geraden Zahlen als Summen von zwei Primzahlen. *Sitz. Heidelberger Akad. Wiss. (Mat.-Natur. Kl.)* **7A**(10) (1916), 1–47.

[Stark 1975]
H. M. Stark. The analytic theory of algebraic numbers. *Bull. Amer. Math. Soc.* **81** (1975), 961–972.

[Stark 1987]
H. M. Stark. *An Introduction to Number Theory*. The MIT Press, Cambridge, 1987.

[Stechkin 1971]
S. B. Stechkin. Lucas's criterion for the primality of numbers of the form $N = h2^n - 1$. *Math. Notes* **10** (1971), 578–584.

[J. Stein 1967]
J. Stein. Computational problems associated with Racah algebra. *J. Comput. Phys.* **1** (1967), 397–405.

[M. Stein and Ulam 1967]
M. Stein and S. Ulam. An observation on the distribution of primes. *Amer. Math. Monthly* **74** (1967), 43. Reprinted in *Stanislaw Ulam: Sets, Numbers, and Universes*, p. 502.

[P. Stein 1974]
P. Stein. Commentary to paper [53]. In W. A. Beyer, J. Mycielski, and G.-C. Rota, editors, *Stanislaw Ulam: Sets, Numbers, and Universes*, pp. 690–695. The MIT Press, 1974.

[Steinitz 1912]
E. Steinitz. Rechteckige Systeme und Moduln in algebraischen Zahlkörpern. I. *Math. Ann.* **71** (1912), 328–354.

[Stepanov and Shparlinski 1991]
S. A. Stepanov and I. E. Shparlinskiy. On the construction of primitive elements and primitive normal bases in a finite field. In A. Pethö, M. E. Pohst, H. C. Williams, and H. G. Zimmer, editors, *Computational Number Theory*, pp. 1–14. Walter de Gruyter, Berlin, 1991.

[Stephens and Williams 1990]
A. J. Stephens and H. C. Williams. An open architecture number sieve. In J. H. Loxton, editor, *Number Theory and Cryptography*, Vol. 154 of *London Mathematical Society Lecture Note Series*, pp. 38–75. Cambridge University Press, 1990.

[Stevin 1585]
S. Stevin. *L'Arithmétique*. Plantin, 1585. Reprinted in E. Crone et al., editors, *The Principal Works of Simon Stevin,* Swets and Zeitlinger, Amsterdam, 1958.

[Stickelberger 1898]
L. Stickelberger. Über eine neue Eigenschaft der Diskriminanten algebraischer Zahlkörper. In *Verhandlungen des ersten internationalen Mathematiker-Kongresses*, pp. 182–193, Leipzig, 1898. Teubner.

[Stieltjes 1889]
T. J. Stieltjes. Sur le devéloppement de log $\Gamma(a)$. *J. Math. Pures Appl.* **5** (1889), 425–444. Reprinted in *Oeuvres Complètes*, Vol. 2, pp. 211–230.

[Stieltjes 1890]
T. J. Stieltjes. Sur la théorie des nombres. *Ann. Fac. Sci. Toulouse* **4** (1890), 1–103. Reprinted in *Oeuvres Complètes*, Vol. II, pp. 279–377.

[Stinson 1990]
D. R. Stinson. Some observations on parallel algorithms for fast exponentiation in $GF(2^n)$. *SIAM J. Comput.* **19** (1990), 711–717.

[Stockmeyer and Meyer 1973]
L. J. Stockmeyer and A. R. Meyer. Word problems requiring exponential time. In *Proc. Fifth Ann. ACM Symp. Theor. Comput.*, pp. 1–9. ACM Press, 1973.

[Strang 1980]
G. Strang. *Linear Algebra and its Applications*. Academic Press, New York, 1980.

[Strassen 1976]
V. Strassen. Einige Resultate über Berechnungskomplexität. *Jahresbericht der Deutsche Math. Verein.* **78** (1976), 1–8.

[Strassen 1983]
V. Strassen. The computational complexity of continued fractions. *SIAM J. Comput.* **12** (1983), 1–27.

[Straus 1964]
E. G. Straus. Solution to problem 5125. *Amer. Math. Monthly* **71** (1964), 807–808.

[Strong 1988]
R. Strong. Some asymptotic results on finite vector spaces. *Adv. in Appl. Math.* **9** (1988), 167–199.

[Sturtivant and Frandsen 1993]
C. Sturtivant and G. S. Frandsen. The computational efficacy of finite field arithmetic. *Theoret. Comput. Sci.* **112** (1993), 291–309.

[Subbarao 1989]
M. V. Subbarao. Addition chains—some results and problems. In R. A. Mollin, editor, *Number Theory and Applications*, Vol. 265 of *NATO ASI Series C: Mathematical and Physical Sciences*, pp. 555–574. Kluwer, 1989.

[Suryanarayana 1974]
D. Suryanarayana. On $\Delta(x, n) = \varphi(x, n) - x\varphi(n)/n$. *Proc. Amer. Math. Soc.* **44** (1974), 17–21.

[Sutherland and Mead 1977]
I. E. Sutherland and C. A. Mead. Microelectronics and computer science. *Scientific American* **237**(3) (1977), 210–229.

[Sutton 1937]
C. S. Sutton. An investigation of the average distribution of twin prime numbers. *J. Math. Phys.* **16** (1937), 1–42.

[Svoboda 1962]
A. Svoboda. The numerical system of residue classes. In W. Hoffman, editor, *Digital Information Processors*. Interscience, New York, 1962.

[Swan 1962]
R. Swan. Factorization of polynomials over finite fields. *Pacific J. Math.* **12** (1962), 1099–1106.

[Sweeney 1963]
D. W. Sweeney. On the computation of Euler's constant. *Math. Comp.* **17** (1963), 170–178. Corrigendum in **17** (1963), 488.

[Swift 1975]
J. D. Swift. Table of Carmichael numbers to 10^9. *Math. Comp.* **29** (1975), 338–339.

[Swinnerton-Dyer 1967]
P. Swinnerton-Dyer. The use of computers in number theory. In J. T. Schwartz, editor, *Mathematical Aspects of Computer Science, 1966*, Vol. 19 of *Proc. Symp. Appl. Math.*, pp. 111–116. Amer. Math. Soc., 1967.

[Sylvester 1840]
J. J. Sylvester. A method of determining by mere inspection the derivatives from two equations of any degree. *Philosophical Magazine* **16** (1840), 132–135. Reprinted in *Collected Mathematical Papers,* Vol. 1, pp. 54–57.

[Sylvester 1883]
J. J. Sylvester. On the number of fractions in their lowest terms whose numerators and denominators are limited not to exceed a certain number. *Johns Hopkins Univ. Circulars* **II** (1883), 44–45. Reprinted in *Collected Mathematical Papers*, Vol. 3, pp. 672–676.

[Sylvester 1888]
J. J. Sylvester. On certain inequalities relating to prime numbers. *Nature* **38** (1888), 259–262. Reprinted in *Collected Mathematical Papers*, Vol. 4, pp. 592–603.

[Szabo and Tanaka 1967]
N. S. Szabo and R. I. Tanaka. *Residue Arithmetic and its Applications to Computer Technology*. McGraw-Hill, New York, 1967.

[Szekeres 1974]
G. Szekeres. On the number of divisors of $x^2 + x + A$. *J. Number Theory* **6** (1974), 434–442.

[Szele 1947]
T. Szele. Über die endlichen Ordnungszahlen, zu denen nur eine Gruppe gehört. *Comment. Math. Helvetici* **20** (1947), 265–267.

[Szep 1947]
J. Szep. On finite groups which are necessarily commutative. *Comment. Math. Helvetici* **20** (1947), 223–224.

[Szymiczek 1965]
K. Szymiczek. On prime numbers p, q, and r such that pq, pr, and qr are pseudoprimes. *Colloq. Math.* **13** (1965), 259–263.

[Szymiczek 1967]
K. Szymiczek. On pseudoprimes which are products of distinct primes. *Amer. Math. Monthly* **74** (1967), 35–37.

[N. Takagi 1991]
N. Takagi. A radix-4 modular multiplication hardware algorithm efficient for iterative modular multiplication. In P. Kornerup and D. W. Matula, editors, *Proc. 10th IEEE Symp. Comp. Arith.*, pp. 35–42. IEEE Computer Society Press, 1991.

[N. Takagi 1992]
N. Takagi. A radix-4 modular multiplication hardware algorithm for modular exponentiation. *IEEE Trans. Comput.* **41** (1992), 949–956.

[N. Takagi 1993]
N. Takagi. A modular multiplication algorithm with triangle additions. In E. Swartzlander, Jr., M. J. Irwin, and G. Jullien, editors, *Proc. 11th IEEE Symp. Comp. Arith.*, pp. 272–276. IEEE Press, 1993.

[N. Takagi and Yajima 1992]
N. Takagi and S. Yajima. Modular multiplication hardware algorithms with a redundant representation and their application to RSA cryptosystems. *IEEE Trans. Comput.* **41** (1992), 887–891.

[T. Takagi 1920]
T. Takagi. Über eine Theorie des relativ Abel'schen Zahlkörpers. *Journal of the College of Science, Imperial University of Tokyo* **41** (1920), 1–133. Reprinted with corrigenda in *Collected Papers*, pp. 73–167.

[Tamarkine and Friedmann 1906]
J. Tamarkine and A. Friedmann. Sur les congruences du second degré et les nombres de Bernoulli. *Math. Ann.* **62** (1906), 409–412.

[Tanaka 1980]
M. Tanaka. A numerical investigation on cumulative sum of the Liouville function. *Tokyo J. Math.* **3** (1980), 187–189.

[Tanner and Wagstaff 1987]
J. W. Tanner and S. S. Wagstaff, Jr. New congruences for the Bernoulli numbers. *Math. Comp.* **48** (1987), 341–350.

[Tanner and Wagstaff 1989]
J. W. Tanner and S. S. Wagstaff, Jr. New bound for the first case of Fermat's last theorem. *Math. Comp.* **53** (1989), 743–750.

[Tarry 1907]
G. Tarry. Théorie des tables à triple entrée pour la recherche des facteurs premiers des nombres. *Assoc. Française pour l'Avancement des Sciences; Comptes Rendus* **36** (1907), 32–42.

[Taton and Flad 1963]
R. Taton and J.-P. Flad. *Le Calcul Mécanique*. Presses Universitaires de France, Paris, 1963.

[Taussky 1956]
O. Taussky. Some computational problems in algebraic number theory. In *Numerical Analysis*, Vol. 6 of *Proc. Symp. Appl. Math.*, pp. 187–193. 1956.

[Taylor and Wiles 1995]
R. Taylor and A. Wiles. Ring-theoretic properties of certain Hecke algebras. *Ann. Math.* **141** (1995), 553–572.

[Templer 1980]
M. Templer. On the primality of $k! + 1$ and $2 * 3 * 5 * \cdots * p + 1$. *Math. Comp.* **34** (1980), 303–304.

[Terrill and Sweeny 1946]
H. M. Terrill and L. Sweeny. Two constants connected with the theory of prime numbers. *J. Franklin Inst.* **239** (1946), 242–243.

[Thiong Ly 1989a]
J. A. Thiong Ly. A deterministic algorithm for factorizing polynomials over extensions $GF(p^m)$ of $GF(p)$, p a small prime. *J. Info. Optim. Sci.* **10** (1989), 337–344.

[Thiong Ly 1989b]
J. A. Thiong Ly. Note for computing the minimum polynomial of elements in large finite fields. In G. Cohen and J. Wolfmann, editors, *Coding Theory and Applications*, Vol. 388 of *Lecture Notes in Computer Science*, pp. 185–192, 1989.

[Thomas, Keller, and Larsen 1986]
J. J. Thomas, J. M. Keller, and G. N. Larsen. The calculation of multiplicative inverses over $GF(P)$ efficiently where P is a Mersenne prime. *IEEE Trans. Comput.* **C-35** (1986), 478–482.

[Thoro 1964a]
D. Thoro. The Euclidean algorithm I. *Fibonacci Quart.* **2** (1964), 53–56.

[Thoro 1964b]
D. Thoro. The Euclidean algorithm II. *Fibonacci Quart.* **2** (1964), 135–137.

[Thurber 1973a]
E. G. Thurber. The Scholz-Brauer problem on addition chains. *Pacific J. Math.* **49** (1973), 229–242.

[Thurber 1973b]
E. G. Thurber. On addition chains $\ell(mn) \leq \ell(n) - b$ and lower bounds for $c(r)$. *Duke Math. J.* **40** (1973), 907–913.

[Thurber 1976]
E. G. Thurber. Addition chains and solutions of $\ell(2n) = \ell(n)$ and $\ell(2^n - 1) = n + \ell(n) - 1$. *Discrete Math.* **16** (1976), 279–289.

[Thurber 1993]
E. G. Thurber. Addition chains – an erratic sequence. *Discrete Math.* **122** (1993), 287–305.

[Thurston 1943]
H. S. Thurston. The solution of p-adic equations. *Amer. Math. Monthly* **50** (1943), 142–148.

[Titchmarsh 1930]
E. C. Titchmarsh. A divisor problem. *Rend. Circ. Mat. Palermo* **54** (1930), 414–429.

[Titchmarsh 1935]
E. C. Titchmarsh. The zeros of the Riemann zeta-function. *Proc. Roy. Soc. London* **151** (1935), 234–255.

[Titchmarsh 1936]
E. C. Titchmarsh. The zeros of the Riemann zeta-function. *Proc. Roy. Soc. London* **157** (1936), 261–263.

[Titchmarsh 1986]
E. C. Titchmarsh. *The Theory of the Riemann Zeta-function.* Oxford University Press, Oxford, 2nd edition, 1986. Revised by D. R. Heath-Brown.

[Tompa 1991]
M. Tompa. Lecture notes on probabilistic algorithms and pseudorandom generators. Technical Report 91-07-05, Department of Computer Science and Engineering, University of Washington, July 1991.

[Tonelli 1891]
A. Tonelli. Bemerkung über die Auflösung quadratischer Congruenzen. *Göttinger Nachrichten* (1891), 344–346.

[Tonelli 1892]
A. Tonelli. Sulla risoluzione della congruenza $x^2 \equiv c \pmod{p^\lambda}$. *Atti R. Accad. Lincei* **1** (1892), 116–120.

[Tonelli 1893]
A. Tonelli. Sulla risoluzione della congruenza $x^2 \equiv c \pmod{p^\lambda}$. *Atti R. Accad. Lincei* **2** (1893), 259–265.

[Tonkov 1974]
T. Tonkov. On the average length of finite continued fractions. *Acta Arithmetica* **26** (1974), 47–57.

[Tôyama 1955]
H. Tôyama. A note on the different of the composed field. *Kodai Math. Sem. Report* **7** (1955), 43–44.

[Trevisan and Wang 1991]
V. Trevisan and P. Wang. Practical factorization of univariate polynomials over finite fields. In S. M. Watt, editor, *Proc. Int'l. Symp. on Symbolic and Algebraic Computation: ISSAC '91*, pp. 22–31. ACM Press, 1991.

[van Trigt 1978]
C. van Trigt. Worst-case analysis of algorithms. A. Some g.c.d algorithms. *Philips J. Res.* **33** (1978), 66–77.

[Tsai and Chin 1987]
Y. H. Tsai and Y. H. Chin. A study of some addition chain problems. *Internat. J. Comput. Math.* **22** (1987), 117–134.

[Tsangaris and Jones 1992]
P. G. Tsangaris and J. P. Jones. An old theorem on the GCD and its application to primes. *Fibonacci Quart.* **30** (1992), 194–198.

[T'u 1974]
K. C. T'u. The structure of Q-matrices and the reducibility of polynomials over a Galois field. *Acta Math. Sinica* **17** (1974), 46–59. In Chinese. Reviewed in *Math. Reviews* **56** (1978), #327.

[Tuckerman 1971]
B. Tuckerman. The 24th Mersenne prime. *Proc. Nat. Acad. Sci. U. S. A.* **68** (1971), 2319–2320.

[Turán 1950]
P. Turán. Results of number-theory in the Soviet Union. *Mat. Lapok* **1** (1950), 243–267. In Hungarian, with English summary.

[Turck 1921]
J. A. V. Turck. *Origin of Modern Calculating Machines*. Western Society of Engineers, Chicago, 1921. Reprinted by Arno Press, New York, 1972.

[Turing 1936]
A. M. Turing. On computable numbers, with an application to the Entscheidungsproblem. *Proc. Lond. Math. Soc.* **42** (1936), 230–265.

[Turing 1939]
A. M. Turing. Systems of logic based on ordinals. *Proc. Lond. Math. Soc.* **45** (1939), 161–228.

[Turing 1945]
A. M. Turing. A method for the calculation of the zeta-function. *Proc. Lond. Math. Soc.* **48** (1945), 180–197.

[Turing 1953]
A. M. Turing. Some calculations of the Riemann zeta-function. *Proc. Lond. Math. Soc.* **3** (1953), 99–117.

[Urbanek 1980]
F. J. Urbanek. An $O(\log n)$ algorithm for computing the nth element of the solution of a difference equation. *Inform. Process. Lett.* **11** (1980), 66–67.

[Uspensky and Heaslet 1939]
J. V. Uspensky and M. A. Heaslet. *Elementary Number Theory*. McGraw-Hill, New York, 1939.

[Utz 1953]
W. R. Utz. A note on the Scholz-Brauer problem in addition chains. *Proc. Amer. Math. Soc.* **4** (1953), 462–463.

[Vacca 1894]
G. Vacca. Intorno alla prima dimonstrazione di un teorema di Fermat. *Bibliotheca Math.* **8** (1894), 46–48.

[Vacca 1899]
G. Vacca. Sui manoscritti inediti di Leibniz. *Bolletino di Biliografia Storia Sc. Mat.* **2** (1899), 113–116.

[Vahlen 1895]
K. Th. Vahlen. Über Näherungswerthe und Kettenbrüche. *J. Reine Angew. Math.* **115** (1895), 221–233.

[Vaidyanathaswamy 1927]
R. Vaidyanathaswamy. The quadratic reciprocity of polynomials modulo p. *J. Indian Math. Soc.* **17** (1927), 185–196.

[de la Vallée Poussin 1896a]
C.-J. de la Vallée Poussin. Recherches analytiques sur la théorie des nombres premiers (première partie). *Ann. Soc. Sci. Bruxelles* **20** (1896), 183–256.

[de la Vallée Poussin 1896b]
C.-J. de la Vallée Poussin. Recherches analytiques sur la théorie des nombres premiers (deuxième partie). *Ann. Soc. Sci. Bruxelles* **20** (1896), 281–397.

[Vanden Eynden 1972]
C. Vanden Eynden. A proof of Gandhi's formula for the nth prime. *Amer. Math. Monthly* **79** (1972), 625.

[Vandiver 1929]
H. S. Vandiver. Algorithms for the solution of the quadratic congruence. *Amer. Math. Monthly* **36** (1929), 83–86.

[Vandiver 1958]
H. S. Vandiver. The rapid computing machine as an instrument in the discovery of new relations in the theory of numbers. *Proc. Nat. Acad. Sci. U. S. A.* **44** (1958), 459–464.

[Vandiver and Wahlin 1928]
H. S. Vandiver and G. E. Wahlin. *Algebraic Numbers – II.* National Research Council, Washington, 1928. (= *Bull. Nat. Res. Council* **62**). Reprinted along with Dickson et al. [1923] by Chelsea, New York, 1967.

[Vegh 1975]
E. Vegh. A note on addition chains. *J. Combin. Theory. Ser. A* **19** (1975), 117–118.

[Vélu 1978]
J. Vélu. Tests for primality under the Riemann hypothesis. *SIGACT News* **10**(2) (1978), 58–59.

[A. Vinogradov 1963]
A. I. Vinogradov. On the residue in Mertens's formula. *Dokl. Akad. Nauk SSSR* **148** (1963), 262–263. In Russian.

[I. Vinogradov 1918]
I. M. Vinogradov. On the distribution of power residues and non-residues. *Zhurnal Fiz. Mat. Obshchestva pri Permskom Universitet* **1** (1918), 94–98. In Russian. English translation in *Selected Works*, pp. 53–56.

[Viry 1993]
G. Viry. Factorization of multivariate polynomials with coefficients in $\mathbb{F}_p$. *J. Symbolic Comput.* **15** (1993), 371–391.

[Vitányi 1988]
P. M. B. Vitányi. Locality, communication, and interconnect length in multicomputers. *SIAM J. Comput.* **17** (1988), 659–672.

[Vitányi and Meertens 1984]
P. M. B. Vitányi and L. Meertens. Big omega versus the wild functions. *Bull. European Assoc. Theor. Comput. Sci.* (22) (1984), 14–19. Also in *SIGACT News* **16** (4) (1985), 56–59.

[Volger 1985]
H. Volger. Some results on addition/subtraction chains. *Inform. Process. Lett.* **20** (1985), 155–160.

[Volpi 1995]
A. Volpi. p-th roots of integers modulo p^n and for p-adic integers. *ACM SIGSAM Bull.*, Special Issue (June, 1995), 9–15.

[van der Waerden 1970]
B. L. van der Waerden. *Algebra.* Ungar, 7th edition, 1970. 2 volumes.

[Wagon 1986]
S. Wagon. Primality testing. *Math. Intelligencer* **8**(3) (1986), 58–61.

[Wagstaff 1978]
S. S. Wagstaff, Jr. The irregular primes to 125000. *Math. Comp.* **32** (1978), 583–591.

[Wagstaff 1979]
S. S. Wagstaff, Jr. Greatest of the least primes in arithmetic progressions having a given modulus. *Math. Comp.* **33** (1979), 1073–1080.

[Wagstaff 1980]
S. S. Wagstaff, Jr. Large Carmichael numbers. *Math. J. Okayama Univ.* **22** (1980), 33–41.

[Wagstaff 1982]
S. S. Wagstaff, Jr. Pseudoprimes and a generalization of Artin's conjecture. *Acta Arith.* **41** (1982), 141–150.

[Wagstaff 1983]
S. S. Wagstaff, Jr. Divisors of Mersenne numbers. *Math. Comp.* **40** (1983), 385–397.

[Wagstaff 1990]
S. S. Wagstaff, Jr. Some uses of microcomputers in number theory research. *Comput. Math. with Appl.* **19**(3) (1990), 53–58.

[Walfisz 1936]
A. Walfisz. Zur additiven Zahlentheorie. II. *Math. Zeitschrift* **40** (1936), 592–607.

[Walfisz 1963]
A. Walfisz. *Weylsche Exponentialsummen in der neueren Zahlentheorie.* VEB Deutscher Verlag der Wissenschaften, Berlin, 1963.

[Wallace 1964]
C. S. Wallace. A suggestion for a fast multiplier. *IEEE Trans. Elect. Comput.* **EC-13** (1964), 14–17.

[Walter 1991]
C. D. Walter. Fast modular multiplication using 2–power radix. *Internat. J. Comput. Math.* **39** (1991), 21–28.

[Walter 1992]
C. D. Walter. Faster modular multiplication by operand scaling. In J. Feigenbaum, editor, *Advances in Cryptology–CRYPTO '91 Proceedings*, Vol. 576 of *Lecture Notes in Computer Science*, pp. 313–323. Springer-Verlag, 1992.

[Walter 1995]
C. D. Walter. Still faster modular multiplication. *Electronics Letters* **31** (1995), 263–264.

[Walter and Eldridge 1990]
C. D. Walter and S. E. Eldridge. A verification of Brickell's fast modular multiplication algorithm. *Internat. J. Comput. Math.* **33** (1990), 153–169.

[Wan 1990]
D. Wan. Factoring multivariate polynomials over large finite fields. *Math. Comp.* **54** (1990), 755–770.

[C. Wang and Pei 1990]
C. C. Wang and D. Pei. A VLSI design for computing exponentiations in $GF(2^m)$ and its application to generate pseudorandom number sequences. *IEEE Trans. Comput.* **39** (1990), 258–262.

[C. Wang et al. 1985]
C. C. Wang, T. K. Truong, H. M Shao, L. J. Deutsch, J. K. Omura, and I. S. Reed. VLSI architectures for computing multiplications and inverses in $GF(2^m)$. *IEEE Trans. Comput.* **C-34** (1985), 709–717.

[P. S. Wang 1980]
P. S. Wang. The EEZ-GCD algorithm. *ACM SIGSAM Bull.* **14**(2) (1980), 50–60.

[Y. Wang 1959]
Y. Wang. On the least primitive root of a prime. *Acta Math. Sinica* **9** (1959), 432–441. English translation in *Sci. Sinica* **10** (1961), 1–14.

[Y. Wang 1964]
Y. Wang. Estimation and application of character sums. *Shuxue Jinzhan* **7** (1964), 78–83. In Chinese. Reviewed in *Math. Reviews* **37** (1969), #5162.

[Y. Wang 1984]
Y. Wang, editor. *Goldbach Conjecture*. World Scientific, Singapore, 1984.

[Y. Wang, Hsieh, and Yu 1965]
Y. Wang, S.-K. Hsieh, and K.-J. Yu. Two results on the distribution of prime numbers. *Zhongguo Kexue Jishu Daxue Xuebao* **1** (1965), 32–38. In Chinese. Reviewed in *Math. Reviews* **34** (1967), # 7482.

[Ward 1959]
M. Ward. Tests for primality based on Sylvester's cyclotomic numbers. *Pacific J. Math.* **9** (1959), 1269–1272.

[Warlimont 1970]
R. Warlimont. Eine Bemerkung zu einem Ergebnis von N. G. de Bruijn. *Monatsh. Math.* **74** (1970), 273–276.

[Washington 1982]
L. C. Washington. *Introduction to Cyclotomic Fields*. Springer-Verlag, New York, 1982.

[Waterman 1977]
M. S. Waterman. Multidimensional greatest common divisor and Lehmer algorithms. *BIT* **17** (1977), 465–478.

[Watrous 1995]
J. Watrous. A polynomial time algorithm for the Artin-Whaples approximation theorem. In K. Dilcher, editor, *Number Theory. Fourth Conference of the Canadian Number Theory Association*, Vol. 15 of *CMS Conference Proceedings*, pp. 397–407. Amer. Math. Soc., 1995.

[Weber 1990]
K. Weber. An experiment in high-precision arithmetic on shared memory multiprocessors. *ACM SIGSAM Bull.* **24**(2) (1990), 22–40.

[Weber 1995]
K. Weber. The accelerated integer GCD algorithm. *ACM Trans. Math. Software* **21** (1995), 111–122.

[Weil 1948a]
A. Weil. *Sur les Courbes Algébriques et les Variétés qui s'en Déduisent*. Hermann, 1948.

[Weil 1984]
A. Weil. *Number Theory: an Approach Through History; from Hammurapi to Legendre*. Birkhäuser Boston, 1984.

[Weinberger 1975]
P. J. Weinberger. On small zeroes of Dirichlet L-functions. *Math. Comp.* **29** (1975), 319–328.

[Weinstock 1960]
R. Weinstock. Greatest common divisor of several integers and an associated linear Diophantine equation. *Amer. Math. Monthly* **67** (1960), 664–667.

[Weintraub 1977a]
S. Weintraub. Primes in arithmetic progression. *BIT* **17** (1977), 239–243.

[Weintraub 1977b]
S. Weintraub. Seventeen primes in arithmetic progression. *Math. Comp.* **31** (1977), 1030.

[Weintraub 1991]
S. Weintraub. A prime gap of 784. *J. Recreational Math.* **23** (1991), 6–7.

[Weintraub 1993]
S. Weintraub. A prime gap of 864. *J. Recreational Math.* **25** (1993), 42–43.

[Weiss 1983]
A. Weiss. The least prime ideal. *J. Reine Angew. Math.* **338** (1983), 56–94.

[Welch and Scholtz 1979]
L. R. Welch and R. A. Scholtz. Continued fractions and Berlekamp's algorithm. *IEEE Trans. Inform. Theory* **IT-25** (1979), 19–27.

[Welsh 1983]
D. J. A. Welsh. Randomised algorithms. *Disc. Appl. Math.* **5** (1983), 133–145.

[Wertheim 1902]
G. Wertheim. Ein Beitrag zur Beurteilung des Pietro Antonio Cataldi. *Bibliotheca Math.* **3** (1902), 76–83.

[Western 1905]
A. E. Western. Note on Fermat's numbers and the converse of Fermat's theorem. *Proc. Lond. Math. Soc.* **3** (1905), xxi–xxii.

[Western 1922]
A. E. Western. Note on the number of primes of the form $n^2 + 1$. *Proc. Cambridge Phil. Soc.* **21** (1922), 108–109.

[Western 1932]
A. E. Western. On Lucas's and Pepin's tests for the primeness of Mersenne numbers. *J. London Math. Soc.* **7** (1932), 130–137.

[Western 1934]
A. E. Western. Note on the magnitude of the difference between successive primes. *J. London Math. Soc.* **9** (1934), 276–278.

[Western and Miller 1968]
A. E. Western and J. C. P. Miller. *Tables of Indices and Primitive Roots*. Royal Society, Cambridge, 1968.

[White, Hunt, and Dresel 1977]
D. J. White, J. N. Hunt, and L. A. G. Dresel. Uniform Huffman sequences do not exist. *Bull. Lond. Math. Soc.* **9** (1977), 193–198.

[Whiteman 1937]
A. Whiteman. On a theorem of higher reciprocity. *Bull. Amer. Math. Soc.* **43** (1937), 567–572.

[Wigert 1906]
S. Wigert. Sur l'ordre de grandeur du nombre des diviseurs d'un entier. *Arkiv för Matematik, Astronomi och Fisik* **3**(18) (1906–1907), 1–9.

[Wiles 1995]
A. Wiles. Modular elliptic curves and Fermat's last theorem. *Ann. Math.* **142** (1995), 443–551.

[Wilf 1982]
H. S. Wilf. What is an answer? *Amer. Math. Monthly* **89** (1982), 289–292.

[Wilf 1987]
H. S. Wilf. A greeting; and a view of the Riemann hypothesis. *Amer. Math. Monthly* **94** (1987), 3–6.

[Wilkins 1950]
J. E. Wilkins, Jr. Solution to problem 4319. *Amer. Math. Monthly* **57** (1950), 346.

[Willans 1964]
C. P. Willans. On formulae for the Nth prime number. *Math. Gazette* **48** (1964), 413–415.

[Willett 1980]
M. Willett. Arithmetic in a finite field. *Math. Comp.* **35** (1980), 1353–1359.

[F. Williams 1975]
F. C. Williams. Early computers at Manchester University. *Radio Electr. Eng.* **45** (1975), 327–331.

[H. Williams 1971]
H. C. Williams. An algorithm for determining certain large primes. *Congr. Numer.* **3** (1971), 533–556.

[H. Williams 1972]
H. C. Williams. Some algorithms for solving $x^q \equiv N \pmod{p}$. In *Proc. 3rd Southeastern Conference on Combinatorics, Graph Theory, and Computing*, pp. 451–462, Winnipeg, 1972. Utilitas Mathematica. (= *Congr. Numer.* **VII**).

[H. Williams 1977]
H. C. Williams. On numbers analogous to the Carmichael numbers. *Canad. Math. Bull.* **20** (1977), 133–143.

[H. Williams 1978a]
H. C. Williams. Primality testing on a computer. *Ars Combin.* **5** (1978), 127–185.

[H. Williams 1978b]
H. C. Williams. Some primes with interesting digit patterns. *Math. Comp.* **32** (1978), 1306–1310. Corrigendum in **39** (1982), 759.

[H. Williams 1980]
H. C. Williams. A modification of the RSA public-key encryption procedure. *IEEE Trans. Inform. Theory* **IT-26** (1980), 726–729.

[H. Williams 1981]
H. C. Williams. The primality of certain integers of the form $2Ar^n - 1$. *Acta Arith.* **39** (1981), 7–17.

[H. Williams 1982a]
H. C. Williams. The influence of computers in the development of number theory. *Comput. Math. with Appl.* **8** (1982), 75–93.

[H. Williams 1982b]
H. C. Williams. A class of primality tests for trinomials which includes the Lucas-Lehmer test. *Pacific J. Math.* **98** (1982), 477–494.

[H. Williams 1987]
H. C. Williams. Effective primality tests for some integers of the forms $A5^n - 1$ and $A7^n - 1$. *Math. Comp.* **48** (1987), 385–403.

[H. Williams 1988]
H. C. Williams. A note on the primality of $6^{2^n} + 1$ and $10^{2^n} + 1$. *Fibonacci Quart.* **26** (1988), 296–305.

[H. Williams and Dubner 1986]
H. C. Williams and H. Dubner. The primality of $R1031$. *Math. Comp.* **47** (1986), 703–711.

[H. Williams and Holte 1978]
H. C. Williams and R. Holte. Some observations on primality testing. *Math. Comp.* **32** (1978), 905–917.

[H. Williams and Judd 1976a]
H. C. Williams and J. S. Judd. Determination of the primality of N by using factors of $N^2 \pm 1$. *Math. Comp.* **30** (1976), 157–172.

[H. Williams and Judd 1976b]
H. C. Williams and J. S. Judd. Some algorithms for prime testing using generalized Lehmer functions. *Math. Comp.* **30** (1976), 867–886.

[H. Williams and Seah 1979]
H. C. Williams and E. Seah. Some primes of the form $(a^n - 1)/(a - 1)$. *Math. Comp.* **33** (1979), 1337–1342.

[H. Williams and Zarnke 1968]
H. C. Williams and C. R. Zarnke. A report on prime numbers of the forms $M = (6a + 1)2^{2m-1} - 1$ and $M' = (6a - 1)2^{2m} - 1$. *Math. Comp.* **22** (1968), 420–422.

[I. Williams 1976]
I. S. Williams. On a problem of Kurt Mahler concerning binomial coefficients. *Bull. Austral. Math. Soc.* **14** (1976), 299–302.

[K. Williams 1974]
K. S. Williams. Mertens' theorem for arithmetic progressions. *J. Number Theory* **6** (1974), 353–359.

[K. Williams and Hardy 1993]
K. S. Williams and K. Hardy. A refinement of H. C. Williams' qth root algorithm. *Math. Comp.* **61** (1993), 475–483.

[Willoner and Chen 1981]
R. Willoner and I-N. Chen. An algorithm for modular exponentiation. In *Proc. 5th IEEE Symp. Comput. Arith.*, pp. 135–138. IEEE Press, 1981.

[Wilson and Shortt 1980]
T. C. Wilson and J. Shortt. An $O(\log n)$ algorithm for computing general order-k Fibonacci numbers. *Inform. Process. Lett.* **10** (1980), 68–75.

[Wilton 1927]
J. R. Wilton. A note on the coefficients in the expansion of $\zeta(s, x)$ in powers of $s - 1$. *Quart. J. Pure Appl. Math.* **59** (1927), 329–332.

[Wintner 1941]
A. Wintner. Statistics and prime numbers. *Nature* **147** (1941), 208–209.

[Wintner 1944]
A. Wintner. Random factorizations and Riemann's hypothesis. *Duke Math. J.* **11** (1944), 267–275.

[Wirth 1973]
N. Wirth. *Systematic Programming: an Introduction*. Prentice-Hall, Englewood Cliffs, NJ, 1973.

[Wood 1961]
T. C. Wood. Algorithm 35: sieve. *Comm. ACM* **4** (1961), 151.

[Woods and Huenemann 1982]
D. Woods and J. Huenemann. Larger Carmichael numbers. *Comput. Math. with Appl.* **8** (1982), 215–216.

[Wrench 1961]
J. W. Wrench, Jr. Evaluation of Artin's constant and the twin-prime constant. *Math. Comp.* **15** (1961), 396–398.

[Wright 1951]
E. M. Wright. A prime-representing function. *Amer. Math. Monthly* **58** (1951), 616–618. Correction in **59** (1952), 99.

[Wright 1954]
E. M. Wright. A class of representing functions. *J. London Math. Soc.* **29** (1954), 63–71.

[Wright 1955]
E. M. Wright. A non-linear difference-differential equation. *J. Reine Angew. Math.* **194** (1955), 66–87.

[Wright 1961]
E. M. Wright. A functional equation in the heuristic theory of primes. *Math. Gazette* **45** (1961), 15–16.

[Wunderlich 1967]
M. C. Wunderlich. Sieving procedures on a digital computer. *J. Assoc. Comput. Mach.* **14** (1967), 10–19.

[Wunderlich 1974]
M. C. Wunderlich. A probabilistic setting for prime number theory. *Acta Arith.* **26** (1974), 59–81.

[Wunderlich 1976]
M. C. Wunderlich. The prime number theorem for random sequences. *J. Number Theory* **8** (1976), 369–371.

[Wunderlich 1983a]
M. C. Wunderlich. A performance analysis of a simple prime-testing algorithm. *Math. Comp.* **40** (1983), 709–714.

[Wunderlich and Selfridge 1974]
M. C. Wunderlich and J. L. Selfridge. A design for a number theory package with an optimized trial division routine. *Comm. ACM* **17** (1974), 272–276.

[Wyman 1972]
B. F. Wyman. What is a reciprocity law? *Amer. Math. Monthly* **79** (1972), 571–586. Corrigendum: **80** (1973), 281.

[Yacobi 1991a]
Y. Yacobi. Exponentiating faster with addition chains. In I. B. Damgård, editor, *Advances in Cryptology–EUROCRYPT '90 Proceedings*, Vol. 473 of *Lecture Notes in Computer Science*, pp. 222–229. Springer-Verlag, 1991.

[Yamada 1962]
H. Yamada. Real-time computation and recursive functions not real-time computable. *IEEE Trans. Elect. Comput.* **EC-11** (1962), 753–760.

[Yang 1990]
K.-W. Yang. Euclid's algorithm = reverse Gaussian elimination. *Math. Mag.* **63** (1990), 163–164.

[Yao 1976]
A. C. Yao. On the evaluation of powers. *SIAM J. Comput.* **5** (1976), 100–103.

[Yao 1985]
A. C. Yao. On the complexity of maintaining partial sums. *SIAM J. Comput.* **14** (1985), 277–288.

[Yao and Knuth 1975]
A. C. Yao and D. E. Knuth. Analysis of the subtractive algorithm for greatest common divisors. *Proc. Nat. Acad. Sci. U. S. A.* **72** (1975), 4720–4722.

[Yeh, Reed, and Truong 1984]
C.-S. Yeh, I. S. Reed, and T. K. Truong. Systolic multipliers for finite fields $GF(2^m)$. *IEEE Trans. Comput.* **C-33** (1984), 357–360.

[Yorinaga 1978]
M. Yorinaga. Numerical computation of Carmichael numbers. *Math. J. Okayama Univ.* **20** (1978), 151–163.

[Yorinaga 1979a]
M. Yorinaga. Search for absolute pseudoprime numbers. *Sugaku* **31** (1979), 374–376. In Japanese.

[Yorinaga 1979b]
M. Yorinaga. Numerical computation of Carmichael numbers. II. *Math. J. Okayama Univ.* **21** (1979), 183–205.

[Yorinaga 1980]
M. Yorinaga. Carmichael numbers with many prime factors. *Math. J. Okayama Univ.* **22** (1980), 169–184.

[Yorinaga 1981]
M. Yorinaga. Numerical investigations of the Carmichael function and C_k–numbers. *Math. J. Okayama Univ.* **23** (1981), 69–75.

[Young and Buell 1988]
J. Young and D. A. Buell. The twentieth Fermat number is composite. *Math. Comp.* **50** (1988), 261–263.

[Young and Potler 1989]
J. Young and A. Potler. First occurrence prime gaps. *Math. Comp.* **52** (1989), 221–224.

[Yun 1977]
D. Y. Y. Yun. Fast algorithm for rational function integration. In *Information Processing 77*, pp. 493–498. North-Holland, 1977.

[Yun and Zhang 1986]
D. Y. Y. Yun and C. N. Zhang. A fast carry-free algorithm and hardware design for extended integer gcd computation. In B. W. Char, editor, *Proc. 1986 Symposium on Symbolic and Algebraic Computation: SYMSAC '86*, pp. 82–84. ACM, 1986.

[Zagier 1977]
D. Zagier. The first 50 million prime numbers. *Math. Intelligencer* **0** (1977), 7–19.

[Zarnke and Williams 1971]
C. R. Zarnke and H. C. Williams. Computer determination of some large primes. *Congr. Numer.* **3** (1971), 563–570.

[Zassenhaus 1968]
H. Zassanhaus. Über die Fundamentalkonstruktionen der endlichen Körpertheorie. *Jahres. Deutscher Math.-Verein.* **70** (1968), 177–181.

[Zassenhaus 1969]
H. Zassenhaus. On Hensel factorization, I. *J. Number Theory* **1** (1969), 291–311.

[Zassenhaus 1978]
H. Zassenhaus. A remark on the Hensel factorization method. *Math. Comp.* **32** (1978), 287–292.

[Zavrotsky 1961]
A. Zavrotsky. Greatest common divisor finder. April 11, 1961, U. S. patent 2,978,816; filed February 4, 1960.

[Zeugmann 1989]
T. Zeugmann. Improved parallel computations in the ring $\mathbb{Z}/p^\alpha$. *J. Inf. Process. Cybern. EIK* **25** (1989), 543–547.

[Zeugmann 1990]
T. Zeugmann. Computing large polynomial powers very fast in parallel. In B. Rovan, editor, *Mathematical Foundations of Computer Science 1990*, Vol. 452 of *Lecture Notes in Computer Science*, pp. 538–544. Springer-Verlag, 1990.

[Zeugmann 1992]
T. Zeugmann. Highly parallel computations modulo a number having only small prime factors. *Inform. Comput.* **96** (1992), 95–114.

[Zhang 1991]
M. Zhang. Searching for large Carmichael numbers. Unpublished manuscript, dated December 3, 1991.

[Zheng 1994]
Z.-Y. Zheng. On a theorem of h. [sic] yao and d. knuth. *Acta Math. Sinica* **37** (1994), 191–202. In Chinese. Reviewed in *Math. Reviews* 95h:11006.

[Zierler 1959]
N. Zierler. Linear recurring sequences. *J. Soc. Indust. Appl. Math.* **7** (1959), 31–48.

[Zierler 1974]
N. Zierler. A conversion algorithm for logarithms on $GF(2^n)$. *J. Pure Appl. Algebra* **4** (1974), 353–356.

[Zimmer 1972]
H. G. Zimmer. *Computational Problems, Methods, and Results in Algebraic Number Theory*, Vol. 262 of *Lecture Notes in Mathematics*. Springer-Verlag, 1972.

[Zippel 1993]
R. Zippel. *Effective Polynomial Computation*. Kluwer, Boston, 1993.

[Zolotarev 1872]
G. Zolotarev. Nouvelle démonstration de la loi de réciprocité de Legendre. *Nouvelles Annales de Mathématiques* **11** (1872), 354–362.

[Zolotarev 1874]
G. Zolotarev. *Théorie des nombres entiers complexes, avec une application au calcul intégral*. Dissertation, St. Petersburg. 1874. Reviewed in *Jahrb. Fort. Math.* **6** (1876), 117–124.

[Zuras 1993]
D. Zuras. On squaring and multiplying large integers. In E. Swartzlander, Jr., M. J. Irwin, and G. Jullien, editors, *Proc. 11th IEEE Symp. Comp. Arith.*, pp. 260–271. IEEE Press, 1993.

Index to Notation

Index